T0281231

Second Edition

Problems
and
Solutions
on
QUANTUM
MECHANICS

Major American Universities Ph.D. Qualifying Questions and Solutions - Physics

Problems and Solutions on Quantum Mechanics
Second Edition
edited by Swee Cheng Lim, Choy Heng Lai and Leong Chuan Kwek
(NUS, Singapore)

Problems and Solutions on Thermodynamics and Statistical Mechanics
Second Edition
edited by Swee Cheng Lim, Choy Heng Lai and Leong Chuan Kwek
(NUS, Singapore)

Problems and Solutions on Mechanics (Second Edition)
edited by Swee Cheng Lim, Choy Heng Lai and Leong Chuan Kwek
(NUS, Singapore)

Problems and Solutions on Thermodynamics and Statistical Mechanics
edited by Yung-Kuo Lim (NUS, Singapore)

Problems and Solutions on Optics
edited by Yung-Kuo Lim (NUS, Singapore)

Problems and Solutions on Electromagnetism
edited by Yung-Kuo Lim (NUS, Singapore)

Problems and Solutions on Mechanics
edited by Yung-Kuo Lim (NUS, Singapore)

Problems and Solutions on Solid State Physics, Relativity and
Miscellaneous Topics
edited by Yung-Kuo Lim (NUS, Singapore)

Problems and Solutions on Quantum Mechanics
edited by Yung-Kuo Lim (NUS, Singapore)

Problems and Solutions on Atomic, Nuclear and Particle Physics
edited by Yung-Kuo Lim (NUS, Singapore)

More information on this series can also be found at https://www.worldscientific.com/series/mauqqsp

Second Edition

Problems
and
Solutions
on
QUANTUM
MECHANICS

Editors

Swee Cheng Lim

Choy Heng Lai

National University of Singapore, Singapore

Leong Chuan Kwek

CQT, National University of Singapore, and
NIE, Quantum Science and Engineering Center, NTU, Singapore

 World Scientific

EW JERSEY · LONDON · SINGAPORE · BEIJING · SHANGHAI · HONG KONG · TAIPEI · CHENNAI · TOKYO

Published by

World Scientific Publishing Co. Pte. Ltd.

5 Toh Tuck Link, Singapore 596224

USA office: 27 Warren Street, Suite 401-402, Hackensack, NJ 07601

UK office: 57 Shelton Street, Covent Garden, London WC2H 9HE

Library of Congress Control Number: 2022937746

British Library Cataloguing-in-Publication Data
A catalogue record for this book is available from the British Library.

Some new problems created by B. Rajesh, American College, Madurai.

Major American Universities Ph.D. Qualifying Questions and Solutions - Physics
PROBLEMS AND SOLUTIONS ON QUANTUM MECHANICS
Second Edition

Copyright © 2022 by World Scientific Publishing Co. Pte. Ltd.

ISBN 978-981-125-606-6 (hardcover)
ISBN 978-981-125-734-6 (paperback)
ISBN 978-981-125-607-3 (ebook for institutions)
ISBN 978-981-125-608-0 (ebook for individuals)

For any available supplementary material, please visit
https://www.worldscientific.com/worldscibooks/10.1142/12829#t=suppl

Desk Editor: Joseph Ang

Typeset by Diacritech Technologies Pvt. Ltd.
Chennai - 600106, India

Printed in Singapore

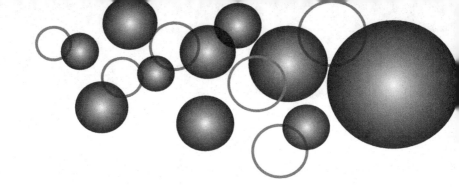

PREFACE

This is the second edition of a former popular series on Problems and Solutions in various topics of physics ranging from Mechanics, Electromagnetism, Optics, Atomic, Nuclear and Particle Physics, Thermodynamics and Statistical Mechanics, Quantum Mechanics, to Solid State Physics, Relativity and Miscellaneous topics.

We have greatly expanded the volumes. Each volume is divided into several subtopics, and there are new interesting problems and solutions. Altogether, we have compiled several thousand problems and solutions, spanning an entire undergraduate physics course. We have also included different genres of problems and questions: qualitative, quantitative, fill-in-the-blanks, and even multiple-choice. These questions are carefully chosen at the level of the PhD Qualifying Examinations at American universities. Moreover, these questions can also serve as an excellent resource for Physics Competitions, like the International Physics Olympiad or the regional Physics Olympiads.

We believe that a good grounding in problem-solving is essential for the study of physics, and we believe that this compendium serves as an invaluable resource for the preparation of graduate study in physics, or competitions. We suggest that the student using the books should first attempt the problems before consulting the solutions. We have tried to elucidate solutions with figures and diagrams that illustrate how to set up the problems.

Lim Swee Cheng, Lai Choy Heng and Kwek Leong Chuan
Editors

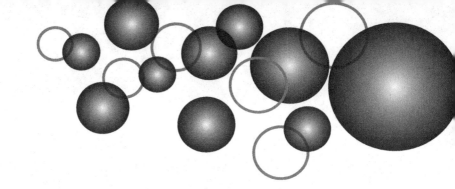

CONTENTS

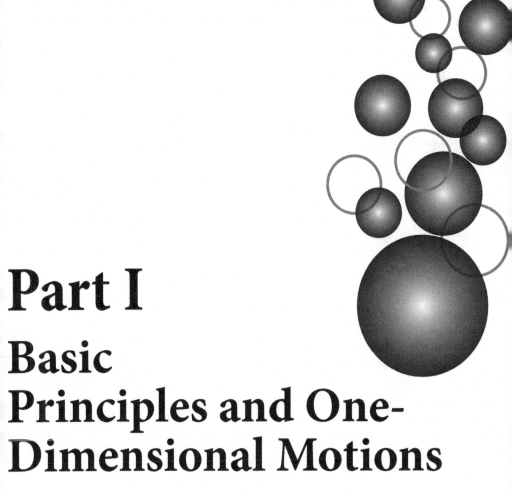

Part I

Basic Principles and One-Dimensional Motions

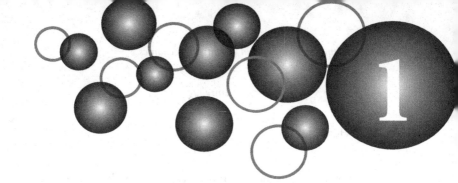

1001

Quantum phenomena are often negligible in the "macroscopic" world. Show this numerically for the following cases:

a. The amplitude of the zero-point oscillation for a pendulum of length $l=1$ m and mass $m=1$ kg.

b. The tunneling probability for a marble of mass $m=5$ g moving at a speed of 10 cm/sec against a rigid obstacle of height $H=5$ cm and width $w=1$ cm.

c. The diffraction of a tennis ball of mass $m=0.1$ kg moving at a speed $v=0.5$ m/sec by a window of size 1×1.5 m².

(*Wisconsin*)

Sol:

a. The theory of the harmonic oscillator gives the average kinetic energy as $\bar{V}=\frac{1}{2}E$, i.e., $\frac{1}{2}m\omega^2A^2=\frac{1}{4}\hbar\omega$, where $\omega=\sqrt{g/l}$ and A is the root-mean-square amplitude of the zero-point oscillation. Hence

$$A=\sqrt{\frac{\hbar}{2m\omega}}\approx 0.41\times10^{-17}\text{ m}.$$

Thus the zero-point oscillation of a macroscopic pendulum is negligible.

b. If we regard the width and height of the rigid obstacle as the width and height of a gravity potential barrier, the tunneling probability is

$$T\approx\exp\left[-\frac{2w}{\hbar}\sqrt{2m\left(mgH-\frac{1}{2}mv^2\right)}\right]$$

$$=\exp\left(-\frac{2mw}{\hbar}\sqrt{2gH-v^2}\right),$$

3

where
$$\frac{2mw}{\hbar}\sqrt{2gH-v^2}\approx 0.9\times 10^{30}.$$

Hence
$$T\approx e^{-0.9\times 10^{30}}\approx 0$$

That is, the tunneling probability for the marble is essentially zero.

c. The de Broglie wavelength of the tennis ball is
$$\lambda=h/p=h/mv=1.3\times 10^{-30}\ \text{cm},$$

and the diffraction angles in the horizontal and the vertical directions are respectively
$$\theta_1\approx\lambda/D=1.3\times 10^{-32}\ \text{rad}\ ,\ \theta_2\approx\lambda/L=9\times 10^{-33}\ \text{rad}.$$

Thus there is no diffraction in any direction.

1002

Express each of the following quantities in terms of $\hbar$, e, c, m = electron mass, M = proton mass. Also give a rough estimate of numerical size for each.

  a. Bohr radius (cm).
  b. Binding energy of hydrogen (eV).
  c. Bohr magneton (choosing your own unit).
  d. Compton wavelength of an electron (cm).
  e. Classical electron radius (cm).
  f. Electron rest energy (MeV).
  g. Proton rest energy (MeV).
  h. Fine structure constant.
  i. Typical hydrogen fine-structure splitting (eV).

(*Berkeley*)

Sol:

  a. $a=\hbar^2/me^2=5.29\times 10^{-9}$ cm.

  b. $E=me^4/2\hbar^2=13.6$ eV.

  c. $\mu_B=e\hbar/2mc=9.27\times 10^{-21}$ erg$\cdot$Gs^{-1}.

  d. $\lambda=2\pi\hbar/mc=2.43\times 10^{-10}$ cm.

e. $r_e = e^2/mc^2 = 2.82 \times 10^{-13}$ cm.

f. $E_e = mc^2 = 0.511$ MeV.

g. $E_p = Mc^2 = 938$ MeV.

h. $\alpha = e^2/\hbar c = 7.30 \times 10^{-3} \approx 1/137$.

i. $\Delta E = e^8 mc^2/8\hbar^2 c^4 = \frac{1}{8}\alpha^4 mc^2 = 1.8 \times 10^{-4}$ eV.

1003

Derive, estimate, guess or remember numerical values for the following, to within one order of magnitude:

a. The electron Compton wavelength.

b. The electron Thomson cross section.

c. The Bohr radius of hydrogen.

d. The ionization potential for atomic hydrogen.

e. The hyperfine splitting of the ground-state energy level in atomic hydrogen.

f. The magnetic dipole moment of $^3\text{Li}^7 (Z=3)$ nucleus.

g. The proton-neutron mass difference.

h. The lifetime of free neutron.

i. The binding energy of a helium-4 nucleus.

j. The radius of the largest stable nucleus.

k. The lifetime of a π^0 meason.

l. The lifetime of a μ^- meason.

(Berkeley)

Sol:

a. $\lambda_e = h/m_e c = 2.43 \times 10^{-2}$ Å.

b. $\sigma = \frac{8\pi}{3} r_e^2 = 6.56 \times 10^{-31}$ m^2.

c. $a = \frac{\hbar^2}{m_e e^2} = 0.53$ Å.

d. $I = \frac{e^2}{2a} = 13.6$ eV.

e. The splitting of the ground-state energy level is

$$\Delta E_f = 13.6 \times \left(\frac{1}{137}\right)^2 \approx 10^{-4} \text{ eV}.$$

The hyperfine splitting of the ground-state energy level is

$$\Delta E_{hf} \approx \Delta E_f / 10^3 \approx 10^{-7} \text{ eV}.$$

f. $\mu = 1.67 \times 10^{-26} \, J \cdot T^{-1}$.

g. $\Delta m = m_p - m_n = -2.3 \times 10^{-30}$ kg.

h. $\tau_n \approx 15$ min $= 9 \times 10^2$ s.

i. $E = 4 \times 7$ MeV $= 28$ MeV.

j. The radius r corresponds to a region of space in which nuclear force is effective. Thus

$$r \approx 1.4 A^{\frac{1}{3}} = 1.4 \times (100)^{\frac{1}{3}} = 6.5 \text{ fm}.$$

k. $\tau = 8.28 \times 10^{-17}$ s.

l. The decay of μ^- is by weak interaction, and so $\tau = 2.2 \times 10^{-6}$ s.

1004

Explain what was learned about quantization of radiation or mechanical system from two of the following experiments:

a. Photoelectric effect.

b. Black body radiation spectrum.

c. Franck–Hertz experiment.

d. Davisson–Germer experiment.

e. Compton scattering.

Describe the experiments selected in detail, indicate which of the measured effects were non-classical and why, and explain how they can be understood as quantum phenomena. Give equations if appropriate.

(Wisconsin)

Sol:

a. Photoelectric Effect

This refers to the emission of electrons observed when one irradiates a metal under vacuum with ultraviolet light. It was found that the magnitude of the electric current thus produced is proportional to the intensity of the striking radiation provided that the frequency of the light is greater than a minimum value characteristic of the metal, while the speed of the electrons does not depend on the light intensity, but on its frequency.

Einstein in 1905 explained these results by assuming light, in its interaction with matter, consisted of corpuscles of energy hv, called photons. When a photon encounters an electron of the metal it is entirely absorbed, and the electon, after receiving the energy hv, spends an amount of work W equal to its binding energy in the metal, and leaves with a kinetic energy

$$\frac{1}{2}mv^2 = hv - W.$$

This quantitative theory of photoelectricity has been completely verified by experiment.

b. Black Body Radiation

A black body is one which absorbs all the radiation falling on it. The spectral distribution of the radiation emitted by a black body can be derived from the general laws of interaction between matter and radiation. The expressions deduced from the classical theory are known as Wien's law and Rayleigh's law. The former is in good agreement with experiment in the short wavelength end of the spectrum only, while the latter is in good agreement with the long wavelength results but leads to divergency in total energy.

Planck in 1900 succeeded in removing the difficulties encountered by classical physics in black body radiation by postulating that energy exchanges between matter and radiation do not take place in a continuous manner but by discrete and indivisible quantities, or quanta, of energy. He showed that by assuming that the quantum of energy was proportional to the frequency, $\varepsilon = hv$, he was able to obtain an expression for the spectrum which is in complete agreement with experiment:

$$E_v = \frac{8\pi h v^3}{c^3} \frac{1}{e^{\frac{hv}{kT}} - 1},$$

where h is a universal constant, now known as Planck's constant.

Planck's hypothesis has been confirmed by a whole array of elementary processes and it directly reveals the existence of discontinuities of physical processes on the microscopic scale, namely quantum phenomena.

c. Franck–Hertz Experiment

The experiment of Franck and Hertz consisted of bombarding atoms with monoenergetic electrons and measuring the kinetic energy of the scattered electrons, from which one deduced by subtraction the quantity of energy absorbed in the collisions by the atoms. Suppose $E_0, E_1, E_2, \ldots$ are the sequence of quantized energy levels of the atoms and T is the kinetic energy of the incident electrons. As long as T is below $\Delta = E_1 - E_0$, the atoms cannot absorb the energy and all collisions are elastic. As soon as $T > E_1 - E_0$, inelastic collisions occur and some atoms go into their first excited states. Similarly, atoms can be excited into the second excited state as soon as $T > E_2 - E_0$, etc. This was exactly what was found experimentally. Thus the Franck–Hertz experiment established the quantization of atomic energy levels.

d.   **Davisson–Germer Experiment**

L. de Broglie, seeking to establish the basis of a unified theory of matter and radiation, postulated that matter, as well as light, exhibited both wave and corpuscular aspects. The first diffraction experiments with matter waves were performed with electrons by Davisson and Germer (1927). The incident beam was obtained by accelerating electrons through an electrical potential. Knowing the parameters of the crystal lattice it was possible to deduce an experimental value for the electron wavelength and the results were in perfect accord with the de Broglie relation $\lambda = h/p$, where h is Planck's constant and p is the momentum of the electrons. Similar experiments were later performed by others with beams of helium atoms and hydrogen molecules, showing that the wavelike structure was not peculiar to electrons.

e.   **Compton Scattering**

Compton observed the scattering of X-rays by free (or weakly bound) electrons and found the wavelength of the scattered radiation exceeded that of the incident radiation. The difference $\Delta\lambda$ varied as a function of the angle θ between the incident and scattered directions:

$$\Delta\lambda = 2\frac{h}{mc}\sin^2\frac{\theta}{2},$$

where h is Planck's constant and m is the rest mass of the electron. Furthermore, $\Delta\lambda$ is independent of the incident wavelength. The Compton effect cannot be explained by any classical wave theory of light and is therefore a confirmation of the photon theory of light.

1005

In the days before Quantum Mechanics, a big theoretical problem was to "stop" an atom from emitting light. Explain. After Quantum Mechanics, a big theoretical problem was to make atoms in excited states emit light. Explain. What does make excited atoms emit light?

(Wisconsin)

Sol: In the days before Quantum Mechanics, according to the Rutherford atomic model electrons move around the nucleus in elliptical orbits. Classical electrodynamics requires radiation to be emitted when a charged particle accelerates. Thus the atom must emit light. This means that the electrons would lose energy continuously and ultimately be captured by the nucleus. Whereas, in actual fact the electrons do not fall towards the nucleus and atoms in ground state are stable and do not emit light. The problem then was to invent a mechanism which could prevent the atom from emitting light. All such attempts ended in failure.

A basic principle of Quantum Mechanics is that, without external interaction, the Hamiltonian of an atom is time-independent. This means that an atom in an excited state (still a stationary state) would stay on and not emit light spontaneously. In reality, however, spontaneous transition of an excited atoms does occur and light is emitted.

According to Quantum Electrodynamics, the interaction of the radiation field and the electrons in an atom, which form two quantum systems, contains a term of the single-photon creation operator a^+, which does not vanish even if there is no photon initially. It is this term that makes atoms in excited states emit light, causing spontaneous transition.

1006

Consider an experiment in which a beam of electrons is directed at a plate containing two slits, labelled *A* and *B*. Beyond the plate is a screen equipped with an array of detectors which enables one to determine where the electrons hit the screen. For each of the following cases draw a rough graph of the relative number of incident electrons as a function of position along the screen and give a brief explanation.

    a. Slit *A* open, slit *B* closed.

    b. Slit *B* open, slit *A* closed.

    c. Both slits open.

d. "Stern–Gerlach" apparatus attached to the slits in such a manner that only electrons with $s_z = \hbar/2$ can pass through A and only electrons with $s_z = -\hbar/2$ can pass through B.

e. Only electrons with $s_z = \hbar/2$ can pass through A and only electrons with $s_z = \hbar/2$ can pass through B.

What is the effect of making the beam intensity so low that only one electron is passing through the apparatus at any time?

(Columbia)

Sol:

    a. The probability detected at the screen is that of the electrons passing through slit A:

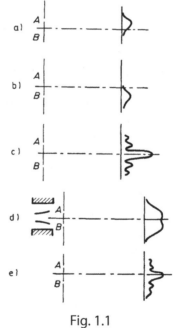

Fig. 1.1

$$I_1 = I_A(x).$$

    b. The probability detected at the screen is that of the electrons passing through slit B:

$$I_2 = I_B(x).$$

c. $I_c = I_{12}(x) = I_1 + I_2 +$ interference term $\neq I_1 + I_2$.

d. The eigenstate of the electrons passing through slit A is different from that of the electrons passing through slit B, and so there is no interference term. The intensity on the screen is just the sum of the intensities of the single-slit cases:

$$I_d = I_1 + I_2.$$

e. Similar to (c), but the intensity is half that in (c):

$$I_e = I_c/2.$$

Because of the self-interference of the wave functions of the electrons, the answers above remain valid even when the incident electron beam intensity is so low that only one electron passes through at a time.

1007

A particle of mass m is subjected to a force $\mathbf{F(r)} = -\nabla V(\mathbf{r})$ such that the wave function $\varphi(\mathbf{p}, t)$ satisfies the momentum-space Schrödinger equation

$$(\mathbf{p}^2/2m - a\nabla_p^2)\varphi(\mathbf{p}, t) = i\partial\varphi(\mathbf{p}, t)/\partial t,$$

where $\hbar = 1$, a is some real constant and

$$\nabla_p^2 = \partial^2/\partial p_x^2 + \partial^2/\partial p_y^2 + \partial^2/\partial p_z^2.$$

Find the force $\mathbf{F(r)}$.

(Wisconsin)

Sol: The coordinate and momentum representations of a wave function are related by

$$\psi(\mathbf{r}, t) = \left(\frac{1}{2\pi}\right)^{\frac{3}{2}} \int \varphi(\mathbf{k}, t)e^{i\mathbf{k}\cdot\mathbf{r}}d\mathbf{k},$$

$$\varphi(\mathbf{k}, t) = \left(\frac{1}{2\pi}\right)^{\frac{3}{2}} \int \psi(\mathbf{r}, t)e^{-i\mathbf{k}\cdot\mathbf{r}}d\mathbf{r},$$

where $\mathbf{k} = \frac{\mathbf{p}}{\hbar}$. Thus (with $\hbar = 1$)

$$\mathbf{P}^2\varphi(\mathbf{P}, t) \to -\nabla^2\psi(\mathbf{r}, t),$$

$$\nabla_p^2\varphi(\mathbf{p}, t) \to -r^2\psi(\mathbf{r}, t),$$

and the Schrödinger equation becomes, in coordinate space,

$$\left(\frac{-\nabla^2}{2m} + ar^2\right)\varphi(\mathbf{r}, t) = i\frac{\partial\varphi(\mathbf{r}, t)}{\partial t}.$$

Hence the potential is

$$V(\mathbf{r}) = ar^2,$$

and the force is

$$F(\mathbf{r}) = -\nabla V(\mathbf{r}) = -\frac{\mathbf{r}}{r}\frac{d}{d\mathbf{r}}V(r) = -2a\mathbf{r}.$$

1008

Consider the one-dimensional time-independent Schrödinger equation for some arbitrary potential $V(x)$. Prove that if a solution $\psi(x)$ has the property that $\psi(x) \to 0$ as $x \to \pm\infty$, then the solution must be nondegenerate and therefore real, apart from a possible overall phase factor.

Hint: Show that the contrary assumption leads to a contradiction.

(Berkeley)

Sol: Suppose that there exists another function $\phi(x)$ which satisfies the same Schrödinger equation with the same energy E as ψ and is such that $\lim_{x\to\infty}\phi(x)=0$. Then

$$\psi''/\psi = -2m(E-V)/\hbar^2,$$
$$\phi''/\phi = -2m(E-V)/\hbar^2,$$

and hence

$$\psi''\phi - \phi''\psi = 0,$$

or

$$\psi'\phi - \phi'\psi = \text{constant}.$$

The boundary conditions at $x \to \infty$ then give

$$\psi'\phi - \phi'\psi = 0,$$

or

$$\frac{\psi'}{\psi} = \frac{\phi'}{\phi}.$$

Integrating we have $\ln \psi = \ln \phi + \text{constant}$, or $\psi = \text{constant} \times \phi$. Therefore, ψ and ϕ represent the same state according to a statistical interpretation of wave function. That is, the solution is nondegenerate.

When $V(x)$ is a real function, ψ^* and ψ satisfy the same equation with the same energy and the same boundary condition $\lim_{x\to\infty}\psi^* = 0$. Hence $\psi^* = c\psi$, or

$\psi = c^* \psi^*$, from which we have $|c|^2 = 1$, or $c = \exp(i\delta)$, where δ is a real number. If we choose $\delta = 0$, then $c = 1$ and ψ is a real function.

1009

Consider a one-dimensional bound particle.

a. Show that

$$\frac{d}{dt} \int_{-\infty}^{\infty} \psi^*(x, t)\, \psi(x, t)\, dx = 0.$$

(ψ need not be a stationary state).

b. Show that, if the particle is in a stationary state at a given time, then it will always remain in a stationary state.

c. If at $t = 0$ the wave function is constant in the region $-a < x < a$ and zero elsewhere, express the complete wave function at a subsequent time in terms of the eigenstates of the system.

(Wisconsin)

Sol:

a. Consider the Schrödinger equation and its complex conjugate

$$i\hbar\, \partial\psi/\partial t = -\frac{\hbar^2}{2m}\, \nabla^2\psi + V\psi, \tag{1}$$

$$-i\hbar\, \partial\psi^*/\partial t = -\frac{\hbar^2}{2m}\, \nabla^2\psi^* + V\psi^*. \tag{2}$$

Taking $\psi^* \times (1) - \psi \times (2)$ we obtain

$$i\hbar\, \frac{\partial}{\partial t}\, (\psi^*\psi) = -\frac{\hbar^2}{2m} \nabla \cdot (\psi^* \nabla\psi - \psi\nabla\psi^*).$$

For the one dimension case we have, integrating over all space,

$$\frac{d}{dt} \int_{-\infty}^{\infty} \psi^*(x, t)\psi(x, t)\, dx = \frac{i\hbar}{2m} \int_{-\infty}^{\infty} \frac{\partial}{\partial x}\left(\psi^* \frac{\partial\psi}{\partial x} - \psi\frac{\partial\psi^*}{\partial x} \right) dx$$

$$= \frac{i\hbar}{2m} [\psi^* \partial\psi/\partial x - \psi\partial\psi^*/\partial x]_{-\infty}^{\infty}.$$

If ψ is a bound state, then $\psi(x \to \pm\infty) = 0$ and hence[1]

$$\frac{d}{dt} \int_{-\infty}^{\infty} \psi^*(x, t)\psi(x, t)\, dx = 0.$$

---

[1] The integral must be unity due to normalization of the wave function. Clearly the derivation is zero.

b. Supposing the particle is in a stationary state with energy E at $t = t_0$, we have

$$\hat{H}\psi(x, t_0) = E\psi(x, t_0),$$

where $\hat{H}$ does not depend on t explicitly. At any later time t, the Schrödinger equation

$$i\hbar\partial\psi(x, t)/\partial t = \hat{H}\psi(x, t)$$

applies. As $\hat{H}$ does not depend on t explicitly, the Schrödinger equation has the formal solution

$$\psi(x, t) = \exp[-i\hat{H}(t - t_0)/\hbar]\,\psi(x, t_0).$$

Multiplying both sides by $\hat{H}$ from the left and noting the commutability between $\hat{H}$ and $\exp[-i(t - t_0)\hat{H}/\hbar]$, we find

$$\hat{H}\psi(x, t) = \exp\left[\frac{-i\hat{H}(t - t_0)}{\hbar}\right]\hat{H}\psi(x, t_0)$$

$$= E\exp\left[\frac{-i\hat{H}(t - t_0)}{\hbar}\right]\hat{H}\psi(x, t_0)$$

$$= E\,\psi(x, t).$$

Hence $\psi(x, t)$ represents a stationary state at any later time t.

c. The wave function given for $t = 0$ can be written as

$$\psi(x, 0) = \begin{cases} C, & |x| < a, \\ 0, & \text{otherwise,} \end{cases}$$

where C is a constant. Normalization $\int_{-a}^{a} \psi^*\,\psi\,dx = 1$ requires that $C = \left(\frac{1}{2a}\right)^{\frac{1}{2}}$.
Suppose the eigenfunction of the bound state is $\langle x|n\rangle$ and $\hat{H}|n\rangle = E_n|n\rangle$. Then

$$1 = \sum_n |n\rangle\langle n|,$$

and

$$|\psi(x, 0)\rangle = \sum_n |n\rangle\langle n|\psi(x, 0)\rangle,$$

$$|\psi(x, t)\rangle = \sum_n |n\rangle\langle n|\psi(x, 0)\rangle \exp\left(-i\frac{E_n}{\hbar}t\right).$$

Hence

$$\psi(x, t) = \sum_n a_n \psi_n(x)\exp\left(-i\frac{E_n}{\hbar}t\right),$$

with

$$a_n = \langle n \mid \psi(x,0) \rangle = \int_{-\infty}^{\infty} \psi_n^*(x)\psi(x,0)dx$$

$$= \sqrt{\frac{1}{2a}} \int_{-a}^{a} \psi_n^*(x)dx.$$

1010

$\psi(x,t)$ is a solution of the Schrödinger equation for a free particle of mass m in one dimension, and

$$\psi(x,0) = A \exp(-x^2/a^2).$$

a. At time $t = 0$ find the probability amplitude in momentum space.

b. Find $\psi(x,t)$.

(Berkeley)

Sol:

a. At time $t = 0$ the probability amplitude in momentum space is

$$\psi(p,0) = \frac{1}{\sqrt{2\pi\hbar}} \int_{-\infty}^{\infty} e^{-ipx/\hbar} \psi(x,0)dx$$

$$= \frac{A}{\sqrt{2\pi\hbar}} \int_{-\infty}^{\infty} \exp(-x^2/a^2 - ipx/\hbar)dx$$

$$= \frac{Aa}{\sqrt{2\hbar}} \exp(-a^2 p^2/4\hbar^2).$$

b. The Schrödinger equation in momentum space for a free particle,

$$i\hbar\partial\psi(p,t)/\partial t = \hat{H}\psi(p,t) = \frac{p^2}{2m}\psi(p,t),$$

gives

$$\psi(p,t) = B \exp\left(\frac{-ip^2 t}{2m\hbar}\right)$$

At time $t = 0$, we have $B = \psi(p,0)$. Hence

$$\psi(p,t) = \frac{Aa}{\sqrt{2\hbar}} \exp\left[-\frac{a^2 p^2}{4\hbar^2} - \frac{ip^2 t}{2m\hbar}\right],$$

$$\psi(x,t) = \frac{1}{(2\pi\hbar)^{1/2}} \int_{-\infty}^{\infty} \exp\left[\frac{ipx}{\hbar}\right] \psi(p,t)\,dp$$

$$= \frac{Aa}{\sqrt{a^2 + \frac{2i\hbar t}{m}}} \exp\left[-\frac{x^2}{\left(a^2 + \frac{2i\hbar t}{m}\right)}\right].$$

We can also expand the wave function as a linear superposition of plane waves and get

$$\psi(x,t) = \frac{1}{(2\pi\hbar)^{1/2}} \int_{-\infty}^{\infty} \psi(p,0) e^{i(kx-\omega t)}\,dp$$

$$= \frac{1}{(2\pi\hbar)^{1/2}} \int_{-\infty}^{\infty} \frac{Aa}{\sqrt{2\hbar}} \exp\left[-\frac{a^2 p^2}{4\hbar^2}\right]$$

$$\times \exp\left[i\left(\frac{p}{\hbar}x - \frac{p^2 t}{2m\hbar}\right)\right] dp$$

$$= \frac{Aa}{2\hbar\sqrt{\pi}} \int_{-\infty}^{\infty} \exp\left[-\frac{a^2 p^2}{4\hbar^2} - \frac{ip^2 t}{2m\hbar} + \frac{ipx}{\hbar}\right] dp$$

$$= \frac{Aa}{\sqrt{a^2 + \frac{2i\hbar t}{m}}} \exp\left[-\frac{x^2}{\left(a^2 + \frac{2i\hbar t}{m}\right)}\right],$$

which agrees with the previous result.

1011

Consider a particle in a state $\psi = \dfrac{|\psi_1> + |\psi_2>}{\sqrt{2}}$ at time $t = 0$, where ψ_1 and ψ_2 are the wave functions of the ground state and first excited state of a particle in an infinite potential well of length L. At what time will the final wave function be orthogonal to the initial state?

Sol: At time t, the wave function is given by $\psi_f = \dfrac{|\psi_1> e^{-iE_1 t/\hbar} + |\psi_2> e^{-iE_2 t/\hbar}}{\sqrt{2}}$

For the orthogonality, $<\psi_i|\psi_f> = 0$, therefore, $\dfrac{<\psi_1|\psi_1> e^{-iE_1 t/\hbar} + <\psi_2|\psi_2> e^{-iE_2 t/\hbar}}{2}$

$= 0$ and all other cross terms are zero, and using the property of Kronecker delta,

$e^{-iE_1 t/\hbar} = -e^{-iE_2 t/\hbar}$, which therefore leads to $t = \dfrac{\hbar}{E_2 - E_1}\cos^{-1}(-1)$ and $E_2 = 4E_1$, so

$t = \dfrac{2mL^2}{3\pi\hbar}.$

1012

A particle of mass m moves in a one-dimensional box of length l with the potential

$$V = \infty, \quad x < 0,$$
$$V = 0, \quad 0 < x < l,$$
$$V = \infty, \quad x > l.$$

At a certain instant, say $t = 0$, the wave function of this particle is known to have the form

$$\psi = \sqrt{30/l^5}\, x(l-x), \qquad 0 < x < l,$$
$$\psi = 0, \qquad\qquad\quad \text{otherwise.}$$

Write down an expression for $\psi(x, t > 0)$ as a series, and expressions for the coefficients in the series.

(*Wisconsin*)

Sol: The eigenfunctions and the corresponding energy eigenvalues are

$$\psi_n(x) = \sqrt{\frac{2}{l}} \sin\left(\frac{\pi x}{l} n\right), \quad E_n = \frac{\hbar^2}{2m}\left(\frac{\pi}{l} n\right)^2, \quad n = 1, 2, 3, \ldots.$$

Thus

$$|\psi\rangle = \sum_n |n\rangle\langle n | \psi\rangle,$$

where

$$\langle n| \psi(t=0)\rangle = \int_0^l \sqrt{\frac{2}{l}}\sin\left(\frac{\pi x}{l} n\right) \cdot \sqrt{\frac{30}{l^5}} x(l-x)\, dx$$

$$= 4\sqrt{15}\left(\frac{1}{n\pi}\right)^3 (1 - \cos n\pi)$$

$$= 4\sqrt{15}[1 - (-1)^n]\,(1/n\pi)^3,$$

and hence

$$\psi(x, t) = \sum_{n=1}^{\infty} \langle n| \psi(t=0)\rangle\, \psi_n(x)\, \exp\left(-i\frac{E_n}{\hbar} t\right)$$

$$= \sum_{n=0}^{\infty} 8\sqrt{\frac{30}{l}} \cdot \frac{1}{(2n+1)^3 \pi^3} \sin\left(\frac{2n+1}{l}\pi x\right) e^{-i\frac{\hbar}{2m}\left(\frac{2n+1}{l}\pi\right)^2 t}.$$

1013

A rigid body with moment of inertia of I_z rotates freely in the $x - y$ plane. Let ϕ be the angle between the x-axis and the rotator axis.

a. Find the energy eigenvalues and the corresponding eigenfunctions.

b. At time $t = 0$ the rotator is described by a wave packet $\psi(0) = A \sin^2 \phi$. Find $\psi(t)$ for $t > 0$.

(Wisconsin)

Sol:

a. The Hamiltonian of a plane rotator is

$$H = -(\hbar^2 / 2I_z) d^2/d\phi^2$$

and so the Schrödinger equation is

$$-(\hbar^2 / 2I_z) d^2\psi / d\phi^2 = E\psi.$$

Setting $\alpha^2 = 2I_z E / \hbar^2$, we write the solution as

$$\psi = A\, e^{i\alpha\phi} + B\, e^{-i\alpha\phi},$$

where A, B are arbitrary constants. For the wave function to be single-valued, i.e. $\psi(\phi) = \psi(\phi + 2\pi)$, we require

$$\alpha \equiv m = 0, \qquad \pm 1, \pm 2, \ldots.$$

The eigenvalues of energy are then

$$E_m = m^2 \hbar^2 / 2I_z, \qquad m = 0, \pm 1, \ldots.$$

and the corresponding eigenfunctions are

$$\psi_m(\phi) = \frac{1}{\sqrt{2\pi}} e^{im\phi}, \qquad m = 0, \pm 1, \ldots,$$

after normalization $\int_0^{2\pi} \psi_m^* \psi_m \, d\phi = 1$.

b. At $t = 0$

$$\psi(0) = A\, \sin^2 \phi = \frac{A}{2}(1 - \cos 2\phi)$$

$$= A/2 - \frac{A}{4}(e^{i2\phi} + e^{-i2\phi}),$$

which corresponds to $m=0$ and $m=\pm2$. The angular speed is given by $E_m=\frac{1}{2}I_z\dot{\phi}^2$, or $\dot{\phi}=\frac{m\hbar}{I_z}$. Hence we have for time t

$$\psi(t)=\frac{A}{2}-\frac{A}{4}[e^{i2(\phi-\hbar t/I_z)}+e^{-i2(\phi+\hbar t/I_z)}].$$

1014

Consider a particle of mass m moving under the influence of potential $V(x)$.

Suppose its energy eigenstate is given by $\psi=\left(\dfrac{r}{\pi}\right)^{1/4}\exp(-rx^2/2)$ and its

energy is $E=\dfrac{\hbar^2 r}{2m}$, find the potential energy of the particle.

Sol:

The time-independent Schrödinger equation is given by $\dfrac{-\hbar^2}{2m}\psi''(x)+V(x)\psi(x)$

$=E\psi(x)$

$\psi'(x)=\left(\dfrac{r}{\pi}\right)^{1/4}\exp(-r\,x^2/2)(-x)=-x\psi(x)r$

$\psi''(x)=r\,[-\psi(x)-x\psi'(x)]=-r\,[\psi(x)+rx(-x\psi(x))]$

So, the equation is given by $\dfrac{-\hbar^2 r}{2m}(-r\psi(x)+x^2r^2\psi(x))+V(x)\psi(x)=\dfrac{\hbar^2 r}{2m}\psi(x)$

$V(x)=\dfrac{\hbar^2}{2m}(r+r^2x^2-r)=\dfrac{\hbar^2 x^2 r^2}{2m}.$

1015

Give the energy levels $E_n^{(a)}$ of the one-dimensional potential in Fig. 1.2(a) as well as the energy levels $E_n^{(b)}$ of the potential in Fig. 1.2(b)

(*Wisconsin*)

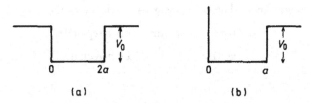

(a)   (b)

Fig. 1.2

Sol:

a. Use coordinate system as shown in Fig. 1.3. The Schrödinger equation is

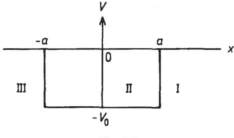

Fig. 1.3

$$-\frac{\hbar^2}{2m}\frac{d^2\psi}{dx^2}+V\psi=E\psi\,.$$

where

$$V=0 \qquad \text{for } x>a\,(\text{region I}),$$
$$V=-V_0 \qquad \text{for } -a<x<a\,(\text{region II}),$$
$$V=0 \qquad \text{for } x<-a\,(\text{region III}).$$

For bound states we require $-V_0<E<0$. Let

$$k^2=\frac{2m(E+V_0)}{\hbar^2}, \quad k'^2=-\frac{2mE}{\hbar^2}\,.$$

The Schrödinger equation becomes

$$\frac{d^2\psi}{dx^2}+k^2\psi=0 \text{ for region II,}$$

and

$$\frac{d^2\psi}{dx^2}-k'^2\psi=0 \text{ for region I and III,}$$

which have solutions

$$\psi=A\sin kx+B\cos kx \qquad \text{for} \quad -a<x<a,$$
$$\psi=Ce^{-k'x}+De^{k'x} \qquad \text{for} \quad x<-a \text{ and } x>a.$$

The requirement that $\psi\to 0$ as $x\to\pm\infty$ demands that

$$\psi=Ce^{-k'x} \qquad \text{for} \quad x>a\,(\text{region I}),$$
$$\psi=De^{k'x} \qquad \text{for} \quad x<a\,(\text{region III}).$$

The boundary conditions that ψ and ψ' be continuous at $x=\pm a$ then give

$$A \sin ka + B \cos ka = Ce^{-k'a},$$
$$-A \sin ka + B \cos ka = De^{-k'a},$$
$$Ak \cos ka - Bk \sin ka = -Ck' e^{-k'a},$$
$$Ak \cos ka + Bk \sin ka = Dk' e^{-k'a};$$

or

$$2A \sin ka = (C-D)e^{-k'a},$$
$$2B \cos ka = (C+D)e^{-k'a},$$
$$2Ak \cos ka = -(C-D)k'e^{-k'a},$$
$$2Bk \sin ka = (C+D)k'e^{-k'a}.$$

For solutions for which not all A, B, C, D vanish, we must have either $A=0, C=D$ giving $k \tan ka = k'$, or $B=0, C=-D$ giving $k \cot ka = -k'$. Thus two classes of solutions are possible, giving rise to bound states. Let $\xi = ka$, $\eta = k'a$.

Class 1:

$$\begin{cases} \xi \tan \xi = \eta, \\ \xi^2 + \eta^2 = \gamma^2, \end{cases}$$

where $\gamma^2 = k^2 a^2 + k'^2 a^2 = \frac{2mV_0 a^2}{\hbar^2}$.

Since ξ and η are restricted to positive values, the energy levels are found from the intersections in the first quadrant of the circle of radius γ with the curve of $\xi \tan \xi$ plotted against ξ, as shown in Fig. 1.4. The number of discrete levels depends on V_0 and a, which determine γ. For small γ only one solution is possible.

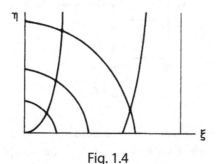

Fig. 1.4

Class 2:

$$\begin{cases} \xi \cot \xi = -\eta, \\ \xi^2 + \eta^2 = \gamma^2. \end{cases}$$

A similar construction is shown in Fig. 1.5. Here the smallest value of $V_0 a^2$ gives no solution while the larger two give one solution each.

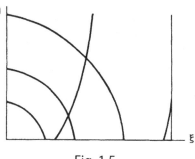

Fig. 1.5

Note that $\xi = 0$, $\eta = 0$ is a solution of $\xi \tan \xi = \eta$ and so no matter how small γ is, there is always a class 1 solution, whereas γ has to be above a minimum for a class 2 solution to exist, given by $\xi \cot \xi = 0$ which has a minimum solution $\xi = \frac{\pi}{2}$, i.e. $\gamma = \frac{\pi}{2}$ or $V_0 a^2 = \frac{\pi^2 \hbar^2}{8m}$.

b.   Use coordinates as shown in Fig. 1.6.

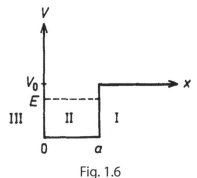

Fig. 1.6

The Schrödinger equation has solutions

$$\psi = A \sin kx + B \cos kx \qquad \text{for} \qquad 0 < x < a,$$
$$\psi = C e^{-k'x} \qquad \text{for} \qquad x > a,$$
$$\psi = 0 \qquad \text{for} \qquad x < 0,$$

satisfying the requirement $\psi \to 0$ as $x \to \infty$. The boundary conditions at $x=0$ and $x=a$ then give $B=0$,

$$A \sin ka = Ce^{-k'a},$$
$$Ak \cos ka = -Ck'e^{-k'a};$$

and finally

$$\xi \cot \xi = -\eta,$$
$$\xi^2 + \eta^2 = \gamma^2,$$

as for the class 2 solutions above.

1016

Consider the one-dimensional problem of a particle of mass m in a potential (Fig. 1.7)

$$V = \infty, \qquad x < 0,$$
$$V = 0, \qquad 0 \le x \le a,$$
$$V = V_0, \qquad x > a.$$

a. Show that the bound state energies $(E < V_0)$ are given by the equation

$$\tan \frac{\sqrt{2mEa}}{\hbar} = -\sqrt{\frac{E}{V_0 - E}}.$$

b. Without solving any further, sketch the ground state wave function.

(Buffalo)

Sol:

a. The Schrödinger equations for the two regions are

$$\psi'' + 2mE\psi/\hbar^2 = 0, \qquad 0 \le x \le a,$$
$$\psi'' - 2m(V_0 - E)\psi/\hbar^2 = 0, \qquad x > a,$$

with respective boundary conditions $\psi = 0$ for $x = 0$ and $\psi \to 0$ for $x \to +\infty$. The solutions for $E < V_0$ are then

$$\psi = \sin(\sqrt{2mE}x/\hbar), \qquad 0 \le x \le a;$$
$$\psi = Ae^{-\sqrt{2m(V_0 - E)}x/\hbar}, \qquad x > a,$$

where A is a constant. The requirement that ψ and $\dfrac{d\psi}{dx}$ are continuous at $x = a$ gives

$$\tan\left(\sqrt{2mEa}\,/\,\hbar\right) = -[E\,/\,(V_0 - E)]^{1/2}.$$

b.   The ground-state wave function is as shown in Fig. 1.7.

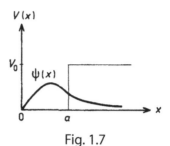

Fig. 1.7

1017

Consider a particle confined to a region, $-a < x < a$ with its wave function given by $\psi(x,t) = \sin\left(\dfrac{\pi x}{a}\right)\exp(-i\omega t)$. Find the potential in the region.

Sol: Using the formula $E\psi(x,t) = i\hbar\dfrac{\partial\psi(x,t)}{\partial t}$

$E\psi(x,t) = i\hbar\sin\left(\dfrac{\pi x}{a}\right)(-i\omega)\exp(-i\omega t)$ so, from this converting this in terms of wave function, $E\psi(x,t) = \hbar\omega\psi(x,t)$

From this we get $E = \hbar\omega$ (1)

The time-independent Schrödinger equation is given by $\dfrac{-\hbar^2}{2m}\psi''(x) + V(x)\psi(x)$

$= E\psi(x)$ where $\psi'(x) = \cos\left(\dfrac{\pi x}{a}\right)\dfrac{\pi}{a}e^{-i\omega t}$ and $\psi''(x) = -\left(\dfrac{\pi}{a}\right)^2\exp(-i\omega t)\sin\left(\dfrac{\pi x}{a}\right)$

and converting this in terms of wave function we get, $\psi''(x) = -\left(\dfrac{\pi}{a}\right)^2\psi(x)$

So plugging this into the earlier equation, we get $\dfrac{\hbar^2\pi^2}{2ma^2}\psi(x) + V(x)\psi(x) = E\psi(x)$

Substituting for E from Equation 1, $V(x) = \hbar\omega - \dfrac{\hbar^2\pi^2}{2ma^2}.$

1018

Consider a free particle confined to a one-dimensional infinite potential well of length L, with potential $V(x) = \begin{cases} 0; 0 < x < L \\ \infty; \text{elsewhere} \end{cases}$. What is the probability of finding an electron in the nth state for the region $x = 0$ to $x = L/n$?

Sol:

The wave function for the nth state is given by $\psi_n(x) = \sqrt{\dfrac{2}{L}} \sin(n\pi x / L)$.

The probability is given by $P = \displaystyle\int_0^{L/n} \psi^* \psi \, dx$

$P = \dfrac{2}{L} \displaystyle\int_0^{L/n} \sin^2(n\pi x / L) dx$ and using the formula for $\sin^2(x) = \dfrac{1 + \cos 2x}{2}$

We get, $P = \dfrac{1}{L} \left[\displaystyle\int_0^{L/n} dx - \displaystyle\int_0^{L/n} \cos(2n\pi x / L) dx \right]$

After applying the limit, it gives $P = \dfrac{1}{L} \left[\left(\dfrac{L}{n} - 0 \right) - (\sin(2\pi) - 0) \right]$

So the probability is given by $P = 1/n$.

1019

Consider the following one-dimensional potential wells:

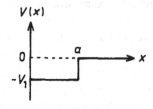

Fig. 1.8

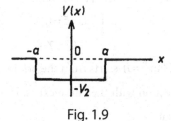

Fig. 1.9

a. Can each well support a bound state for an arbitrarily small depth $V_i (i = 1, 2)$? Explain qualitatively.

b. For $V_1 = V_2$, what is the relationship between the energies of the bound states of the two wells?

c. For continuum states of a given energy, how many independent solutions can each well have?

d. Explain qualitatively how it is possible to have bound states for which the particle is more likely to be outside the well than inside.

(*Wisconsin*)

Sol:

a. For bound states, we must have $-V < E < 0$. Let

$$k^2 = \frac{2m(E+V)}{\hbar^2}, \quad k'^2 = -\frac{2mE}{\hbar^2},$$

where $V = V_1$, V_2 for the two cases, and set $\xi = ka$, $\eta = k'a$, $\gamma = \frac{\sqrt{2mV}}{\hbar} a$.

The discussion in **Problem 1015** shows that for the potential in Fig. 1.8, the solutions are given by

$$\xi \cot \xi = -\eta, \quad \xi^2 + \eta^2 = \gamma^2.$$

The energy levels are given by the intersection of the curve $\xi \cot \xi = -\eta$ with a circle of radius γ with center at the origin (Fig. 1.5) in the first quadrant. As the figure shows, γ must be greater than the value of ξ for which $\xi \cot \xi = 0$, i.e.

$\xi \geq \frac{\pi}{2}$. Hence for a bound state to exist, we require $\frac{a\sqrt{2mV_1}}{\hbar} \geq \frac{\pi}{2}$, or

$$V_1 \geq \frac{\pi^2 \hbar^2}{8ma^2}.$$

For the potential shown in Fig. 1.9, two classes of solutions are possible. One class are the same as those for the case of Fig. 1.5 and are not possible for arbitrarily small V_2. The other class of solutions are given by

$$\begin{cases} \xi \tan \xi = \eta, \\ \xi^2 + \eta^2 = \gamma^2. \end{cases}$$

As the curve of $\xi \tan \xi = \eta$ starts from the origin, γ may be arbitrarily small and yet an intersection with the curve exists. However small V_2 is, there is always a bound state.

b. For $V_1 = V_2$, the bound states of the potential of Fig. 1.8 are also bound states of the potential of Fig. 1.9.

c. For continuum states of a given energy, there is only one independent solution for well 1, which is a stationary-wave solution with $\psi = 0$ at $x = 0$; there are two independent solutions corresponding to traveling waves in $+x$ and $-x$ directions for well 2.

d. Let p_1, p_2 denote respectively the probabilities that the particle is inside and outside the well. Consider, for example, the odd-parity solution

$$\psi = A \sin kx \qquad \text{for} \qquad 0 < x < a,$$
$$\psi = C e^{-k'x} \qquad \text{for} \qquad a < x,$$

where $k = \dfrac{\sqrt{2(mV_i + E)}}{\hbar}\,(i = 1, 2)$, $k' = \dfrac{\sqrt{-2mE}}{\hbar}$, for which

$$\frac{p_1}{p_2} = \frac{\int_0^a A^2 \sin^2 kx\, dx}{\int_a^\infty C^2 e^{-2k'x} dx} = \frac{A^2}{C^2}\frac{k'a}{e^{-2k'a}}\left(1 - \frac{\sin 2ka}{2ka}\right).$$

The continuity of ψ at $x = a$ gives

$$\frac{A}{C} = \frac{e^{-k'a}}{\sin ka}.$$

Setting, as before, $\eta = k'a$, $\xi = ka$, we have

$$\frac{p_1}{p_2} = \frac{\eta}{\sin^2 \xi}\left(1 - \frac{\sin 2\xi}{2\xi}\right) = \frac{2\eta}{1 - \cos 2\xi}\left(1 - \frac{\sin 2\xi}{2\xi}\right).$$

The odd-parity solutions are given by

$$\begin{cases} \xi \cot \xi = -\eta, \\ \xi^2 + \eta^2 = \gamma^2, \end{cases}$$

where $\gamma^2 = \dfrac{2mV_i a^2}{\hbar^2}$ $(i = 1, 2)$.

An analytic solution is possible if $\gamma \to (n + \tfrac{1}{2})\pi$, or

$$V_i a^2 \to \frac{\left(n + \tfrac{1}{2}\right)^2 \pi^2 \hbar^2}{2m}, \qquad (n = 0, 1, 2, \ldots)$$

for which the solution is $\xi \to (n + \tfrac{1}{2})\pi$, $\eta \to 0$, and

$$\frac{p_1}{p_2} \to 0.$$

The particle is then more likely outside the well than inside.

1020

Obtain the binding energy of a particle of mass m in one dimension due to the following short-range potential:

$$V(x)=-V_0\delta(x).$$

(Wisconsin)

Sol: The Schrödinger equation

$$d^2\psi/dx^2+\frac{2m}{\hbar^2}[E-V(x)]\,\psi=0,\ (E<0),$$

on setting

$$k=\sqrt{2m|E|}\big/\hbar,\ U_0=2mV_0\big/\hbar^2,$$

can be written as

$$\psi''(x)-k^2\psi(x)+U_0\delta(x)\,\psi(x)=0.$$

Integrating both sides of the above equation over x from $-\varepsilon$ to ε, where ε is an arbitrarily small positive number, we get

$$\psi'(\varepsilon)-\psi'(-\varepsilon)-k^2\int_{-\varepsilon}^{\varepsilon}\psi\,dx+U_0\psi(0)=0,$$

which becomes, by letting $\varepsilon\rightarrow0$,

$$\psi'(0^+)-\psi'(0^-)+U_0\psi(0)=0. \tag{1}$$

At $x\neq0\,(\delta(x)=0)$ the Schrödinger equation has solutions

$$\psi(x)\sim\exp(-kx)\quad\text{for}\quad x>0,$$
$$\psi(x)\sim\exp(kx)\quad\text{for}\quad x<0.$$

It follows from Eq. (1) that

$$\psi'(0^+)-\psi'(0^-)=-2k\psi(0).$$

A comparison of the two results gives $k=U_0/2$. Hence the binding energy is

$$-E=\hbar^2k^2\big/2m=mV_0^2\big/2\hbar^2.$$

1021

Consider a particle of mass m in the one-dimensional δ function potential

$$V(x)=V_0\,\delta(x).$$

Show that if V_0 is negative there exists a bound state, and that the binding energy is $mV_0^2\big/2\hbar^2$.

(Columbia)

Sol: In the Schrödinger equation

$$d^2\psi/dx^2 + 2m[E - V(x)]\,\psi/\hbar^2 = 0,$$

we set $E < 0$ for a bound state as well as

$$k^2 = 2m\,|E|/\hbar^2, \quad U_0 = 2mV_0/\hbar^2,$$

and obtain

$$d^2\psi/dx^2 - k^2\psi - U_0\delta(x)\psi = 0.$$

Integrating both sides over x from $-\varepsilon$ to $+\varepsilon$, where ε is an arbitrarily small positive number, we obtain

$$\psi'(\varepsilon) - \psi'(-\varepsilon) - k^2\int_{-\varepsilon}^{\varepsilon}\psi\,dx - U_0\psi(0) = 0.$$

With $\varepsilon \to 0^+$, this becomes $\psi'(0^+) - \psi'(0^-) = U_0\psi(0)$. For $x \neq 0$ the Schrödinger equation has the formal solution $\psi(x) \sim \exp(-k|x|)$ with k positive, which gives

$$\psi(x) \sim -k\frac{|x|}{x}e^{-k|x|} = \begin{cases} -ke^{-kx}, & x > 0, \\ ke^{kx}, & x < 0, \end{cases}$$

and hence

$$\psi'(0^+) - \psi'(0^-) = -2k\psi(0) = U_0\psi(0).$$

Thus $k = -U_0/2$, which requires V_0 to be negative. The energy of the bound state is then $E = -\frac{\hbar^2 k^2}{2m} = -mV_0^2/2\hbar^2$ and the binding energy is $E_b = 0 - E = mV_0^2/2\hbar^2$. The wave function of the bound state is

$$\psi(x) = A\,\exp\left(\frac{mV_0}{\hbar^2}|x|\right) = \sqrt{-mV_0/\hbar^2}\,\exp(mV_0\,|x|/\hbar^2),$$

where the arbitrary constant A has been obtained by the normalization $\int_{-\infty}^{0}\psi^2\,dx + \int_{0}^{\infty}\psi^2\,dx = 1$.

1022

A particle of mass m moves non-relativistically in one dimension in a potential given by $V(x) = -a\delta(x)$, where $\delta(x)$ is the usual Dirac delta function. The particle is bound. Find the value of x_0 such that the probability of finding the particle with $|x| < x_0$ is exactly equal to 1/2.

(Columbia)

Sol: For bound states, $E < 0$. The Schrödinger equation

$$\left[-\frac{\hbar^2}{2m}\frac{d^2}{dx^2} - a\delta(x) \right]\psi(x) = E\psi(x)$$

has for $x \neq 0$ the solutions finite at $x = \pm\infty$ as follows,

$$\psi(x) = \begin{cases} A\,e^{kx} & \text{for} & x < 0, \\ A\,e^{-kx} & \text{for} & x > 0, \end{cases}$$

where $k = \dfrac{\sqrt{-2mE}}{\hbar}$ and A is an arbitrary constant. Applying $\lim_{\varepsilon \to 0^+} \int_{-\varepsilon}^{\varepsilon} dx$ to the

Schrödinger equation gives

$$\psi'(0^+) - \psi'(0^-) = -\frac{2ma}{\hbar^2}\psi(0) \tag{1}$$

since

$$\int_{-\varepsilon}^{\varepsilon} \psi(x)\delta(x)\,dx = \psi(0), \quad \lim_{\varepsilon \to 0} \int_{-\varepsilon}^{\varepsilon} \psi(x)\,dx = 0$$

for finite $\psi(0)$. Substitution of $\psi(x)$ in (1) gives

$$k = \frac{ma}{\hbar^2}\,.$$

Hence

$$\psi(x) = A\,\exp\left(-\frac{ma|x|}{\hbar^2} \right).$$

On account of symmetry, the probabilities are

$$P(|x| < x_0) = 2|A|^2 \int_0^{x_0} e^{-2kx}\,dx = \frac{|A|^2}{k}\left(1 - e^{-2kx_0}\right),$$

$$P(-\infty < x < \infty) = 2|A|^2 \int_0^{\infty} e^{-2kx}\,dx = \frac{|A|^2}{k}.$$

As it is given

$$1 - e^{-2kx_0} = \frac{1}{2},$$

we have

$$x_0 = \frac{1}{2k}\ln 2 = \frac{\hbar^2}{2ma}\ln 2\,.$$

1023

A particle of mass m moving in one dimension is confined to the region $0 < x < L$ by an infinite square well potential. In addition, the particle experiences a delta function potential of strength λ located at the center of the well (Fig. 1.10). The Schrödinger equation which describes this system is, within the well,

$$-\frac{\hbar^2}{2m}\frac{\partial^2 \psi(x)}{\partial x^2}+\lambda\delta(x-L/2)\psi(x)=E\psi(x), \quad 0<x<L.$$

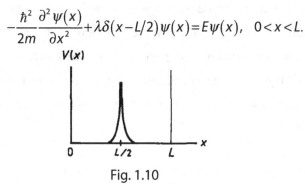

Fig. 1.10

Find a transcendental equation for the energy eigenvalues E in terms of the mass m, the potential strength λ, and the size L of the system.

(Columbia)

Sol: Applying $\lim_{\varepsilon\to 0}\int_{L/2-\varepsilon}^{L/2+\varepsilon}dx$ to both sides of the Schrödinger equation, we get

$$\psi'(L/2+\varepsilon)-\psi'(L/2-\varepsilon)=\left(2m\lambda/\hbar^2\right)\psi(L/2), \qquad (1)$$

since

$$\int_{\frac{L}{2}-\varepsilon}^{\frac{L}{2}+\varepsilon} \psi(x)\delta\left(x-\frac{L}{2}\right)dx=\psi\left(\frac{L}{2}\right), \ \lim_{\varepsilon\to 0}\int_{\frac{L}{2}-\varepsilon}^{\frac{L}{2}+\varepsilon}\psi(x)dx=0.$$

Subject to the boundary conditions $\psi(0)=\psi(L)=0$, the Schrödinger equation has solutions for $x\neq\frac{L}{2}$:

$$\psi(x)=\begin{cases}A_1\,\sin(kx), & 0\leq x\leq L/2-\varepsilon \\ A_2\,\sin[k(x-L)], & L/2+\varepsilon\leq x\leq L,\end{cases}$$

where $k=\dfrac{\sqrt{2mE}}{\hbar}$ and ε is an arbitrarily small positive number. The continuity of the wave function at $L/2$ requires $A_1\sin(kL/2)=-A_2\sin(kL/2)$, or $A_1=-A_2$.

Substituting the wave function in (1), we get

$$A_2 k \cos(kL/2) - A_1 k \cos(kL/2) = (2m\lambda A_1/\hbar^2) \sin(kL/2),$$

whence $\tan\dfrac{kL}{2} = -\dfrac{k\hbar^2}{m\lambda}$, or $\tan\dfrac{\sqrt{2mEL}}{2\hbar} = -\sqrt{\dfrac{2E}{m}}\dfrac{\hbar}{\lambda}$, which is the transcendental equation for the energy eigenvalue E.

1024

An infinitely deep one-dimensional square well potential confines a particle to the region $0 \le x \le L$. Sketch the wave function for its lowest energy eigenstate. If a repulsive delta function potential, $H' = \lambda \delta(x - L/2)(\lambda > 0)$, is added at the center of the well, sketch the new wave function and state whether the energy increases or decreases. If it was originally E_0, what does it became when $\lambda \to \infty$?

(*Wisconsin*)

Sol: For the square well potential the eigenfunction corresponding to the lowest energy state and its energy value are respectively

$$\phi_0(x) = \sqrt{2/L} \sin(\pi x/L),$$

$$E_0 = \frac{\pi^2 \hbar^2}{2mL^2}.$$

A sketch of this wave function is shown in Fig. 1.11

With the addition of the delta potential $H' = \lambda \delta(x - L/2)$, the Schrödinger equation becomes

$$\psi'' + \left[k^2 - \alpha\delta(x - L/2) \right] \psi = 0,$$

where $k^2 = 2mE/\hbar^2$, $\alpha = 2m\lambda/\hbar^2$. The boundary conditions are

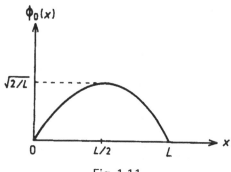

Fig. 1.11

$$\psi(0) = \psi(L) = 0, \tag{1}$$

$$\psi'\left[\left(\frac{L}{2}\right)^{+}\right] - \psi'\left[\left(\frac{L}{2}\right)^{-}\right] = \alpha\psi(L/2), \tag{2}$$

$$\psi\left[\left(\frac{L}{2}\right)^{+}\right] = \psi\left[\left(\frac{L}{2}\right)^{-}\right]. \tag{3}$$

Note that (2) arises from taking $\lim_{\varepsilon \to 0} \int_{\frac{L}{2}-\varepsilon}^{\frac{L}{2}+\varepsilon} dx$ over both sides of the Schrödinger equation and (3) arises from the continuity of $\psi(x)$ at $x = \frac{L}{2}$.

The solutions for $x \neq \frac{L}{2}$ satisfying (1) are

$$\psi = \begin{cases} A_1 \sin(kx), & 0 \leq x \leq L/2, \\ A_2 \sin[k(x-L)], & L/2 \leq x \leq L. \end{cases}$$

Let $k = k_0$ for the ground state. Condition (3) requires that $A_1 = -A_2 = A$, say, and the wave function for the ground state becomes

$$\psi_0(x) = \begin{cases} A \sin(k_0 x), & 0 \leq x \leq L/2, \\ -A \sin[k_0(x-L)], & L/2 \leq x \leq L. \end{cases}$$

Condition (2) then shows that k_0 is the smallest root of the transcendental equation

$$\cot(kL/2) = -\frac{m\lambda}{k\hbar^2}.$$

As $\cot\left(\frac{kL}{2}\right)$ is negative, $\pi/2 \leq k_0 L/2 \leq \pi$, or $\pi/L \leq k_0 \leq 2\pi/L$. The new ground-state wave function is shown Fig. 1.12. The corresponding energy is $E = \hbar^2 k_0^2/2m \geq E_0 = \frac{\pi^2 \hbar^2}{2mL^2}$, since $k_0 \geq \frac{\pi}{2}$. Thus the energy of the new ground state increases.

Furthermore, if $\lambda \to +\infty$, $k_0 \to 2\pi/L$ and the new ground-state energy $E \to 4E_0$.

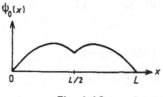

Fig. 1.12

1025

Consider a particle of mass m confined to a one-dimensional infinite square well. Assume the particle is in the ground state initially. If one of the wall at $x = L$ is suddenly shifted to $x = 2L$, find the probability of finding the particle in the ground state of the new potential.

Sol: The initial wave function is given by $\psi_i = \sqrt{\dfrac{2}{L}} \sin(\pi x / L)$ and the final wave function is given by $\psi_f = \sqrt{\dfrac{1}{L}} \sin(\pi x / 2L)$, so the probability amplitude is given by $P_{amp} = \int \psi_i^* \psi_f dx$ therefore the probability amplitude is given by

$$\frac{\sqrt{2}}{L} \int_0^L \sin(\pi x / L) \sin(\pi x / 2L) d\flat$$

where the limits are common to both systems. And using the formula

$$\int \sin(mx)\sin(nx) = \frac{\sin[(m-n)x]}{2(m-n)} - \frac{\sin[(m+n)x]}{2(m+n)} \quad \text{where } m = \pi/L \text{ and } n = \pi/2L$$

So $P_{amp} = \dfrac{\sqrt{2}}{L} \left(\dfrac{\sin[(\pi / 2L)x]}{\pi / L} - \dfrac{\sin[(3\pi / 2L)x]}{3\pi / L} \right)_0^L = P_{amp} = \sqrt{2}\left(\dfrac{1}{\pi} + \dfrac{1}{3\pi} \right) = 4\sqrt{2} / 3\pi$

And the probability is given by $P = |P_{amp}|^2 = \dfrac{32}{9\pi^2}$.

1026

An approximate model for the problem of an atom near a wall is to consider a particle moving under the influence of the one-dimensional potential given by

$$V(x) = -V_0 \delta(x), \qquad x > -d,$$
$$V(x) = \infty, \qquad x < -d,$$

where $\delta(x)$ is the so-called "delta function".

a. Find the modification of the bound-state energy caused by the wall when it is far away. Explain also how far is "far away".

b. What is the exact condition on V_0 and d for the existence of at least one bound state?

(*Buffalo*)

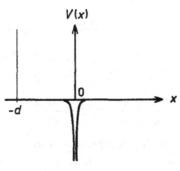

Fig. 1.13

Sol:

a. The potential is as shown in Fig. 1.13. In the Schrödinger equation

$$\psi'' + (2m/\hbar^2)[E + V_0\delta(x)]\psi = 0, \quad x > -d,$$

let $k = \sqrt{-2mE}/\hbar$, where $E < 0$. This has the formal solutions

$$\psi(x) = \begin{cases} ae^{kx} + be^{-kx} & \text{for} \quad -d < x < 0, \\ e^{-kx} & \text{for} \quad x > 0, \end{cases}$$

as $\psi(x)$ is finite for $x \to \infty$. The continuity of the wave function and the discontinuity of its derivative at $x = 0$ (Eq. (1) of **Problem 1020**), as well as the requirement $\psi(x = -d) = 0$, give

$$a + b = 1,$$

$$-k - (a - b)k = -2mV_0/\hbar^2,$$

$$ae^{-kd} + be^{kd} = 0.$$

Solving these we find

$$a = -\frac{e^{2kd}}{1 - e^{2kd}}, \quad b = \frac{1}{1 - e^{2kd}},$$

$$k = \frac{mV_0}{\hbar^2}(1 - e^{-2kd}).$$

The wall is "far away" from the particle if $kd \gg 1$, for which $k \approx mV_0/\hbar^2$. A better approximation is $k \approx (mV_0/\hbar^2)\left[1-\exp(-2mV_0d/\hbar^2)\right]$, which gives the bound-state energy as

$$E = -\frac{\hbar^2 k^2}{2m} \approx -\frac{\hbar^2}{2m}\left(\frac{mV_0}{\hbar^2}\right)^2\left[1-\exp\left(-\frac{2mV_0d}{\hbar^2}\right)\right]^2$$
$$\approx -\frac{mV_0^2}{2\hbar^2}\left[1-2\exp\left(-\frac{2mV_0d}{\hbar^2}\right)\right]$$
$$\approx -\frac{mV_0^2}{2\hbar^2}+\frac{mV_0^2}{\hbar^2}\exp\left(-\frac{2mV_0d}{\hbar^2}\right).$$

The second term in the last expression is the modification of energy caused by the wall. Thus for the modification of energy to be small we require $d \gg 1/k = \hbar^2/mV_0$. This is the meaning of being "far away"

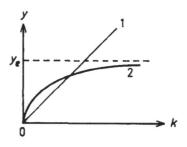

Fig. 1.14

b. Figure 1.14 shows line 1 representing $y=k$ and curve 2 representing $y = y_e\left[1-\exp(-2kd)\right]$, where $y_e = mV_0/\hbar^2$. The condition for the equation

$$k = mV_0\left[1-\exp(-2kd)\right]/\hbar^2$$

to have a solution is that the slope of curve 2 at the origin is greater than that of line 1:

$$\left.\frac{dy}{dk}\right|_{k=0} = 2mV_0d/\hbar > 1.$$

Hence if $V_0d > \frac{\hbar^2}{2m}$, there is one bound state.

1027

The wave function of the ground state of a harmonic oscillator of force constant k and mass m is

$$\psi_0(x) = (\alpha/\pi)^{1/4}e^{-\alpha x^2/2}, \quad \alpha = m\omega_0/\hbar, \quad \omega_0^2 = k/m.$$

Obtain an expression for the probability of finding the particle outside the classical region.

(*Wisconsin*)

Sol: The particle is said to be outside the classical region if $E < V(x)$. For the ground state, $E = \hbar\omega_0/2$ and the nonclassical region is $\frac{1}{2}\hbar\omega_0 < \frac{1}{2}m\omega_0^2 x^2$, i.e.,

$$x^2 > \frac{\hbar}{\omega_0 m} = \frac{1}{\alpha}, \text{ or } \begin{cases} x > \sqrt{\dfrac{1}{\alpha}}, \\ x < -\sqrt{\dfrac{1}{\alpha}}. \end{cases}$$

The probability of finding the particle in this nonclassical region is therefore

$$P = \int_{|x| > \sqrt{\frac{1}{a}}} \psi_0^2(x)\,dx$$

$$= \int_{-\infty}^{-\sqrt{1/\alpha}} \sqrt{\frac{\alpha}{\pi}} e^{-\alpha x^2}\,dx + \int_{\sqrt{1/\alpha}}^{\infty} \sqrt{\frac{\alpha}{\pi}} e^{-\alpha x^2}\,dx$$

$$= 2\int_{\sqrt{1/\alpha}}^{\infty} \sqrt{\frac{\alpha}{\pi}} e^{-\alpha x^2}\,dx$$

$$= 2\int_{1}^{\infty} \frac{1}{\sqrt{\pi}} e^{-t^2}\,dt \approx 16\%.$$

1028

Consider a linear harmonic oscillator and let ψ_0 and ψ_1 be its real, normalized ground and first excited state energy eigenfunctions respectively. Let $A\psi_0 + B\psi_1$ with A and B real numbers be the wave function of the oscillator at some instant of time. Show that the average value of x is in general different from zero. What values of A and B maximize $\langle x \rangle$ and what values minimize it?

(*Wisconsin*)

Sol: The orthonormal condition

$$\int (A\psi_0 + B\psi_1)^2\,dx = 1$$

gives $A^2 + B^2 = 1$. Generally A and B are not zero, so the average value of x,

$$\langle x \rangle = \int x(A\psi_0 + B\psi_1)^2\,dx = 2AB\,\langle \psi_0 | x | \psi_1 \rangle$$

is not equal to zero. Rewriting the above as

$$\langle x \rangle = \left[1-(A^2+B^2+2AB)\right]\langle \psi_0 \mid x \mid \psi_1 \rangle$$

$$= \left[1-(A-B)^2\right]\langle \psi_0 \mid x \mid \psi_1 \rangle$$

and considering $f = AB = A(1-A^2)^{\frac{1}{2}}$, which has extremums at $A=\pm\frac{1}{\sqrt{2}}$, we see that if $A=B=1/\sqrt{2}$, $\langle x \rangle$ is maximized; if $A=-B=1/\sqrt{2}$, $\langle x \rangle$ is minimized.

1029

Show that the minimum energy of a simple quantum harmonic oscillator is $\hbar\omega/2$ if $\Delta x \Delta p = \hbar/2$, where $(\Delta p)^2 = \langle (p-\langle p \rangle)^2 \rangle$.

(Wisconsin)

Sol: For a harmonic oscillator, $\langle x \rangle = \langle p \rangle = 0$, and so

$$(\Delta x)^2 = \langle x^2 \rangle, \quad (\Delta p)^2 = \langle p^2 \rangle.$$

Then the Hamiltonian of a harmonic oscillator, $H = p^2/2m + m\omega^2 x^2/2$, gives the average energy as

$$\langle H \rangle = \langle p^2 \rangle / 2m + m\omega^2 \langle x^2 \rangle /2 = \langle \Delta p \rangle^2 /2m + m\omega^2 \langle \Delta x \rangle^2 /2.$$

As for a, b real positive we have $\left(\sqrt{a}-\sqrt{b}\right)^2 \geq 0$, or $a+b \geq 2\sqrt{ab}$,

$$\langle H \rangle_{min} = \langle \Delta p \rangle \langle \Delta x \rangle \omega = \hbar\omega /2.$$

1030

An electron is confined in the ground state of a one-dimensional harmonic oscillator such that $\sqrt{\langle (x-\langle x \rangle)^2 \rangle} = 10^{-10}$ m. Find the energy (in eV) required to excite it to its first excited state.

[Hint: The virial theorem can help.]

(Wisconsin)

Sol: The virial theorem for a one-dimensional harmonic oscillator states that $\langle T \rangle = \langle V \rangle$. Thus $E_0 = \langle H \rangle = \langle T \rangle + \langle V \rangle = 2\langle V \rangle = m_e\omega^2 \langle x^2 \rangle$, or, for the ground state,

$$\frac{\hbar\omega}{2} = m_e\omega^2 \langle x^2 \rangle,$$

giving

$$\omega = \frac{\hbar}{2m_e\langle x^2 \rangle}.$$

As $\langle x \rangle = 0$ for a harmonic oscillator, we have

$$\sqrt{\langle (x - \langle x \rangle)^2 \rangle} = \sqrt{\langle x^2 \rangle - \langle x \rangle^2} = \sqrt{\langle x^2 \rangle} = 10^{-10} \text{ m}.$$

The energy required to excite the electron to its first excited state is therefore

$$\Delta E = \hbar \omega = \frac{\hbar^2}{2m_e \langle x^2 \rangle} = \frac{\hbar^2 c^2}{2m_e c^2 \langle x^2 \rangle}$$

$$= \frac{(6.58 \times 10^{-16})^2 \times (3 \times 10^8)^2}{2 \times 0.51 \times 10^{-20}} = 3.8 \text{ eV}.$$

1031

The wave function at time $t = 0$ for a particle in a harmonic oscillator potential $V = \frac{1}{2} k x^2$, is of the form

$$\psi(x, 0) = A e^{-(\alpha x)^2/2} \left[\cos \beta H_0(\alpha x) + \frac{\sin \beta}{2\sqrt{2}} H_2(\alpha x) \right],$$

where β and A are real constants, $\alpha^2 \equiv \sqrt{mk}/\hbar$, and the Hermite polynomials are normalized so that

$$\int_{-\infty}^{\infty} e^{-\alpha^2 x^2} [H_x(\alpha x)]^2 \, dx = \frac{\sqrt{\pi}}{\alpha} 2^n n!.$$

a. Write an expression for $\psi(x, t)$.

b. What are the possible results of a measurement of the energy of the particle in this state, and what are the relative probabilities of getting these values?

c. What is $\langle x \rangle$ at $t = 0$? How does $\langle x \rangle$ change with time?

(Wisconsin)

Sol:

a. The Schrödinger equation for the system is

$$i\hbar \partial_t \psi(x, t) = \hat{H} \psi(x, t),$$

where $\psi(x, t)$ takes the given value $\psi(x, 0)$ at $t = 0$. As $\hat{H}$ does not depend on t explicitly,

$$\psi_n(x, t) = \psi_n(x) e^{-iE_n t/\hbar},$$

where $\psi_n(x)$ is the energy eigenfunction satisfying

$$\hat{H}\psi_n(x) = E_n\psi_n(x).$$

Expanding $\psi(x, 0)$ in terms of $\psi_n(x)$:

$$\psi(x,\ 0) = \sum_n a_n\psi_n(x),$$

where

$$a_n = \int \psi_n^*(x)\psi(x,\ 0)\ dx\ .$$

Thus

$$\psi(x,\ t) = \sum_n a_n\psi_n(x,\ t) = \sum_n a_n\psi_n(x)e^{-iE_nt/\hbar}.$$

For a harmonic oscillator,

$$\psi_n(x) = N_n e^{-\alpha^2 x^2/2} H_n(\alpha x),$$

so that

$$a_n = \int N_n e^{-\alpha^2 x^2/2} H_n(\alpha x)\cdot A e^{-\alpha^2 x^2/2}$$
$$\times\left[\cos\beta H_0(\alpha x) + \frac{\sin\beta}{2\sqrt{2}} H_2(\alpha x)\right]dx.$$

As the functions $\exp\left(-\frac{1}{2}x^2\right)H_n(x)$ are orthogonal, all $a_n = 0$ except

$$a_0 = AN_0\sqrt{\frac{\pi}{\alpha^2}}\cos\beta,$$

$$a_2 = AN_2\sqrt{\frac{\pi}{\alpha^2}}2\sqrt{2}\sin\beta.$$

Hence

$$\psi(x, t) = A\sqrt{\frac{\pi}{\alpha^2}}\left[N_0\cos\beta\psi_0(x)e^{-iE_0t/\hbar}\right.$$
$$\left. +2\sqrt{2}N_2\sin\beta\psi_2(x)e^{-iE_2t/\hbar}\right].$$
$$= A\left(\frac{\pi}{\alpha^2}\right)^{\frac{1}{4}}\left[\cos\beta\psi_0(x)e^{-i\frac{E_0t}{\hbar}} + \sin\beta\psi_2(x)\,e^{-i\frac{E_2t}{\hbar}}\right],$$

as N_n are given by $\int[\psi_n(x)]^2 dx = 1$ to be $N_0 = \left(\frac{\alpha^2}{\pi}\right)^{\frac{1}{4}}$, $N_2 = \frac{1}{2\sqrt{2}}\left(\frac{\alpha^2}{\pi}\right)^{\frac{1}{4}}$.

b.   The observable energy values for this state are $E_0 = \hbar\omega/2$ and $E_2 = 5\hbar\omega/2$, and the relative probability of getting these values is

$$P_0/P_2 = \cos^2\beta/\sin^2\beta = \cot^2\beta.$$

c. As $\psi(x, 0)$ is a linear combination of only $\psi_0(x)$ and $\psi_2(x)$ which have even parity,

$$\psi(-x,\ 0)=\psi(x,\ 0).$$

Hence for $t=0$,

$$\langle x\rangle= \int \psi(x,0)\, x\psi(x,0)\, dx=0.$$

It follows that the average value of x does not change with time.

1032

a. For a particle of mass m in a one-dimensional harmonic oscillator potential $V = m\omega^2 x^2/2$, write down the most general solution to the time-dependent Schrödinger equation, $\psi(x, t)$, in terms of harmonic oscillator eigenstates $\phi_n(x)$.

b. Using (a) show that the expectation value of x, $\langle x\rangle$, as a function of time can be written as $A\cos\omega t + B\sin\omega t$, where A and B are constants.

c. Using (a) show explicitly that the time average of the potential energy satisfies $\langle V\rangle=\frac{1}{2}\langle E\rangle$ for a general $\psi(x, t)$.

Note the equality

$$\sqrt{\frac{m\omega}{\hbar}}x\phi_n = \sqrt{\frac{n+1}{2}}\phi_{n+1} + \sqrt{\frac{n}{2}}\phi_{n-1}.$$

(Wisconsin)

Sol:

a. From the time-dependent Schrödinger equation

$$i\hbar\frac{\partial}{\partial t}\psi(x,t)=-\frac{\hbar^2}{2m}\frac{\partial^2 \psi(x,t)}{\partial x^2}+m\omega^2 x^2/2$$

as $\hat{H}$ does not depend on time explicitly, we get

$$\psi(x,t)=e^{-iHt/\hbar}\psi(x,0).$$

We can expand $\psi(x, 0)$ in terms of $\phi_n(x)$:

$$\psi(x, 0) = \sum_n a_n \phi_n(x),$$

where

$$a_n = \langle \phi_n(x) | \psi(x, 0) \rangle,$$

and $\phi_n(x)$ are the eigenfunctions of

$$H\phi_n(x) = E_n \phi_n(x), \quad \text{with} \quad E_n = \left(n + \frac{1}{2} \right) \hbar\omega.$$

Hence

$$\psi(x, t) = \sum_n a_n \phi_n(x) e^{-iE_n t/\hbar}.$$

b. Using the given equality we have

$$\langle x \rangle = \int \psi^*(x, t) x \psi(x, t) dx$$

$$= \sum_{n, n'} a_n^* a_{n'} \, e^{-i(E_{n'} - E_n)t/\hbar} \int \phi_n^*(x) x \phi_{n'}(x) \, dx$$

$$= \sum_{n, n'} a_n^* a_{n'} \, e^{-i(E_{n'} - E_n)t/\hbar} \left(\sqrt{\frac{n'+1}{2}} \delta_{n, n'+1} + \sqrt{\frac{n'}{2}} \delta_{n, n'-1} \right) \sqrt{\frac{\hbar}{m\omega}}$$

$$= \sum_n a_n^* \left(a_{n-1} \sqrt{\frac{n}{2}} e^{i\omega t} + a_{n+1} \sqrt{\frac{n+1}{2}} e^{-i\omega t} \right) \sqrt{\frac{\hbar}{m\omega}}$$

$$= A \cos\omega t + B \sin\omega t,$$

where

$$A = \sqrt{\frac{\hbar}{m\omega}} \sum_n a_n^* \left(a_{n-1} \sqrt{\frac{n}{2}} + a_{n+1} \sqrt{\frac{n+1}{2}} \right),$$

$$B = \sqrt{\frac{\hbar}{m\omega}} \sum_n i a_n^* \left(a_{n-1} \sqrt{\frac{n}{2}} - a_{n+1} \sqrt{\frac{n+1}{2}} \right),$$

and we have used $E_{n+1} - E_n = \hbar\omega$.

c. The time average of the potential energy can be considered as the time average of the ensemble average of the operator $\hat{V}$ on $\psi(x, t)$. It is sufficient to take time average over one period $T = 2\pi/\omega$. Let $\langle A \rangle$ and $\bar{A}$ denote the time average and ensemble average of an operator A respectively.

As

$$V|\psi\rangle = \frac{1}{2}\hbar\omega \cdot \frac{m\omega}{\hbar}x^2|\psi\rangle$$

$$= \frac{1}{2}\hbar\omega\sqrt{\frac{m\omega}{\hbar}}x\sum a_n\sqrt{\frac{m\omega}{\hbar}}x\phi_n(x)e^{i\frac{E_n t}{\hbar}}$$

$$= \frac{1}{2}\hbar\omega\sqrt{\frac{m\omega}{\hbar}}x\sum_{n=0}^{\infty}a_n\left(\sqrt{\frac{n+1}{2}}|n+1\rangle\right.$$

$$\left. + \sqrt{\frac{n}{2}}|n-1\rangle\right)e^{-i\omega(n+1/2)t}$$

$$= \frac{1}{2}\hbar\omega\sum_{n=0}^{\infty}a_n\left[\sqrt{\frac{(n+1)(n+2)}{2}}|n+2\rangle\right.$$

$$\left. + \left(n+\frac{1}{2}\right)|n\rangle + \sqrt{\frac{n(n-1)}{2}}|n-2\rangle\right]e^{-i\frac{E_n t}{\hbar}},$$

we have

$$\bar{V} = \langle\psi|V|\psi\rangle$$

$$= \frac{1}{2}\hbar\omega\sum_{n=0}^{\infty}a_n^*a_n\left(n+\frac{1}{2}\right) + \frac{1}{2}\hbar\omega\sum_{n=0}^{\infty}a_{n+2}^*$$

$$\times a_n\sqrt{\frac{(n+1)(n+2)}{4}}e^{i2\omega t} + \frac{1}{2}\hbar\omega\sum_{n=0}^{\infty}a_n$$

$$\times a_{n+2}^*\sqrt{\frac{(n+1)(n+2)}{4}}e^{-i2\omega t}$$

$$= \frac{1}{2}\hbar\omega\sum_{n=0}^{\infty}a_n^*a_n\left(n+\frac{1}{2}\right)$$

$$+ \frac{1}{2}\hbar\omega\sum_{n=0}^{\infty}|a_{n+2}^*a_n|\sqrt{(n+1)(n+2)}\cos(2\omega t+\delta_n),$$

where δ_n is the phase of $a_{n+1}^*a_n$. Averaging $\bar{V}$ over a period, as the second term becomes zero, we get

$$\langle V\rangle = \frac{1}{T}\int_0^T \bar{V}\,dt = \frac{1}{2}\hbar\omega\sum_{n=0}^{\infty}a_n^*a_n\left(n+\frac{1}{2}\right).$$

On the other hand,

$$\bar{E} = \langle\psi|H|\psi\rangle = \hbar\omega\sum_{n=0}^{\infty}a_n^*a_n\left(n+\frac{1}{2}\right),$$

and $\langle E\rangle = \bar{E}$. Therefore $\langle V\rangle = \langle E\rangle/2$.

1033

Consider a particle of mass m in the one-dimensional potential

$$V(x) = m\omega^2 x^2/2, \qquad |x| > b;$$
$$V(x) = V_0, \qquad |x| < b,$$

where $V_0 \gg \hbar^2/mb^2 \gg \hbar\omega$, i.e. a harmonic oscillator potential with a high, thin, nearly impenetrable barrier at $x = 0$ (see Fig 1.15).

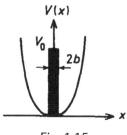

Fig. 1.15

a. What is the low-lying energy spectrum under the approximation that the barrier is completely impenetrable?

b. Describe qualitatively the effect on the spectrum of the finite penetrability of the barrier.

(MIT)

Sol:

a. For the low-lying energy spectrum, as the barrier is completely impenetrable, the potential is equivalent to two separate halves of a harmonic oscillator potential and the low-lying eigenfunctions must satisfy the condition $\psi(x) = 0$ at $x = 0$. The low-lying energy spectrum thus corresponds to that of a normal harmonic oscillator with odd quantum numbers $2n+1$, for which $\psi_n(x) = 0$, at $x = 0$ and $E_n = (2n + 3/2)\hbar\omega, n = 0, 1, 2, \ldots$ with a degeneracy of 2. Thus only the odd-parity wave functions are allowed for the low-lying levels.

b. There will be a weak penetration of the barrier. Obviously the probability for the particle to be in $|x| < b$, where the barrier exists, becomes less than that for the case of no potential barrier, while the probability outside the barrier becomes relatively larger. A small portion of the even-parity solutions is mixed into the particle states, while near the origin the probability distribution of even-parity states is greater than that of odd-parity states. Correspondingly, a small portion of the energy $E'_n = (2n+1/2)\hbar\omega$ is mixed into the energy for the case (a). Since $\langle\psi|$ barrier potential $|\psi\rangle > 0$, the energy levels will shift upwards. The level shifts for the even-parity states are greater than for odd-parity states. Furthermore, the energy shift is smaller for greater energies for states of the same parity.

1034

The Hamiltonian for a harmonic oscillator can be written in dimensionless units $(m = \hbar = \omega = 1)$ as

$$\hat{H} = \hat{a}^+\hat{a} + 1/2,$$

where

$$\hat{a} = (\hat{x} + i\hat{p})/\sqrt{2}, \quad \hat{a}^+ = (\hat{x} - i\hat{p})/\sqrt{2}.$$

One unnormalized energy eigenfunction is

$$\psi_a = (2x^3 - 3x)\,exp(-x^2/2).$$

Find two other (unnormalized) eigenfunctions which are closest in energy to ψ_a.

(MIT)

Sol: In the Fock representation of harmonic oscillation, $\hat{a}$ and $\hat{a}^+$ are the annihilation and creation operators such that

$$\hat{a}\psi_n = \sqrt{n}\,\psi_{n-1}, \quad \hat{a}^+\psi_n = \sqrt{n+1}\,\psi_{n+1}, \quad \hat{a}\hat{a}^+\psi_n = (n+1)\psi_n,$$

$$E_n = \left(n + \frac{1}{2}\right)\hbar\omega, \quad n = 0, 1, 2, \ldots.$$

As

$$\hat{a}\hat{a}^+\psi_a = \frac{1}{2}\left(x + \frac{d}{dx}\right)\left(x - \frac{d}{dx}\right)(2x^3 - 3x)e^{-\frac{x^2}{2}}$$

$$= \frac{1}{2}\left(x + \frac{d}{dx}\right)(4x^4 - 12x^2 + 3)e^{-\frac{x^2}{2}}$$

$$= 4(2x^3 - 3x)\,e^{-\frac{x^2}{2}} = (3+1)\psi_a,$$

we have $n = 3$. Hence the eigenfunctions closest in energy to ψ_a have $n = 2, 4$, the unnormalized wave functions being

$$\psi_2 = \frac{1}{\sqrt{3}}\hat{a}\psi_a = \frac{1}{\sqrt{6}}\left(x+\frac{d}{dx}\right)(2x^3-3x)\,e^{-x^2/2}$$

$$\sim (2x^2-1)\,e^{-x^2/2},$$

$$\psi_4 = \frac{1}{2}\hat{a}^+\psi_a = \frac{1}{2\sqrt{2}}\left(x-\frac{d}{dx}\right)(2x^3-3x)\,e^{-x^2/2}$$

$$\sim (4x^4-12x^2+3)\,e^{-x^2/2},$$

where the unimportant constants have been omitted.

1035

A particle of mass "m" is in a one-dimensional infinite potential well with an initial wave function $\psi(x)=\sqrt{\dfrac{30}{a^5}}x(a-x)$ between 0 and a. Outside 0 and a, the wave function would be zero.
Find the initial energy of the particle.

Sol:

Normalized wave function,

$$\psi_{(x)} = \sqrt{\frac{30}{a^5}}x(a-x)$$

To find energy, we apply Schrödinger equation,

$$\frac{-\hbar^2}{2m}\frac{d^2\psi}{dx^2}+v(x)\psi = \in\psi$$

$$\frac{-\hbar^2}{2m}\times\sqrt{\frac{30}{a^5}}\times-2 = \in\psi$$

$$\int_0^a \frac{\hbar^2}{m}\sqrt{\frac{30}{a^5}}\times\underbrace{\sqrt{\frac{30}{a^5}}x(a-x)}_{\psi_{(x)}}dx = \int_0^a \in\psi^*\,\psi\,dx$$

$$\Rightarrow \frac{5\hbar^2}{ma^2} = \in$$

1036

Consider the one-dimensional motion of a particle of mass μ in the potential

$$V(x) = V_0(x/a)^{2n},$$

where n is a positive integer and $V_0 > 0$. Discuss qualitatively the distribution of energy eigenvalues and the parities, if any, of the corresponding eigenfunctions. Use the uncertainty principle to get an order-of-magnitude estimate for the lowest energy eigenvalue. Specialize this estimate to the cases $n = 1$ and $n \to \infty$. State what $V(x)$ becomes in these cases and compare the estimates with your previous experience.

(*Buffalo*)

Sol: Since the potential $V(x) \to \infty$ as $x \to \infty$, there is an infinite number of bound states in the potential and the energy eigenvalues are discrete. Also, the mth excited state should have m nodes in the region of $E > V(x)$ given by $k\Delta x \approx (m+1)\pi$. Δx Increases slowly as m increases. From the virial theorem $2\overline{T} \propto 2n\overline{V}$, we have

$$k^2 \propto (\Delta x)^{2n} \propto [(m+1)\pi / k]^{2n},$$

and so

$$E \propto k^2 \propto (m+1)^{2n/(n+1)}.$$

Generally, as n increases, the difference between adjacent energy levels increases too. Since $V(-x) = V(x)$, the eigenstates have definite parities. The ground state and the second, fourth, . . . excited states have even parity while the other states have odd parity.

The energy of the particle can be estimated using the uncertainty principle

$$p_x \sim \hbar/2b,$$

where

$$b = \sqrt{(\Delta x)^2}.$$

Thus

$$E \sim \frac{1}{2\mu}(\hbar/2b)^2 + V_0(b/a)^{2n}.$$

For the lowest energy let $dE/db = 0$ and obtain

$$b = (h^2 a^{2n}/8\mu n V_0)^{1/2(n+1)}.$$

Hence the lowest energy is

$$E \sim [(n+1)V_0/a^{2n}]\,(\hbar^2 a^{2n}/8\mu n V_0)^{n/(n+1)}.$$

For $n = 1$, $V(x)$ is the potential of a harmonic oscillator,

$$V(x) = V_0 x^2/a^2 = \mu\omega^2 x^2/2.$$

In this case E equals $\hbar\omega/2$, consistent with the result of a precise calculation. For $n=\infty$, $V(x)$ is an infinite square-well potential, and

$$E = \hbar^2/8\mu a^2,$$

to be compared with the accurate result $\hbar^2\pi^2/2\mu a^2$.

1037

Consider a particle in one dimension with Hamiltonian

$$H = p^2/2m + V(x),$$

where $V(x) \le 0$ for all x, $V(\pm\infty)=0$, and V is not everywhere zero. Show that there is at least one bound state. (One method is to use the Rayleigh-Ritz variational principle with a trial wave function

$$\psi(x) = (b/\pi)^{1/4} \exp(-bx^2/2).$$

However, you may use any method you wish.)

(*Columbia*)

Sol: Method 1: Assume a potential $V(x)=f(x)$ as shown in Fig. 1.16. We take a square-well potential $V'(x)$ in the potential $V(x)$ such that

$$V'(x) = -V_0, \qquad |x| < a,$$
$$V'(x) = 0, \qquad |x| > a,$$
$$V'(x) \ge f(x) \qquad \text{for all } x.$$

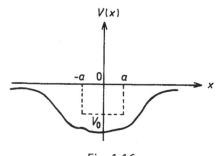

Fig. 1.16

We know there is at least a bound state $\varphi(x)$ in the well potential $V'(x)$ for which

$$\langle \varphi(x)| H' | \varphi(x)\rangle = \langle \varphi | p^2/2m + V'(x) | \varphi\rangle$$
$$= E_0 < 0.$$

We then have

$$\langle\varphi|H|\varphi\rangle=\langle\varphi|p^2/2m+f(x)|\varphi\rangle$$
$$\leq\langle\varphi|p^2/2m+V'(x)|\varphi\rangle$$
$$=E_0<0.$$

Let $\cdots\psi_{n-1}(x)$, $\psi_n(x)$, ... denote the eigenfunctions of H, and expand

$$\varphi(x)=\sum_n C_n\psi_n(x).$$

As

$$\langle\varphi|H|\varphi\rangle=\sum_n^\infty |C_n|^2\langle\psi_n|H|\psi_n\rangle<0,$$

there is at least an eigenfunction $\psi_i(x)$ satisfying the inequality

$$\langle\psi_i|H|\psi_i\rangle<0.$$

Hence there exists at least one bound state in $V(x)$.

Method 2: Let the wave function be

$$\psi(x)=(b/\pi)^{1/4}\exp(-bx^2/2),$$

where b is an undetermined parameter. We have

$$\langle H\rangle=\sqrt{b/\pi}\int_{-\infty}^\infty e^{-bx^2/2}\left(-\frac{\hbar^2}{2m}\frac{d^2}{dx^2}\right)e^{-bx^2/2}\,dx+\langle V\rangle$$
$$=\hbar^2 b/4m+\langle V\rangle,$$

where

$$\langle V\rangle=(b/\pi)^{1/2}\int_{-\infty}^\infty V(x)\exp(-bx^2)\,dx,$$

and thus

$$\frac{\partial\langle H\rangle}{\partial b}=\frac{\hbar^2}{4m}+\frac{1}{2b}\langle V\rangle-\left(\frac{b}{\pi}\right)^{1/2}\int_{-\infty}^{+\infty}x^2 V(x)e^{-bx^2}\,dx$$
$$=\frac{\hbar^2}{4m}+\frac{1}{2b}\langle V\rangle-\langle x^2 V\rangle=0,$$

giving

$$b=\frac{\langle V\rangle}{2\left[\langle x^2 V\rangle-\frac{\hbar^2}{4m}\right]}.$$

Substitution in the expression for $\langle H\rangle$ yields

$$\overline{E} = \langle H \rangle = \frac{\left[2\langle x^2 V \rangle - \dfrac{\hbar^2}{4m} \right] \langle V \rangle}{2 \left[\langle x^2 V \rangle - \dfrac{\hbar^2}{4m} \right]}.$$

As $V(x) \leq 0$ for all x, $V(\pm\infty) = 0$, and V is not everywhere zero, we have $\langle V \rangle < 0$, $\langle x^2 V \rangle < 0$ and hence $\overline{E} < 0$, $b > 0$.

In fact, under the condition that the total energy has a certain negative value (which must be greater than $\langle V \rangle$ to make $\langle T \rangle$ positive), whatever the form of V a particle in it cannot move to infinity and must stay in a bound state.

1038

The wave function for a particle of mass M in a one-dimensional potential $V(x)$ is given by the expression

$$\psi(x, t) = \alpha x \, \exp(-\beta x) \, \exp(i\gamma t / \hbar), \quad x > 0,$$
$$= 0, \quad\quad\quad\quad\quad\quad\quad\quad\quad\; x < 0,$$

where α, β and γ are all positive constants.

a. Is the particle bound? Explain.

b. What is the probability density $\rho(E)$ for a measurement of the total energy E of the particle?

c. Find the lowest energy eigenvalue of $V(x)$ in terms of the given quantities.

(MIT)

Sol:

a. The particle is in a bound state because the wave function $\psi(x, t)$ satisfies

$$\lim_{x \to -\infty} \psi(x, t) = 0,$$
$$\lim_{x \to +\infty} \psi(x, t) = \lim_{x \to +\infty} \alpha x e^{-\beta x} e^{i\gamma t/\hbar} = 0.$$

b, c. Substituting the wave function for $x > 0$ in the Schrödinger equation

$$i\hbar \frac{\partial}{\partial t} \psi(x, t) = \left[-\frac{\hbar^2}{2M} \frac{\partial^2}{\partial x^2} + V(x) \right] \psi(x, t)$$

gives

$$-\gamma x = -\frac{\hbar^2}{2M}(\beta^2 x - 2\beta) + V(x)x,$$

whence the potential for $x > 0$:

$$V(x) = -\gamma + \frac{\hbar^2}{2M}(\beta^2 - 2\beta/x).$$

As the stationary wave function of the particle in $V(x)$ satisfies

$$-\frac{\hbar^2}{2M}\left(\frac{d^2}{dx^2} - \beta^2 + 2\beta/x\right)\psi_E(x) = (E+\gamma)\,\psi_E(x),\ (x>0)$$

or

$$\frac{d^2}{dx^2}\psi_E(x) + \frac{2M}{\hbar^2}(E' + e^2/x)\psi_E(x) = 0$$

by setting

$$E' = E + \gamma - \beta^2\hbar^2/2M,\ e^2 = \beta\hbar^2/M,$$

and $\psi_E(x)\xrightarrow{x\to 0}0$, we see that the above equation is the same as that satisfied by the radial wave function of a hydrogen atom with $l = 0$. The corresponding Bohr radius is $a = \hbar^2/Me^2 = 1/\beta$, while the energy levels are

$$E'_n = -Me^4/2\hbar^2 n^2 = -\beta^2\hbar^2/2Mn^2,\ n = 1, 2, \ldots.$$

Hence

$$E_n = -\gamma + (\beta^2\hbar^2/2M)(1 - 1/n^2),\ n = 1, 2, \ldots,$$

and consequently the lowest energy eigenvalue is $E_1 = -\gamma$ with the wave function

$$\psi(x, t) = \alpha x\,\exp(-\beta x)\,\exp(i\gamma t/\hbar) \propto \psi_{E_1}(x)\,\exp(-iE_1 t/\hbar).$$

The probability density $\rho(E) = \psi^*\psi = \psi^*_{E_1}\psi_{E_1}$ is therefore

$$\rho(E) = \begin{cases} 1 & \text{for } E = -\gamma, \\ 0 & \text{for } E \neq -\gamma. \end{cases}$$

1039

A particle of mass m is released at $t = 0$ in the one-dimensional double square well shown in Fig. 1.13 in such a way that its wave function at $t = 0$ is just one sinusoidal loop ("half a sine wave") with nodes just at the edges of the left half of the potential as shown.

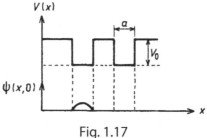

Fig. 1.17

a. Find the average value of the energy at $t=0$ (in terms of symbols defined above).

b. Will the average value of the energy be constant for times subsequent to the release of the particle? Why?

c. Is this a state of definite energy? (That is, will a measurement of the energy in this state always give the same value?) Why?

d. Will the wave function change with time from its value at $t=0$? If "yes", explain how you would attempt to calculate the change in the wave function. If "no", explain why not.

e. Is it possible that the particle could escape from the potential well (from the whole potential well, from both halves)? Explain.

(*Wisconsin*)

Sol:

a. The normalized wave function at $t=0$ is $\psi(x,0)=\sqrt{\frac{2}{a}}\sin\frac{\pi x}{a}$. Thus

$$\left\langle \hat{H} \right\rangle_{t=0} = -V_0 - \frac{\hbar^2}{2m}\frac{2}{a}\int_0^a \sin\left(\frac{\pi x}{a}\right)\frac{d^2}{dx^2}\sin\left(\frac{\pi x}{a}\right)dx$$

$$= \frac{\hbar^2\pi^2}{2ma} - V_0.$$

b. $\left\langle \hat{H} \right\rangle$ is a constant for $t>0$ since $\partial\left\langle \hat{H} \right\rangle/\partial t = 0$.

c. It is not a state of definite energy, because the wave function of the initial state is the eigenfunction of an infinitely deep square well potential with width a, and not of the given potential. It is a superposition state of the different energy eigenstates of the given potential. Therefore different measurements of the energy in this state will not give the same value, but a group of energies according to their probabilities.

d. The shape of the wave function is time dependent since the solution satisfying the given conditions is a superposition state:

$$\psi(x, 0) = \sqrt{\frac{2}{a}} \sin\left(\frac{\pi x}{a}\right) = \sum_n c_n \psi_n(x),$$

$$\psi(x, t) = \sum_n c_n \psi_n(x) e^{-iE_n t/\hbar}.$$

The shape of $\psi(x, t)$ will change with time because E_n changes with n.

e. The particle can escape from the whole potential well if the following condition is satisfied: $\hbar^2 \pi^2 / 2ma > V_0$. That is to say, if the width of the potential well is small enough (i.e., the kinetic energy of the particle is large enough), the depth is not very large (i.e., the value of V_0 is not very large), and the energy of the particle is positive, the particle can escape from the whole potential well.

1040

A free particle of mass m moves in one dimension. At time $t = 0$ the normalized wave function of the particle is

$$\psi(x, 0, \sigma_x^2) = (2\pi\sigma_x^2)^{-1/4} \exp(-x^2 / 4\sigma_x^2),$$

where $\sigma_x^2 = \langle x^2 \rangle$.

a. Compute the momentum spread $\sigma_p = \sqrt{\langle p^2 \rangle - \langle p \rangle^2}$ associated with this wave function.

b. Show that at time $t > 0$ the probability density of the particle has the form

$$|\psi(x, t)|^2 = |\psi(x, 0, \sigma_x^2 + \sigma_p^2 t^2 / m^2)|^2.$$

c. Interpret the results of parts (a) and (b) above in terms of the uncertainty principle.

(Columbia)

Sol:

    a. As

$$\langle p \rangle = \int_{-\infty}^{\infty} \psi^* \left(-i\hbar \frac{d}{dx}\right) \psi \, dx$$

$$= -i\hbar \int_{-\infty}^{\infty} \frac{1}{(2\pi\sigma_x^2)^{1/2}} \left(-\frac{x}{2\sigma_x^2}\right) e^{-\frac{x^2}{2\sigma_x^2}} dx = 0,$$

$$\langle p^2 \rangle = \int_{-\infty}^{\infty} \psi^* \left(-\hbar^2 \frac{d^2}{dx^2} \right) \psi \, dx$$

$$= -\hbar^2 \int_{-\infty}^{\infty} \frac{1}{\left(2\pi\sigma_x^2\right)^{1/2}} \left\{ -\frac{1}{2\sigma_x^2} + \frac{x^2}{4\sigma_x^4} \right\}$$

$$\times \, e^{-\frac{x^2}{2\sigma_x^2}} dx = \hbar^2 \, / \, 4\sigma_x^2,$$

$$\sigma_p = \sqrt{\langle p^2 \rangle} = \frac{\hbar}{2\sigma_x}.$$

b. By Fourier transform,

$$\psi(p, 0) = \frac{1}{\left(2\pi\hbar\right)^{1/2}} \int e^{-ipx/\hbar} \, \psi(x, 0) dx$$

$$= \frac{1}{\left(2\pi\hbar\right)^{1/2}} \int e^{-ipx/\hbar} \, \frac{1}{\left(2\pi\sigma_x^2\right)^{1/4}}$$

$$\times \, \exp\left(-x^2 \, / \, 4\sigma_x^2\right) dx$$

$$= \left[\left(2\pi\sigma_x^2\right)^{1/4} \Big/ \sqrt{2\pi\hbar} \right] \exp\left[-\sigma_x^2 p^2 \, / \, \hbar^2 \right]$$

Then

$$\psi(p, t) = \psi(p, 0) e^{-iEt/\hbar},$$

where

$$E = \frac{p^2}{2m}$$

for a free particle. By inverse Fourier transformation

$$\psi(x, t) = \int e^{ipx/\hbar} \psi(p, t) dp = \frac{\left(2\pi\sigma_x^2\right)^{1/4}}{\sqrt{2}\left(\pi\hbar\right)} \int e^{ipx/\hbar} \exp\left(-i\frac{p^2}{2m\hbar} t \right)$$

$$\times \exp\left(-\frac{\sigma_x^2 p^2}{\hbar^2} \right) dp = \left(\frac{\sigma_x^2}{2\pi} \right)^{1/4} \frac{1}{\left(\sigma_x^2 + i\dfrac{\hbar t}{2m} \right)^{1/2}}$$

$$\times \exp\left[-\frac{x^2}{4\left(\sigma_x^2 + i\dfrac{\hbar t}{2m} \right)} \right].$$

$$|\psi(x,t)|^2 = \frac{1}{\sqrt{2\pi\left(\sigma_x^2 + \frac{\sigma_p^2 t^2}{m^2}\right)}} \exp\left[-\frac{x^2}{2} \frac{1}{\sigma_x^2 + \frac{\sigma_p^2 t^2}{m^2}}\right]$$

$$= |\psi(x, 0, \sigma_x^2 + \sigma_p^2 t^2 m^{-2})|^2 .$$

c. Discussion:

⟨*i*⟩ The results indicate the width of the Gaussian wave packet at time *t* (which was originally σ_x at $t = 0$) is

$$\sqrt{\sigma_x^2 + \sigma_p^2 t^2 / m^2},$$

where $\sigma_p^2 = \hbar^2/4\sigma_x^2$.

⟨*ii*⟩ As $\sigma_x \sigma_p = \hbar/2$, the uncertainty principle is satisfied.

1041

A particle of mass *m* moves in one dimension under the influence of a potential $V(x)$. Suppose it is in an energy eigenstate $\psi(x) = \left(\gamma^2/\pi\right)^{1/4} \exp\left(-\gamma^2 x^2/2\right)$ with energy $E = \hbar^2 \gamma^2/2m$.

a. Find the mean position of the particle.

b. Find the mean momentum of the particle.

c. Find $V(x)$.

d. Find the probability $P(p)\, dp$ that the particle's momentum is between *p* and $p + dp$.

(*Wisconsin*)

Sol:

a. The mean position of the particle is

$$\langle x \rangle = \int_{-\infty}^{\infty} \psi^*(x) x \psi(x)\, dx = \frac{\gamma}{\sqrt{\pi}} \int_{-\infty}^{\infty} x e^{-\gamma^2 x^2}\, dx = 0 .$$

b. The mean momentum is

$$\langle p \rangle = \int_{-\infty}^{\infty} \psi^*(x) \frac{\hbar}{i} \left(\frac{d}{dx} \psi(x) \right) dx$$

$$= \frac{\gamma \hbar}{i \sqrt{\pi}} \int_{-\infty}^{\infty} e^{-\gamma^2 x^2/2} \frac{d}{dx} \left(e^{-\gamma^2 x^2/2} \right) dx = 0.$$

c. The Schrödinger equation

$$\left(-\frac{\hbar^2}{2m} \frac{d^2}{dx^2} + V(x) \right) \psi(x) = E\psi(x)$$

can be written as

$$-\frac{\hbar^2}{2m} \frac{d^2}{dx^2} \psi(x) = \left[E - V(x) \right] \psi(x).$$

As

$$-\frac{\hbar^2}{2m} \frac{d^2}{dx^2} e^{-\gamma^2 x^2/2} = -\frac{\hbar^2}{2m} \left(-\gamma^2 + \gamma^4 x^2 \right) e^{-\gamma^2 x^2/2},$$

we have

$$E - V(x) = -\frac{\hbar^2}{2m} \left(-\gamma^2 + \gamma^4 x^2 \right),$$

or

$$V(x) = \frac{\hbar^2}{2m} \left(\gamma^4 x^2 - \gamma^2 \right) + \frac{\hbar^2 \gamma^2}{2m} = \frac{\hbar^2 \gamma^4 x^2}{2m}.$$

d. The Schrödinger equation in momentum representation is

$$\left(\frac{p^2}{2m} - \frac{\hbar^4 \gamma^4}{2m} \frac{d^2}{dp^2} \right) \psi(p) = E\psi(p).$$

Letting

$$\psi(p) = N e^{-ap^2}$$

and substituting it into the above equation, we get

$$\frac{p^2}{2m} e^{-ap^2} - \frac{\hbar^4 \gamma^4}{2m} \left(-2a + 4a^2 p^2 \right) e^{-ap^2} = E e^{-ap^2},$$

or

$$4a^2 p^2 - 2a = \frac{1}{\hbar^2 \gamma^2} \left(\frac{p^2}{\hbar^2 \gamma^2} - 1 \right).$$

As the parameter a is independent of p, the above relation can be satisfied by $a = 1/2\hbar^2\gamma^2$. Hence

$$\psi(p) = N\exp\left(-p^2/2\hbar^2\gamma^2\right).$$

This is the eigenfunction of the state with energy $\hbar^2\gamma^2/2m$ in the momentum representation. Normalization gives $N = \left(1/\hbar^2\gamma^2\pi\right)^{1/4}$. Thus the probability that the particle momentum is between p and $p + dp$ is

$$P(p)dp = \left|\psi(p)\right|^2 dp = \left(\frac{1}{\hbar^2\gamma^2\pi}\right)^{1/2}\exp\left(-\frac{p^2}{\hbar^2\gamma^2}\right)dp.$$

Note that $\psi(p)$ can be obtained directly by the Fourier transform of $\psi(x)$:

$$\psi(p) = \int \frac{dx}{(2\pi\hbar)^{1/2}} e^{-ip\cdot x/\hbar}\left(\frac{\gamma}{\pi}\right)^{1/4} e^{-\gamma^2 x^2/2}$$

$$= \int \frac{dx}{(2\pi\hbar)^{1/2}}\left(\frac{\gamma^2}{\pi}\right)^{1/4}\exp\left[\left(\frac{p}{\sqrt{2\hbar\gamma}} - \frac{i\gamma x}{\sqrt{2}}\right)^2\right]\exp\left(-\frac{p^2}{2\hbar^2\gamma^2}\right)$$

$$= \left(\frac{1}{\hbar^2\gamma^2\pi}\right)^{1/4} e^{-p^2/2\hbar^2\gamma^2}.$$

1042

In one dimension, a particle of mass m is in the ground state of a potential which confines the particle to a small region of space. At time $t = 0$, the potential suddenly disappears, so that the particle is free for time $t > 0$. Give a formula for the probability per unit time that the particle arrives at time t at an observer who is a distance L away.

(Wisconsin)

Sol: Let $\psi_0(x)$ be the wave function at $t = 0$. Then

$$\psi(x,t) = \left\langle x\left|\exp\left(\frac{-ip^2 t}{2m\hbar}\right)\right|\psi_0(x)\right\rangle$$

$$= \int_{-\infty}^{+\infty}\left\langle x\left|e^{-ip^2 t/(2m\hbar)}\right|x'\right\rangle dx'\left\langle x'|\psi_0\right\rangle$$

$$= \int_{-\infty}^{+\infty}\left\langle x\left|e^{-ip^2 t/(2m\hbar)}\right|x'\right\rangle\psi_0(x')dx',$$

where

$$\left\langle x \left| \exp\left(-i\frac{p^2 t}{2m\hbar} \right) \right| x' \right\rangle$$

$$= \iint_{-\infty}^{+\infty} \langle x|p' \rangle dp' \left\langle p' \left| \exp\left(-i\frac{p^2 t}{2m\hbar} \right) \right| p \right\rangle dp \langle p|x' \rangle$$

$$= \int_{-\infty}^{+\infty} \frac{1}{2\pi\hbar} \exp\left[i\frac{px}{\hbar} - i\frac{px'}{\hbar} - i\frac{p^2 t}{2m\hbar} \right] dp$$

$$= \frac{1}{2\pi\hbar} \int_{-\infty}^{+\infty} \exp\left\{ -i\left[p\sqrt{\frac{t}{2m\hbar}} + \frac{(x'-x)\sqrt{m}}{\sqrt{2\hbar t}} \right]^2 \right\} dp \cdot \exp\left[i(x'-x)^2 \frac{m}{2\hbar t} \right]$$

$$= \frac{1}{2\pi\hbar} \sqrt{\frac{2m\hbar}{t}} \exp\left[i(x'-x)^2 \frac{m}{2\hbar t} \right] \int_{-\infty}^{+\infty} e^{-iq^2} dq$$

$$= \frac{1-i}{2} \sqrt{\frac{m}{\pi\hbar t}} \exp\left[i(x'-x)^2 \frac{m}{2\hbar t} \right].$$

Thus

$$\psi(x,t) = \frac{1-i}{2} \sqrt{\frac{m}{\pi\hbar t}} \int_{-\infty}^{+\infty} \exp\left[i(x'-x)^2 \frac{m}{2\hbar t} \right] \cdot \psi_0(x') dx'.$$

Represent the particle as a Gaussian wave packet of dimension a:

$$\psi_0(x) = \left(\pi a^2 \right)^{-1/4} \exp\left(-x^2 / 2a^2 \right).$$

The last integral then gives

$$\psi(x,t) = \frac{(1-i)}{2} \sqrt{\frac{m}{\pi\hbar t}} \frac{1}{\pi^{1/4} a^{1/2}}$$

$$\times \exp\left\{ i\frac{mx^2}{2ht} \left[1 - \frac{m}{2ht\left(\frac{m}{2\hbar t} + \frac{i}{2a^2} \right)} \right] \right\}$$

$$\times \frac{1}{\sqrt{\frac{m}{2\hbar t} + \frac{i}{2a^2}}} \int_{-\infty}^{+\infty} e^{i\xi^2} d\xi$$

$$= \frac{1}{a^{1/2}\pi^{1/4}} \sqrt{\frac{m}{m+i\frac{ht}{a^2}}} \exp\left[-\frac{mx^2}{2a^2} \frac{1}{m+i\frac{ht}{a^2}} \right],$$

whence the current density

$$j = Re\left(\psi^* \frac{p_x}{m} \psi\right) = \frac{\hbar^2 xt}{\sqrt{\pi}a^5 m^2} \frac{1}{\left(1+\frac{\hbar^2 t^2}{m^2 a^4}\right)^{3/2}}$$

$$\times \exp\left[-\frac{x^2}{a^2} \frac{1}{1+\left(\frac{\hbar t}{ma^2}\right)^2}\right].$$

By putting $x = L$, we get the probability per unit time that the particle arrives at the observer a distance L away.

1043

A free particle of mass m moves in one dimension. The initial wave function of the particle is $\psi(x,0)$.

a. Show that after a sufficiently long time t the wave function of the particle spreads to reach a unique limiting form given by

$$\psi(x,t) = \sqrt{m/\hbar t}\, \exp(-i\pi/4)\exp\left(imx^2/2\hbar t\right)\varphi(mx/\hbar t),$$

where φ is the Fourier transform of the initial wave function:

$$\varphi(k) = (2\pi)^{-1/2} \int \psi(x,0)\exp(-ikx)dx.$$

b. Give a plausible physical interpretation of the limiting value of $|\psi(x,t)|^2$.

Hint: Note that when $\alpha \to \infty$,

$$\exp(-i\alpha u^2) \to \sqrt{\pi/\alpha}\, \exp(-i\pi/4)\delta(u).$$

(Columbia)

Sol:

a. The Schrödinger equation is

$$\left[i\hbar\partial/\partial t + \left(h^2/2m\right)d^2/dx^2\right]\psi(x,t) = 0.$$

By Fourier transform, we can write

$$\psi(k,t) = \frac{1}{\sqrt{2\pi}} \int_{-\infty}^{\infty} dx e^{-ikx}\psi(x,t),$$

and the equation becomes

$$\left(i\hbar \frac{\partial}{\partial t} - \frac{\hbar^2 k^2}{2m} \right) \psi(k, t) = 0 \,.$$

Integration gives

$$\psi(k, t) = \psi(k, 0) \exp\left(-i \frac{k^2 \hbar t}{2m} \right),$$

where

$$\psi(k, 0) = \frac{1}{\sqrt{2\pi}} \int_{-\infty}^{\infty} dx e^{-ikx} \psi(x, 0) \equiv \varphi(k) \,.$$

Hence

$$\psi(k, t) = \varphi(k) \exp\left(-i \frac{k^2 \hbar t}{2m} \right),$$

giving

$$\psi(x, t) = \frac{1}{\sqrt{2\pi}} \int_{-\infty}^{\infty} dk e^{ikx} \psi(k, t)$$

$$= \frac{1}{\sqrt{2\pi}} \int_{-\infty}^{\infty} dk \varphi(k) \exp\left(ikx - i \frac{k^2 \hbar t}{2m} \right)$$

$$= \frac{1}{\sqrt{2\pi}} \exp\left(i \frac{mx^2}{2\hbar t} \right) \int_{-\infty}^{\infty} dk$$

$$\times \exp\left[-i \frac{\hbar t}{2m} \left(k - \frac{mx}{\hbar t} \right)^2 \right] \varphi(k).$$

With $\xi = k - mx/\hbar t$, this becomes

$$\psi(x, t) = \frac{1}{\sqrt{2\pi}} \exp\left(i \frac{mx^2}{2\hbar t} \right) \int_{-\infty}^{\infty} d\xi$$

$$\times \exp\left(-i \frac{\hbar t}{2m} \xi^2 \right) \varphi\left(\xi + \frac{mx}{\hbar t} \right).$$

For $\alpha \to \infty$,

$$e^{-i\alpha u^2} \to \sqrt{\frac{\pi}{\alpha}} \exp\left(-i \frac{\pi}{4} \right) \delta(u),$$

and so after a long time t $(t \to \infty)$,

$$\exp\left(-i \frac{\hbar t}{2m} \xi^2 \right) \to \sqrt{\frac{2\pi m}{\hbar t}} \delta(\xi) \exp\left(-i \frac{\pi}{4} \right),$$

and

$$\psi(x,t)=\frac{1}{\sqrt{2\pi}}\exp\left(i\frac{mx^2}{2\hbar t}\right)\int_{-\infty}^{\infty}d\xi\sqrt{\frac{2\pi m}{\hbar t}}$$

$$\times\delta(\xi)\varphi\left(\xi+\frac{mx}{\hbar t}\right)\exp\left(-i\frac{\pi}{4}\right)$$

$$=\frac{1}{\sqrt{2\pi}}\exp\left(i\frac{mx^2}{2\hbar t}\right)\sqrt{\frac{2\pi m}{\hbar t}}\exp\left(-i\frac{\pi}{4}\right)\varphi\left(\frac{mx}{\hbar t}\right)$$

$$=\sqrt{\frac{m}{\hbar t}}\exp\left(-i\frac{\pi}{4}\right)\exp\left(i\frac{mx^2}{2\hbar t}\right)\varphi\left(\frac{mx}{\hbar t}\right).$$

b.

$$|\psi(x,t)|^2=\frac{m}{\hbar t}\left|\varphi\left(\frac{mx}{\hbar t}\right)\right|^2.$$

Because $\varphi(k)$ is the Fourier transform of $\psi(x,0)$, we have

$$\int_{-\infty}^{+\infty}|\varphi(k)|^2\,dk=\int_{-\infty}^{+\infty}|\psi(x,0)|^2\,dx.$$

On the other hand, we have

$$\int_{-\infty}^{+\infty}|\psi(x,t)|^2\,dx=\int_{-\infty}^{+\infty}\frac{m}{\hbar t}\left|\varphi\left(\frac{mx}{\hbar t}\right)\right|^2\,dx$$

$$=\int_{-\infty}^{+\infty}|\varphi(k)|^2\,dk=\int_{-\infty}^{+\infty}|\psi(x,0)|^2\,dx,$$

which shows the conservation of total probability. For the limiting case of $t\to\infty$, we have

$$|\psi(x,t)|^2\to 0\cdot|\varphi(0)|^2=0,$$

which indicates that the wave function of the particle will diffuse infinitely.

1044

The one-dimensional quantum mechanical potential energy of a particle of mass m is given by

$$V(x)=V_0\delta(x),\qquad -a<x<\infty,$$
$$V(x)=\infty,\qquad x<-a,$$

as shown in Fig. 1.18. At time $t=0$, the wave function of the particle is completely confined to the region $-a<x<0$. [Define the quantities $k=\sqrt{2mE}/\hbar$ and $\alpha=2mV_0/\hbar^2$]

a. Write down the normalized lowest-energy wave function of the particle at time $t=0$.

b. Give the boundary conditions which the energy eigenfunctions

$$\psi_k(x)=\psi_k^{I}(x) \quad \text{and} \quad \psi_k(x)=\psi_k^{II}(x)$$

must satisfy, where the region I is $-a<x<0$ and the region II $x\geq0$.

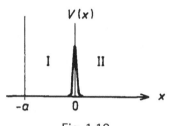

Fig. 1.18

c. Find the (real) solutions for the energy eigenfunctions in the two regions (up to an overall constant) which satisfy the boundary conditions.

d. The $t=0$ wave function can be expressed as an integral over energy eigenfunctions:

$$\psi(x)=\int_{-\infty}^{\infty}f(k)\psi_k(x)dk.$$

Show how $f(k)$ can be determined from the solutions $\psi_k(x)$.

e. Give an expression for the time development of the wave function in terms of $f(k)$. What values of k are expected to govern the time behavior at large times?

(Wisconsin)

Sol:

a. The required wave function $\psi(x)$ must satisfy the boundary conditions $\psi(-a)=\psi(0)=0$. A complete orthonormal set of wave functions defined in $-a<x<0$ and satisfying the Schrödinger equation consists of

$$\phi_n(x)=\begin{cases}\sqrt{\dfrac{2}{a}}\sin\left(\dfrac{n\pi x}{a}\right), & -a<x<0,\\[2mm]0, & \text{outside}[-a,\,0],\end{cases}$$

where $n = 1, 2, \ldots,$ with

$$\langle \phi_n \mid H \mid \phi_m \rangle = E_n \delta_{mn}, \quad E_n = (\hbar^2/2m)(n\pi/a)^2 .$$

The normalized lowest-energy wave function is given by $n = 1$ as

$$\psi_n(x) = \begin{cases} \sqrt{\dfrac{2}{a}} \sin\left(\dfrac{\pi x}{a}\right), & -a < x < 0, \\ 0, & \text{outside}[-a, 0]. \end{cases}$$

b. The Schrödinger equation for $x > -a$ is

$$-\frac{\hbar^2}{2m} \frac{d^2}{dx^2} \psi + V_0 \delta(x)\psi = E\psi ,$$

or

$$\psi''(x) + k^2 \psi(x) = \alpha \delta(x)\psi(x)$$

with

$$k^2 = \frac{2mE}{\hbar^2}, \quad \alpha = \frac{2mV_0}{\hbar^2} .$$

The boundary conditions and the discontinuity condition to be satisfied are

$$\psi^{\mathrm{I}}(-a) = 0, \quad \psi^{\mathrm{I}}(0) = \psi^{\mathrm{II}}(0), \quad \psi^{\mathrm{II}}(+\infty) = \text{finite},$$

$$\psi^{\mathrm{II}'}(0) - \psi^{\mathrm{I}'}(0) = \alpha \psi^{\mathrm{I}}(0).$$

The last equation is obtained by integrating the Schrödinger equation over a small interval $[-\epsilon, \epsilon]$ and letting $\epsilon \to 0$ (see **Problem 1020**).

c. In both the regions I and II, the wave equation is

$$\psi''(x) + k^2 \psi(x) = 0 ,$$

whose real solutions are sinusoidal functions. The solutions that satisfy the boundary conditions are

$$\psi_k(x) = \begin{cases} \psi_k^{\mathrm{I}}(x) = c_k \sin k(x+a), & -a < x < 0, \\ \psi_k^{\mathrm{II}}(x) = c_k \sin k(x+a) + A_k \sin kx, & x \geq 0, \\ 0, & x < -a. \end{cases}$$

The discontinuity and normalization conditions then give

$$A_k = \frac{c_k \alpha}{k} \sin ka,$$

$$c_k = \left\{ \frac{\pi}{2} \left[1 + \frac{\alpha \sin 2ka}{k} + \left(\frac{\alpha \sin ka}{k} \right)^2 \right] \right\}^{-\frac{1}{2}} .$$

d. Expand the wave function $\psi(x)$ in terms of $\psi_k(x)$,

$$\psi(x) = \int_{-\infty}^{\infty} f(k)\psi_k(x)\,dk,$$

and obtain

$$\int_{-a}^{\infty} \psi_{k'}^*(x)\psi(x)\,dx = \int \int f(k)\psi_k(x)\psi_{k'}^*(x)\,dk\,dx$$
$$= \int_{-\infty}^{\infty} f(k)\delta(k-k')\,dk = f(k'),$$

or

$$f(k) = \int_{-a}^{\infty} \psi_k^*(x)\psi(x)\,dx.$$

e. As

$$\psi(x,0) = \int_{-\infty}^{\infty} f(k)\psi_k(x)\,dk,$$

we have

$$\psi(x,t) = \int_{-\infty}^{+\infty} f(k)\psi_k(x)e^{-iE_k t/\hbar}\,dk.$$

At time $t=0$, the particle is in the ground state of an infinitely deep square well potential of width a, it is a wave packet. When $t>0$, since the $\delta(x)$ potential barrier is penetrable, the wave packet will spread over to the region $x>0$. Quantitatively, we compute first

$$f(k) = \int_{-a}^{0} c_k \sin k(x+a) \cdot \sqrt{\frac{2}{a}} \sin\frac{\pi x}{a}\,dx$$
$$= \sqrt{\frac{1}{2a}} \int_{-a}^{0} ck\left\{\cos\left[\left(k-\frac{\pi}{a}\right)x + ka\right]\right.$$
$$\left. -\cos\left[\left(k+\frac{\pi}{a}\right)x + ka\right]\right\}dx$$
$$= \frac{\pi}{a} \cdot \sqrt{\frac{2}{a}} \frac{\sin ka}{k^2 - \left(\frac{\pi}{a}\right)^2} c_k$$

and then

$$\psi(x,t) = \frac{\pi}{a}\sqrt{\frac{2}{a}} \int_{-\infty}^{-\infty} c_k \frac{\sin ka}{k^2 - \left(\frac{\pi}{a}\right)^2}$$
$$\times \left\{ \begin{array}{c} \sin k(x+a) \\ \sin k(x+a) + \frac{\alpha}{k}\sin ka \sin kx \end{array} \right\} e^{-iE_k t/\hbar}\,dk,$$

where $E_k = \hbar^2 k^2/2m$. In the last expression the upper and lower rows are for regions I and II respectively.

When $t \to \infty$, the oscillatory factor $\exp(-iE_k t/\hbar)$ changes even more rapidly, while the other functions of the integrand behave quite normally ($k = \pi/a$ is not a pole). Thus $\psi(x,t)$ tend to zero for any given x. When t is very large, component waves of small wave number k play the principal role. At that time the particle has practically escaped from the region $[-a, 0]$.

1045

The radioactive isotope $_{83}\text{Bi}^{212}$ decays to $_{81}\text{Tl}^{208}$ by emitting an alpha particle with the energy $E = 6.0$ MeV .

a. In an attempt to calculate the lifetime, first consider the finite potential barrier shown in Fig. 1.19. Calculate the transition probability T for a particle of mass m incident from the left with energy E in the limit $T \ll 1$.

b. Using the above result, obtain a rough numerical estimate for the lifetime of the nucleus Bi^{212}. Choose sensible barrier parameters to approximate the true alpha particle potential.

(CUS)

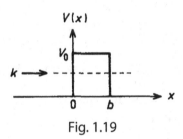

Fig. 1.19

Sol:

a. If $T \ll 1$, the incident wave is reflected at $x = 0$ as if the potential barrier were infinitely thick. We thus have

$$\psi(x) = e^{ikx} + (t_1 - 1)e^{-ikx}, \qquad x < 0,$$
$$\psi(x) = t_1 e^{-k'x}, \qquad 0 < x < b,$$

where t_1 is the amplitude transmission coefficient and

$$k' = \sqrt{\frac{2m(V_0 - E)}{\hbar^2}}, \qquad k = \sqrt{\frac{2mE}{\hbar^2}} .$$

The continuity of $\psi'(x)$ at $x = 0$ gives

$$ik(2 - t_1) = -k't_1, \text{ or } t_1 = \frac{2k}{k + ik'}.$$

Consider the reflection at b. We have

$$\psi(x) = t_1 e^{-k'b}[e^{-k'(x-b)} + (t_2 - 1)e^{k'(x-b)}], \qquad 0 < x < b,$$

$$\psi(x) = t_1 t_2 e^{-k'b} e^{ik(x-b)}, \qquad x > b,$$

and so

$$-k'(2 - t_2) = ikt_2, \quad \text{or} \quad t_2 = 2ik'/(k + ik').$$

Hence the transition probability is given by

$$T = t_1 t_2 e^{-k'b}$$

to be

$$|T|^2 = \frac{16k^2 k'^2}{(k^2 + k'^2)^2} e^{-2k'b} = \frac{16E(V_0 - E)}{V_0^2} e^{-2k'b}.$$

b. To estimate the rate of α-decay of $_{83}\text{Bi}^{212}$, we treat, in first approximation, the Coulomb potential experienced by the α-particle in the $_{81}\text{Tl}$ nucleus as a rectangular potential barrier. As shown in Fig. 1.20, the width of the barrier r_0 can be taken to be

$$r_0 = \frac{2Ze^2}{E} = \frac{2(83 - 2)}{6} \frac{e^2}{\hbar c} \frac{\hbar c}{\text{MeV}}$$

$$= \frac{162}{6} \times \frac{1}{137} \times 6.58 \times 10^{-22} \times 3 \times 10^{10}$$

$$= 3.9 \times 10^{-12} \text{cm}.$$

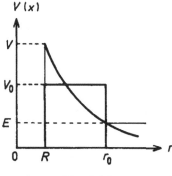

Fig. 1.20

The radius of the nucleus of Tl is

$$R = 1 \times 10^{-13} \times 208^{\frac{1}{3}} = 6 \times 10^{-13} \text{cm},$$

corresponding to a Coulomb potential height of

$$V = \frac{2Ze^2}{r_0} \cdot \frac{r_0}{R} = 39 \text{ MeV}.$$

An α-particle, moving with speed v in the nucleus of Tl, makes $\frac{v}{2R}$ collisions per second with the walls. Hence the lifetime τ of $_{83}\text{Bi}^{212}$ is given by

$$\tau |T|^2 \frac{v}{2R} \approx 1,$$

or

$$\tau \approx \frac{2R}{v|T|^2}.$$

Taking for the rectangular potential barrier a height $V_0 \approx \frac{1}{2}(39-6)+6 = 22.5 \text{ MeV}$, $b = r_0 - R = 33 \times 10^{-13}$ cm (see Fig. 1.20), $v = \sqrt{\frac{2E}{m}} = \sqrt{\frac{2 \times 6}{940}} c \approx 0.1 \, c$,

we find

$$2k'b = \frac{2\sqrt{2mc^2(V_0 - E)}}{\hbar c} \, b = \frac{2\sqrt{2 \times 940 \times 16.5} \times 33 \times 10^{-13}}{6.58 \times 10^{-22} \times 3 \times 10^{10}} = 59,$$

$$\tau = \frac{2 \times 6 \times 10^{-13}}{3 \times 10^9} \times \frac{22.5^2}{16 \times 6 \times (22.5-6)} \times e^{59}$$

$$= 5.4 \times 10^3 \text{ s}.$$

1046

A particle of mass m is in an infinite potential well extending from 0 to a. If the initial wave function of the particle is $\psi_{(x)} = x(a-x)$ apart from normalization, what is the wave function after time $t = T$sec?

Sol:

$$\int \psi^2 dx = 1$$

$$A^2 \int x^2 (a-x)^2 = 1$$

$$A^2 \int_0^a x^2 (a^2 - 2ax + x^2) dx = 1$$

$$A^2 \frac{a^5}{30} = 1 \qquad\qquad A = \sqrt{\frac{30}{a^5}}.$$

$$\psi_{(x,0)} = \int_0^a \sqrt{\frac{30}{a^5}} x(a-x) \sqrt{\frac{2}{a}} \sin \frac{n\pi x}{a}$$

$$= \sqrt{\frac{30}{a^5}} \sqrt{\frac{2}{a}} \int_0^a x(a-x) \sin \frac{n\pi x}{a} dx$$

$$= \sqrt{\frac{30}{a^5}} \sqrt{\frac{2}{a}} \int_0^a xa \sin \frac{n\pi x}{a} - \int_0^a x^2 \sin \frac{n\pi x}{a}$$

$$= \sqrt{\frac{30}{a^5}} \sqrt{\frac{2}{a}} a \left[-x \cos \frac{n\pi x}{a} \times \frac{a}{n\pi} + \sin \frac{n\pi x}{a} \left(\frac{a}{n\pi} \right)^2 \right]$$

$$- \left[-x^2 \cos \frac{n\pi x}{a} \left(\frac{a}{n\pi} \right) + 2x \sin \frac{n\pi x}{a} \times \left(\frac{a}{n\pi} \right)^2 + 2 \cos \frac{n\pi x}{a} \times \left(\frac{a}{n\pi} \right)^3 \right]_0^a$$

$$= \sqrt{\frac{30}{a^5}} \sqrt{\frac{2}{a}} \left[(-1)^n \frac{a^{\cancel{2}^0}}{n\pi} + 0 \right] - \left[\frac{a^3}{n\pi} \cancel{(-1)^0} + 1 - \frac{2a^3}{\cancel{(n\pi)^2}} \times 0 + 2 \left(\frac{a}{n\pi} \right)^3 (-1)^n \right]$$

$$= \sqrt{\frac{60}{a^6}} \left[2(-1)^n - 1 \left(\frac{a}{n\pi} \right)^3 \right] = \frac{\sqrt{120}}{n^3 \pi^3} \left[1 - (-1)^n \right] = \frac{8\sqrt{15}}{n^3 \pi^3} \left[1 - (-1)^n \right].$$

$$\psi_{(x,0)} = \Sigma C_n \psi_{(x,t)}.$$

$$\psi_{(x,t)} = \frac{8\sqrt{15}}{\pi^3} \sum_{n=\text{odd}} \frac{1}{n^3} \sin \frac{n\pi x}{a} e^{\frac{-iET}{\hbar}}.$$

1047

Consider a one-dimensional square-well potential (see Fig. 1.21)

$$V(x) = 0, \qquad x < 0,$$
$$V(x) = -V_0, \qquad 0 < x < a,$$
$$V(x) = 0, \qquad x > a,$$

where V_0 is positive. If a particle with mass m is incident from the left with nonrelativistic kinetic energy E, what is its probability for transmission through the potential? For what values of E will this probability be unity?

(Columbia)

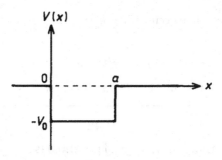

Fig. 1.21

Sol: Let the wave function be

$$\psi(x) = e^{ikx} + Re^{-ikx}, \qquad x < 0,$$

$$\psi(x) = Se^{ikx}, \qquad x > a,$$

$$\psi(x) = Ae^{ik'x} + Be^{-ik'x}, \qquad 0 < x < a,$$

where

$$k = \frac{\sqrt{2mE}}{\hbar}, \ k' = \frac{\sqrt{2m(E + V_0)}}{\hbar}.$$

The constants R, S, A, B are to be determined from the boundary conditions that $\psi(x)$ and $\psi'(x)$ are both continuous at $x = 0$ and $x = a$, which give

$$\begin{cases} 1 + R = A + B, \\ k(1 - R) = k'(A - B), \\ Ae^{ik'a} + Be^{-ik'a} = Se^{ika}, \\ k'(Ae^{ik'a} - Be^{-ik'a}) = kSe^{ika}. \end{cases}$$

Hence

$$S = \frac{4kk'e^{-ika}}{(k + k')^2 e^{-ik'a} - (k - k')^2 e^{ik'a}}$$

and the probability for transmission is

$$P = \frac{j_t}{j_i} = |S|^2$$

$$= \frac{4k^2 k'^2}{4(kk' \cos k'a)^2 + (k^2 - k'^2)^2 \sin^2(k'a)}.$$

Resonance transmission occurs when $k'a = n\pi$, i.e., when the kinetic energy E of the incident particle is

$$E = n^2 \pi^2 \hbar^2 / 2ma^2 - V_0 .$$

The probability for transmission, P, then becomes unity.

1048

Consider a one-dimensional square-well potential (see Fig. 1.22):

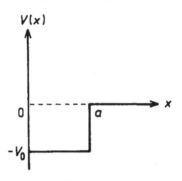

Fig. 1.22

$$V(x) = \infty, \qquad x < 0,$$
$$V(x) = -V_0, \qquad 0 < x < a,$$
$$V(x) = 0, \qquad x > a.$$

a. For $E < 0$, find the wave function of a particle bound in this potential. Write an equation which determines the allowed values of E.

b. Suppose a particle with energy $E > 0$ is incident upon this potential. Find the phase relation between the incident and the outgoing wave.

(Columbia)

Sol: The Schrödinger equations for the different regions are

$$\left[-\frac{\hbar^2}{2m} \frac{d^2}{dx^2} - V_0 - E \right] \psi(x) = 0, \quad 0 < x < a,$$

$$\left[-\frac{\hbar^2}{2m} \frac{d^2}{dx^2} - E \right] \psi(x) = 0, \quad x > a.$$

a. $E < 0$.

⟨i⟩ Consider first the case of $V_0 < -E$, for which the wave function is

$$\psi(x) = \begin{cases} 0, & x < 0, \\ A\sinh(kx), & 0 < x < a, \\ Be^{-k'x}, & x > a, \end{cases}$$

where

$$k = \sqrt{\frac{2m(-V_0 - E)}{\hbar^2}}, \quad k' = \sqrt{\frac{2m(-E)}{\hbar^2}}.$$

The continuity conditions of the wave function give

$$A\sinh(ka) = Be^{-k'a},$$
$$Ak\cosh(ka) = -Bk'e^{-k'a}$$

and hence

$$k\coth(ka) = -k'.$$

As $\coth x > 0$ for $x > 0$, there is no solution for this case.

⟨ii⟩ For $V_0 > -E$, $ik \to k$, $k = \sqrt{2m(V_0 + E)}/\hbar$, and the equation determining the energy becomes $k\cot(ka) = -k'$. The wave function is

$$\psi(x) = \begin{cases} 0, & x < 0, \\ A\sin(kx), & 0 < x < a, \\ Be^{-k'x}, & x > a. \end{cases}$$

From the continuity and normalization of the wave function we get

$$A = \left[\frac{2}{\frac{1}{k'}\sin^2(ka) + a - \frac{1}{2k}\sin(2ka)}\right]^{1/2},$$

$$B = \left[\frac{2}{\frac{1}{k'}\sin^2(ka) + a - \frac{1}{2k}\sin(2ka)}\right]^{1/2} e^{k'a}\sin(ka).$$

b. $E > 0$. The wave function is

$$\psi(x) = \begin{cases} 0, & x < 0, \\ A\sin(kx), & 0 < x < a, \\ B\sin(k'x + \varphi), & x > a, \end{cases}$$

where

$$k = \sqrt{\frac{2m(E + V_0)}{\hbar^2}}, \quad k' = \sqrt{2mE/\hbar^2}.$$

As $\partial \ln \psi / \partial \ln x$ is continuous at $x = a$,

$$(ka)\cot(ka) = (k'a)\cot(k'a + \varphi),$$

whence

$$\varphi = \text{arccot}\left(\frac{k}{k'}\cot(ka)\right) - k'a.$$

For $x > a$,

$$\psi(x) = \frac{B}{2i}e^{-ik'x - i\varphi} - \frac{B}{2i}e^{ik'x + i\varphi},$$

where

$$\varphi_{\text{inc}}(x) \propto e^{-ik'x - i\varphi},$$

$$\varphi_{\text{out}}(x) \propto e^{ik'x + i\varphi}.$$

Hence the phase shift of the outgoing wave in relation to the incident wave is

$$\delta = 2\varphi = 2\left[\text{arccot}\left(\frac{k}{k'}\cot(ka)\right) - k'a\right].$$

1049

Consider a one-dimensional system with potential energy (see Fig. 1.23)

$$V(x) = V_0, \qquad x > 0,$$
$$V(x) = 0, \qquad x < 0,$$

where V_0 is a positive constant. If a beam of particles with energy E is incident from the left (i.e., from $x = -\infty$), what fraction of the beam is transmitted and what fraction reflected? Consider all possible values of E.

(Columbia)

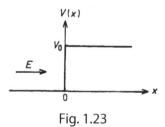

Fig. 1.23

Sol: For $x < 0$, the Schrödinger equation is

$$\frac{d^2}{dx^2}\psi + \frac{2mE}{\hbar^2}\psi = 0,$$

whose solution has the form

$$\psi(x) = e^{ikx} + re^{-ikx},$$

where

$$k = \sqrt{\frac{2mE}{\hbar^2}}.$$

For $x > 0$, the equation is

$$\frac{d^2}{dx^2}\psi + \frac{2m(E - V_0)}{\hbar^2}\psi = 0.$$

⟨i⟩ If $E < V_0$, write the above as

$$\frac{d^2}{dx^2}\psi - \frac{2m(V_0 - E)}{\hbar^2}\psi = 0.$$

As $\psi(x)$ must be finite for $x \to \infty$, the solution has the form

$$\psi(x) = te^{-k'x},$$

where

$$k' = \sqrt{\frac{2m(V_0 - E)}{\hbar^2}}.$$

The continuity conditions then give

$$1 + r = t,$$

$$ik - ikr = -tk',$$

whence $r = (k' + ik)/(ik - k') = (1 - ik'/k)/(1 + ik'/k)$. Therefore the fraction reflected is $R = j_{\text{ref}} / j_{\text{inc}} = |r|^2 = 1$, the fraction transmitted is $T = 1 - R = 0$.

⟨ii⟩ $E > V_0$. For $x > 0$, we have

$$\left[\frac{d^2}{dx^2} + \frac{2m(E - V_0)}{\hbar^2}\right]\psi(x) = 0.$$

where

$$\psi(x) = te^{ik'x}, \quad k' = \sqrt{\frac{2m(E - V_0)}{\hbar^2}}.$$

Noting that there are only outgoing waves for $x \to \infty$, we have $1 + r = t$, $ik - ikr = ik't$, and thus $r = (k' - k)/(k' + k)$. Hence the fraction reflected is $R = [(k' - k)/(k' + k)]^2$, the fraction transmitted is $T = 1 - R = 4kk'/(k + k')^2$.

1050

A particle of mass m and momentum p is incident from the left on the potential step shown in Fig. 1.24.

Calculate the probability that the particle is scattered backward by the potential if

a. $p^2/2m < V_0$,

b. $p^2/2m > V_0$.

(Columbia)

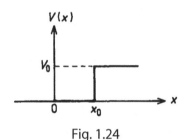

Fig. 1.24

Sol: The Schrödinger equations are

$$\left(\frac{d^2}{dx^2} + \frac{2mE}{\hbar^2}\right)\psi(x) = 0 \quad \text{for } x < x_0,$$

$$\left[\frac{d^2}{dx^2} + \frac{2m}{\hbar^2}(E - V_0)\right]\psi(x) = 0 \quad \text{for } x > x_0.$$

a.   If $E < V_0$, we have

$$\psi(x) = \begin{cases} e^{ik(x-x_0)} + re^{-ik(x-x_0)}, & x < x_0, \\ te^{-k'(x-x_0)}, & x > x_0, \end{cases}$$

where

$$k = \sqrt{\frac{2mE}{\hbar^2}},$$

$$k' = \sqrt{\frac{2m(V_0 - E)}{\hbar^2}},$$

the condition that $\psi(x)$ is finite for $x \to \infty$ having been made use of. The continuity conditions give $1 + r = t$, $ik - ikr = -k't$, whence $r = (k' + ik)/(ik - k')$. The probability of reflection is $R = j_r / j_i = |r|^2 = 1$.

b. If $E > V_0$. We have

$$\psi(x) = \begin{cases} e^{ik(x-x_0)} + re^{-ik(x-x_0)}, & x < x_0, \\ te^{ik'(x-x_0)}, & x > x_0, \end{cases}$$

where

$$k = \sqrt{\frac{2mE}{\hbar^2}},$$

$$k' = \sqrt{\frac{2m(E - V_0)}{\hbar^2}},$$

noting that there is only outgoing wave for $x > x_0$. The continuity conditions give $1 + r = t$, $ik - ikr = ik't$, and hence $r = (k - k')/(k + k')$. The probability of reflection is then $R = |r|^2 = [(k - k')/(k + k')]^2$.

1051

Find the reflection and transmission coefficients for the one-dimensional potential step shown in Fig. 1.25 if the particles are incident from the right.

(Wisconsin)

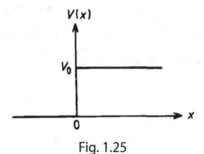

Fig. 1.25

Sol: As the particles are incident from the right we must have $E > V_0$. And there are both incident and reflected waves in the region $x > 0$. The Schrödinger equation for $x > 0$,

$$\psi''(x) + k_1^2 \psi(x) = 0,$$

where $k_1 = \sqrt{2m(E - V_0)}/\hbar$, has solutions of the form

$$\psi = \exp(-ik_1 x) + R\exp(ik_1 x).$$

There are only transmitted waves in the region $x < 0$, where the Schrödinger equation is

$$\psi''(x) + k_2^2 \psi(x) = 0$$

with $k_2 = \sqrt{2mE}/\hbar$, and has the solution

$$\psi(x) = S \exp(-ik_2 x).$$

Using the continuity conditions of the wave function at $x = 0$, we get $1 + R = S$. From the continuity of the first derivative of the wave function, we get $k_1(1 - R) = k_2 S$. Hence $R = (k_1 - k_2)/(k_1 + k_2)$, giving the reflection coefficient

$$|R|^2 = \left| \frac{k_1 - k_2}{k_1 + k_2} \right|^2 = \frac{V_0^2}{(\sqrt{E} + \sqrt{E - V_0})^4},$$

and the transmission coefficient

$$|S|^2 = 1 - |R|^2 = 1 - \frac{V_0^2}{(\sqrt{E} + \sqrt{E - V_0})^4}.$$

1052

Consider, quantum mechanically, a stream of particles of mass m, each moving in the positive x direction with kinetic energy E toward a potential jump located at $x = 0$. The potential is zero for $x \leq 0$ and $3E/4$ for $x > 0$. What fraction of the particles are reflected at $x = 0$?

(Buffalo)

Sol: The Schrödinger equations are

$$\psi'' + k^2 \psi = 0 \qquad \text{for} \quad x \leq 0,$$
$$\psi'' + (k/2)^2 \psi = 0 \qquad \text{for} \quad x > 0,$$

where $k = \sqrt{2mE}/\hbar$. As for $x < 0$ there will also be reflected waves, the solutions are of the form

$$\psi = \exp(ikx) + r \exp(-ikx), \qquad x \leq 0,$$
$$\psi = t \exp(ikx/2), \qquad x > 0.$$

From the continuity conditions of the wave function at $x = 0$, we obtain $1 + r = t$, $k(1 - r) = kt/2$, and hence $r = 1/3$. Thus one-ninth of the particles are reflected at $x = 0$.

1053

Consider a particle beam approximated by a plane wave directed along the x-axis from the left and incident upon a potential $V(x) = \gamma\delta(x)$, $\gamma > 0$, $\delta(x)$ is the Dirac delta function.

a. Give the form of the wave function for $x < 0$.

b. Give the form of the wave function for $x > 0$.

c. Give the conditions on the wave function at the boundary between the regions.

d. Calculate the probability of transmission.

(Berkeley)

Sol:

a. For $x < 0$, there are incident waves of the form $\exp(ikx)$ and reflected waves of the form $R\exp(-ikx)$. Thus

$$\psi(x) = \exp(ikx) + R\exp(-ikx), \qquad x < 0.$$

b. For $x > 0$, there only exist transmitted waves of the form $S\exp(ikx)$. Thus

$$\psi(x) = S\exp(ikx), \quad x > 0.$$

c. The Schrödinger equation is

$$-\frac{\hbar^2}{2m}\frac{d^2}{dx^2}\psi(x) + \gamma\delta(x)\psi(x) = E\psi(x)$$

and its solutions satisfy (**Problem 1020**)

$$\psi'(0^+) - \psi'(0^-) = \frac{2m\gamma}{\hbar^2}\psi(0).$$

As the wave function is continuous at $x = 0$, $\psi(0^+) = \psi(0^-)$.

d. From (a), (b) and (c) we have $1 + R = S$, $ikS - ik(1 - R) = 2m\gamma S / \hbar^2$, giving $S = 1/(1 + im\gamma/\hbar^2 k)$. Hence the transmission coefficient is

$$T = |S|^2 = \left(1 + \frac{m^2\gamma^2}{\hbar^4 k^2}\right)^{-1} = \left(1 + \frac{m\gamma^2}{2E\hbar^2}\right)^{-1},$$

where $E = \hbar^2 k^2 / 2m$.

1054

Consider a one-dimensional problem of a particle of mass m incident upon a potential of a shape shown in Fig. 1.26. Assume that the energy E at $x \to -\infty$ is greater than V_0, where V_0 is the asymptotic value of the potential as $x \to \infty$.

Show that the sum of reflected and transmitted intensities divided by the incident intensity is one.

(Princeton)

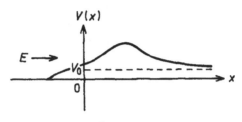

Fig. 1.26

Sol: As $E > V_0$ we may assume the asymptotic forms

$$\psi \to e^{ikx} + re^{-ikx} \quad \text{for} \quad x \to -\infty,$$

$$\psi \to te^{i\beta x} \quad \text{for} \quad x \to +\infty,$$

where r, t, k, β are constants. The incident intensity is defined as the number of particles incident per unit time: $I = \hbar k/m$. Similarly, the reflected and transmitted intensities are respectively

$$R = |r|^2 \, \hbar k / m, \quad T = |t|^2 \, \hbar \beta / m.$$

Multiplying the Schrödinger equation by ψ^*,

$$-\frac{\hbar^2}{2m} \psi^* \nabla^2 \psi + \psi^* V \psi = E \psi^* \psi,$$

and the conjugate Schrödinger equation by ψ,

$$-\frac{\hbar^2}{2m} \psi \nabla^2 \psi^* + \psi^* V \psi = E \psi^* \psi,$$

and taking the difference of the two equations, we have

$$\psi^* \nabla^2 \psi - \psi \nabla^2 \psi^* = \nabla \cdot (\psi^* \nabla \psi - \psi \nabla \psi^*) = 0.$$

This means that

$$f(x) = \psi^* d\psi/dx - \psi d\psi^*/dx$$

is a constant. Then equating $f(+\infty)$ and $f(-\infty)$, we find

$$k(1 - |r|^2) = \beta |t|^2.$$

Multiplying both sides by $\frac{\hbar}{m}$ gives

$$I = R + T.$$

1055

A Schrödinger equation in one dimension reads

$$(-\partial^2/\partial x^2 - 2\operatorname{sech}^2 x)\psi = \varepsilon\psi$$

$(\hbar = 1, m = 1/2)$.

a. Show that $\exp(ikx)(\tanh x + \text{const})$ is a solution for a particular value of the constant. Calculate the S-matrix (transmission and reflection coefficients) for this problem.

b. The wave function sech x happens to satisfy the Schrödinger equation. Calculate the energy of the corresponding bound state and give a simple argument that it must be the ground state of the potential.

c. Outline how you might have proceeded to estimate the ground-state energy if you did not know the wave function.

(Buffalo)

Sol:

a. Letting the constant in the given solution ψ be K and substituting ψ in the Schrödinger equation, we obtain

$$k^2(\tanh x + K) - 2(ik + K)\operatorname{sech}^2 x = \varepsilon(\tanh x + K).$$

This equation is satisfied if we set $K = -ik$ and $\varepsilon = k^2$. Hence

$$\psi(x) = e^{ikx}(\tanh x - ik)$$

is a solution of the equation and the corresponding energy is k^2. Then as $\tanh x \to 1$ for $x \to \infty$ and $\tanh(-x) = -\tanh x$ we have

$$\psi = (1 - ik)e^{ikx} \quad \text{as} \quad x \to \infty,$$
$$\psi = -(1 + ik)e^{ikx} \quad \text{as} \quad x \to -\infty.$$

Since $V(x) \le 0$, $\varepsilon > 0$, the transmission coefficient is $T = 1$ and the reflection coefficient is $R = 0$ as the particle travels through $V(x)$. So the S-matrix is

$$\begin{pmatrix} 1 & -(1-ik)/(1+ik) \\ -(1-ik)/(1+ik) & 0 \end{pmatrix}.$$

b. Letting $\psi = \text{sech}\, x$ in the Schrödinger equation we have $-\psi = \varepsilon \psi$. Hence $\varepsilon = -1$. Because sech x is a non-node bound state in the whole coordinate space, it must be the ground state.

c. We might proceed by assuming a non-node bound even function with a parameter and obtain an approximate value of the ground state energy by the variational method.

1056

A monoenergetic parallel beam of nonrelativistic neutrons of energy E is incident onto the plane surface of a plate of matter of thickness t. In the matter, the neutrons move in a uniform attractive potential V. The incident beam makes an angle θ with respect to the normal to the plane surface as shown in Fig. 1.27.

a. What fraction of the incident beam is reflected if t is infinite?

b. What fraction of the incident beam is reflected if V is repulsive and $V = E$? Consider t finite.

(CUS)

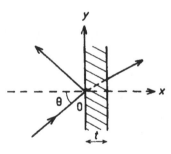

Fig. 1.27

Sol:

a. Let k_0 be the wave number of an incident neutron, given by $k_0^2 = \dfrac{2mE}{\hbar^2}$. For $x < 0$, the wave function is

$$\psi_1(x,\, y) = e^{ik_0 x \cos\theta + ik_0 y \sin\theta}$$
$$+ R\, e^{-ik_0 x \cos\theta + ik_0 y \sin\theta}.$$

With t infinite and the potential negative, for $x > 0$ the Schrödinger equation is

$$-\frac{\hbar^2}{2m}\nabla^2\psi = (E+V)\psi .$$

Assuming a solution

$$\psi_2(x,y) = Te^{ik_x x + ik_y y}$$

and substituting it the equation, we obtain $k_x^2 + k_y^2 = 2m(E+V)/\hbar^2$.

The boundary conditions at $x = 0$

$$\psi_1(0,y) = \psi_2(0,y),$$

$$\left.\frac{\partial\psi_1}{\partial x}\right|_{x=0} = \left.\frac{\partial\psi_2}{\partial x}\right|_{x=0},$$

then give

$$e^{ik_0 y\sin\theta} + Re^{ik_0 y\sin\theta} = Te^{ik_y y},$$

$$ik_0\cos\theta e^{ik_0 y\sin\theta} - Rik_0\cos\theta e^{ik_0 y\sin\theta} = Tik_x e^{ik_y y},$$

As the potential does not vary with y, $k_y = k_0\sin\theta$ and the above become

$$1 + R = T,$$

$$k_0(1-R)\cos\theta = k_x T,$$

which give $R = (k_0\cos\theta - k_x)/(k_0\cos\theta + k_x)$. The probability of reflection is then $P = |R|^2 = (k_0\cos\theta - k_x)^2/(k_0\cos\theta + k_x)^2$, with $k_x^2 = 2m(E+V)/\hbar^2 - k_0^2\sin^2\theta$, $k_0^2 = 2mE/\hbar^2$.

b. For $x < 0$ the wave function has the same form as that in (a). For $0 < x < t$, $E - V = 0$, and the Schrödinger equation is

$$-\left(\hbar^2/2m\right)\nabla^2\psi = 0 .$$

As the potential is uniform in y we assume $\psi = \exp(ik'y)\exp(kx)$, where $k' = k_0\sin\theta$. Substitution gives $-k'^2 + k^2 = 0$, or $k = \pm k'$. Hence the wave function for $0 < x < t$ is

$$\psi_2(x,y) = \left(ae^{k'x} + be^{-k'x}\right)e^{ik'y} .$$

Writing $\psi(x,y) = \phi(x)e^{ik'y}$, we have for the three regions

$$x < 0,\ \phi_1(x) = e^{ik_x x} + re^{-ik_x x},$$

$$0 < x < t,\ \phi_2(x) = ae^{k'x} + be^{-k'x},$$

$$x > t,\ \phi_3(x) = ce^{ik_x x},$$

with

$$k_x = \sqrt{\frac{2mE}{\hbar^2}}\cos\theta, \ k' = \sqrt{2mE/\hbar^2}\sin\theta \ .$$

The boundary conditions

$$\phi_1(0) = \phi_2(0), \ \ \phi_2(t) = \phi_3(t),$$

$$\left.\frac{d\phi_1}{dx}\right|_{x=0} = \left.\frac{d\phi_2}{dx}\right|_{x=0}, \ \left.\frac{d\phi_2}{dx}\right|_{x=t} = \left.\frac{d\phi_3}{dx}\right|_{x=t}$$

give

$$1 + r = a + b,$$

$$ik_x(1 - r) = k'(a - b),$$

$$c \ \exp(ik_x t) = a \ \exp(k't) + b \ \exp(-k't),$$

$$ik_x c \ \exp(ik_x t) = k'a \ \exp(k't) - k'b \ \exp(-k't),$$

whose solution is

$$r = \left[\frac{ik_x}{k'}\left(\frac{a/b+1}{a/b-1}\right) - 1\right]\left[\frac{ik_x}{k'}\left(\frac{a/b+1}{a/b-1}\right) - 1\right]^{-1},$$

$$a/b = \left[(k' + ik_x)/(k' - ik_x)\right]\cdot e^{-2k't} \ .$$

Hence

$$r = \frac{e^{2k't} - 1}{1 - \beta^2 e^{2k't}}$$

with

$$\beta = \frac{k' - ik_x}{k' + ik_x},$$

and the fraction of neutrons reflected is

$$|R|^2 = |r|^2 = \frac{e^{2k't} + e^{-2k't} - 2}{e^{2k't} + e^{-2k't} - 2\cos 4\theta}.$$

Alternative Sol: The solution can also be obtained by superposition of infinite amplitudes, similar to the case of a Fabry–Perot interferometer in optics (see Fig. 1.28).

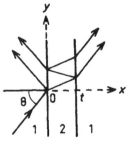

Fig. 1.28

We need only consider the x-component of the waves. Let T_{12}, R_{12} denote the coefficients of amplitude transmission and reflection as a wave goes from medium 1 to medium 2, respectively. Let T_{21}, R_{21} denote coefficients of amplitude transmission and reflection from medium 2 to medium 1. Take the amplitude of incident wave as 1. Then the amplitude of the wave that is transmitted to medium 2 is the sum

$$T = T_{12}e^{-\delta} \cdot T_{21} + T_{12}e^{-\delta} \cdot R_{21}e^{-2\delta}R_{21} \cdot T_{21}$$
$$+ T_{12}e^{-\delta} \cdot (R_{21}e^{-2\delta}R_{21})^2 \cdot T_{21} + \cdots$$
$$= T_{12}e^{-\delta}T_{21}[1 + R_{21}^2 e^{-2\delta} + (R_{21}^2 e^{-2\delta})^2 + \cdots]$$
$$= T_{12}T_{21}e^{-\delta}\frac{1}{1 - R_{21}^2 e^{-2\delta}}.$$

In the above $\exp(-\delta)$ is the attenuation coefficient of a wave in medium 2, where

$$\delta = k't \text{ with } k' = \sqrt{2mE(1 - \cos^2\theta)}\Big/\hbar = \sqrt{2mE}\,\sin\theta/\hbar.$$

From (a) we have the coefficients of transmission and reflection

$$R_{12} = (k_{1x} - k_{2x})/(k_{1x} + k_{2x}),$$
$$T_{12} = 1 + R_{12} = 2k_1/(k_{1x} + k_{2x}).$$

As

$$k_{1x} = \sqrt{2mE}\,\cos\theta/\hbar,$$
$$k_{2x} = ik' = i\sqrt{2mE}\,\sin\theta/\hbar,$$

we find

$$T_{12} = 2k_{1x}/(k_{1x} + k_{2x}) = \frac{2\cos\theta}{\cos\theta + i\sin\theta} = 2\cos\theta e^{-i\theta},$$

$$T_{21} = 2k_{2x}/(k_{1x} + k_{2x}) = \frac{2i\sin\theta}{\cos\theta + i\sin\theta} = 2i\sin\theta e^{-i\theta},$$

$$R_{21} = (k_{2x} - k_{1x})/(k_{1x} + k_{2x}) = \frac{i\sin\theta - \cos\theta}{i\sin\theta + \cos\theta} = e^{-2i\theta},$$

and hence

$$T = T_{12}T_{21}e^{-\delta}\frac{1}{1 - R_{21}^2 e^{-2\delta}}$$

$$= \frac{4i\,\cos\theta\,\sin\theta e^{-2i\theta} \cdot e^{-k't}}{1 - e^{-4i\theta}e^{-2k't}}$$

$$= \frac{2i\,\sin 2\theta e^{-2i\theta}e^{-k't}}{1 - e^{-4i\theta}e^{-2k't}}.$$

The transmissivity is therefore

$$|T|^2 = \frac{4 \sin^2 2\theta e^{-2k't}}{(1-e^{-2k't}\cos^4\theta)^2 + (e^{-2k't}\sin 4\theta)^2}$$

$$= \frac{4 \sin^2 2\theta e^{-2k't}}{1 + e^{-4k't} - 2e^{-2k't} \cos 4\theta}$$

$$= \frac{4 \sin^2 2\theta}{e^{2k't} + e^{-2k't} - 2\cos 4\theta},$$

and the reflectivity is

$$1 - |T|^2 = 1 - \frac{4 \sin^2 2\theta}{e^{2k't} + e^{-2k't} - 2 \cos 4\theta}$$

$$= \frac{e^{2k't} + e^{-2k't} - 2}{e^{2k't} + e^{-2k't} - 2 \cos 4\theta},$$

where $k' = \sqrt{2mE} \sin\theta / \hbar$.

1057

Find the wave function for a particle moving in one dimension in a constant imaginary potential $-iV$ where $V \ll E$.

Calculate the probability current and show that an imaginary potential represents absorption of particles. Find an expression for the absorption coefficient in terms of V.

(Wisconsin)

Sol: The Schrödinger equation is

$$i\hbar \partial\psi/\partial t = (\hat{p}^2/2m - iV)\psi.$$

Supposing $\psi = \exp(-iEt/\hbar) \exp(ikx)$, we have $k^2 = (2mE/\hbar^2)(1+iV/E)$. As $V \ll E$,

$$k \approx \pm\sqrt{\frac{2mE}{\hbar^2}}\left(1 + i\frac{V}{2E}\right),$$

and hence

$$\psi_\pm(x, t) = \exp\left(\mp\sqrt{\frac{2mE}{\hbar^2}}\frac{V}{2E}x\right) \cdot \exp\left(\pm i\sqrt{\frac{2mE}{\hbar^2}}x\right)\exp\left(-\frac{iEt}{\hbar}\right),$$

where ψ_+ and ψ_- refer to the exponentially attenuated right- and left-traveling waves respectively. The probability current is

$$j = Re\left(\psi * \frac{\hat{p}}{m}\psi\right) = Re\left(\psi * \frac{\hbar k}{m}\psi\right) = Re\left[\frac{\hbar k}{m}\exp\left(\mp\sqrt{\frac{2mE}{\hbar^2}}\frac{V}{E}x\right)\right]$$

$$\approx \pm\frac{\hbar}{m}\sqrt{\frac{2mE}{\hbar^2}}\exp\left(\mp\sqrt{\frac{2mE}{\hbar^2}}\frac{V}{E}x\right).$$

These are the exponentially attenuated currents in the respective directions. The absorption coefficient is then

$$\mu = \left|-\frac{1}{j}\frac{dj}{dx}\right| = \left|-\frac{d\ln j}{dx}\right| = \sqrt{\frac{2mE}{\hbar^2}}\frac{V}{E}.$$

The imaginary potential iV is responsible for the absorption of the particle, since the exponent in j would be imaginary. Hence there would be no absorption if V were real.

1058

Let the solution to the one-dimensional free-particle time-dependent Schrödinger equation of definite wavelength λ be $\psi(x,t)$ as described by some observer O in a frame with coordinates (x,t). Now consider the same particle as described by wave function $\psi'(x',t')$ according to observer O' with coordinates (x',t') related to (x,t) by the Galilean transformation

$$x' = x - \upsilon t,$$
$$t' = t.$$

a. Do $\psi(x,t)$, $\psi'(x',t')$ describe waves of the same wavelength?

b. What is the relationship between $\psi(x,t)$ and $\psi'(x',t')$ if both satisfy the Schrödinger equation in their respective coordinates?

(Berkeley)

Sol:

a. The one-dimensional time-dependent Schrödinger equation for a free particle

$$i\hbar\partial_t\psi(x,t) = \left(-\hbar^2/2m\right)\partial_x^2\psi(x,t)$$

has a solution corresponding to a definite wavelength λ

$$\psi_\lambda(x,t) = \exp\left[i(kx - \omega t)\right]$$

with

$$\lambda = 2\pi / k = 2\pi\hbar / p, \quad \omega = \hbar k^2 / 2m.$$

As the particle momentum p is different in the two reference frames, the wavelength λ is also different.

b. Applying the Galilean transformation and making use of the Schrödinger equation in the (x', t') frame we find

$$i\hbar\partial_t \psi'(x', t') = i\hbar\partial_t \psi'(x - \upsilon t, t)$$

$$= i\hbar\left[\partial_t' \psi'(x', t') - \upsilon\partial_x' \psi'(x', t')\right]$$

$$= -\frac{\hbar^2}{2m}\partial_x'^2 \psi'(x', t') - i\hbar\upsilon\partial_x' \psi'(x', t')$$

$$= -\frac{\hbar^2}{2m}\partial_x^2 \psi'(x - \upsilon t, t) - i\hbar\upsilon\partial_x \psi'(x - \upsilon t, t). \tag{1}$$

Considering

$$i\hbar\partial_t\left[e^{i(kx-\omega t)}\psi'(x', t')\right]$$

$$= i\hbar e^{i(kx-\omega t)}\left(\partial_t \psi' - i\omega\psi'\right)$$

and

$$-\frac{\hbar^2}{2m}\partial_x^2\left[e^{i(kx-\omega t)}\psi'(x', t')\right]$$

$$= -\frac{\hbar^2}{2m}e^{i(kx-\omega t)}\left(-k^2\psi' + 2ik\partial_x \psi' + \partial_x^2 \psi'\right)$$

$$= i\hbar e^{i(kx-\omega t)}\left(\partial_t \psi' - i\omega\psi'\right),$$

making use of Eq. (1) and the definitions of k and ω, we see that

$$-\frac{\hbar^2}{2m}\partial_x^2\left[e^{im\upsilon x/\hbar}e^{-im\upsilon^2 t/2\hbar}\psi'(x - \upsilon t, t)\right]$$

$$= i\hbar\partial_t\left[e^{im\upsilon x/\hbar}e^{-im\upsilon^2 t/2\hbar}\psi'(x - \upsilon t, t)\right].$$

This is just the Schrödinger equation that $\psi(x, t)$ satisfies. Hence, accurate to a phase factor, we have the relation

$$\psi(x, t) = \psi'(x - \upsilon t, t)\exp\left\{\frac{i}{\hbar}\left[m\upsilon x - \frac{m\upsilon^2}{2}t\right]\right\}.$$

1059

A particle of mass m bound in a one-dimensional harmonic oscillator potential of frequency ω and in the ground state is subjected to an impulsive force $p\delta(t)$.

Find the probability it remains in its ground state.

(Wisconsin)

Sol: The particle receives an instantaneous momentum p at $t=0$ and its velocity changes to p/m instantaneously. The duration of the impulse is, however, too short for the wave function to change. Hence, in the view of a frame K' moving with the particle, the latter is still in the ground state of the harmonic oscillator $\psi_0(x')$. But in the view of a stationary frame K, it is in the state $\psi_0(x')\exp(-ipx/\hbar)$. We may reasonably treat the position of the particle as constant during the process, so that at the end of the impulse the coordinate of the particle is the same for both K and K'. Hence the initial wave function in K is

$$\psi_0' = \psi_0(x)\exp(-ipx/\hbar).$$

Thus, the probability that the particle remains in its ground state after the impulse is

$$P_0 = \left|\left\langle \psi_0 \left| \exp\left(-i\frac{px}{\hbar}\right)\right|\psi_0\right\rangle\right|^2$$

$$= \frac{\alpha^2}{\pi}\left|\int_{-\infty}^{+\infty} e^{-\alpha^2 x^2 - i\frac{px}{\hbar}}dx\right|^2$$

$$= \exp\left(-\frac{p^2}{2\alpha^2\hbar^2}\right)\left|\frac{\alpha}{\sqrt{\pi}}\int_{-\infty}^{\infty}\exp\left[-\alpha^2\left(x+\frac{ip}{2\alpha^2\hbar}\right)^2\right]dx\right|^2$$

$$= \exp\left(\frac{-p^2}{2m\omega\hbar}\right),$$

where $\alpha = \sqrt{m\omega/\hbar}$.

1060

An idealized ping pong ball of mass m is bouncing in its ground state on a recoilless table in a one-dimensional world with only a vertical direction.

a. Prove that the energy depends on m, g, h according to:

$\varepsilon = Kmg(m^2g/h^2)^\alpha$ and determine α.

b. By a variational calculation estimate the constant K and evaluate ε for $m = 1$ gram in ergs.

Sol:

a. By the method of dimensional analysis, if we have

$$[\varepsilon] = \frac{[m]^{1+2\alpha}[g]^{1+\alpha}}{[h]^{2\alpha}},$$

or

$$\frac{[m][L]^2}{[T]^2} = \frac{[m][L]^{1-3\alpha}}{[T]^2},$$

then $\alpha = -\frac{1}{3}$. Thus, provided $\alpha = -\frac{1}{3}$, the expression gives the energy of the ball.

b. Take the x coordinate in the vertical up direction with origin at the table. The Hamiltonian is

$$H = \frac{p^2}{2m} + mgx = -\frac{\hbar^2}{2m}\frac{d^2}{dx^2} + mgx,$$

taking the table surface as the reference point of gravitational potential. Try a ground state wave function of the form $\psi = x\exp(-\lambda x^2/2)$, where λ is to be determined. Consider

$$\langle H \rangle = \frac{\int \psi^* H \psi dx}{\int \psi^* \psi dx}$$

$$= \frac{\int_0^\infty xe^{-\lambda x^2/2}\left(-\frac{\hbar^2}{2m}\frac{d^2}{dx^2} + mgx\right)xe^{-\lambda x^2/2}dx}{\int_0^\infty x^2 e^{-\lambda x^2}dx}$$

$$= \frac{3\hbar^2}{4m}\lambda + \frac{2mg}{\sqrt{\pi}\lambda^{1/2}}.$$

To minimize $\langle H \rangle$, take $\dfrac{d\langle H \rangle}{d\lambda} = 0$ and obtain $\lambda = \left(\dfrac{4m^2 g}{3\sqrt{\pi}\hbar^2}\right)^{\frac{2}{3}}$.

The ground state energy is then

$$\langle H \rangle = 3(3/4\pi)^{1/3}mg(m^2 g/\hbar^2)^{-1/3}.$$

giving

$$K = 3(3/4\pi)^{1/3}.$$

Numerically

$$\varepsilon = 3mg\left(\frac{3\hbar^2}{4\pi m^2 g}\right)^{\frac{1}{3}}$$

$$= 3\times 980\times \left(\frac{3\times 1.054^2\times 10^{-54}}{4\pi\times 980}\right)^{\frac{1}{3}}$$

$$= 1.9\times 10^{-16}\,\text{erg}.$$

1061

The following theorem concerns the energy eigenvalues $E_n (E_1 < E_2 < E_3 < \ldots)$ of the Schrödinger equation in one dimension:

Theorem: If the potential $V_1(x)$ gives the eigenvalues E_{1n} and the potential $V_2(x)$ gives the eigenvalues E_{2n} and $V_1(x) \le V_2(x)$ for all x, then $E_{1n} \le E_{2n}$.

a. Prove this theorem.

Hint: Consider a potential $V(\lambda, x)$, where $V(0, x) = V_1(x)$ and $V(1, x) = V_2(x)$ and $\partial V/\partial \lambda \ge 0$ (for all x), and calculate $\partial E_n/\partial \lambda$.

b. Now consider the potential (Fig. 1.29)

$$U(x) = kx^2/2, \qquad |x| < a,$$
$$U(x) = ka^2/2, \qquad |x| \ge a.$$

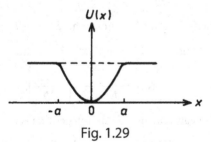

Fig. 1.29

We want to determine the number of bound states that this potential can hold. Assume this number N is $\gg 1$. It may be helpful to draw a qualitative picture of the wave function for the highest bound state.

Choose a solvable comparison potential and use the theorem above to determine either a rigorous upper bound to *N* or a rigorous lower bound to *N*. (Both can be done but you are asked for only one.)

(*Berkeley*)

Sol:

a. Define $V(\lambda, x) = \lambda V_2(x) + (1-\lambda)V_1(x)$. Obviously $V(0, x) = V_1(x)$, $V(1, x) = V_2(x)$, $\partial V / \partial \lambda = V_2(x) - V_1(x) \geq 0$. The Hamiltonian is then

$$\hat{H}(\lambda) = \hat{p}^2 / 2m + V(\lambda, x),$$

and the eigenequation is

$$\hat{H}(\lambda)|n, \lambda\rangle = E_n(\lambda)|n, \lambda\rangle,$$

where $E_n(\lambda) = \langle n, \lambda | \hat{H}(\lambda) | n, \lambda \rangle$. As

$$\partial E_n(\lambda) / \partial \lambda = \frac{\partial}{\partial \lambda}\left[\langle n, \lambda | \hat{H}(\lambda) | n, \lambda \rangle\right]$$

$$= \langle n\lambda | \partial V(\lambda) / \partial \lambda | n\lambda \rangle$$

$$= \int \frac{\partial V}{\partial \lambda} |\psi_n(x, \lambda)|^2 \, dx \geq 0,$$

we have $E_{1n} = E_n(0) \leq E_n(1) = E_{2n}$, and the theorem is proved. Note that we have used $\langle n\lambda | n\lambda \rangle = 1$.

b. Let $V(x) = kx^2 / 2$. Then $V(x) \geq U(x)$. If E_n is an energy level for the potential $U(x)$, then $E_n \leq (n + 1/2)\hbar\omega$, where $\omega = \sqrt{k/m}$. For a bound state, $E_n \leq ka^2 / 2$. Solving $(N + 1/2)\hbar\omega \leq ka^2 / 2$, we find

$$N \leq \frac{m\omega a^2}{2\hbar} - \frac{1}{2} = \left[\frac{m\omega a^2}{2\hbar}\right],$$

where $[A]$ indicates the maximum integer that is less than A.

We now choose for $V(x)$ a square well of finite depth,

$$V(x) = ka^2 / 2, \qquad |x| > a,$$
$$V(x) = 0, \qquad |x| \leq a.$$

The number of bound states of $U(x)$ is less than that of $V(x)$, which for the latter is $\left[2m\omega a^2 / \pi\hbar\right] + 1$. We can take the upper bound to the number of bound states of $U(x)$ as $\left[2m\omega a^2/\pi\hbar\right]$ as for $N \gg 1$ the term 1 can be neglected. Taken together, we get that the number of bound states is between $\left[m\omega a^2/2\hbar\right]$ and $\left[2m\omega a^2/\pi\hbar\right]$.

1062

For electronic states in a one-dimensional system, a simple model Hamiltonian is

$$H = \sum_{n=1}^{N} E_0 |n\rangle\langle n| + \sum_{n=1}^{N} W\{|n\rangle\langle n+1| + |n+1\rangle\langle n|\},$$

where $|n\rangle$ are an orthonormal basis, $\langle n|n'\rangle = \delta_{nn'}$; E_0 and W are parameters. Assume periodic boundary conditions so that $|N+j\rangle = |j\rangle$. Calculate the energy levels and wave functions.

(Wisconsin)

Sol: From the fact that $|n\rangle$ form a complete set of orthonormal functions, where $n = 1, 2, 3, \ldots, N$, and

$$\hat{H} = \sum_{n=1}^{N} E_0 |n\rangle\langle n| + \sum_{n=1}^{N} W\{|n\rangle\langle n+1| + |n+1\rangle\langle n|\},$$

or

$$\hat{H} = E_0 + W\left(A + A^+\right),$$

with

$$A \equiv \sum_{n=1}^{N} |n\rangle\langle n+1|, \quad A^+ \equiv \sum_{n=1}^{N} |n+1\rangle\langle n|,$$

and

$$A|n\rangle = |n-1\rangle, \quad A^+|n\rangle = |n+1\rangle,$$
$$AA^+ = A^+A = 1, \quad \text{or} \quad A^+ = A^{-1},$$

we know that $\hat{H}, A$ and A^+ have the same eigenvectors. Hence we only need to find the eigenvectors and eigenvalues of the operator A^+ to solve the problem. As

$$A_{k'k} = \langle k'|A|k\rangle = \delta_{k',k-1},$$

We have

$$
A = \begin{pmatrix}
0 & 1 & 0 & . & . & 0 \\
0 & 0 & 1 & . & . & 0 \\
0 & 0 & 0 & 1 & . & . \\
. & . & . & . & . & . \\
. & . & . & . & 0 & 1 \\
1 & 0 & 0 & . & . & 0
\end{pmatrix}_{N \times N}
$$

and so

$$
A - \lambda = \begin{pmatrix}
-\lambda & 1 & 0 & . & . & 0 \\
0 & -\lambda & 1 & . & . & 0 \\
0 & 0 & -\lambda & 1 & . & . \\
. & . & . & . & . & . \\
. & . & . & . & -\lambda & 1 \\
1 & 0 & 0 & . & . & -\lambda
\end{pmatrix}_{N \times N} ,
$$

i.e.,

$$
\det(A - \lambda I) = (-\lambda)^N + (-1)^{N+1} = (-1)^N \left(\lambda^N - 1 \right) = 0,
$$

giving

$$
\lambda_j = e^{i\theta_j}, \ \theta_j = \frac{2\pi}{N} j, \ j = 0,\ 1,\ 2,\ ...,\ N-1.
$$

If $|E\rangle$ are the same eigenvectors of the operators A and A^+, i.e.,

$$
A \big| E_j \big\rangle = \lambda_j \big| E_j \big\rangle, \ A^+ \big| E_j \big\rangle = \frac{1}{\lambda_j} \big| E_j \big\rangle,
$$

then

$$
\hat{H} \big| E_j \big\rangle = \left[E_0 + W \left(\lambda_j + \frac{1}{\lambda_j} \right) \right] \big| E_j \big\rangle
$$

$$
= \left(E_0 + 2W \cos\theta_j \right) \big| E_j \big\rangle.
$$

Hence the eigenvalues of $\hat{H}$ are

$$
E_j = E_0 + 2W \cos\theta_j, \ \text{with} \ \theta_j = \frac{2\pi}{N} j, (j = 0,\ 1,\ 2,\ ...,\ N-1).
$$

The corresponding eigenfunctions can be obtained from the matrix equations

$$
\left(A - \lambda_j \right) \big| E_j \big\rangle = 0.
$$

Thus

$$|E_i\rangle = \frac{1}{\sqrt{N}} \begin{pmatrix} 1 \\ e^{i\theta_j} \\ e^{i2\theta_j} \\ \vdots \\ e^{i(N-1)\theta_j} \end{pmatrix},$$

or

$$|E_j\rangle = \frac{1}{\sqrt{N}} \sum_{n=1}^{N} e^{i(n-1)\theta_j} |n\rangle.$$

1063

Give a brief discussion of why there are energy bands in a crystalline solid. Use the ideas of quantum mechanics but do not attempt to carry out any complicated calculations. You should assume that anyone reading your discussion understands quantum mechanics but does not understand anything about the theory of solids.

(*Wisconsin*)

Sol: A crystal may be regarded as an infinite, periodic array of potential wells, such as the lattice structure given in **Problem 1065**. Bloch's theorem states that the solution to the Schrödinger equation then has the form $u(x)\exp(iKx)$, where K is a constant and $u(x)$ is periodic with the periodicity of the lattice. The continuity conditions of $u(x)$ and $du(x)/dx$ at the well boundaries limit the energy of the propagating particle to certain ranges of values, i.e., energy bands. An example is given in detail in **Problem 1065**.

1064

A particle of mass m moves in one dimension in a periodic potential of infinite extent. The potential is zero at most places, but in narrow regions of width b separated by spaces of length $a(b \ll a)$ the potential is V_0, where V_0 is a large positive potential.

[One may think of the potential as a sum of Dirac delta functions:

$$V(x) = \sum_{n=-\infty}^{\infty} V_0 b\, \delta(x - na).$$

Alternatively one can arrive at the same answer in a somewhat more messy way by treating the intervals as finite and then going to the limit.]

a. What are the appropriate boundary conditions to apply to the wave function, and why?

b. Let the lowest energy of a wave that can propagate through this potential be $E_0 = \hbar^2 k_0^2 / 2m$ (this defines k_0). Write down a transcendental equation (not a differential equation) that can be solved to give k_0 and thus E_0.

c. Write down the wave function at energy E_0 valid in the region $0 \le x \le a$ (For uniformity, let us choose normalization and phase such that $\psi(x=0)=1$). What happens to the wave function between $x=a$ and $x=a+b$?

d. Show that there are ranges of values of E, greater than E_0, for which there is no eigenfunction. Find (exactly) the energy at which the first such gap begins.

(*Berkeley*)

Sol:

a. The Schrödinger equation is

$$\left[-\frac{\hbar^2}{2m}\frac{d^2}{dx^2} + \sum_{n=-\infty}^{\infty} V_0 b \delta(x-na) \right] \psi(x) = E\psi(x).$$

Integrating it from $x=a-\varepsilon$ to $x=a+\varepsilon$ and letting $\varepsilon \to 0$, we get

$$\psi'(a^+) - \psi'(a^-) = 2\Omega\psi(a),$$

where $\Omega = mV_0 b / \hbar^2$. This and the other boundary condition

$$\psi(a^+) - \psi(a^-) = 0$$

apply to the wave function at $x=na$, where $n=-\infty, \ldots, -1, 0, 1, \ldots, +\infty$.

b. For $x \ne na$, there are two fundamental solutions to the Schrödinger equation:

$$u_1(x) = e^{ikx}, \; u_2(x) = e^{-ikx},$$

the corresponding energy being

$$E = \hbar^2 k^2 / 2m.$$

Let

$$\psi(x) = Ae^{ikx} + Be^{-ikx}, \; 0 \le x \le a.$$

According to Bloch's Theorem, in the region $a \le x \le 2a$

$$\psi(x) = e^{iKa}\left[Ae^{ik(x-a)} + Be^{-ik(x-a)}\right],$$

where K is the Bloch wave number. The boundary conditions give

$$e^{iKa}(A+B) = Ae^{ika} + Be^{-ika},$$

$$ike^{iKa}(A-B) = ik\left(Ae^{ika} - Be^{-ika}\right)$$
$$+ 2\Omega\left(Ae^{ika} + Be^{-ika}\right).$$

For nonzero solutions of A and B we require

$$\begin{vmatrix} e^{iKa} - e^{ika} & e^{iKa} - e^{-ika} \\ ike^{iKa} - (ik+2\Omega)e^{ika} & -ike^{iKa} + (ik-2\Omega)e^{-ika} \end{vmatrix} = 0,$$

or

$$\cos ka + \frac{\Omega}{k}\sin ka = \cos Ka,$$

which determines the Bloch wave number K. Consequently, the allowed values of k are limited to the range given by

$$\left|\cos ka + \frac{\Omega}{k}\sin ka\right| \le 1,$$

or

$$\left(\cos ka + \frac{\Omega}{k}\sin ka\right)^2 \le 1.$$

k_0 is the minimum of k that satisfy this inequality,

c. For $E = E_0$,

$$\psi(x) = Ae^{ik_0 x} + Be^{-ik_0 x}, \quad 0 \le x \le a,$$

where $k_0 = \sqrt{\frac{2mE_0}{\hbar}}$.

Normalization $\psi(x=0) = 1$ gives

$$\psi(x) = 2i \, A \sin k_0 x + e^{-ik_0 x}, \quad 0 \le x \le a.$$

The boundary conditions at $x = a$ give

$$e^{iKa} = 2i \, A \sin k_0 a + e^{-ik_0 a},$$

or

$$2iA = \left(e^{iKa} - e^{-ik_0 a}\right) / \sin k_0 a.$$

So

$$\psi(x)=\left(e^{iKa}-e^{-ik_0a}\right)\frac{\sin k_0 x}{\sin k_0 a}+e^{-ik_0 x}; \quad 0\le x\le a.$$

For $x \in [a, a+b]$, the wave function has the form $\exp(\pm k_1, x)$, where

$$k_1=\sqrt{2m(V_0-E_0)}\,/\,\hbar\,.$$

d. For $ka=n\pi+\delta$, where δ is a small positive number, we have

$$\left|\cos ka+\frac{\Omega}{k}\sin ka\right|$$

$$=\left|\cos(n\pi+\delta)+\frac{\Omega}{k}\sin(n\pi+\delta)\right|$$

$$\approx\left|1-\frac{\delta^2}{2}+\frac{\Omega}{k}\delta\right|\le1.$$

When δ is quite small, the left side $\approx 1+\Omega\delta\,/\,k>1$. Therefore in a certain region of $k>n\pi\,/\,a$, there is no eigenfunction. On the other hand, $ka=n\pi$ corresponds to eigenvalues. So the energy at which the first energy gap begins satisfies the relation $ka=\pi$, or $E=\pi^2\hbar^2\,/\,2ma^2$.

1065

We wish to study particle-wave propagation in a one-dimensional periodic potential constructed by iterating a "single-potential" $V(x)$ at intervals of length *l*. $V(x)$ vanishes for $|x|\ge l\,/\,2$ and is symmetric in *x* (i.e., $V(x)=V(-x)$). The scattering properties of $V(x)$ can be summarized as follows:

If a wave is incident from the left, $\psi_+(x)=\exp(ikx)$ for $x<-l\,/\,2$, it produces a transmitted wave $\psi_+(x)=\exp(ikx)$ for $x>l\,/\,2$ and a reflected wave $\psi_-(x)=\exp(-ikx)$ for $x<-l\,/\,2$. Transmitted and reflected coefficients are given by

$$T(x)=\frac{\psi_+\left(|x|\right)}{\psi_+\left(-|x|\right)}=\frac{1}{2}e^{2ik|x|}\left[e^{2i\delta_e}+e^{2i\delta_0}\right],\ |x|\ge\frac{l}{2},$$

$$R(x)=\frac{\psi_-\left(-|x|\right)}{\psi_+\left(-|x|\right)}\approx\frac{1}{2}e^{2ik|x|}\left[e^{2i\delta_e}-e^{2i\delta_0}\right],\ |x|\ge\frac{l}{2},$$

and δ_e and δ_0 are the phase shifts due to the potential $V(x)$. Take these results as given. Do not derive them.

Now consider an infinite periodic potential $V_\infty(x)$ constructed by iterating the potential $V(x)$ with centers separated by a distance l (Fig. 1.30). Call the points at which $V_\infty(x) = 0$ "interpotential points". We shall attempt to construct waves propagating in the potential $V_\infty(x)$ as superpositions of left- and right-moving waves ϕ_+ and ϕ_-.

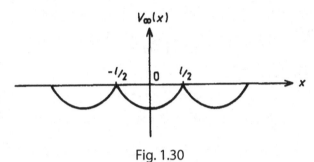

Fig. 1.30

a. Write recursion relations which relate the amplitudes of the right- and left-moving waves at the nth interpotential point, $\phi_\pm^n$, to the amplitudes at the $(n-1)$th and $(n+1)$th interpotential points, $\phi_\pm^{n-1}$ and $\phi_\pm^{n+1}$.

b. Obtain a recursion relation for ϕ_- or ϕ_+ alone by eliminating the other from part (a).

c. Obtain an expression for the ratio of amplitudes of ϕ_+ to ϕ_- at successive interpotential points.

d. Find the condition on k, δ_e and δ_0 such that traveling waves are allowed.

e. Use this result to explain why it is "normal" for conduction by electrons in metals to be allowed only for bands of values of energy.

(MIT)

Sol: For the wave incident from the left, the potential being $V(x)$, let

$$\psi_{t+} = t\phi + e^{ikx}, \quad t = \frac{1}{2}\left(e^{i2\delta_e} + e^{i2\delta_0}\right);$$

$$\psi_{r-} = r\phi - e^{-ikx}, \quad r = \frac{1}{2}\left(e^{i2\delta_e} - e^{i2\delta_0}\right).$$

For the wave incident from the right, let the transmission and reflection coefficients be t' and r' respectively. It can be shown that $t'=t$, $r'=-r^{*}t/t^{*}$. In the periodic potential, the transmission and reflection coefficients at adjacent interpotential points have relations $t_{n}=t_{n-1}$ and $r_{n}=r_{n-1}\exp(i2kl)$. So the transmission coefficient can be denoted by a single notation t.

a.   The waves at adjacent interpotential points are as shown in Fig. 1.31. Obviously, only the reflection term of ϕ_{-}^{n} and the transmission term of ϕ_{+}^{n-1} contribute to ϕ_{+}^{n}:

$$\phi_{+}^{n}=r'_{n-1}\phi_{-}^{n}+t\phi_{+}^{n-1}. \tag{1}$$

Similarly,

$$\phi_{-}^{n}=r_{n}\phi_{+}^{n}+t'\phi_{-}^{n+1}. \tag{2}$$

Thus we have

$$\phi_{+}^{n}=e^{i2kl}r'_{n}\phi_{-}^{n}+t\phi_{+}^{n-1}, \tag{3}$$

$$\phi_{-}^{n}=r_{n}\phi_{+}^{n}+t\phi_{-}^{n+1} \tag{4}$$

$$\begin{array}{ccc} \phi_{+}^{n-1} & \phi_{+}^{n} & \phi_{+}^{n+1} \\ \rightarrow & \rightarrow & \rightarrow \\ \leftarrow & \leftarrow & \leftarrow \\ \phi_{-}^{n-1} & \phi_{-}^{n} & \phi_{-}^{n+1} \end{array}$$

Fig. 1.31

b.   With n replaced by $n+1$, Eq. (1) gives

$$\phi_{+}^{n+1}=r'_{n}\phi_{-}^{n+1}+t\phi_{+}^{n}. \tag{5}$$

Equations (3), (4), (5) then give

$$\phi_{+}^{n}=\frac{t\left(\phi_{+}^{n-1}+e^{i2kl}\phi_{+}^{n+1}\right)}{1+t^{2}e^{i2kl}-r_{n}r'_{n}e^{i2kl}}.$$

Let $r_{0}=r$.   Then $r_{n}=r\exp(i2nkl)$.   Assume $r'_{n}=-r^{*}_{n}t/t^{*}$.   Then $r'_{n}=r'\exp(-i2nkl)$. Hence

$$\phi_{+}^{n}=\frac{t\left(\phi_{+}^{n-1}+e^{i2kl}\phi_{+}^{n+1}\right)}{1+t^{2}e^{i2kl}-rr'e^{i2kl}}. \tag{6}$$

Similarly,

$$\phi_{-}^{n}=\frac{t\left(\phi_{-}^{n+1}+e^{i2kl}\phi_{-}^{n-1}\right)}{1+t^{2}e^{i2kl}-rr'e^{i2kl}}.$$

c. As the period of the potential is l, if $\psi(x)$ is the wave function in the region $\left[x_{n-1}, x_n\right]$, then $\psi(x-l)\exp(i\delta)$ is the wave function in the region $\left[x_n, x_{n+1}\right]$. Thus

$$\begin{cases} \phi_+^{n+1} = e^{i(\delta-kl)}\phi_+^n, \\ \phi_-^{n+1} = e^{i(\delta+kl)}\phi_-^n. \end{cases} \tag{7}$$

Let $c_n = \phi_+^n / \phi_-^n$. From (4) and (5) we obtain respectively

$$1 = r_n c_n + t\frac{\phi_-^{n+1}}{\phi_-^n}, \tag{8}$$

$$\frac{\phi_+^{n+1}}{\phi_-^n} = r_n' \frac{\phi_-^{n+1}}{\phi_-^n} + t c_n. \tag{9}$$

Using (7), (9) can be written as

$$c_n e^{-i2kl}\frac{\phi_-^{n+1}}{\phi_-^n} = r_n'\frac{\phi_-^{n+1}}{\phi_-^n} + t c_n,$$

or, using (8),

$$c_n e^{-i2kl}\left(1 - r_n c_n\right) = r_n'\left(1 - r_n c_n\right) + t^2 c_n,$$

i.e.,

$$r_n c_n^2 + \left(t^2 e^{i2kl} - r_n r_n' e^{i2kl} - 1\right)c_n + r_n' e^{i2kl} = 0.$$

Solving for c_n we have

$$c_n = \frac{\left(1 + rr'e^{i2kl} - t^2 e^{i2kl}\right) \pm \sqrt{\Delta}}{2r_n},$$

where

$$\Delta = (t^2 e^{i2kl} - r_n r_n' e^{i2kl} - 1)^2 - 4r_n r_n' e^{i2kl}$$
$$= (t^2 e^{i2kl} - rr'e^{i2kl} - 1)^2 - 4rr'e^{i2kl}.$$

d. The necessary condition for a stable wave to exist in the infinite periodic field is

$$\phi_+^{n+1} / \phi_+^n = e^{i\delta_1},$$

where δ_1 is real and independent of n. If this were not so, when $n \to \infty$ one of ϕ_+^n and $\phi_+^{(-n)}$ would be infinite. From (7), we see that $\delta_1 = \delta - kl$. From (6), we obtain

$$1 + t^2 e^{i2kl} - rr'e^{i2kl} = t\left(\frac{\phi_+^{n-1}}{\phi_+^n} + e^{i2kl}\frac{\phi_+^{n+1}}{\phi_+^n}\right)$$
$$= t\left[e^{i(kl-\delta)} + e^{i(kl+\delta)}\right].$$

Substituting $r'=-r^*t/t^*$ in the above equation and using $rr^*+tt^*=1$, we obtain

$$te^{ikl}+t^*e^{-ikl}=2tt^*\cos\delta,$$

which means

$$\left|\frac{te^{ikl}+t^*e^{-ikl}}{2tt^*}\right|\leq 1,$$

or, using the definition of t,

$$\left|\frac{\cos(2\delta_e+kl)+\cos(2\delta_0+kl)}{1+\cos[2(\delta_e-\delta_0)]}\right|\leq 1,$$

i.e.,

$$\left|\frac{\cos(\delta_e+\delta_0+kl)}{\cos(\delta_e-\delta_0)}\right|\leq 1.$$

In general, only some of the values of k satisfy the above inequality, i.e., only energy values in certain regions are allowed while the others are forbidden. Thus we obtain the band structure of energy levels.

e. In metals, the distribution of positive ions is regular and so the conduction electrons move in a periodic potential, (d) Shows that the electron waves can only have certain k values, corresponding to bands of electron energies.

1066

You are given a real operator $\hat{A}$ satisfying the quadratic equation

$$\hat{A}^2 - 3\hat{A} + 2 = 0.$$

This is the lowest-order equation that $\hat{A}$ obeys.

a. What are the eigenvalues of $\hat{A}$?

b. What are the eigenstates of $\hat{A}$?

c. Prove that $\hat{A}$ is an observable.

(Buffalo)

Sol:

a. As $\hat{A}$ satisfies a quadratic equation it can be represented by a 2×2 matrix. Its eigenvalues are the roots of the quadratic equation $\lambda^2-3\lambda+2=0$, $\lambda_1=1$, $\lambda_2=2$.

b. $\hat{A}$ is represented by the matrix

$$\hat{A} = \begin{pmatrix} 1 & 0 \\ 0 & 2 \end{pmatrix}.$$

The eigenvalue equation

$$\begin{pmatrix} 1 & 0 \\ 0 & 2 \end{pmatrix} \begin{pmatrix} a \\ b \end{pmatrix} = \lambda \begin{pmatrix} a \\ b \end{pmatrix}$$

then gives $a = 1$, $b = 0$ for $\lambda = 1$ and $a = 0$, $b = 1$ for $\lambda = 2$. Hence the eigen-

states of $\hat{A}$ are $\begin{pmatrix} 1 \\ 0 \end{pmatrix}$ and $\begin{pmatrix} 0 \\ 1 \end{pmatrix}$.

c. Since $\hat{A} = \hat{A}^+$, $\hat{A}$ is Hermitian and hence an observable.

1067

If $|\psi_q\rangle$ is any eigenstate of the electric charge operator Q corresponding to eigenvalue q, that is to say,

$$Q|\psi_q\rangle = q|\psi_q\rangle,$$

the "charge conjugation" operator C applied to $|\psi_q\rangle$ leads to an eigenstate $|\psi_{-q}\rangle$ of Q corresponding to eigenvalue $-q$:

$$C|\psi_q\rangle = |\psi_{-q}\rangle.$$

a. Find the eigenvalues of the operator $CQ + QC$.

b. Can a state simultaneously be an eigenstate of C and of Q?

(Chicago)

Sol:

a. Let

$$|\psi\rangle = \sum_q c_q |\psi_q\rangle.$$

Then

$$(CQ + QC)|\psi_q\rangle = qC|\psi_q\rangle + Q|\psi_{-q}\rangle = q|\psi_{-q}\rangle - q|\psi_{-q}\rangle = 0.$$

Thus the eigenvalue of the operator $CQ + QC$ is zero.

b. As C is the charge conjugation transformation, $CQC^{-1} = -Q$, or $CQ + QC = 0$, i.e., C and Q do not commute (they anticommute) they cannot have common eigenstates. (Unless $q = 0$, in which case it is quite meaningless to introduce charge conjugation.)

1068

A quantum-mechanical system is known to possess only two energy eigen-states denoted $|1\rangle$ and $|2\rangle$. The system also includes three other observables (besides the energy), known as P, Q and R. The states $|1\rangle$ and $|2\rangle$ are normalized but they are not necessarily eigenstates of P, Q or R.

Determine as many of the eigenvalues of P, Q and R as possible on the basis of the following sets of "experimental data". [Warning: one data set is unphysical.]

   a. $\langle 1|P|1\rangle=1/2$, $\langle 1|P^2|1\rangle=1/4$.

   b. $\langle 1|Q|1\rangle=1/2$, $\langle 1|Q^2|1\rangle=1/6$.

   c. $\langle 1|R|1\rangle=1$, $\langle 1|R^2|1\rangle=5/4$, $\langle 1|R^3|1\rangle=7/4$.

(MIT)

Sol: We are given three observables P, Q, R, which satisfy the Hermiticity $\langle n|P|m\rangle=\langle m|P|n\rangle^*$, and that the mechanical system has a complete set of energy eigenstates $|1\rangle$ and $|2\rangle$.

   a.   The completeness of the two states and the "experimental data", give

$$P|1\rangle=\frac{1}{2}|1\rangle+\alpha|2\rangle,$$

where α is a constant to be determined. The orthogonality of the eigenstates and the Hermiticity of P give

$$\langle 1|P|2\rangle=\frac{1}{2}\langle 1|2\rangle+\alpha^2\langle 2|2\rangle=\alpha^*.$$

So we have

$$P|2\rangle=\alpha^*|1\rangle+\beta|2\rangle,$$

where β is to be determined. Then

$$P^2|1\rangle=P(P|1\rangle)$$

$$=\frac{1}{2}P|1\rangle+\alpha P|2\rangle=(1/4+\alpha^*\alpha)|1\rangle+(\alpha/2+\alpha\beta)|2\rangle.$$

As $P^2|1\rangle=\frac{1}{4}$ according to experiment, $\alpha^*\alpha=0$ and hence $\alpha=0$.

Therefore,

$$P|1\rangle = \frac{1}{2}|1\rangle ,$$

i.e., at least one of the eigenvalues of P is 1/2.

b. Let

$$Q|1\rangle = \frac{1}{2}|1\rangle + \gamma|2\rangle ,$$

where γ is to be determined. By a similar procedure, we get $\gamma^* \gamma = 1/6 - 1/4 < 0$. So this data set is unphysical and the eigenvalue of Q could not be determined.

c. As $\langle 1|R|1\rangle = 1$, we can write

$$R|1\rangle = |1\rangle + \lambda|2\rangle ,$$

where λ is to be determined. Then

$$\langle 1|R|2\rangle = \langle 1|2\rangle + \lambda^* \langle 2|2\rangle = \lambda^* ,$$

showing that

$$R|2\rangle = \lambda^*|1\rangle + \eta|2\rangle .$$

where η is to be determined. Consider

$$R^2|1\rangle = R|1\rangle + \lambda R|2\rangle$$

$$= |1\rangle + \lambda|2\rangle + \lambda\lambda^*|1\rangle + \lambda\eta|2\rangle .$$

Then as

$$\langle 1|R^2|1\rangle = 1 + \lambda\lambda^* = \frac{5}{4} ,$$

we have

$$\lambda\lambda^* = \frac{1}{4} ,$$

and so

$$\lambda = \frac{1}{2}\exp(i\delta),$$

and

$$R|1\rangle = |1\rangle + \frac{1}{2}e^{i\delta}|2\rangle ,$$

$$R|2\rangle = \frac{1}{2}e^{-i\delta}|1\rangle + \eta|2\rangle ,$$

$$R^2|1\rangle = \frac{5}{4}|1\rangle + \frac{1}{2}(1+\eta)e^{i\delta}|2\rangle .$$

It follows also

$$R^3|1\rangle = \frac{5}{4}R|1\rangle + \frac{1}{2}(1+\eta)e^{i\delta}R|2\rangle$$

$$= \frac{5}{4}|1\rangle + \frac{5}{8}e^{i\delta}|2\rangle + \frac{1}{4}(1+\eta)|1\rangle$$

$$+ \frac{1}{2}(1+\eta)\eta e^{i\delta}|2\rangle.$$

Experimentally $\langle 1|R^3|1\rangle = \frac{7}{4}$. Thus $\frac{5}{4} + \frac{1}{4}(1+\eta) = \frac{7}{4}$, giving $\eta = 1$. Hence on the bases $|1\rangle$ and $|2\rangle$ the matrix of R is

$$\begin{pmatrix} 1 & \frac{1}{2}e^{-i\delta} \\ \frac{1}{2}e^{i\delta} & 1 \end{pmatrix}.$$

To find the eigenvalues of R solve

$$\begin{vmatrix} 1-\lambda & \frac{1}{2}e^{-i\delta} \\ \frac{1}{2}e^{i\delta} & 1-\lambda \end{vmatrix} = 0,$$

i.e., $(1-\lambda)^2 - \frac{1}{4} = (1-\lambda-\frac{1}{2})(1-\lambda+\frac{1}{2}) = 0$, and obtain the eigenvalues of $R = \frac{1}{2}, \frac{3}{2}$,

---

1069

For a charged particle in a magnetic field, find the commutation rules for the operators corresponding to the components of the velocity.

(Berkeley)

Sol: Suppose the magnetic field arises from a vector potential **A**. Then the velocity components of the particle are

$$\hat{v}_i = \hat{P}_i/m - qA_i/mc .$$

Hence

$$\left[\hat{v}_i, \hat{v}_j\right] = \frac{1}{m^2}\left[\hat{P}_i - \frac{q}{c}A_i, \hat{P}_j - \frac{q}{c}A_j\right]$$

$$= \frac{q}{m^2 c}\{[\hat{p}_j, A_i] - [\hat{p}_i, A_j]\} = \frac{i\hbar q}{mc^2}\left(\frac{\partial A_j}{\partial x_i} - \frac{\partial A_i}{\partial x_j}\right)$$

$$= \frac{i\hbar q}{m^2 c}\sum_{k=1}^{3}\varepsilon_{ijk}B_k ,$$

where ε_{ijk} is the Levi-Civita density, use having been made of the correspondence rule $\hat{p}_i \to \dfrac{\hbar}{i}\dfrac{\partial}{\partial x_i}$.

1070

Using the coordinate-momentum commutation relation prove that

$$\sum_n (E_n - E_0)\,|\langle n|x|0\rangle|^2 = \text{constant},$$

where E_n is the energy corresponding to the eigenstate $|n\rangle$. Obtain the value of the constant. The Hamiltonian has the form $H = p^2/2M + V(x)$.

(Berkeley)

Sol: As

$$H = p^2/2M + V(x),$$

we have

$$[H, x] = \frac{1}{2M}\left[p^2, x\right] = -i\hbar p/M,$$

and so

$$[[H, x], x] = -\frac{i\hbar}{M}[p, x] = -\hbar^2/M.$$

Hence

$$\langle m|[[H, x], x]|m\rangle = -\frac{\hbar^2}{M}.$$

On the other hand,

$$\langle m|[[H, x], x]|m\rangle = \langle m|\,Hx^2 - 2xHx + x^2H\,|m\rangle$$

$$= 2E_m\langle m|x^2|m\rangle - 2\langle m|xHx|m\rangle$$

$$= 2E_m\sum_n |\langle m|x|n\rangle|^2 - 2\sum_n E_n |\langle m|x|n\rangle|^2$$

$$= 2\sum_n (E_m - E_n)|\langle m|x|n\rangle|^2.$$

In the above we have used

$$H|m\rangle = E_m|m\rangle,$$

$$\langle m|x^2|m\rangle = \sum_n \langle m|x|n\rangle\langle n|x|m\rangle$$

$$= \sum_n |\langle m|x|n\rangle|^2$$

$$\langle m|xHx|m\rangle = \sum_n \langle m|xH|n\rangle\langle n|x|m\rangle$$

$$= \sum_n E_n \langle m|x|n\rangle\langle n|x|m\rangle$$

$$= \sum_n E_n |\langle m|x|n\rangle|^2 .$$

Equating the two results and setting $m = 0$, we obtain

$$\sum_n (E_n - E_0)|\langle n|x|0\rangle|^2 = \hbar^2/2M .$$

1071

a. Given a Hermitian operator A with eigenvalues a_n and eigenfunctions $u_n(x)$ $[n = 1, 2, \ldots, N; 0 \le x \le L]$, show that the operator $\exp(iA)$ is unitary.

b. Conversely, given the matrix U_{mn} of a unitary operator, construct the matrix of a Hermitian operator in terms of U_{mn}.

c. Given a second Hermitian operator B with eigenvalues b_m and eigenfunctions $v_m(x)$, construct a representation of the unitary operator V that transforms the eigenvectors of B into those of A.

(*Chicago*)

Sol:

a. As $A^+ = A$, A being Hermitian,

$$\{\exp(iA)\}^+ = \exp(-iA^+) = \exp(-iA) = \{\exp(iA)\}^{-1} .$$

Hence $\exp(iA)$ is unitary.

b. Let

$$C_{mn} = U_{mn} + U_{nm}^* = U_{mn} + (U^+)_{mn} ,$$

i.e.,

$$C = U + U^+ .$$

As $U^{++} = U$, $C^+ = C$. Therefore $C_{mn} = U_{mn} + U_{nm}^*$ is the matrix representation of a Hermitian operator.

c. The eigenkets of a Hermitian operator form a complete and orthonormal set. Thus any $|u_m\rangle$ can be expanded in the complete set $|v_n\rangle$:

$$|u_m\rangle = \sum_k |vk\rangle\langle v_k | u_m\rangle \equiv \sum_k |v_k\rangle V_{km},$$

which defines V_{km},

$$V_{km} = \int_0^L v_k^*(x) u_m(x) dx.$$

Similarly,

$$|v_n\rangle = \sum_j |u_j\rangle\langle u_j | v_n\rangle$$

$$= \sum_j |u_j\rangle |v_n | u_j|^*$$

$$= \sum_j |u_j\rangle V_{nj}^*$$

$$= \sum_j |u_j\rangle \tilde{V}_{jn}^*$$

$$= \sum_j |u_j\rangle V_{jn}^+.$$

Hence

$$|u_m\rangle = \sum_j \sum_k |u_j\rangle V_{jk}^+ V_{km} = \sum_j |u_j\rangle \delta_{jm},$$

or

$$V^+V = 1,$$

i.e.,

$$V^+ = V^{-1},$$

showing that V is unitary. Thus V is a unitary operator transforming the eigenvectors of B into those of A

1072

Consider a one-dimensional oscillator with the Hamiltonian

$$H = p^2/2m + m\omega^2 x^2/2.$$

a. Find the time dependence of the expectation values of the "initial position" and "initial momentum" operators

$$x_0 = x \cos\omega t - (p/m\omega) \sin\omega t,$$

$$p_0 = p \cos\omega t + m\omega x \sin\omega t.$$

b. Do these operators commute with the Hamiltonian?

c. Do you find your results for (a) and (b) to be compatible? Discuss.

d. What are the motion equations of the operators in the Heisenberg picture?

e. Compute the commutator $[p_0, x_0]$. What is its significance for measurement theory?

(Princeton)

Sol:

a. Making use of the relation

$$\frac{df}{dt} = \frac{1}{i\hbar}[f, H] + \frac{\partial f}{\partial t},$$

we have

$$\frac{d\langle x_0\rangle}{dt} = \frac{1}{i\hbar}[\langle x\rangle\cos\omega t, H] - \omega\langle x\rangle\sin\omega t$$

$$-\frac{1}{i\hbar}\left[\frac{\langle p\rangle}{m\omega}\sin\omega t, H\right] - \frac{\langle p\rangle}{m}\cos\omega t$$

$$= \frac{1}{i\hbar}\overline{[x, H]}\cos\omega t - \omega\langle x\rangle\sin\omega t$$

$$-\frac{1}{m\omega}\frac{1}{i\hbar}\overline{[p, H]}\sin\omega t - \frac{1}{m}\langle p\rangle\cos\omega t = 0,$$

$$\frac{d\langle p_0\rangle}{dt} = \frac{1}{i\hbar}[\langle p\rangle\cos\omega t, H] - \langle p\rangle\omega\sin\omega t$$

$$+\frac{1}{i\hbar}[m\omega\langle x\rangle\sin\omega t, H] + m\omega^2\langle x\rangle\cos\omega t$$

$$= \frac{1}{i\hbar}\overline{[p, H]}\cos\omega t - \omega\langle p\rangle\sin\omega t$$

$$+\frac{m\omega}{i\hbar}\overline{[x, H]}\sin\omega t + m\omega^2\langle x\rangle\cos\omega t = 0.$$

Thus the expectation values of these operators are independent of time.

b. Consider

$$[x_0, H] = [x, H]\cos\omega t - \frac{[p, H]}{mw}\sin\omega t$$

$$= \frac{i\hbar p}{m}\cos\omega t + i\hbar\omega x\sin\omega t,$$

$$[p_0, H]=[p, H]\cos\omega t+m\omega[x, H]\sin\omega t$$
$$=-i\hbar m\omega^2 x\cos\omega t+i\hbar\omega p\sin\omega t.$$

Thus the operators x_0, p_0 do not commute with H.

c. The results of (a) and (b) are still compatible. For while the expressions for x_0 and p_0 contain t explicitly, their non-commutation with H does not exclude their being conserved. In fact

$$\frac{dx_0}{dt}=\frac{1}{i\hbar}[x_0,H]+\frac{\partial x_0}{\partial t}=0,$$
$$\frac{dp_0}{dt}=\frac{1}{i\hbar}[p_0,H]+\frac{\partial p_0}{\partial t}=0,$$

showing that they are actually conserved.

d. In the Heisenberg picture, the motion equation of an operator is

$$dA/dt=\frac{1}{i\hbar}[A,H]+\frac{\partial A}{\partial t}.$$

Thus the motion equations of x_0 and p_0 are respectively

$$dx_0/dt=0, \quad dp_0/dt=0.$$

e. Using the expressions for x_0 and p_0, we have

$$[p_0, x_0]=[p\cos\omega t+m\omega x\sin\omega t, x_0]$$
$$=[p, x_0]\cos\omega t+[x, x_0]m\omega\sin\omega t$$
$$=\left[p, x\cos\omega t-\frac{p}{m\omega}\sin\omega t\right]\cos\omega t$$
$$+\left[x, x\cos\omega t-\frac{p}{m\omega}\sin\omega t\right]m\omega\sin\omega t.$$
$$=[p, x]\cos^2\omega t-[x, p]\sin^2\omega t$$
$$=-[x, p]=-i\hbar,$$

as $[x, x]=[p, p]=0, [x, p]=i\hbar.$

In general, if two observables A and B satisfy the equation

$$[A, B]=i\hbar,$$

then their root-mean-square deviations $\Delta A, \Delta B$, when they are measured simultaneously, must satisfy the uncertainty principle

$$\Delta A\cdot\Delta B\geq\frac{\hbar}{2}.$$

In the present case, the simultaneous measurements of position and momentum in the same direction must result in

$$\Delta x \cdot \Delta p \geq \frac{\hbar}{2}.$$

The relation shows

$$\sqrt{\Delta x_0^2} \sqrt{\Delta p_0^2} \geq \hbar / 2.$$

It is a relation between possible upper limits to the precision of the two quantities when we measure them simultaneously.

Part II
Central Potentials

2001

An electron is confined in a three-dimensional infinite potential well. The sides parallel to the x-, y-, and z-axes are of length L each.

  a. Write the appropriate Schrödinger equation.

  b. Write the time-independent wave function corresponding to the state of the lowest possible energy.

  c. Give an expression for the number of states, N, having energy less than some given E. Assume $N \gg 1$.

(Wisconsin)

Sol:

  a. The Schrödinger equation is

$$i\hbar \partial \psi (\mathbf{r},\, t)/\partial t = -(\hbar^2/2m)\nabla^2 \psi (\mathbf{r},\, t),\ 0 \le x, y, z \le L,$$
$$\psi = 0, \qquad\qquad\qquad\qquad \text{otherwise.}$$

  b. By separation of variables, we can take that the wave function to be the product of three wave functions each of a one-dimensional infinite well potential. The wave function of the lowest energy level is

$$\psi_{111}(x, y, z) = \psi_1(x)\psi_1(y)\psi_1(z),$$

  where

$$\psi_1(x) = \sqrt{\frac{2}{L}}\sin\left(\frac{\pi}{L}x\right),\ \text{etc.}$$

  Thus

$$\psi_{111}(x, y, z) = \left(\frac{2}{L}\right)^{3/2}\sin\left(\frac{\pi x}{L}\right)\sin\left(\frac{\pi y}{L}\right)\sin\left(\frac{\pi z}{L}\right).$$

  The corresponding energy is $E_{111} = 3\hbar^2\pi^2/2mL^2$.

c. For a set of quantum numbers n_x, n_y, n_z for the three dimensions, the energy is

$$E = \frac{\hbar^2 \pi^2}{2mL^2}(n_x^2 + n_y^2 + n_z^2).$$

Hence the number N of states whose energy is less than or equal to E is equal to the number of sets of three positive integers n_x, n_y, n_z satisfying the inequality

$$n_x^2 + n_y^2 + n_z^2 \leq \frac{2mL^2}{\hbar^2 \pi^2} E.$$

Consider a Cartesian coordinate system of axes n_x, n_y, n_z. The number N required is numerically equal to the volume in the first quadrant of a sphere of radius $(2mL^2 E/\hbar^2 \pi^2)^{1/2}$, provided $N \geq 1$. Thus

$$N = \frac{1}{8} \cdot \frac{4\pi}{3}\left(\frac{2mL^2}{\hbar^2 \pi^2}E\right)^{3/2} = \frac{4\pi}{3}\left(\frac{mL^2}{2\hbar^2 \pi^2}E\right)^{3/2}.$$

2002

A 'quark' (mass $= m_p/3$) is confined in a cubical box with sides of length 2 fermis $= 2 \times 10^{-15}$ m. Find the excitation energy from the ground state to the first excited state in MeV.

(Wisconsin)

Sol: The energy levels in the cubical box are given by

$$E_{n_1 n_2 n_3} = \frac{\hbar^2 \pi^2}{2ma^2}(n_1^2 + n_2^2 + n_3^2), \quad n_i = 1, 2, \ldots$$

Thus the energy of the ground state is $E_{111} = 3\hbar^2 \pi^2/2ma^2$, that of the first excited state is $E_{211} = 6\hbar^2 \pi^2/2ma^2 = 3\hbar^2 \pi^2/ma^2$. Hence the excitation energy from the ground state to the first excited state is

$$\Delta E = 3\hbar^2 \pi^2/2ma^2 = \frac{1.5\pi^2 \hbar^2 c^2}{mc^2 a^2}$$

$$= \frac{1.5\pi^2 (6.58 \times 10^{-22})^2 \times (3 \times 10^8)^2}{\left(\frac{938}{3}\right) \times (2 \times 10^{-15})^2} = 461 \text{ MeV}.$$

2003

A NaCl crystal has some negative ion vacancies, each containing one electron. Treat these electrons as moving freely inside a volume whose dimensions are on the order of the lattice constant. The crystal is at room temperature. Give a numerical estimate for the longest wavelength of electromagnetic radiation absorbed strongly by these electrons.

(MIT)

Sol: The energy levels of an electron in a cubical box of sides a are given by

$$E_{nmk} = (\pi^2 \hbar^2 / 2ma^2)(n^2 + m^2 + k^2),$$

where n, m and k are positive integers. Taking $a \sim 1\text{Å}$, the ground state energy is $E_{111} = 3\hbar^2\pi^2/2ma^2 \approx 112$ eV. For a crystal at room temperature, the electrons are almost all in the ground state. The longest wavelength corresponds to a transition from the ground state to the nearest excited state:

$$\Delta E = E_{211} - E_{111} = \frac{3\pi^2\hbar^2}{ma^2} = 112 \text{ eV},$$

for which

$$\lambda = \frac{c}{v} = \frac{hc}{\Delta E} = 110 \text{ Å}.$$

2004

An electron is confined to the interior of a hollow spherical cavity of radius R with impenetrable walls. Find an expression for the pressure exerted on the walls of the cavity by the electron in its ground state.

(MIT)

Sol: For the ground state, $l=0$, and if we set the radial wave function as $R(r) = \chi(r)/r$, then $\chi(r)$ is given by

$$\frac{d^2\chi}{dr^2} + \frac{2\mu E}{\hbar^2}\chi = 0 \quad \text{for} \quad r < R,$$

$$\chi = 0 \quad \text{for} \quad r \geq R,$$

where μ is the electron rest mass. $R(r)$ is finite at $r=0$, so that $\chi(0)=0$.

The solutions satisfying this condition are

$$\chi_n = \sqrt{\frac{2}{R}}\sin\frac{n\pi r}{R}, \quad (n=1,2,3,\ldots)$$

for which

$$E_n = \frac{\pi^2\hbar^2}{2\mu R^2}n^2.$$

The average force F acting radially on the walls by the electron is given by

$$F=\left\langle -\frac{\partial\hat{V}}{\partial R}\right\rangle = -\left\langle\frac{\partial\hat{H}}{\partial R}\right\rangle = -\frac{\partial}{\partial R}\langle\hat{H}\rangle = -\frac{\partial E}{\partial R}.$$

As the electron is in the ground state, $n=1$ and

$$F=-\partial E_1/\partial R = \pi^2\hbar^2/\mu R^3.$$

The pressure exerted on the walls is

$$p = F/4\pi R^2 = \pi\hbar^2/4\mu R^5.$$

2005

A particle of mass m is constrained to move between two concentric impermeable spheres of radii $r=a$ and $r=b$. There is no other potential. Find the ground state energy and normalized wave function.

(MIT)

Sol: Let the radial wave function of the particle be $R(r)=\chi(r)/r$. Then $\chi(r)$ satisfies the equation

$$\frac{d^2\chi(r)}{dr^2}+\left\{\frac{2m}{\hbar^2}[E-V(r)]-\frac{l(l+1)}{r^2}\right\}\chi(r)=0.$$
$$(a\le r\le b)$$

For the ground state, $l=0$, so that only the radial wave function is non-trivial. Since $V(r)=0$, letting $K^2=2mE/\hbar^2$, we reduce the equation to

$$\chi''+K^2\chi=0,$$

with

$$\chi|_{r=a}=\chi|_{r=b}=0.$$

$\chi(a)=0$ requires the solution to have the form

$$\chi(r)=A\,\sin[K(r-a)].$$

Then from $\chi(b)=0$, we get the possible values of K:

$$K = n\pi/(b-a), \quad (n=1,2,\ldots)$$

For the particle in the ground state, i.e., $n=1$, we obtain the energy

$$E = \hbar^2 K^2/2m = \hbar^2\pi^2/2m(b-a)^2.$$

From the normalization condition

$$\int_a^b R^2(r)r^2\,dr = \int_a^b \chi^2(r)\,dr = 1,$$

we get $A=\sqrt{2/(b-a)}$. Hence for the ground state, the normalized radial wave function is

$$R(r) = \sqrt{\frac{2}{b-a}}\frac{1}{r}\sin\frac{\pi(r-a)}{b-a},$$

and the normalized wave function is

$$\psi(\mathbf{r}) = \frac{1}{\sqrt{4\pi}}\sqrt{\frac{2}{b-a}}\frac{1}{r}\sin\frac{\pi(r-a)}{b-a}.$$

2006

a. For a simple quantum harmonic oscillator with $H=(p^2/m+kx^2)/2$, show that the energy of the ground state has the lowest value compatible with the uncertainty principle.

b. The wave function of the state where the uncertainty principle minimum is realized is a Gaussian function $\exp(-\alpha x^2)$. Making use of this fact, but without solving any differential equation, find the value of α.

c. Making use of raising or lowering operators, but without solving any differential equation, write down the (non-normalized) wave function of the first excited state of the harmonic oscillator.

d. For a three-dimensional oscillator, write down, in polar coordinates, the wave functions of the degenerate first excited state which is an eigenstate of l_z.

(Berkeley)

Sol:

a. The ground state of the harmonic oscillator has even parity, so that

$$\bar{x} = \langle 0|x|0\rangle = 0, \quad \bar{p} = \langle 0|p|0\rangle = 0;$$

and so

$$\overline{\Delta p^2} = \overline{p^2}, \qquad \overline{\Delta x^2} = \overline{x^2}.$$

The uncertainty principle requires

$$\overline{\Delta p^2} \cdot \overline{\Delta x^2} \geq \frac{\hbar^2}{4}.$$

It follows that

$$\overline{E} = \frac{\overline{p^2}}{2m} + \frac{k}{2}\overline{x^2}$$

$$\geq \sqrt{\frac{k}{m}} \sqrt{\overline{p^2} \cdot \overline{x^2}}$$

$$\geq \frac{\hbar}{2}\sqrt{\frac{k}{m}} = \hbar\omega/2 = E_0,$$

as $\sqrt{\frac{k}{m}} = \omega$. Thus the energy of the ground state has the lowest value compatible with the uncertainty principle.

b. Using the given wave function we calculate

$$\overline{x^2} = \int_{-\infty}^{\infty} e^{-2\alpha x^2} x^2 \, dx \Big/ \int_{-\infty}^{\infty} e^{-2\alpha x^2} \, dx = 1/4\alpha,$$

$$\overline{p^2} = -\hbar^2 \int_{-\infty}^{\infty} e^{-\alpha x^2} \frac{d^2}{dx^2} e^{-\alpha x^2} \, dx \Big/ \int_{-\infty}^{\infty} e^{-2\alpha x^2} \, dx = \hbar^2 \alpha,$$

and hence

$$E = \frac{\hbar^2}{2m}\alpha + \frac{k}{2}\cdot\frac{1}{4\alpha}.$$

From $\frac{dE}{d\alpha} = 0$ we see that when $\alpha = \sqrt{km}/2\hbar = m\omega/2\hbar$ the energy is minimum. Therefore $\alpha = m\omega/2\hbar$.

c. In the Fock representation of harmonic oscillation we define

$$\hat{a} = i(\hat{p} - im\omega\hat{x})/\sqrt{2m\hbar\omega},$$

$$\hat{a}^+ = -i(\hat{p} + im\omega\hat{x})/\sqrt{2m\hbar\omega}.$$

Then $[\hat{a}, \hat{a}^+] = 1$,

$$H = (\hat{a}^+\hat{a} + 1/2)\hbar\omega.$$

Denoting the ground state wave function by $|0\rangle$. As $H|0\rangle = \frac{1}{2}\hbar\omega|0\rangle$, the last equation gives $\hat{a} + \hat{a}|0\rangle = 0$. It also gives

$$H(\hat{a}^+|0\rangle) = \frac{1}{2}\hbar\omega\hat{a}^+|0\rangle + \hbar\omega\hat{a}^+\hat{a}\hat{a}^+|0\rangle$$

$$= \frac{1}{2}\hbar\omega\hat{a}^+|0\rangle + \hbar\omega\hat{a}^+(\hat{a}^+\hat{a}+1)|0\rangle$$

$$= \frac{3}{2}\hbar\omega(\hat{a}^+|0\rangle).$$

Hence

$$|1\rangle = \hat{a}^+|0\rangle = \frac{-i}{\sqrt{2m\hbar\omega}}\left(-i\hbar\frac{\partial}{\partial x} + im\omega x\right)\exp\left(-\frac{m\omega}{2\hbar}x^2\right)$$

$$= \sqrt{\frac{2m\omega}{\hbar}}x\,\exp\left(-\frac{m\omega}{2\hbar}x^2\right)$$

in the coordinate representation.

d. For a 3-dimensional oscillator, the wave function is

$$\psi_{n_1 n_2 n_3}(\mathbf{r}) = \psi_{n_1}(x)\psi_{n_2}(y)\psi_{n_3}(z).$$

For the ground state, $(n_1,n_2,n_3)=(0,0,0)$. For the first excited states, $(n_1, n_2, n_3)=(1, 0, 0); (0, 1, 0); (0, 0, 1)$.

$$\psi_{100}(\mathbf{r}) = N_0^2 N_1 2\alpha x \exp\left(-\frac{1}{2}\alpha^2 r^2\right),$$

$$\psi_{010}(\mathbf{r}) = N_0^2 N_1 2\alpha y \exp\left(-\frac{1}{2}\alpha^2 r^2\right),$$

$$\psi_{001}(\mathbf{r}) = N_0^2 N_1 2\alpha z \exp\left(-\frac{1}{2}\alpha^2 r^2\right).$$

Expanding x, y, z in spherical harmonics and recombining the wave functions, we get the eigenstates of l_z

$$\psi_m(\mathbf{r}) = N_m \exp\left(-\frac{1}{2}\alpha^2 r^2\right) r Y_{1m}(\theta,\varphi).$$

where

$$N_m^{-1} = \sqrt{2}/\alpha^2.$$

Note that here $\alpha = \sqrt{m\omega/\hbar}$, which is the usual definition, different from that given in (b).

2007

The diagram (Fig. 2.1) shows the six lowest energy levels and the associated angular momenta for a spinless particle moving in a certain three-dimensional central potential. There are no "accidental" degeneracies in this energy spectrum. Give the number of nodes (changes in sign) in the radial wave function associated with each level.

(MIT)

Sol: The radial wave function of a particle in a three-dimensional central potential can be written as $R(r) = \chi(r)/r$. With a given angular quantum number l, the equation satisfied by $\chi(r)$ has the form of a one-dimensional Schrödinger equation. Hence, if an energy spectrum has no "accidental" degeneracies, the role of the nodes in the radial wave function of the particle is the same as that in the one-dimensional wave function. For bound states, Sturm's theorem remains applicable, i.e., $\chi(r)$ obeys Sturm's theorem: the radial wave function of the ground state has no node, while that of the nth excited state has n nodes. Thus, for a bound state of energy E_n, which has quantum number $n = n_r + l + 1$, the radial wave has n_r nodes.

For angular quantum number $l = 0$, the numbers of nodes for the three energy levels (ordered from low to high energy) are 0, 1 and 2.

Similarly, for $l = 1$, the numbers of nodes are 0 and 1; for $l = 2$, the number of nodes is 0.

Thus, the numbers of nodes in the energy levels shown in Fig. 2.1 are 0, 1, 0, 0, 1, 2, from low to high energy.

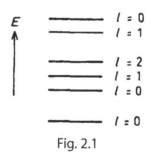

Fig. 2.1

2008

A particle of mass m and charge q is bound to the origin by a spherically symmetric linear restoring force. The energy levels are equally spaced at intervals $\hbar\omega_0$ above the ground state energy $E_0 = 3\hbar\omega_0/2$. The states can be described alternatively in a Cartesian basis (three one-dimensional harmonic oscillators) or in a spherical basis (central field, separated into angular and radial motions).

a. In the Cartesian basis, table the occupation numbers of the various states of the oscillators for the ground and first three excited levels. Determine the total degeneracy of each of these levels.

b. In the spherical basis, write down (do not solve) the radial equation of motion.

(Note that in spherical coordinates $\nabla^2 = \frac{1}{r^2}\frac{\partial}{\partial r}\left(r^2\frac{\partial}{\partial r}\right) - \frac{L^2}{r^2}$, where L^2 is the operator of total orbital angular momentum squared in units of $\hbar^2$.)

Identify the effective potential and sketch it. For a given angular momentum, sketch the "ground state" radial wave function (for a given l value) and also the radial wave functions for the next two states of the same l.

c. For the four levels of part (a), write down the angular momentum content and the parity of the states in each level. Compare the total degeneracies with the answers in (a).

d. Does the second excited state $(E_2 = 7\hbar\omega_0/2)$ have a linear Stark effect? Why or why not? Compare similarities and differences between this oscillator level and the second excited level $(n = 3)$ of the nonrelativistic hydrogen atom.

(Berkeley)

Sol:

a.

Table 2. 1

Energy level	Occupation Numbers	Degeneracy
E_0	$\lvert 0,0,0 \rangle$	1
E_1	$\lvert 1,0,0 \rangle, \lvert 0,1,0 \rangle, \lvert 0,0,1 \rangle$	3
E_2	$\lvert 2,0,0 \rangle, \lvert 0,2,0 \rangle, \lvert 0,0,2 \rangle$	6
	$\lvert 1,1,0 \rangle, \lvert 1,0,1 \rangle, \ \lvert 0,1,1 \rangle$	
E_3	$\lvert 3,0,0 \rangle, \lvert 0,3,0 \rangle, \lvert 0,0,3 \rangle$	10
	$\lvert 2,1,0 \rangle, \lvert 0,2,1 \rangle, \lvert 1,0,2 \rangle$	
	$\lvert 1,2,0 \rangle, \lvert 0,1,2 \rangle, \lvert 2,0,1 \rangle$	
	$\lvert 1,1,1 \rangle$	

b. Let

$$\psi(\mathbf{r}) = R(r)Y_{lm}(\theta, \varphi).$$

The radial wave function $R(r)$ satisfies the equation

$$\frac{1}{r^2}\frac{d}{dr}\left(r^2\frac{d}{dr}R\right) + \left[\frac{2m}{\hbar^2}\left(E - \frac{m}{2}\omega^2 r^2\right) - \frac{l(l+1)}{r^2}\right]R = 0,$$

so that the effective potential is

$$V_{\text{eff}} = m\omega^2 r^2/2 + \hbar^2 l(l+1)/2mr^2,$$

which is sketched in Fig. 2.2, where $r_0 = [\hbar^2 l(l+1)/m^2\omega^2]^{1/4}$. The shapes of the radial wave functions of the three lowest states for a given l are shown in Fig. 2.3.

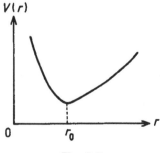

Fig. 2.2

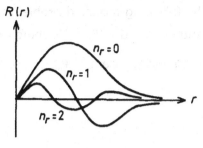

Fig. 2.3

Note that the number of nodes of a wave function is equal to n_r.

c.

Table 2.2

E	l	m	P	D
E_0	0	0	+	1
E_1	1	$0, \pm 1$	−	3
E_2	2	$0, \pm 1 \pm 2$	+	6
	0	0		
E_3	3	$0, \pm 1, \pm 2, \pm 3$	−	10
	1	$0, \pm 1$		

Note: P = parity, D = degeneracy.

d. The second excited state does not have a linear Stark effect because x is an operator of odd parity while all the degenerate states for E_2 have even parity, with the result that the matrix elements of H' in the subspace of the energy level E_2 are all zero.

On the other hand, for the second excited level of the hydrogen atom, $n = 3$, its degenerate states have both even and odd parities, so that linear Stark effect exists.

2009

a. A nonrelativistic particle of mass m moves in the potential

$$V(x,y,z) = A(x^2 + y^2 + 2\lambda xy) + B(z^2 + 2\mu z),$$

where $A > 0, B > 0, |\lambda| < 1, \mu$ is arbitrary. Find the energy eigenvalues.

b. Now consider the following modified problem with a new potential V_{new}: for $z > -\mu$ and any x and y, $V_{\text{new}} = V$, where V is the same as in part (a) above; for $z < -\mu$ and any x and y, $V_{\text{new}} = +\infty$. Find the ground state energy.

(CUS)

Sol:

a. We choose two new variables μ, t defined by

$$\mu = \frac{1}{\sqrt{2}}(x+y), \quad t = \frac{1}{\sqrt{2}}(x-y),$$

or

$$x = (\mu+t)/\sqrt{2}, \quad y = \frac{1}{\sqrt{2}}(\mu-t),$$

and write the potential as

$$V(x, y, z) = A\frac{1}{2}\left[(\mu^2+t^2+2\mu t)+\frac{1}{2}(\mu^2+t^2-2\mu t)\right.$$

$$\left. +2\lambda \cdot \frac{1}{2}(\mu^2-t^2)\right]+B(z^2+2\mu z)$$

$$= A\left[(1+\lambda)\mu^2+(1-\lambda)t^2\right]+B(z^2+2\mu z)$$

and the differentials as

$$\frac{\partial}{\partial x} = \frac{\partial}{\partial \mu}\frac{1}{\sqrt{2}}+\frac{\partial}{\partial t}\cdot\frac{1}{\sqrt{2}},$$

$$\frac{\partial^2}{\partial x^2} = \left(\frac{\partial^2}{\partial \mu^2}\frac{1}{\sqrt{2}}+\frac{\partial^2}{\partial \mu \partial t}\frac{1}{\sqrt{2}}\right)\frac{1}{\sqrt{2}}$$

$$+\left(\frac{\partial^2}{\partial t \partial \mu}\frac{1}{\sqrt{2}}+\frac{\partial^2}{\partial t^2}\frac{1}{\sqrt{2}}\right)\frac{1}{\sqrt{2}};$$

$$\frac{\partial}{\partial y} = \frac{\partial}{\partial \mu}\frac{1}{\sqrt{2}}-\frac{\partial}{\partial t}\frac{1}{\sqrt{2}},$$

$$\frac{\partial^2}{\partial y^2} = \left(\frac{\partial^2}{\partial \mu^2}\frac{1}{\sqrt{2}}-\frac{\partial^2}{\partial \mu \partial t}\frac{1}{\sqrt{2}}\right)\frac{1}{\sqrt{2}}$$

$$-\frac{1}{\sqrt{2}}\left(\frac{\partial^2}{\partial t \partial \mu}\frac{1}{\sqrt{2}}-\frac{\partial^2}{\partial t^2}\frac{1}{\sqrt{2}}\right),$$

$$\nabla^2 = \frac{\partial^2}{\partial \mu^2}+\frac{\partial^2}{\partial t^2}+\frac{\partial^2}{\partial z^2}.$$

Then Schrödinger equation becomes

$$\left(\frac{\partial^2}{\partial\mu^2}+\frac{\partial^2}{\partial t^2}+\frac{\partial^2}{\partial z^2}\right)\phi(\mu, t, z)+\frac{2m}{\hbar^2}[E-V(\mu, t, z)]\phi(\mu, t, z)=0.$$

Let $\phi(\mu, t, z)=U(\mu)T(t)Z(z)$. The above equation can be separated into

$$\frac{\partial^2}{\partial\mu^2}U(\mu)+\frac{2m}{\hbar^2}[E_1-A(1+\lambda)\mu^2]U(\mu)=0,$$

$$\frac{\partial^2}{\partial t^2}T(t)+\frac{2m}{\hbar^2}[E_2-A(1-\lambda)t^2]T(t)=0,$$

$$\frac{\partial^2}{\partial z^2}Z(z)+(2m/\hbar^2)[E_3-B(z^2+2\mu z)]Z(z)=0,$$

with

$$E_1+E_2+E_3=E.$$

By setting $z'=z+\mu, E_3'=E_3+B\mu^2$, all the above three equations can be reduced to that for a harmonic oscillator. Thus the energy eigenvalues are

$$E_1=\left(n_1+\frac{1}{2}\right)\hbar\omega_1, \quad \omega_1=\sqrt{\frac{2A}{m}(1+\lambda)};$$

$$E_2=\left(n_2+\frac{1}{2}\right)\hbar\omega_2, \quad \omega_2=\sqrt{\frac{2A}{m}(1-\lambda)};$$

$$E_3=\left(n_3+\frac{1}{2}\right)\hbar\omega_3-B\mu^2, \quad \omega_3=\sqrt{\frac{2B}{m}}.$$

$$(n_1, n_2, n_3=0, 1, 2, 3, \ldots)$$

b. With a new potential V_{new} such that for $z<-\mu, V_{new}=\infty$ and for $z>-\mu, V_{new}$ is the same as that in (a), the wave function must vanish for $z\rightarrow-\mu$. The Z-equation has solution

$$Z \sim H_{n_3}(\zeta)e^{-\zeta^2/2},$$

where $\zeta=(2mB/\hbar^2)^{1/4}(z+\mu)$, $H_{n_3}(\zeta)$ is the n_3th Hermite polynomial and has the parity of n_3. Hence n_3 must be an odd integer. The ground state is the state for $n_1=n_2=0$ and $n_3=1$, with the corresponding energy

$$E=\hbar(\omega_1+\omega_2)/2+3\hbar\omega_3/2-B\mu^2.$$

2010

Find the root mean squared (RMS) value of the momentum $\hat{p}$ for the first three states for a particle trapped in a 1D quantum harmonic oscillator.

Sol:

(i) $\left\langle 0 \middle| p^2 \middle| 0 \right\rangle$

$$p = i\sqrt{\frac{m\omega\hbar}{2}}\left(a^+ - a\right)$$

$$p^2 = \frac{-m\omega\hbar}{2}\left(a^+ - a\right)^2$$

$$= \frac{-m\omega\hbar}{2}\left\langle 0 \middle| a^+a^{+^{0}} - a^+a^{0} - aa^+ + aa^{0} \middle| 0 \right\rangle$$

$$= \frac{m\omega\hbar}{2}\left\langle 0 \middle| aa^+ \middle| 0 \right\rangle = \frac{m\omega\hbar}{2}\sqrt{1}\left\langle 0 \middle| a \middle| 1 \right\rangle$$

$$= \frac{m\omega\hbar}{2}\sqrt{1}\sqrt{1}\left\langle 0 \middle| 0 \right\rangle = \frac{m\omega\hbar}{2}.$$

(ii) $\left\langle 1 \middle| p^2 \middle| 1 \right\rangle$

$$= \frac{-m\omega\hbar}{2}\left\langle 1 \middle| a^+a^{+^{0}} - a^+a - aa^+ + aa^{0} \middle| 1 \right\rangle$$

$$= \frac{-m\omega\hbar}{2}\left\{-\left\langle 1 \middle| a^+a \middle| 1 \right\rangle - \left\langle 1 \middle| aa^+ \middle| 1 \right\rangle\right\}$$

$$= \frac{-m\omega\hbar}{2}\left\{-\sqrt{1}\left\langle 1 \middle| a^+ \middle| 0 \right\rangle - \sqrt{2}\left\langle 1 \middle| a \middle| 2 \right\rangle\right\}$$

$$= \frac{-m\omega\hbar}{2}\left\{-\sqrt{1}\sqrt{1}\left\langle 1 \middle| 1 \right\rangle - \sqrt{2}\sqrt{2}\left\langle 1 \middle| 1 \right\rangle\right\} = \frac{3m\omega\hbar}{2}.$$

(iii) $\left\langle 2 \middle| p^2 \middle| 2 \right\rangle$

$$= \frac{-m\omega\hbar}{2}\left\langle 2 \middle| a^+a^{+^{0}} - a^+a - aa^+ + aa^{0} \middle| 2 \right\rangle$$

$$= \frac{-m\omega\hbar}{2}\left\{-\left\langle 2 \middle| a^+a \middle| 2 \right\rangle - \left\langle 2 \middle| aa^+ \middle| 2 \right\rangle\right\}$$

$$= \frac{-m\omega\hbar}{2}\left\{-\sqrt{2}\left\langle 2 \middle| a^+ \middle| 1 \right\rangle - \sqrt{3}\left\langle 2 \middle| a \middle| 3 \right\rangle\right\}$$

$$= \frac{-m\omega\hbar}{2}\left\{-\sqrt{2}\left\langle 2 \middle| 2 \right\rangle - 3\left\langle 2 \middle| 2 \right\rangle\right\}$$

$$= \frac{5m\omega\hbar}{2}.$$

2011

Assume that the eigenstates of a hydrogen atom isolated in space are all known and designated as usual by

$$\psi_{nlm}(r, \theta, \phi) = R_{nl}(r)Y_{lm}(\theta, \phi).$$

Suppose the nucleus of a hydrogen atom is located at a distance d from an infinite potential wall which, of course, tends to distort the hydrogen atom.

a. Find the explicit form of the ground state wave function of this hydrogen atom as d approaches zero.

b. Find all other eigenstates of this hydrogen atom in half-space, i.e. $d \rightarrow 0$, in terms of the R_{nl} and Y_{lm}.

(Buffalo)

Sol:

a. Choose a coordinate system with origin at the center of the nucleus and z-axis perpendicular to the wall surface as shown in Fig. 2.4. As $d \rightarrow 0$, the solutions of the Schrödinger equation are still $R_{nl}Y_{lm}$ in the half-space $z > 0$ i.e., $0 < \theta < \pi/2$, but must satisfy the condition $\psi = 0$ at $\theta = \pi/2$ where $V = \infty$. That is, only solutions satisfying $l + m =$ odd integer are acceptable. As $|m| \leq l$, the first suitable spherical harmonic is

$$Y_{10} = \sqrt{3/4\pi} \, \cos\theta.$$

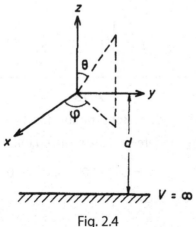

Fig. 2.4

Since $n \geq l+1$, the ground state wave function is $R_{21}Y_{10}$.

b. All the other eigenstates have wave functions $R_{nl}Y_{lm}$ where $l+m =$ odd integer. For a given l, we have $m = l-1, 1-3, \ldots, -1+1$ and hence a degeneracy l.

2012

A harmonic oscillator with constant "k" in ground state suddenly changes its constant to $2k$. What is the probability that the oscillator remains in ground state after this sudden change?

Sol:

$$\psi_i = \left(\frac{\alpha_i}{\sqrt{\pi}}\right)^{1/2} e^{-\alpha_i^2 x^2/2} \qquad \alpha_i = \sqrt{\frac{m\omega_i}{\hbar}} \qquad \omega_i = \sqrt{\frac{k}{m}} \qquad \alpha_f = \sqrt{2}\,\alpha_i$$

$$\psi_f = \left(\frac{\alpha_f}{\sqrt{\pi}}\right)^{1/2} e^{-\alpha_f^2 x^2/2} \qquad \alpha_f = \sqrt{\frac{m\omega_f}{\hbar}} \qquad \omega_f = \sqrt{\frac{2k}{m}}$$

Probability amplitude $= \langle \psi_i | \psi_f \rangle$

$$= \int_0^\alpha \left(\frac{\alpha_i}{\sqrt{\pi}}\right)^{1/2} e^{-\alpha_i^2 x^2/2} \times \left(\frac{\sqrt{2}\alpha_i}{\sqrt{\pi}}\right)^{1/2} e^{-2\alpha_i^2 x^2/2}$$

$$= \left(\sqrt{2}\right)^{1/2} \frac{\alpha_i}{\sqrt{\pi}} \int_0^\alpha e^{-3\alpha^2 x^2/2}$$

$$= \left(\sqrt{2}\right)^{1/2} \frac{\alpha_i}{\sqrt{\pi}} \times \sqrt{\frac{\pi}{3\alpha^2}} = \left(\frac{\sqrt{2}}{3}\right)^{1/2}$$

Probability $= |\text{Prob amp}|^2 = \left|\left(\frac{\sqrt{2}}{3}\right)^{1/2}\right|^2 = \frac{\sqrt{2}}{3}$.

2013

The ground state energy and Bohr radius for the hydrogen atom are $E_0 = -e^2/2a_0$, $a_0 = \hbar^2/me^2$, where m is the reduced mass of the system.

[$m_e = 9.11 \times 10^{-28}$ g, $m_p = 1.67 \times 10^{-24}$ g, $e = 4.80 \times 10^{-10}$ e.s.u., $\hbar = 1.05 \times 10^{-27}$ erg sec.]

a. Compute the ground state energy and Bohr radius of the positronium.

b. What is the degeneracy of the positronium ground state due to electron spin? Write down the possible spin wave functions which have definite values of the total spin together with the corresponding eigenvalues.

c. The ground state of positronium can decay by annihilation into photons. Calculate the energy and angular momentum released in this process and prove that there must be at least two photons in the final state.

(Buffalo)

Sol:

a. The reduced mass m of the positronium is given by $\frac{1}{m}=\frac{1}{m_e}+\frac{1}{m_e}$, i.e., $m=m_e/2$. Its use in the formulas gives

$$E_0=\frac{-e^2}{2a_0}=-\frac{mc^2}{2}\left(\frac{e^2}{\hbar c}\right)^2=\frac{-0.51\times10^6}{4\times137^2}=-6.8\ \text{eV},$$

$$a_0=1.05\times10^{-8}\ \text{cm}.$$

b. The degeneracy of the positronium ground state is 4. Denote the positron by 1 and the electron by 2, and let the spin eigenstates in the z direction of a single particle be α and β, corresponding to eigenvalues $\hbar/2$ and $-\hbar/2$ respectively. Then the ground eigenstates with definite total spin $S=s_1+s_2$ and z-component of the total spin $S_z=s_{1z}+s_{2z}$ are

$$\alpha(1)\,\alpha(2), \qquad\qquad S=\hbar, \qquad S_z=\hbar.$$

$$\frac{1}{\sqrt{2}}[\alpha(1)\beta(2)+\beta(1)\alpha(2)], \qquad S=\hbar, \qquad S_z=0.$$

$$\beta(1)\beta(2), \qquad\qquad S=\hbar, \qquad S_z=-\hbar.$$

$$\frac{1}{\sqrt{2}}[\alpha(1)\beta(2)-\beta(1)\alpha(2)], \qquad S=0, \qquad S_z=0.$$

c. The energy released in the annihilation process mostly comes from the rest masses of the electron and positron, $\Delta E=2m_ec^2=1.02$ MeV. The released angular momentum depends on the state of the positronium before the annihilation. For the ground state $S=0$, no angular momentum is released. For the state $S=\hbar$, the angular momentum released is $\Delta J=\sqrt{l(l+1)}\hbar=\sqrt{2}\hbar$. There must be at least two photons in the final state of an annihilation, for if there were only one photon produced in the annihilation of a positronium, the energy and momentum of the system could not both be conserved. This can be shown by a simple argument. If the single photon has energy E, it must have a momentum E/c at the same time. Thus the momentum of the photon cannot be zero in any reference frame. But in the positronium's rest frame the momentum is zero throughout the annihilation. Hence we have to conclude that there must be at least two photons in the final state of an annihilation whose momenta cancel out in the rest frame of the positronium.

2014

Consider an electron moving in a spherically symmetric potential $V = kr$, where $k > 0$.

a. Use the uncertainty principle to estimate the ground state energy.

b. Use the Bohr-Sommerfeld quantization rule to calculate the ground state energy.

c. Do the same using the variational principle and a trial wave function of your own choice.

d. Solve for the energy eigenvalue and eigenfunction exactly for the ground state.

(Hint: Use Fourier transforms.)

e. Write down the effective potential for nonzero angular momentum states.

(Berkeley)

Sol:

a. The uncertainty principle states that

$$\Delta p \Delta r \geq \frac{\hbar}{2},$$

where

$$\Delta p = \left[\overline{(p - \bar{p})^2} \right]^{1/2} = \left[\overline{(p^2 - 2p\bar{p} + \bar{p}^2)} \right]^{1/2}$$
$$= (\overline{p^2} - \bar{p}^2)^{1/2},$$

$$\Delta r = (\overline{r^2} - \bar{r}^2)^{1/2}.$$

The potential is spherically symmetric, so we can take $\bar{p} = 0$, i.e. $\Delta p \approx \sqrt{\overline{p^2}}$.

For an estimate of the energy

$$E = \frac{\overline{p^2}}{2m} + k\bar{r},$$

we shall also take

$$\Delta r \sim \bar{r}.$$

Then

$$E \approx \frac{(\Delta p)^2}{2m} + k\Delta r \geq \frac{(\Delta p)^2}{2m} + \frac{k\hbar}{2\Delta p}.$$

For the ground state energy E, we have

$$\frac{\partial E}{\partial \Delta p} = \frac{\Delta p}{m} - \frac{k\hbar}{2(\Delta p)^2} = 0,$$

giving

$$\Delta p = \left(\frac{mk\hbar}{2} \right)^{1/3}$$

and

$$E = \frac{3}{2}\left(\frac{k^2\hbar^2}{4m} \right)^{1/3}.$$

b. The Bohr-Sommerfeld quantization rule gives

$$\oint P_r \, dr = n_r h, \qquad \oint P_\phi \, d\phi = n_\phi h.$$

Choose polar coordinates such that the particle is moving in the plane $\theta = \pi / 2$. The ground state is given by $n_r = 0, n_\phi = 1$, and the orbit is circular with radius a. The second integral gives

$$P_\phi = I\omega = ma^2\omega = \hbar.$$

The central force is

$$F_r = -\frac{dV}{dr} = -k = -m\omega^2 a.$$

Combining we have $a = (\hbar^2/mk)^{1/3}$, and hence

$$E_0 = P_\phi^2/2ma^2 + ka = \frac{3}{2}(k^2\hbar^2/m)^{1/3}.$$

c. The notion in the ground state does not depend on θ and ϕ. Take a trial wave function $\psi = \exp(-\lambda r)$ and evaluate

$$\bar{H} = \frac{\langle \psi | \hat{H} | \psi \rangle}{\langle \psi | \psi \rangle},$$

where

$$\hat{H} = -\frac{\hbar^2}{2m}\nabla^2 + kr.$$

As

$$\left\langle \psi \,|\, \hat{H} \,|\, \psi \right\rangle \propto -\frac{\hbar^2}{2m} \int_0^\infty e^{-\lambda r} \frac{1}{r^2} \frac{d}{dr}\left(r^2 \frac{d}{dr} e^{-\lambda r} \right) r^2 \, dr$$
$$+ k \int_0^\infty r^3 e^{-2\lambda r} \, dr$$
$$= \frac{\hbar^2}{2m} \lambda \int_0^\infty \left(\frac{2}{r} - \lambda \right) e^{-2\lambda r} r^2 \, dr + \frac{k 3!}{(2\lambda)^4}$$
$$= \frac{\hbar^2}{8m\lambda} + \frac{3k}{8\lambda^4},$$

$$\left\langle \psi \,|\, \psi \right\rangle \propto \int_0^\infty e^{-2\lambda r} r^2 \, dr = \frac{2}{(2\lambda)^3} = \frac{1}{4\lambda^3},$$

we have

$$\bar{H} = \frac{\hbar^2 \lambda^2}{2m} + \frac{3k}{2\lambda}.$$

For stable motion, $\bar{H}$ is a minimum. Then taking

$$\frac{\partial \bar{H}}{\partial \lambda} = 0,$$

we find

$$\lambda = \left(\frac{3mk}{2\hbar^2} \right)^{1/3},$$

and hence

$$\bar{H} = \frac{9}{4}\frac{k}{\lambda} = \frac{3}{2}\left(\frac{9}{4}\frac{k^2 \hbar^2}{m} \right)^{1/3}.$$

d. The Schrödinger equation for the radial motion can be written as

$$-\frac{\hbar^2}{2m}\frac{d^2}{dr^2}\chi + (kr - E)\chi = 0,$$

where $\chi = rR$, R being the radial wave function. For the ground state, the angular wave function is constant. By the transformation

$$y = \left(\frac{2mk}{\hbar^2} \right)^{1/3} \left(r - \frac{E}{k} \right),$$

the Schrödinger equation becomes the Airy equation

$$\frac{d^2 \chi(y)}{dy^2} - y\chi(y) = 0,$$

whose solutions are $Ai(-x)$ and $Ai(x)$, where $x=-|y|$, for $y<0$ and $y>0$ respectively. The boundary conditions that $R(r)$ and $R'(r)$ be continuous at $r=\frac{E}{k}$, i.e. $y=0$, are satisfied automatically as $Ai(x)=Ai(-x)$, $Ai'(x)=Ai'(-x)$ for $x\rightarrow 0$. The condition that $R(r)$ is finite at $r\rightarrow 0$ requires that $Ai(-x)=rR(r)\rightarrow 0$ as $r\rightarrow 0$. The first zero of $Ai(-x)$ occurs at $x=x_0\approx 2.35$. Hence the ground state energy is

$$E_0=\left(\frac{\hbar^2}{2mk}\right)^{\frac{1}{3}}kx_0,$$

and the ground state eigenfunction is

$$R(r)=\frac{1}{r}Ai(-x)\quad\text{with}\quad x=\left(\frac{2mk}{\hbar^2}\right)^{\frac{1}{3}}\left(\frac{E_0}{k}-r\right).$$

e. The effective potential for nonzero angular momentum is
$$V_{\text{eff}}=kr+\hbar^2 l(l+1)/2mr^2.$$

2015

The interactions of heavy quarks are often approximated by a spin-independent nonrelativistic potential which is a linear function of the radial variable r, where r is the separation of the quarks: $V(r)=A+Br$. Thus the famous "charmonium" particles, the ψ and ψ', with rest energies 3.1 GeV and 3.7 GeV $(1\,\text{GeV}=10^9\text{ eV})$, are believed to be the $n=0$ and $n=1$ bound states of zero orbital angular momentum of a "charm" quark of mass $m_c=1.5\text{ GeV}/c^2$ (i.e. $E=1.5\text{ GeV}$) and an anti-quark of the same mass in the above linear potential. Similarly, the recently discovered upsilon particles, the Υ and Υ', are believed to be the $n=0$ and $n=1$ zero orbital angular momentum bound states of a "bottom" quark and anti-quark pair in the same potential. The rest mass of bottom quark is $m_b=4.5\text{ GeV}/c^2$. The rest energy of Υ is 9.5 GeV.

a. Using dimensional analysis, derive a relation between the energy splitting of the ψ and ψ' and that of the Υ and Υ', and thereby evaluate the rest energy of the Υ'. (Express all energies in units of GeV)

b. Call the $n=2$, zero orbital angular momentum charmonium particle the ψ'' Use the WKB approximation to estimate the energy splitting of the ψ' and the ψ'' in terms of the energy splitting of the ψ and the ψ', and thereby give a numerical estimate of the rest energy of the ψ''.

(Princeton)

Sol: In the center-of-mass system of a quark and its antiquark, the equation of relative motion is

$$\left[-\frac{\hbar^2}{2\mu}\nabla_r^2+V(r)\right]\psi(\mathbf{r})=E_R\psi(\mathbf{r}),\qquad \mu=m_q/2,$$

where E_R is the relative motion energy, m_q is the mass of the quark. When the angular momentum is zero, the above equation in spherical coordinates can be simplified to

$$\left[-\frac{\hbar^2}{2\mu}\frac{1}{r^2}\frac{d}{dr}\left(r^2\frac{d}{dr}\right)+V(r)\right]R(r)=E_RR(r).$$

Let $R(r)=\chi_0(r)/r$. Then $\chi_0(r)$ satisfies

$$\frac{d^2\chi_0}{dr^2}+\frac{2\mu}{\hbar^2}[E_R-V(r)]\chi_0=0,$$

i.e.,

$$\frac{d^2\chi_0}{dr^2}+\frac{2\mu}{\hbar^2}[E_R-A-Br]\chi_0=0.$$

a. Suppose the energy of a bound state depends on the principal quantum number n, which is a dimensionless quantity, the constant B in $V(r)$, the quark reduced mass μ, and $\hbar$, namely

$$E=f(n)B^x\mu^y\hbar^z.$$

As,

$$[E]=[M][L]^2[T]^{-2},$$
$$[B]=[M][L][T]^{-2},\qquad [\mu]=[M]$$
$$[\hbar]=[M][L]^2[T]^{-1},$$

we have

$$x=z=\frac{2}{3},\ y=-\frac{1}{3}$$

and hence

$$E = f(n)(B\hbar)^{2/3}(\mu)^{-1/3},$$

where $f(n)$ is a function of the principal quantum number n. Then

$$\Delta E_\psi = E_{\psi'} - E_\psi = f(1)\frac{(B\hbar)^{2/3}}{\mu_c^{1/3}} - f(0)\frac{(B\hbar)^{2/3}}{\mu_c^{1/3}}$$

$$= \frac{(B\hbar)^{2/3}}{\mu_c^{1/3}}[f(1) - f(0)],$$

and similarly

$$\Delta E_\Upsilon = \frac{(B\hbar)^{2/3}}{\mu_b^{1/3}}[f(1) - f(0)].$$

Hence

$$\frac{\Delta E_\Upsilon}{\Delta E_\psi} = \left(\frac{\mu_c}{\mu_b}\right)^{1/3} = \left(\frac{1}{3}\right)^{1/3}.$$

As

$$E_{\Upsilon'} - E_\Upsilon \approx 0.42 \text{ GeV},$$
$$E_{\Upsilon'} = E_\Upsilon + 0.42 = 9.5 + 0.42 \approx 9.9 \text{ GeV}.$$

b. Applying the WKB approximation to the equation for χ_0 we obtain the Bohr-Sommerfeld quantization rule

$$2\int_0^a \sqrt{2\mu(E_R - A - Br)}\, dr = (n + 3/4)\hbar \text{ with } a = \frac{E_R - A}{B},$$

which gives, writing E_n for E_R,

$$E_n = A + \frac{\left[3\left(n + \dfrac{3}{4}\right)B\hbar/4\right]^{2/3}}{(2\mu)^{1/3}}.$$

Application to the energy splitting gives

$$E_{\psi'} - E_\psi = \frac{(B\hbar)^{2/3}}{(2\mu_c)^{1/3}}\left[\left(\frac{21}{16}\right)^{2/3} - \left(\frac{9}{16}\right)^{2/3}\right],$$

$$E_{\psi''} - E_{\psi'} = \frac{(B\hbar)^{2/3}}{(2\mu_c)^{1/3}}\left[\left(\frac{33}{16}\right)^{2/3} - \left(\frac{21}{16}\right)^{2/3}\right],$$

and hence

$$\frac{E_{\psi''} - E_{\psi'}}{E_{\psi'} - E_\psi} = \frac{(33)^{2/3} - (21)^{2/3}}{(21)^{2/3} - (9)^{2/3}} \approx 0.81.$$

Thus

$$E_{\psi''} - E_{\psi'} = 0.81 \times (E_{\psi'} - E_{\psi}) = 0.81 \times (3.7 - 3.1)$$
$$\approx 0.49 \text{ GeV},$$

and

$$E_{\psi''} = 3.7 + 0.49 \approx 4.2 \text{ GeV}.$$

2016

Find the acceleration operator for a particle under the Hamiltonian $H = AP^2 + Bx^2$, where A and B are real constants.

Sol:

Ehrenfest theorem

$$V_x = \frac{dx}{dt} = \frac{1}{i\hbar}[x_1, H] + \frac{\partial x}{\partial t}$$

$$= \frac{1}{i\hbar}[x_1, AP^2 + Bx^2] + \frac{\partial x}{\partial t}$$

$$= \frac{1}{i\hbar}\{[x, AP^2] + [x, Bx^2]\} + \frac{\cancel{\partial x}^{\,0}}{\cancel{\partial t}}$$

$$= \frac{1}{\cancel{i\hbar}}(A2P\,\cancel{i\hbar} + 0 + 0) = 2AP.$$

$$a_x = \frac{dV_x}{dt} = \frac{1}{i\hbar}[V_x, H] + \frac{\partial V_x}{\partial t}$$

$$= \frac{1}{i\hbar}[2AP, AP^2 + Bx^2] + \frac{\partial V_x}{\partial t}$$

$$= \frac{1}{i\hbar}\{[2AP, AP^2] + [2AP, Bx^2]\} + \frac{\partial V_x}{\partial t}$$

$$= \frac{1}{i\hbar}2BA[Px^2] + \frac{\cancel{\partial V_x}^{\,0}}{\cancel{\partial t}}$$

$$= \frac{1}{\cancel{i\hbar}}2BA - 2x\,\cancel{i\hbar} = -4ABx.$$

2017

Prove that in any single-particle bound energy eigenstate the following relation is satisfied in nonrelativistic quantum mechanics for a central field-of-force potential $V(r)$,

$$|\psi(0)|^2 = \frac{m}{2\pi}\left\langle\frac{dV(r)}{dr}\right\rangle - \frac{1}{2\pi}\left\langle\frac{L^2}{r^3}\right\rangle,$$

where $\psi(0)$ is the wave function at the origin, m the particle mass, and L^2 the square of the orbital angular momentum operator (let $\hbar = 1$). Give a classical interpretation of this equation for the case of a state with angular momentum $\neq 0$.

(Columbia)

Sol:

a. In the field of central force, the Schrödinger equation is

$$-\frac{1}{2m}\left[\frac{1}{r^2}\frac{\partial}{\partial r}\left(r^2\frac{\partial}{\partial r}\right) - \frac{\hat{l}^2}{r^2}\right]\psi + V(r)\psi = E\psi.$$

Let

$$\psi(r, \theta, \varphi) = R(r)\, Y_{lm}(\theta, \varphi) \equiv \frac{1}{r}u(r)Y_{lm}(\theta, \varphi),$$

where $u(r) = rR(r)$, and we have for the radial motion

$$u'' + \left\{2m[E - V(r)] - \frac{l(l+1)}{r^2}\right\}u = 0, \quad (r > 0).$$

Multiplying the two sides of the above with $u'(r)$ and integrating from $r = 0$ to $r = \infty$, we get

$$\int u'(r)u''(r)dr + \int\left\{2m[E - V(r)] - \frac{l(l+1)}{r^2}\right\}\left(\frac{1}{2}u^2(r)\right)' dr = 0.$$

For the eigenstates we may assume $u'(\infty) = 0$, $u(\infty) = u(0) = 0$. With $u'(0) = [R(r) + rR'(r)]_{r=0} = R(0)$, partial integration gives

$$-\frac{1}{2}R^2(0) + \frac{1}{2}\left[2m\int R^2\frac{dV(r)}{dr}r^2 dr - \int\frac{2l(l+1)}{r^3}R^2 r\, dr\right] = 0.$$

Hence

$$|\psi(0)|^2 = \frac{1}{4\pi} R^2(0)$$

$$= \frac{m}{2\pi} \int [R(r)Y_{lm}(\theta,\varphi)] \frac{dV(r)}{dr} [R(r)Y_{lm}(\theta,\varphi)] r^2 dr d\Omega$$

$$- \frac{1}{2\pi} \int [R(r)Y_{lm}(\theta,\varphi)] \frac{\hat{L}^2}{r^3} [R(r)Y_{lm}(\theta,\varphi)] r^2 dr d\Omega,$$

or

$$|\psi(0)|^2 = \frac{m}{2\pi} \left\langle \frac{dV(r)}{dr} \right\rangle - \frac{1}{2\pi} \left\langle \frac{\hat{L}^2}{r^3} \right\rangle.$$

b. For $l \neq 0, |\psi(\dot{0})|^2 = 0$, and so

$$\left\langle \frac{dV(r)}{dr} \right\rangle = \frac{1}{m} \left\langle \frac{\hat{L}^2}{r^3} \right\rangle.$$

Its corresponding classical expression is

$$\frac{d}{dr} V(r) = \frac{1}{m} \frac{L^2}{r^3}.$$

Here $F_r = -\frac{dV(r)}{dr}$ is the centripetal force, and

$$\frac{1}{m} \frac{L^2}{r^3} = m \frac{|\mathbf{r} \times \mathbf{v}|^2}{r^3} = m \frac{1}{r} (v \sin\angle \mathbf{r},\mathbf{v})^2 = m \frac{v_t^2}{r},$$

$$= ma_r,$$

where v_t is the tangential velocity along the spherical surface of $\mathbf{r}$, is mass multiplied by the centripetal acceleration $a_r = \frac{v_t^2}{r}$. The equation thus expresses Newton's second law of motion.

2018

A spinless particle of mass m is subject (in 3 dimensions) to a spherically symmetric attractive square-well potential of radius r_0.

a. What is the minimum depth of the potential needed to achieve two bound states of zero angular momentum?

b. With a potential of this depth, what are the eigenvalues of the Hamiltonian that belong to zero total angular momentum? (If necessary you may express part of your answer through the solution of a transcendental equation.)

c. If the particle is in the ground state, sketch the wave function in the coordinate basis and the corresponding coordinate probability distribution. Explain carefully the physical significance of the latter.

d. Predict the result of a (single) measurement of the particle kinetic energy in terms of this wave function. You may express your prediction through one-dimensional definite integrals.

e. On the basis of the uncertainty principle, give a qualitative connection between parts (c) and (d) above.

(Berkeley)

Sol:

a. The attractive potential may be represented by $V = -V_0$, where V_0 is a positive constant. For bound states $0 > E > -V_0$. Thus for $l = 0$ the radial wave function $R(r) = \chi(r)/r$ satisfies the equations

$$-\frac{\hbar^2}{2m}\frac{d^2\chi}{dr^2} - V_0\chi = E\chi, \qquad (0 < r < r_0)$$

$$-\frac{\hbar^2}{2m}\frac{d^2\chi}{dr^2} = E\chi, \qquad (r_0 < r < \infty)$$

with $\chi(0) = 0$ and $\chi(\infty)$ finite. To suit these conditions the wave function may be chosen as follows:

$$\chi(r) = \sin \alpha r, \qquad 0 < r < r_0,$$
$$\chi(r) = B \exp(-\beta r), \qquad r_0 < r < \infty,$$

where $\alpha = \frac{1}{\hbar}\sqrt{2m(E+V_0)}$, $\beta = \frac{1}{\hbar}\sqrt{-2mE}$.

From the boundary condition that at $r = r_0$, χ and χ' should be continuous we get $-\alpha \cot \alpha r_0 = \beta$. Defining $\xi = \alpha r_0$, $\eta = \beta r_0$, we have

$$\xi^2 + \eta^2 = 2mV_0 r_0^2 / \hbar^2,$$
$$-\xi \cot \xi = \eta.$$

Each set of the positive numbers ξ, η satisfying these equations gives a bound state. In Fig. 2.5 curve 1 represents $\eta = -\xi \cot \xi$ and curve 2, $\xi^2 + \eta^2 = 2^2$, for example. As shown in the figure, for a given value of V_0, to have two intersections in the quadrant we require

$$\frac{2mV_0 r_0^2}{\hbar^2} \geq \left(\frac{3\pi}{2}\right)^2,$$

or

$$V_0 \geq \frac{9\pi^2 \hbar^2}{8mr_0^2},$$

which is the minimum potential depth needed to achieve two bound states of zero angular momentum.

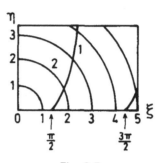

Fig. 2.5

b. With a potential of depth given above, one intersection occurs at $\eta = 0$, for which $\beta = \sqrt{-2mE} = 0$, i.e., $E = 0$. The other intersection occurs at

$$\xi^2 + \eta^2 = \left(\frac{3\pi}{2}\right)^2, \text{ i.e., } \xi = \frac{3\pi}{2}\sqrt{1 - \left(\frac{2\beta r_0}{3\pi}\right)^2}, \text{ and } -\xi \cot \xi = \eta, \text{ i.e.,}$$

$$-\frac{1}{r_0}\left(\frac{3\pi}{2}\right)\sqrt{1 - \left(\frac{2\beta r_0}{3\pi}\right)^2} \cot\left[\frac{3\pi}{2}\sqrt{1 - \left(\frac{2\beta r_0}{3\pi}\right)^2}\right] = \beta.$$

Solving for β, we get the second eigenvalue of the Hamiltonian,

$$E = -\frac{\beta^2 \hbar^2}{2m}.$$

c. Setting the normalized ground state wave function as

$$\chi(r) = A \sin \alpha r, \qquad\qquad 0 \leq r \leq r_0,$$
$$\chi(r) = A \sin a r_0 \exp\left[\beta(r_0 - r)\right], \qquad\qquad r > r_0,$$

we have

$$4\pi \int_0^\infty R^2 r^2 \, dr = 4\pi \int_0^\infty u^2 \, dr$$

$$= 4\pi A^2 \int_0^{r_0} \sin^2 \alpha r \, dr + 4\pi A^2 \sin^2 \alpha r e^{2\beta r_0}$$

$$\times \int_{r_0}^\infty e^{-2\beta r} \, dr = 1,$$

or

$$\frac{1}{4\pi A^2} = \frac{1}{2\alpha}(\alpha r_0 - \sin \alpha r_0 \cos \alpha r_0) + \frac{1}{2\beta} \sin^2 \alpha r_0.$$

The wave function and probability distribution are shown in Figs 2.6(a) and 2.6(b) respectively.

It can be seen that the probability of finding the particle is very large for $r < r_0$ and it attenuates exponentially for $r > r_0$, and we can regard the particle as being bound in the square-well potential.

d. The kinetic energy of the particle, $E_T = p^2 / 2m$, is a function dependent solely on the momentum p; thus the probability of finding a certain value of the kinetic energy by a single measurement is the same as that of finding the corresponding value of the momentum p, $|\psi(\mathbf{p})|^2$. Here $\psi(\mathbf{p})$ is the Fourier transform of the ground state coordinate wave function,

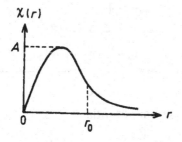

Fig. 2.6(a)

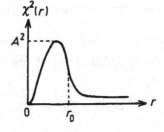

Fig. 2.6(b)

$$|\psi(\mathbf{p})|^2 = |\psi(\mathbf{p})|^2 = \left| \frac{1}{(2\pi\hbar)^{3/2}} \int \frac{1}{\sqrt{4\pi}} \frac{\chi(r)}{r} e^{-i\mathbf{p}\cdot\mathbf{r}/\hbar} \, d\mathbf{r} \right|^2$$

$$= \frac{1}{4\pi(2\pi\hbar)^3} \left| \int_0^\infty \int_0^\pi \int_0^{2\pi} \frac{\chi(r)}{r} e^{-ipr\cos\theta/\hbar} \, \sin\theta \, d\theta \, d\varphi \, r^2 \, dr \right|^2$$

$$= \frac{1}{2\pi^2\hbar^3} \left| \int_0^\infty \frac{\chi(r)}{r} \frac{\sin\left(\frac{pr}{\hbar}\right)}{\left(\frac{pr}{\hbar}\right)} r^2 \, dr \right|^2 .$$

The integration can be effected when the expression for $\chi(r)$ in (c) is substituted in the integrand. The average kinetic energy $\overline{E}_T$ is

$$\overline{E}_T = \left\langle \psi_1 \left| \frac{p^2}{2m} \right| \psi_1 \right\rangle = E_1 - \langle \psi_1 | V | \psi_1 \rangle$$

$$= E_1 + V_0 \int_0^{r_0} A^2 \sin^2\alpha r \cdot \frac{1}{r^2} \cdot 4\pi r^2 \, dr$$

$$= E_1 + 2\pi \, V_0 A^2 \left(r_0 - \frac{\sin 2\alpha r_0}{2\alpha} \right).$$

e. From the above we see that the wave function in space coordinates in (c) gives the space probability distribution, whereas the wave function in p-space in (d) gives the momentum probability distribution. The product of uncertainities of one simultaneous measurement of the position and the momentum must satisfy the uncertainty principle

$$\Delta p \Delta r \geq \hbar / 2.$$

That is to say, the two complement each other.

2019

a. Given a one-dimensional potential (Fig. 2.7)

$$V = -V_0, \qquad |x| < a,$$
$$V = 0, \qquad |x| > a,$$

show that there is always at least one bound state for attractive potentials $V_0 > 0$. (You may solve the eigenvalue condition by graphical means.)

b. Compare the Schrödinger equation for the above one-dimensional case with that for the radial part $U(r)$ of the three-dimensional wave function when $L = 0$,

$$\psi(\mathbf{r}) = r^{-1} U(r) Y_{LM}(\Omega),$$

where $\psi(\mathbf{r})$ is the solution of the Schrödinger equation for the potential

$$V = -V_0, \qquad r < a,$$
$$V = 0, \qquad r > a.$$

Why is there not always a bound state for $V_0 > 0$ in the three-dimensional case?

(MIT)

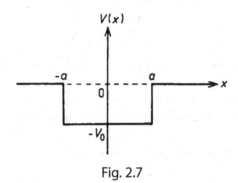

Fig. 2.7

Sol:

a. For the bound state of a particle, $E < 0$. For $|x| > a, V = 0$ and the Schrödinger equation

$$\frac{d^2\psi}{dx^2} + \frac{2mE}{\hbar^2}\psi = 0$$

has solutions

$$\psi(x) = \begin{cases} Ae^{-k'x}, & x > a, \\ Be^{k'x}, & x < -a, \end{cases}$$

where

$$k' = \sqrt{-\frac{2mE}{\hbar^2}}.$$

For $|x| < a$, the Schrödinger equation

$$\frac{d^2\psi}{dx^2} + \frac{2m}{\hbar^2}(V_0 + E)\psi = 0$$

has solutions

$$\psi(x) \sim \cos kx, \text{ (even parity)}$$
$$\psi(x) \sim \sin kx, \text{ (odd parity)}$$

where

$$k = \sqrt{\frac{2m(V_0+E)}{\hbar^2}},$$

provided $E > -V_0$. Here we need only consider states of even parity which include the ground state.

The continuity of the wave function and its derivative at $x = \pm a$ requires $k \tan ka = k'$. Set $ka = \xi$, $k'a = \eta$. Then the following equations determine the energy levels of the bound states:

$$\xi \tan \xi = \eta,$$

$$\xi^2 + \eta^2 = 2mV_0a^2 / \hbar^2.$$

These equations must in general be solved by graphical means. In Fig. 2.8(a), curve 1 is a plot of $\eta = \xi \tan \xi$, and curve 2 plots $\xi^2 + \eta^2 = 1$. The dashed line 3 is the asymptotic curve for the former with $\xi = \pi / 2$. Since curve 1 goes through the origin, there is at least one solution no matter how small is the value of V_0a^2. Thus there is always at least one bound state for a one-dimensional symmetrical square-well potential.

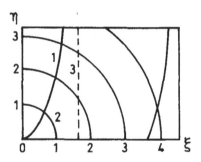

Fig. 2.8(a)

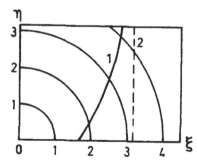

Fig. 2.8(b)

b. For $r > a$, the radial Schrödinger equation

$$\frac{d^2U}{dr^2} + \frac{2mE}{\hbar^2}U = 0,$$

has solution $U(r) = A \exp(-\kappa' r)$, where

$$\kappa' = \sqrt{-\frac{2mE}{\hbar^2}}.$$

For $r < a$, the equation is

$$\frac{d^2U}{dr^2} + \frac{2m}{\hbar^2}(V_0 + E)\, U = 0$$

and the solution that satisfies the boundary condition $U(r) \xrightarrow{r \to 0} 0$ is $U(r) = B \sin \kappa r$, where

$$\kappa = \sqrt{\frac{2m(V_0 + E)}{\hbar^2}}.$$

The continuity of the wave function and its derivative at $r = a$, requires $\kappa \cot \kappa a = -\kappa'$. Setting $\kappa a = \xi$, $\kappa' a = \eta$ we get

$$\xi \cot \xi = -\eta.$$
$$\xi^2 + \eta^2 = 2mV_0 a^2 / \hbar^2.$$

These are again to be solved by graphical means. Fig. 2.8(b) shows curve 1 which is a plot of $\xi \cot \xi = -\eta$, and the dashed line 2 which is its asymptotic if $\xi = \pi$. It can be seen that only when

$$\xi^2 + \eta^2 = 2mV_0 a^2 / \hbar^2 \geq \left(\frac{\pi}{2}\right)^2,$$

or

$$V_0 a^2 \geq \pi^2 \hbar^2 / 8m,$$

can the equations have a solution. Hence, unlike the one-dimensional case, only when $V_0 a^2 \geq \pi^2 \hbar^2 / 8m$ can there be a bound state.

2020

a. Consider a particle of mass m moving in a three-dimensional square well potential $V(|\mathbf{r}|)$. Show that for a well of fixed radius R, a bound state exists only if the depth of the well has at least a certain minimum value. Calculate that minimum value.

b. The analogous problem in one dimension leads to a different answer. What is that answer?

c. Can you show that the general nature of the answers to (a) and (b) above remains the same for a well of arbitrary shape? For example, in the one-dimensional case (b)

$$V(x) = \lambda f(x) < 0, \qquad a \le x \le b,$$
$$V(x) = 0, \qquad x < a \text{ or } x > b,$$

consider various values of λ while keeping $f(x)$ unchanged.

(CUSPEA)

Sol:

a. Suppose that there is a bound state $\psi(r)$ and that it is the ground state $(l=0)$, so that $\psi(\mathbf{r}) = \psi(r)$. The eigenequation is

$$-\frac{\hbar^2}{2m} \frac{1}{r^2} \frac{d}{dr}\left(r^2 \frac{d}{dr}\psi(r)\right) + V(r)\psi(r) = E\psi(r),$$

where $E < 0$, and

$$V(r) = 0, \qquad r > R,$$
$$V(r) = -V_0, \qquad 0 < r < R,$$

with $V_0 > 0$, as shown in Fig. 2.9. The solution is

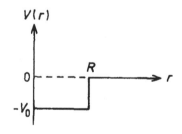

$$V(r)$$

$$0 \quad \dashleftarrow \quad R$$

$$-V_0$$

Fig. 2.9

$$\psi(r) = \begin{cases} A\sin(kr)/r, & r < R, \quad k = \sqrt{\dfrac{2m(E+V_0)}{\hbar^2}}, \\[4mm] Be^{-k'r}/r, & r > R, \quad k' = \sqrt{\dfrac{-2mE}{\hbar^2}}, \end{cases}$$

where A and B are normalization constants. The continuity of ψ and ψ' at $r = R$, or equivalently

$$\left[\ln(r\psi(r))\right]'_{r=R^-} = \left[\ln(r\psi(r))\right]'_{r=R^+},$$

gives

$$k \cot (kR) = -k',$$

while the definitions of κ, κ' require

$$k^2 + k'^2 = \frac{2mV_0}{\hbar^2}.$$

These equations can be solved graphically as in **Problem 2018**. In a similar way, we can show that for there to be at least a bound state we require

$$\frac{2mV_0R^2}{\hbar^2} \geq \left(\frac{\pi}{2}\right)^2,$$

i.e.,

$$V_0 \geq \frac{\pi^2 \hbar^2}{8mR^2}.$$

b. If the potential is a one-dimensional rectangular well potential, no matter how deep the well is, there is always a bound state. The ground state is always symmetric about the origin which is the center of the well. The eigenequation is

$$\frac{d^2}{dx^2}\psi(x) + \frac{2m}{\hbar^2}[E - V(x)]\psi(x) = 0,$$

where, as shown in the Fig. 2.10,

$$V(x) = -V_0, \quad (V_0 > 0), \quad |x| < R/2,$$
$$V(x) = 0, \quad\quad\quad\quad\; |x| > R/2.$$

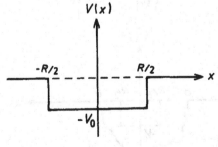

Fig. 2.10

For bound states, we require $0 > E > -V_0$. As $V(x) = V(-x)$, the equation has solution

$$\psi(x) = \begin{cases} A\cos(kx), & |x| < \dfrac{R}{2}, \\[2mm] Be^{-k'|x|}, & |x| > \dfrac{R}{2}, \end{cases}$$

where k, k' have the same definitions as in (a). The continuity of ψ and ψ' at $x = \dfrac{R}{2}$ gives

$$\tan(kR/2) = k'/k.$$

or

$$\sec^2(kR/2) = \frac{V_0}{E + V_0},$$

i.e.,

$$\cos(kR/2) = \pm\sqrt{\frac{E + V_0}{V_0}}.$$

Since $V_0 > -E > 0$, there is always a bound-state solution for any V_0.

c. For a one-dimensional potential well of arbitrary shape, we can always define a rectangular potential well $V_s(x)$ such that

$$V_s(x) = -V_0, \qquad |x| < R/2,$$
$$V_s(x) = 0, \qquad |x| > R/2,$$

and $-V_0 \geq V(x)$ always (see Fig. 2.11). From (b) we see that there always exists a $|\psi_0(x)\rangle$ which is a bound eigenstate of $V_s(x)$ for which

$$\left\langle \psi_0 \left| \frac{p^2}{2m} + V_s(x) \right| \psi_0 \right\rangle < 0.$$

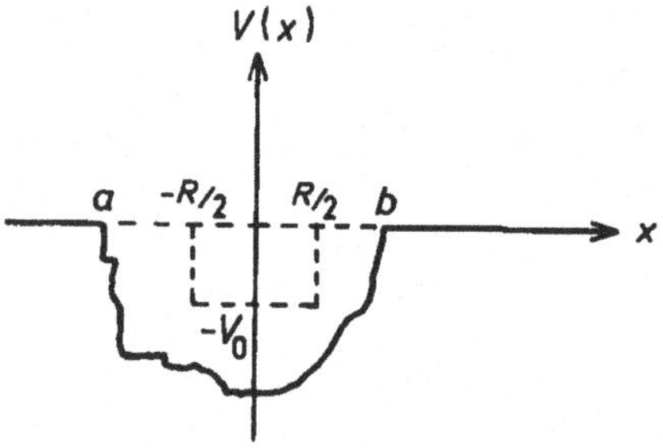

Fig. 2.11

Since

$$\left\langle \psi_0 \left| \frac{p^2}{2m} + V(x) \right| \psi_0 \right\rangle < \left\langle \psi_0 \left| \frac{p^2}{2m} + V_s(x) \right| \psi_0 \right\rangle,$$

we have

$$\left\langle \psi_0 \left| \frac{p^2}{2m} + V(x) \right| \psi_0 \right\rangle < 0.$$

This means that there is always a bound state for a one-dimensional well of any shape.

2021

Calculate Green's function for a nonrelativistic electron in the potential

$$V(x,y,z) = \infty, \quad x \leq 0, \quad (\text{any } y, z)$$
$$V(x,y,z) = 0, \quad x > 0, \quad (\text{any } y, z)$$

and evaluate $|G(\mathbf{r}, \mathbf{r}', t)|^2$. Describe the evolution in time of the pattern of probability and interpret physically the reason for this behavior.

(Berkeley)

Sol: The potential in this problem can be replaced by the boundary condition $G(\mathbf{r}, \mathbf{r}', t) = 0$ and $x = 0$. The boundary problem can then be solved by the method of images. Suppose at $\mathbf{r}''$ is the image of the electron at $\mathbf{r}'$ about $x = 0$. Then

$$(i\hbar \partial_t - H) G(\mathbf{r}, \mathbf{r}', t) = \delta(t) \left[\delta(\mathbf{r} - \mathbf{r}') - \delta(\mathbf{r} - \mathbf{r}'') \right]. \qquad (1)$$

The Green's function is zero for $x \leq 0$ and for $x \geq 0$ is equal to the $x > 0$ part of the solution of (1). Let

$$G(\mathbf{r}, \mathbf{r}', t) = \frac{1}{(2\pi)^4} \int d^3 k \int_{-\infty}^{\infty} e^{i\mathbf{k}\cdot\mathbf{r} - i\omega t} \overline{G}(\mathbf{k}, \mathbf{r}', \omega) d\omega. \qquad (2)$$

We have $i\hbar \partial_t G = \hbar w G$ and $H = \frac{\hbar^2 k^2}{2m}$, and the substitution of (2) in (1) gives

$$\overline{G}(\mathbf{k}, \mathbf{r}', \omega) = \frac{1}{\hbar\omega - \frac{\hbar^2 k^2}{2m}} (e^{-i\mathbf{k}\cdot\mathbf{r}'} - e^{-i\mathbf{k}\cdot\mathbf{r}''}), \qquad (3)$$

Re-substituting (3) in (2) gives

$$G(\mathbf{r}, \mathbf{r}', t) = \frac{1}{(2\pi)^4} \int_\Gamma d^3 k \int e^{i\mathbf{k}\cdot\mathbf{r} - i\omega t} \frac{(e^{-i\mathbf{k}\cdot\mathbf{r}'} - e^{-i\mathbf{k}\cdot\mathbf{r}''})}{\hbar\omega - \frac{\hbar^2 k^2}{2m}} d\omega. \qquad (4)$$

We first integrate with respect to ω. The path Γ is chosen to satisfy the causality condition.

Causality requires that when $t < 0$, $G(\mathbf{r}, \mathbf{r}, t) = 0$. First let the polar point of ω shift a little, say by $-i\varepsilon$, where ε is a small positive number. Finally letting $\varepsilon \to 0$, we get

$$
G(\mathbf{r}, \mathbf{r}', t) = \frac{-i}{\hbar(2\pi)^3} \int \exp\left(i\mathbf{k} \cdot \mathbf{r} - i\frac{\hbar k^2}{2m}t\right)(e^{-i\mathbf{k}\cdot\mathbf{r}'} - e^{-i\mathbf{k}\cdot\mathbf{r}''})d^3k
$$

$$
= \frac{1}{\hbar}\left[\frac{m}{2\pi\hbar t}\right]^{3/2}\left\{\exp\left[\frac{im(\mathbf{r} - \mathbf{r}')^2}{2\hbar t}\right] - \exp\left[\frac{im(\mathbf{r} - \mathbf{r}'')^2}{2\hbar t}\right]\right\}. \tag{5}
$$

Hence when both x and t are greater than zero, the Green's function is given by (5); otherwise, it is zero. When $x > 0$ and $t > 0$,

$$
|G(\mathbf{r}, \mathbf{r}', t)|^2 = \frac{1}{\hbar^2}\left[\frac{m}{2\pi\hbar t}\right]^3
$$

$$
\times\left\{2 - 2Re\,\exp\left(\frac{im}{2\hbar t}[\mathbf{r}'^2 - \mathbf{r}''^2 - 2\mathbf{r}\cdot(\mathbf{r}' - \mathbf{r}'')]\right)\right\}.
$$

If the potential $V(x, y, z)$ were absent, the Green's function for the free space, $|G(\mathbf{r}, \mathbf{r}', t)|^2$, would be proportional to t^{-3}. But because of the presence of the reflection wall the interference term occurs.

2022

An electron moves above an impenetrable conducting surface. It is attracted toward this surface by its own image charge so that classically it bounces along the surface as shown in Fig. 2.12.

a. Write the Schrödinger equation for the energy eigenstates and energy eigenvalues of the electron. (Call y the distance above the surface.) Ignore inertial effects of the image.

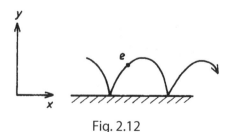

Fig. 2.12

b. What is the *x* and *z* dependence of the eigenstates?

c. What are the remaining boundary conditions?

d. Find the ground state and its energy.

[Hint: they are closely related to those for the usual hydrogen atom).

e. What is the complete set of discrete and/or continuous energy eigenvalues?

(Columbia)

Sol:

a. Figure 2.13 shows the electron and its image. Accordingly the electric energy for the system is $V(r) = \frac{1}{2}\sum_i q_i V_i = \frac{1}{2}\left[e \cdot \frac{-e}{2y} + (-e) \cdot \frac{e}{2y} \right] = -e^2/4y$. The Schrödinger equation is then

$$\left(-\frac{\hbar^2}{2m}\nabla^2 - \frac{e^2}{4y} \right)\psi(x,\,y,\,z) = E\psi(x,\,y,\,z).$$

b. Separating the variables by assuming solutions of the type

$$\psi(x,\,y,\,z) \equiv \psi_n(y)\phi(x,\,z) \equiv \psi_n(y)\phi_x(x)\phi_z(z),$$

we can write the above equation as

$$-\frac{\hbar^2}{2m}\frac{d^2}{dy^2}\psi_n(y) - \frac{e^2}{4y}\psi_n(y) = E_y\psi_n(y), \qquad (1)$$

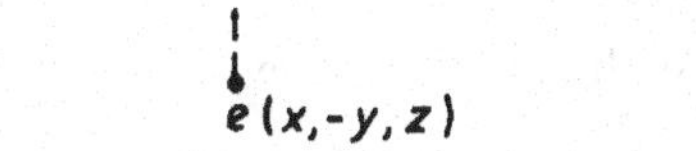

Fig. 2.13

$$-\frac{h^2}{2m}\frac{d^2}{dx^2}\phi_x(x) = \frac{p_x^2}{2m}\phi_x(x),$$

$$-\frac{\hbar^2}{2m}\frac{d^2}{dz^2}\phi_z(z) = \frac{p_z^2}{2m}\phi_z(x),$$

with

$$E_y + \frac{p_x^2}{2m} + \frac{p_z^2}{2m} = E.$$

Note that since $V(y) = -\frac{e^2}{4y}$ depends on y only, p_x and p_z are constants of the motion. Hence

$$\phi(x,z) \equiv \phi_x(x)\phi_z(z) \sim e^{i(p_x x + p_z z)/\hbar},$$

and

$$\psi(x,y,z) \equiv \psi_n(y)e^{i(p_x x + p_z z)/\hbar}.$$

c. The remaining boundary condition is $\psi(x,y,z) = 0$ for $y \le 0$.

d. Now consider a hydrogen-like atom of nuclear charge Z. The Schrödinger equation in the radial direction is

$$-\frac{\hbar^2}{2m}\frac{1}{r^2}\frac{d}{dr}\left(r^2\frac{dR}{dr}\right) - \frac{Ze^2}{r}R + \frac{l(l+1)\hbar^2}{2mr^2}R = ER.$$

On setting $R = \chi / r$, the above becomes

$$-\frac{\hbar^2}{2m}\frac{d^2\chi}{dr^2} - \frac{Ze^2}{r}\chi + \frac{l(l+1)\hbar^2}{2mr^2}\chi = E\chi.$$

In particular, when $l = 0$ we have

$$-\frac{\hbar^2}{2m}\frac{d^2\chi}{dr^2} - \frac{Ze^2}{r}\chi = E\chi, \tag{2}$$

which is identical with (1) with the replacements $r \to y, Z \to \frac{1}{4}$. Hence the solutions of (1) are simply y multiplied by the radial wave functions of the ground state of the atom. Thus

$$\psi_1(y) = yR_{10}(y) = 2y\left(\frac{Z}{a}\right)^{3/2}e^{-Zy/a},$$

where $a = \frac{\hbar^2}{me^2}$. With $Z = \frac{1}{4}$, we have

$$\psi_1(y) = 2y\left(\frac{me^2}{4\hbar^2}\right)^{3/2}\exp\left[-\frac{me^2 y}{4\hbar^2}\right].$$

Note that the boundary condition in (c) is satisfied by this wave function. The ground-state energy due to y motion is similarly obtained:

$$E_y = -\frac{Z^2 me^4}{2\hbar^2} = -\frac{me^4}{32\hbar^2}.$$

e. The complete energy eigenvalue for quantum state n is

$$E_{n,\,p_x,\,p_z} = -\frac{me^4}{32\hbar^2 n^2} + \frac{1}{2m}(p_x^2 + p_z^2), \quad (n=1,\,2,\,3,\,\ldots)$$

with wave function

$$\psi_{n,\,p_x,\,p_z}(\mathbf{r}) = A y R_{n0}(y)\exp\left[\frac{i}{\hbar}(p_x x + p_z z)\right],$$

where A is the normalization constant.

2023

A nonrelativistic electron moves in the region above a large flat grounded conductor. The electron is attracted by its image charge but cannot penetrate the conductor's surface.

a. Write down the appropriate Hamiltonian for the three-dimensional motion of this electron. What boundary conditions must the electron's wave function satisfy?

b. Find the energy levels of the electron.

c. For the state of lowest energy, find the average distance of the electron above the conductor's surface.

(Columbia)

Sol:

a. Take Cartesian coordinates with the origin on and the z-axis perpendicular to the conductor surface such that the conductor occupies the half-space $z \le 0$. As in **Problem 2022**, the electron is subject to a potential $V(z) = -\dfrac{e^2}{4z}$. Hence the Hamiltonian is

$$H = \frac{1}{2m}(p_x^2 + p_y^2 + p_z^2) - \frac{e^2}{4z}$$

$$= -\frac{\hbar^2}{2m}\left(\frac{\partial^2}{\partial x^2} + \frac{\partial^2}{\partial y^2} + \frac{\partial^2}{\partial z^2}\right) - \frac{e^2}{4z}.$$

The wave function of the electron satisfies the boundary condition $\psi(x,y,z) = 0$ for $z \le 0$.

b. As shown in **Problem 2022** the energy eigenvalues are

$$E_n = \frac{1}{2m}(p_x^2 + p_y^2) - \frac{me^4}{32\hbar^2}\frac{1}{n^2}. \quad (n = 1,\,2,\,3\,\ldots)$$

c. The ground state has energy

$$E = \frac{1}{2m}(p_x^2 + p_y^2) - \frac{me^4}{32\hbar^2},$$

and wave function

$$\psi_{100}(x, y, z) = \frac{Az}{a'^{3/2}} e^{-z/a'},$$

where

$$a' = \frac{4\hbar^2}{me^2},$$

and A is the normalization constant. Hence

$$
\begin{aligned}
\langle z \rangle &= \frac{\int \psi_{100}^* \, z \, \psi_{100} \, dx \, dy \, dz}{\int \psi_{100}^* \psi_{100} \, dx \, dy \, dz} \\[2mm]
&= \frac{\int_0^\infty z^3 e^{-2z/a'} dz}{\int_0^\infty z^2 e^{-2z/a'} dz} \\[2mm]
&= \frac{3!}{\left(\frac{2}{a'}\right)^4} \cdot \frac{\left(\frac{2}{a'}\right)^3}{2!} \\[2mm]
&= \frac{3}{2} a' = \frac{6\hbar^2}{me^2}.
\end{aligned}
$$

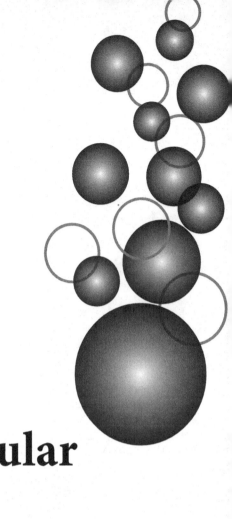

Part III
Spin and Angular Momentum

3001

Consider four Hermitian 2×2 matrices I, σ_1, σ_2, and σ_3, where I is the unit matrix, and the others satisfy $\sigma_i \sigma_j + \sigma_j \sigma_i = 2\delta_{ij}$. You must prove the following without using a specific representation or form for the matrices.

a. Prove that $Tr(\sigma_i) = 0$.

b. Show that the eigenvalues of σ_i are ± 1 and that $\det(\sigma_i) = -1$.

c. Show that the four matrices are linearly independent and therefore that any 2×2 matrix can be expanded in terms of them.

d. From (c) we know that

$$M = m_0 I + \sum_{i=1}^{3} m_i \sigma_i ,$$

where M is any 2×2 matrix. Derive an expression for m_i ($i = 0, 1, 2, 3$).

(Buffalo)

Sol:

a. As

$$\sigma_i \sigma_j = -\sigma_j \sigma_i \qquad (i \neq j), \qquad \sigma_j \sigma_j = I,$$

we have

$$\sigma_i = \sigma_i \sigma_j \sigma_j = -\sigma_j \sigma_i \sigma_j$$

and thus

$$Tr(\sigma_i) = -Tr(\sigma_j \sigma_i \sigma_j) = -Tr(\sigma_i \sigma_j \sigma_j) = -Tr(\sigma_i).$$

Hence $Tr(\sigma_i) = 0$.

b. Suppose σ_i has eigenvector ϕ and eigenvalue λ_i, i.e.

$$\sigma_i \phi = \lambda_i \phi.$$

Then

$$\sigma_i\sigma_i\phi=\sigma_i\lambda_i\phi=\lambda_i\sigma_i\phi=\lambda_i^2\phi\,.$$

On the other hand,

$$\sigma_i\sigma_i\phi=I\phi=\phi\,.$$

Hence

$$\lambda_i^2=1,$$

or

$$\lambda_i=\pm1\,.$$

As

$$Tr(\sigma_i)=\lambda_1+\lambda_2=0,$$

the two eigenvalues of σ_i are $\lambda_1=+1$, $\lambda_2=-1$, and so $\mathrm{Det}(\sigma_i)=\lambda_1\lambda_2=-1$.

c. If I, σ_i, $i=1,2,3$, were linearly dependent, then four constants m_0, m_i could be found such that

$$m_0I+\sum_{i=1}^{3}m_i\sigma_i=0\,.$$

Multiplying by σ_j from the right and from the left we have

$$m_0\sigma_j+\sum_{i=1}^{3}m_i\sigma_i\sigma_j=0,$$

and

$$m_0\sigma_j+\sum_{i=1}^{3}m_i\sigma_j\sigma_i=0\,.$$

Adding the above two equations gives

$$2m_0\sigma_j+\sum_{i\neq j}m_i(\sigma_i\sigma_j+\sigma_j\sigma_i)+2m_jI=0,$$

or

$$m_0\sigma_j+m_jI=0\,.$$

Thus

$$Tr(m_0\sigma_j+m_jI)=m_0Tr(\sigma_j)+2m_j=0\,.$$

As $Tr(\sigma_j)=0$, we would have $m_j=0$, and so $m_0=0$. Therefore the four matrices I and σ_i are linearly independent and any 2×2 matrix can be expanded in terms of them.

d. Given M any 2×2 matrix, we can write

$$M = m_0 I + \sum_{i=1}^{3} m_i \sigma_i.$$

To determine the coefficients m_0, m_i, take the trace of the above:

$$Tr(M) = 2m_0,$$

or

$$m_0 = \frac{1}{2} Tr(M).$$

Consider

$$\sigma_j M = m_0 \sigma_j + \sum_{i=1}^{3} m_i \sigma_j \sigma_i,$$

and

$$M\sigma_j = m_0 \sigma_j + \sum_{i=1}^{3} m_i \sigma_i \sigma_j.$$

Adding the last two equations gives

$$\sigma_j M + M\sigma_j = 2m_0 \sigma_j + 2m_j I.$$

Thus

$$Tr(\sigma_j M + M\sigma_j) = 2Tr(\sigma_j M) = 4m_j,$$

or

$$m_j = \frac{1}{2} Tr(\sigma_j M).$$

3002

The three matrix operators for spin one satisfy $s_x s_y - s_y s_x = i s_z$ and cyclic permutations. Show that

$$s_z^3 = s_z, \quad (s_x \pm i s_y)^3 = 0.$$

(Wisconsin)

Sol: The matrix forms of the spin angular momentum operator $(s=1)$ in the (s^2, s_z) representation, in which s^2, s_z are diagonal, are

$$s_x = \frac{1}{\sqrt{2}} \begin{pmatrix} 0 & 1 & 0 \\ 1 & 0 & 1 \\ 0 & 1 & 0 \end{pmatrix}, \quad s_y = \frac{1}{\sqrt{2}} \begin{pmatrix} 0 & -i & 0 \\ i & 0 & -i \\ 0 & i & 0 \end{pmatrix},$$

$$s_z = \begin{pmatrix} 1 & 0 & 0 \\ 0 & 0 & 0 \\ 0 & 0 & -1 \end{pmatrix}.$$

Direct calculation gives

$$s_z^3 = \begin{pmatrix} 1 & 0 & 0 \\ 0 & 0 & 0 \\ 0 & 0 & -1 \end{pmatrix}^3 = \begin{pmatrix} 1 & 0 & 0 \\ 0 & 0 & 0 \\ 0 & 0 & -1 \end{pmatrix} = s_z,$$

$$(s_x + i s_y)^3 = \left[\frac{1}{\sqrt{2}} \begin{pmatrix} 0 & 2 & 0 \\ 0 & 0 & 2 \\ 0 & 0 & 0 \end{pmatrix} \right]^3 = 0,$$

$$(s_x - i s_y)^3 = \left[\frac{1}{\sqrt{2}} \begin{pmatrix} 0 & 0 & 0 \\ 2 & 0 & 0 \\ 0 & 2 & 0 \end{pmatrix} \right]^3 = 0.$$

3003

Three matrices M_x, M_y, M_z, each with 256 rows and columns, are known to obey the commutation rules $[M_x, M_y] = iM_z$ (with cyclic permutations of x, y and z). The eigenvalues of the matrix M_x are ± 2, each once; $\pm 3/2$, each 8 times; ± 1, each 28 times; $\pm 1/2$, each 56 times; and 0, 70 times. State the 256 eigenvalues of the matrix $M^2 = M_x^2 + M_y^2 + M_z^2$.

(Wisconsin)

Sol: M^2 commutes with M_x. So we can select the common eigenstate $|M, M_x\rangle$. Then

$$M^2 | M, M_x \rangle = m(m+1) | M, M_x \rangle,$$
$$M_x | M, M_x \rangle = m_x | M, M_x \rangle.$$

For the same m, m_x can have the values $+m$, $m-1, \ldots, -m$, while M^2 has eigenvalue $m(m+1)$. Thus

m	m_x	$M^2 = m(m+1)$	
2	$\begin{array}{l}\pm 2 \\ \pm 1 \\ 0\end{array}\right\}$ each once	6	$5 \times 1 = 5$ times
3/2	$\begin{array}{l}\pm 3/2 \\ \pm 1/2\end{array}\right\}$ each 8 times	$\dfrac{15}{4}$	$4 \times 8 = 32$ times
1	$\begin{array}{l}\pm 1 \\ 0\end{array}\right\}$ each 27 times	2	$3 \times 27 = 81$ times
1/2	$\pm 1/2$ each 48 times	$\dfrac{3}{4}$	$2 \times 48 = 96$ times
0	0, each 42 times	0	$1 \times 42 = 42$ times
			Total 256 eigenvalues

3004

A certain state $|\psi\rangle$ is an eigenstate of $\hat{\mathbf{L}}^2$ and $\hat{L}_z$:

$$\hat{\mathbf{L}}^2|\psi\rangle = l(l+1)\hbar^2|\psi\rangle, \quad \hat{L}_z|\psi\rangle = m\hbar|\psi\rangle.$$

For this state calculate $\left\langle \hat{L}_x \right\rangle$ and $\left\langle \hat{L}_x^2 \right\rangle$.

(MIT)

Sol: As $\hat{L}_z$ is a Hermitian operator, we have

$$\hat{L}_z|\psi\rangle = m\hbar|\psi\rangle \to \langle\psi|\hat{L}_z = m\hbar\langle\psi|.$$

Then

$$\left\langle \hat{L}_x \right\rangle = \left\langle \psi \left| \frac{1}{i\hbar}\left[\hat{L}_y, \hat{L}_z\right] \right| \psi \right\rangle = \frac{1}{i\hbar}\langle\psi|\hat{L}_y\hat{L}_z - \hat{L}_z\hat{L}_y|\psi\rangle$$

$$= \frac{m\hbar}{i\hbar}\left(\langle\psi|\hat{L}_y|\psi\rangle - \langle\langle\psi|\hat{L}_y|\psi\rangle\right) = 0.$$

Considering the symmetry with respect to x, y, we have

$$\left\langle \hat{L}_x^2 \right\rangle = \left\langle \hat{L}_y^2 \right\rangle = \frac{1}{2}\left\langle \hat{L}_x^2 + \hat{L}_y^2 \right\rangle = \frac{1}{2}\left\langle \hat{\mathbf{L}}^2 - \hat{L}_z^2 \right\rangle,$$

and so

$$\left\langle \hat{L}_x^2 \right\rangle = \frac{1}{2}\left\langle \psi \left| \hat{\mathbf{L}}^2 - \hat{L}_z^2 \right| \psi \right\rangle = \frac{1}{2}[l(l+1)-m^2]\hbar^2.$$

It can also be calculated using the "raising" and "lowering" operators.

3005

The spin functions for a free electron in a basis where $\hat{s}_z$ is diagonal can be written as $\begin{pmatrix} 1 \\ 0 \end{pmatrix}$ and $\begin{pmatrix} 0 \\ 1 \end{pmatrix}$ with eigenvalues of $\hat{s}_z$ being $+1/2$ and $-1/2$ respectively. Using this basis find a normalized eigenfunction of $\hat{s}_y$ with eigenvalue $-1/2$.

(MIT)

Sol: In the diagonal representation of $\hat{s}^2$, $\hat{s}_z$, we can represent $\hat{s}_y$ by

$$\hat{s}_y = \frac{1}{2}\begin{pmatrix} 0 & -i \\ i & 0 \end{pmatrix}, \quad (\hbar = 1).$$

Let the required eigenfunction of $\hat{s}_y$ be $\sigma_y = \begin{pmatrix} a \\ b \end{pmatrix}$. Then as

$$\frac{1}{2}\begin{pmatrix} 0 & -i \\ i & 0 \end{pmatrix}\begin{pmatrix} a \\ b \end{pmatrix} = -\frac{1}{2}\begin{pmatrix} a \\ b \end{pmatrix},$$

we have $a = ib$, and so $\sigma_y = b\begin{pmatrix} i \\ 1 \end{pmatrix}$.

Normalization

$$\sigma_y^+ \sigma_y = b^2(-i, 1)\begin{pmatrix} i \\ 1 \end{pmatrix} = 2b^2 = 1,$$

gives $b = \frac{1}{\sqrt{2}}$. Hence

$$\sigma_y = \frac{1}{\sqrt{2}}\begin{pmatrix} i \\ 1 \end{pmatrix}.$$

3006

Consider a spinless particle represented by the wave function

$$\psi = K(x+y+2z)e^{-\alpha r},$$

where $r = \sqrt{x^2 + y^2 + z^2}$, and K and α are real constants.

a. What is the total angular momentum of the particle?

b. What is the expectation value of the z-component of angular momentum?

c. If the z-component of angular momentum, L_z, were measured, what is the probability that the result would be $L_z = +\hbar$?

d. What is the probability of finding the particle at θ, ϕ and in solid angle $d\Omega$? Here θ, ϕ are the usual angles of spherical coordinates.

You may find the following expressions for the first few spherical harmonics useful:

$$Y_0^0 = \sqrt{\frac{1}{4\pi}}, \qquad Y_1^{\pm 1} = \mp\sqrt{\frac{3}{8\pi}}\sin\theta\, e^{\pm i\phi},$$

$$Y_1^0 = \sqrt{\frac{3}{4\pi}}\cos\theta, \quad Y_2^{\pm 1} = \mp\sqrt{\frac{15}{8\pi}}\sin\theta\,\cos\theta\, e^{\pm i\phi}.$$

(CUS)

Sol: The wave function may be rewritten in spherical coordinates as

$$\psi = Kr(\cos\phi\sin\theta + \sin\phi\sin\theta + 2\cos\theta)e^{-\alpha r},$$

its angular part being

$$\psi(\theta, \phi) = K'(\cos\phi\sin\theta + \sin\phi\sin\theta + 2\cos\theta),$$

where K' is the normalization constant such that

$$K'^2 \int_\theta^\pi d\theta \int_0^{2\pi} \sin\theta\,(\cos\phi\sin\theta + \sin\phi\sin\theta + 2\cos\theta)^2\, d\phi = 1.$$

Since

$$\cos\phi = \frac{1}{2}(e^{i\phi} + e^{-i\phi}), \quad \sin\phi = \frac{1}{2i}(e^{i\phi} - e^{-i\phi}),$$

we have

$$\psi(\theta, \phi) = K'\left[\frac{1}{2}(e^{i\phi} + e^{-i\phi})\sin\theta + \frac{1}{2i}(e^{i\phi} - e^{-i\phi})\sin\theta + \cos 2\theta\right],$$

$$= K'\left[-\frac{1}{2}(1-i)\sqrt{\frac{8\pi}{3}}Y_1^1 + \frac{1}{2}(1+i)\sqrt{\frac{8\pi}{3}}Y_1^{-1} + 2\sqrt{\frac{4\pi}{3}}Y_1^0\right].$$

The normalization condition and the orthonormality of Y_l^m then give

$$K'^2\left[\frac{1}{2}\cdot\frac{8\pi}{3}+\frac{1}{2}\cdot\frac{8\pi}{3}+4\cdot\frac{4\pi}{3}\right]=1,$$

or

$$K'=\sqrt{\frac{1}{8\pi}},$$

and thus

$$\psi(\theta,\phi)=\left[\sqrt{\frac{1}{8\pi}}-\frac{1}{2}(1-i)\sqrt{\frac{8\pi}{3}}Y_1^1\right.$$
$$\left.+\frac{1}{2}(1+i)\sqrt{\frac{8\pi}{3}}Y_1^{-1}+2\sqrt{\frac{4\pi}{3}}Y_1^0\right].$$

a. The total angular momentum of the particle is

$$\sqrt{\langle\mathbf{L}^2\rangle}=\sqrt{l(l+1)}\hbar=\sqrt{2}\hbar.$$

as the wave function corresponds to $l=1$.

b. The z-component of the angular momentum is

$$\langle\psi^*|L_z|\psi\rangle=K'^2\left[\frac{1}{2}\cdot\frac{8\pi}{3}\cdot\hbar(Y_1^1)^2+\frac{1}{2}\cdot\frac{8\pi}{3}(-\hbar)(Y_1^{-1})^2\right.$$
$$\left.+4\cdot\frac{4\pi}{3}(0)(Y_1^0)^2\right]$$
$$=\frac{1}{8\pi}\left[\frac{1}{2}\cdot\frac{8\pi}{3}(+\hbar)+\frac{1}{2}\cdot\frac{8\pi}{3}(-\hbar)\right]=0.$$

c. The probability of finding $L_z=+\hbar$ is

$$P=\left|\langle L_z=+\hbar\,|\,\psi(\theta,\phi)\rangle\right|^2$$
$$=\frac{1}{8\pi}\cdot\frac{1}{2}\cdot\frac{8\pi}{3}=\frac{1}{6}.$$

d. The probability of finding the particle in the solid angle $d\Omega$ at θ,φ is

$$\int\psi^*(\theta,\varphi)\psi(\theta,\varphi)d\Omega=\frac{1}{8\pi}[\sin\theta\,(\sin\phi+\cos\phi)+2\cos\theta]^2\,d\Omega.$$

3007

Consider a particle in spherically symmetric potential with orbital angular momentum $l=\hbar$ and $s=\frac{1}{2}\hbar$.

Find the

  a. Eigen energy values associated with L-S coupling scheme

  b. Number of degenerate states in each energy level

Sol:

a. The Eigen functions of different angular momentum components are given by

$$J^2\varphi = \hbar^2 j(j+1)\varphi, \ L^2\varphi = \hbar^2 l(l+1)\varphi, \text{ and } S^2\varphi = \hbar^2 s(s+1)\varphi, \tag{1}$$

where j, l and s are quantum numbers and φ is the angular component of the wave function.

In L-S coupling, the coupling energy is given by the equation

$$H = a(L.S) = \frac{1}{2}a(2L.S + L^2 + S^2 - L^2 - S^2) = \frac{1}{2}a(J^2 - L^2 - S^2) \tag{2}$$

as $J = L + S$

the possible values of j are given by $j = (l + s, \ \ldots\ldots l - s)$

therefore $j = \dfrac{3}{2} and \dfrac{1}{2}$ $\tag{3}$

from (1), (2) and (3)

$$E_{so} = \frac{\hbar^2}{2}a[j(j+1) - l(l+1) - s(s+1)]$$

$$\text{For } j = \frac{3}{2}, \ E_{so} = \frac{\hbar^2}{2}a$$

$$j = \frac{1}{2}, \ E_{so} = -\hbar^2 a$$

b. Degeneracy of the energy states $D = 2j + 1$

$$D = \begin{cases} 4, j = \dfrac{3}{2} \\[2mm] 2, j = \dfrac{1}{2} \end{cases}$$

3008

Consider the orbital angular momentum $L = (L_x, L_y, L_z)$ in Cartesian coordinates (x, y, z) and determine the commutators $[[[L_x, L_y], L_x], L_x]$, $[[[L_x, L_y], L_x], L_y]$, $[[[L_x, L_y], L_x], L_z]$. Finally, determine the action of $L^2 = (L_x^2 + L_y^2 + L_z^2)$ on the combination $x[L_y, z] - y[L_x, z] + z[L_x, y] = i\hbar(x^2 + y^2 + z^2)$.

Sol:

To solve this problem, we need to consider the commutation rules of the angular momentum. In particular, we know that

$$L \times L = i\hbar L$$

from which we get

$$[L_x, L_y] = i\hbar L_z$$
$$[L_z, L_x] = i\hbar L_y$$
$$[L_y, L_z] = i\hbar L_x$$

The above relations can be used to simplify the first commutator requested by the text.

$$[[[L_x, L_y], L_x], L_x] = i\hbar[L_z, L_x], L_x] = -\hbar^2[L_y, L_x] = i\hbar^3 L_z$$

When the last L_x is interchanged with L_y, we get

$$[[[L_x, L_y], L_x], L_y] = i\hbar[L_z, L_x], L_y] = -\hbar^2[L_y, L_y] = 0$$

Also, when the last L_x is interchanged with L_z, we get

$$[[[L_x, L_y], L_x], L_z] = i\hbar[L_z, L_x], L_z] = -\hbar^2[L_y, L_z] = -i\hbar^3$$

As for the second point, from the relation

$$[L_x, y] = [yp_z - zp_y, y] = i\hbar z$$

and its cyclic permutations, we find

$$x[L_y, z] - y[L_x, z] + z[L_x, y] = i\hbar(x^2 + y^2 + z^2).$$

Therefore, we see that the combination $x[L_y, z] - y[L_x, z] + z[L_x, y] = i\hbar(x^2 + y^2 + z^2)$ is directly proportional to the square of the distance from the origin of coordinates (r^2) and is independent of the angles, that is, when projected on the angular variables, it is proportional to the spherical harmonic $Y_{0,0}(\theta, \varphi)$. Since the spherical harmonics $Y_{l,m}$ are eigenfunctions of L^2 with eigenvalues $\hbar^2 l(l+1)$, we find

$$L^2(x[L_y, z] - y[L_x, z] + z[L_x, y]) = 0$$

3009

Suppose an electron is in a state described by the wave function

$$\psi = \frac{1}{\sqrt{4\pi}}(e^{i\phi}\sin\theta + \cos\theta)\,g(r),$$

where

$$\int_0^\infty |g(r)|^2\, r^2\, dr = 1,$$

nd ϕ, θ are the azimuth and polar angles respectively.

a. What are the possible results of a measurement of the z-component L_z of the angular momentum of the electron in this state?

b. What is the probability of obtaining each of the possible results in part (a)?

c. What is the expectation value of L_z?

(Wisconsin)

Sol:

a. As

$$Y_{10} = \sqrt{\frac{3}{4\pi}}\cos\theta, \ Y_{1,\pm1} = \mp\sqrt{\frac{3}{8\pi}}\sin\theta e^{\pm i\phi},$$

the wave function can be written as

$$\psi = \sqrt{\frac{1}{3}}(-\sqrt{2}Y_{11} + Y_{10})g(r).$$

Hence the possible values of L_z are $+h$, 0.

b. Since

$$\int |\psi|^2 \, dr = \frac{1}{4\pi}\int_0^\infty |g(r)|^2 \, r^2 \, dr \int_0^\pi d\theta \int_0^{2\pi}(1+\cos\phi\sin2\theta)\sin\theta \, d\phi$$

$$= \frac{1}{2}\int_0^\pi \sin\theta \, d\theta = 1,$$

the given wave function is normalized. The probability density is then given by $P = |\psi|^2$. Thus the probability of $L_z = +\hbar$ is $\left(\sqrt{\frac{2}{3}}\right)^2$ or 2/3 and that of $L_z = 0$ is $\left(\frac{1}{\sqrt{3}}\right)^2$ or 1/3.

c.

$$\int \psi^* L_z \psi r^2 \sin\theta \, d\theta \, d\phi \, dr = \int \left[\sqrt{\frac{1}{3}}(-\sqrt{2}Y_{11} + Y_{10})\right]^*$$

$$\times \hat{L}_z\left[\sqrt{\frac{1}{3}}(-\sqrt{2}Y_{11} + Y_{10})\right]$$

$$\times |g(r)|^2 \, r^2 dr \sin\theta \, d\theta \, d\phi$$

$$= \frac{2}{3}\hbar \int_0^\pi d\theta \int_0^{2\pi} Y_{11}^2 \, d\phi = \frac{2}{3}\hbar.$$

3010

If $U(\beta, \hat{y})$ refers to a rotation through an angle β about the y-axis, show that the matrix elements

$$\langle j, m | U(\beta, \hat{y}) | j, m' \rangle, \quad -j \leq m, m' \leq j,$$

are polynomials of degree $2j$ with respect to the variables $\sin(\beta/2)$ and $\cos(\beta/2)$. Here $|j, m\rangle$ refers to an eigenstate of the square and z-component of the angular momentum:

$$\hat{j}^2 | j, m \rangle = j(j+1)\hbar^2 | j, m \rangle,$$

$$\hat{j}_z | j, m \rangle = m\hbar | j, m' \rangle$$

(Wisconsin)

Sol: We use the method of mathematical induction. If $j = 0$, then $m = m' = 0$ and the statement is obviously correct. If $j = 1/2$, let

$$\hat{j}_y = \frac{\hbar}{2}\sigma_y \equiv \frac{\hbar}{2} \begin{pmatrix} 0 & -i \\ i & 0 \end{pmatrix}.$$

Consider Pauli's matrices σ_k, where $k = x$, y or z. Since

$$\sigma_x^2 = \sigma_y^2 = \sigma_z^2 = \begin{pmatrix} 1 & 0 \\ 0 & 1 \end{pmatrix},$$

the unit matrix, we have for $\alpha = \text{constant}$

$$\exp(\pm i\alpha\sigma_k) = 1 \pm \frac{i\alpha\sigma_k}{1!} + \frac{(\pm i\alpha\sigma_k)^2}{2!} + \frac{(\pm i\alpha\sigma_k)^3}{3!} + \cdots$$

$$= \left(1 - \frac{\alpha^2}{2!} + \frac{\alpha^4}{4!} - \cdots\right)$$

$$\pm i\sigma_k \left(\frac{\alpha}{1!} - \frac{\alpha^3}{3!} + \frac{\alpha^5}{5!} - \cdots\right)$$

$$= \cos\alpha \pm i\sigma_k \sin\alpha.$$

Thus

$$U(\beta, \hat{y}) \equiv \exp(-i\beta\hat{j}_y / \hbar) = \exp\left(-i\frac{\beta}{2}\sigma_y\right) = \cos\frac{\beta}{2} - i\sigma_y \sin\frac{\beta}{2}$$

$$= \cos\frac{\beta}{2} - i\frac{2}{\hbar}\hat{j}_y \sin\frac{\beta}{2},$$

and

$$\left\langle \frac{1}{2},m\left|U\right|\frac{1}{2},m'\right\rangle=\left\langle \frac{1}{2},m\left|\exp\left(-i\beta\hat{j}_y/\hbar\right)\right|\frac{1}{2},m'\right\rangle$$

$$=\delta_{mm'}\cos\frac{\beta}{2}-\frac{i2}{\hbar}\left\langle \frac{1}{2},m\left|\hat{j}_y\right|\frac{1}{2},m'\right\rangle\sin\frac{\beta}{2}.$$

As the matrix elements of $\hat{j}_y$ in the second term are independent of β,

$$\langle 1/2,m|U|1/2,m'\rangle$$

is a linear homogeneous form of $\cos(\beta/2)$ and $\sin(\beta/2)$ and the statement is correct also.

If the statement is correct for j, i.e.,

$$\left\langle j,m|U|j,m'\right\rangle=\left\langle j,m\left|\exp\left[-i\beta\hat{j}_y/\hbar\right]\right|j,m\right\rangle$$

$$=\sum_{n=0}^{2j}A_n\left(\cos\frac{\beta}{2}\right)^{2j-n}\left(\sin\frac{\beta}{2}\right)^n,$$

where A_n depends on j, m, m', i.e.,

$$A_n=A_n(j,m,m'),$$

we shall prove that the statement is also correct for $j+1/2$. Let $\hat{\mathbf{J}}=\hat{\mathbf{j}}+\hat{\mathbf{j}}_1$, where the quantum numbers of $\hat{\mathbf{j}}$ and $\hat{\mathbf{j}}_1$ are j and $1/2$ respectively. We can expand

$$|J,m\rangle=|j+1/2,m\rangle$$

in the coupling representation using terms of the uncoupling representation:

$$\left|j+\frac{1}{2},m\right\rangle=C_1\left|j,m+\frac{1}{2}\right\rangle\left|\frac{1}{2},-\frac{1}{2}\right\rangle$$

$$+C_2\left|j,m-\frac{1}{2}\right\rangle\left|\frac{1}{2},\frac{1}{2}\right\rangle,$$

where C_1 and C_2 (independent of β) are Clebsch-Gordan coefficients. Applying the expansion to

$$\left\langle j+\frac{1}{2},m\left|\exp\left[-i\beta(\hat{j}_y+\hat{j}_{1y})/\hbar\right]\right|j+\frac{1}{2},m'\right\rangle,$$

we reduce the procedure to calculating the matrix elements

$$\left\langle \frac{1}{2},\mp\frac{1}{2}\right|\left\langle j,m\pm\frac{1}{2}\left|\exp\left[-i\beta(\hat{j}_y+\hat{j}_{1y})/\hbar\right]\right|j,m'\pm\frac{1}{2}\right\rangle\left|\frac{1}{2},\mp\frac{1}{2}\right\rangle,$$

$$\left\langle \frac{1}{2},\pm\frac{1}{2}\right|\left\langle j,m\mp\frac{1}{2}\left|\exp\left[-i\beta(\hat{j}_y+\hat{j}_{1y})/\hbar\right]\right|j,m'\mp\frac{1}{2}\right\rangle\left|\frac{1}{2},\pm\frac{1}{2}\right\rangle.$$

For example,

$$\left\langle \frac{1}{2}, -\frac{1}{2} \middle| \left\langle j, m+\frac{1}{2} \middle| \exp\left[-i\beta(\hat{j}_y + \hat{j}_{1y})/\hbar\right] \middle| j, m'+\frac{1}{2} \right\rangle \middle| \frac{1}{2}, -\frac{1}{2} \right\rangle$$

$$= \left\langle \frac{1}{2}, -\frac{1}{2} \middle| \exp\left[-i\beta \hat{j}_{1y}/\hbar\right] \middle| \frac{1}{2}, -\frac{1}{2} \right\rangle$$

$$\times \left\langle j, m+\frac{1}{2} \middle| \exp\left[-i\beta \hat{j}_y/\hbar\right] \middle| j, m'+\frac{1}{2} \right\rangle$$

$$= \left(a_1 \cos\frac{\beta}{2} + b_1 \sin\frac{\beta}{2} \right) \sum_{n=0}^{2j} A_n \left(\cos\frac{\beta}{2} \right)^{2j-n} \left(\sin\frac{\beta}{2} \right)^n$$

$$= \sum_{l=0}^{2(i+\frac{1}{2})} B_n \left(j+\frac{1}{2}, m, m' \right) \left(\cos\frac{\beta}{2} \right)^{2(j+\frac{1}{2})-l} \left(\sin\frac{\beta}{2} \right)^l.$$

Thus the statement is also valid for $j+1/2$. That is to say, the matrix elements

$$\left\langle j, m \middle| U(\beta, \hat{y}) \middle| j, m' \right\rangle = \left\langle j, m \middle| \exp\left(-i\beta \hat{j}_y/\hbar\right) \middle| j, m' \right\rangle,$$

are polynomials of degree $2j$ with respect to the variables $\cos(\beta/2)$ and $\sin(\beta/2)$

.

3011

An operator f describing the interaction of two spin-1/2 particles has the form

$$f = a + b\sigma_1 \cdot \sigma_2,$$

where a and b are constants, σ_1 and σ_2 are Pauli matrices. The total spin angular momentum is $\mathbf{J} = \mathbf{j}_1 + \mathbf{j}_2 = \frac{\hbar}{2}(\sigma_1 + \sigma_2)$.

a. Show that f, $\mathbf{J}^2$ and J_z can be simultaneously measured.

b. Derive the matrix representation for f in the $\left| J, M, j_1, j_2 \right\rangle$ basis. (Label rows and columns of your matrix).

c. Derive the matrix representation for f in the $\left| j_1, j_2, m_1, m_2 \right\rangle$ basis.

(Wisconsin)

Sol:

a. f, $\mathbf{J}^2$ and J_z can be measured simultaneously if each pair of them commute. We know that $\mathbf{J}^2$ and J_z commute, also that either commutes with a, a constant.

From definition,

$$\mathbf{J}^2 = \frac{\hbar^2}{4}(\sigma_1^2 + \sigma_2^2 + 2\sigma_1 \cdot \sigma_2),$$

or

$$\sigma_1 \cdot \sigma_2 = \frac{2\mathbf{J}^2}{\hbar^2} - \frac{1}{2}(\sigma_1^2 + \sigma_2^2).$$

Now for each particle

$$\sigma^2 = \sigma_x^2 + \sigma_y^2 + \sigma_z^2 = 3I,$$

I being the unit matrix, so

$$\sigma_1 \cdot \sigma_2 = \frac{2\mathbf{J}^2}{\hbar^2} - 3.$$

Hence

$$[\mathbf{J}^2, f] = [\mathbf{J}^2, a] + b[\mathbf{J}^2, \sigma_1 \cdot \sigma_2]$$

$$= b\left[\mathbf{J}^2, \frac{2\mathbf{J}^2}{\hbar^2} - 3\right] = 0,$$

$$[J_z, f] = [J_z, a] + b\left[J_z, \frac{2\mathbf{J}^2}{\hbar^2} - 3\right] = 0.$$

Therefore, f, $\mathbf{J}^2$ and J_z can be measured simultaneously

b. In the $\left| J, M, j_1, j_2 \right\rangle$ basis,

$$\left\langle J, M, j_1, j_2 \middle| f \middle| J', M', j_1, j_2 \right\rangle = a\delta_{JJ'}\delta_{MM'} + b\left\langle J, M, j_1, j_2 \middle| \sigma_1 \cdot \sigma_2 \right.$$
$$\left| J', M', j_1, j_2 \right\rangle$$

$$= a\delta_{JJ'}\delta_{MM'} + b\left[\frac{2}{\hbar^2}J'(J'+1)\hbar^2 - 3\right]$$

$$\times \delta_{JJ'}\delta_{MM'}$$

$$= [a + 2bJ(J+1) - 3b]\delta_{JJ'}\delta_{MM'},$$

where J, M are row labels, J', M' are column labels.

c. Denote the state of $J=0$ and the state of $J=1$ and $J_z = M$ as χ_0 and χ_{1M}, respectively. Since $j_1 = j_2 = 1/2$, we can denote the state $\left| j_1, j_2, m_1, m_2 \right\rangle$ simply as $\left| m_1, m_2 \right\rangle$. Then as

$$\begin{cases} \chi_0 = \dfrac{1}{\sqrt{2}}\left\{\left|\dfrac{1}{2}, -\dfrac{1}{2}\right\rangle - \left|-\dfrac{1}{2}, \dfrac{1}{2}\right\rangle\right\}, \\[2mm] \chi_{10} = \dfrac{1}{\sqrt{2}}\left\{\left|\dfrac{1}{2}, -\dfrac{1}{2}\right\rangle + \left|-\dfrac{1}{2}, \dfrac{1}{2}\right\rangle\right\}, \\[2mm] \chi_{1,\pm 1} = \left|\pm\dfrac{1}{2}, \pm\dfrac{1}{2}\right\rangle, \end{cases}$$

we have

$$
\begin{cases}
\left|\pm\dfrac{1}{2},\ \pm\dfrac{1}{2}\right\rangle = \chi_{1,\pm1}, \\[2mm]
\left|\dfrac{1}{2},\ -\dfrac{1}{2}\right\rangle = \dfrac{1}{\sqrt{2}}(\chi_0 + \chi_{10}), \\[2mm]
\left|-\dfrac{1}{2},\ \dfrac{1}{2}\right\rangle = \dfrac{1}{\sqrt{2}}(-\chi_0 + \chi_{10}).
\end{cases}
$$

Using the above expressions and the result of (b) we can write the matrix elements $\langle m_1, m_2 | f | m_1', m_2' \rangle$ in the basis $| j_1, j_2, m_1, m_2 \rangle$ as follows:

$\begin{array}{c} m_1', m_2' \\ m_1, m_2 \end{array}$	$\dfrac{1}{2}, \dfrac{1}{2}$	$\dfrac{1}{2}, -\dfrac{1}{2}$	$-\dfrac{1}{2}, \dfrac{1}{2}$	$-\dfrac{1}{2}, -\dfrac{1}{2}$
$\dfrac{1}{2}, \dfrac{1}{2}$	$a+b$	0	0	0
$\dfrac{1}{2}, -\dfrac{1}{2}$	0	$a-b$	$2b$	0
$-\dfrac{1}{2}, \dfrac{1}{2}$	0	$2b$	$a-b$	0
$-\dfrac{1}{2}, -\dfrac{1}{2}$	0	0	0	$a+b$

3012

Consider the following two-particle wave function in position space:

$\psi(\mathbf{r}_1, \mathbf{r}_2) = f(r_1^2)\, g(r_2^2)\, [\alpha(\mathbf{a}\cdot\mathbf{r}_1)(\mathbf{b}\cdot\mathbf{r}_2) + \beta(\mathbf{b}\cdot\mathbf{r}_1)(\mathbf{a}\cdot\mathbf{r}_2) + \gamma(\mathbf{a}\cdot\mathbf{b})(\mathbf{r}_1\cdot\mathbf{r}_2)]$, where $\mathbf{a}$ and $\mathbf{b}$ are arbitrary constant vectors, f and g are arbitrary functions, and α, β and γ are constants.

a. What are the eigenvalues of the squared angular momentum for each particle ($\mathbf{L}_1^2$ and $\mathbf{L}_2^2$)?

b. With an appropriate choice of α, β and γ, $\psi_2(\mathbf{r}_1, \mathbf{r}_2)$ can also be an eigenfunction of the total angular momentum squared $\mathbf{J}^2 = (\mathbf{L}_1 + \mathbf{L}_2)^2$. What are the possible values of the total angular momentum squared and what are the appropriate values of α, β and γ for each state?

(MIT)

Sol:

a. We first note that

$$\nabla f(r) = f'(r)\frac{\mathbf{r}}{r},$$

or

$$\mathbf{r} \times \nabla f(r) = f'(r)\frac{\mathbf{r} \times \mathbf{r}}{r} = 0,$$

and that

$$\mathbf{r} \times \nabla (\mathbf{a} \cdot \mathbf{r}) = \mathbf{r} \times \mathbf{a},$$
$$(\mathbf{r} \times \nabla) \cdot (\mathbf{r} \times \mathbf{a}) = -2\mathbf{a} \cdot \mathbf{r}.$$

As $\mathbf{L} = \mathbf{r} \times \mathbf{p} = -i\hbar \mathbf{r} \times \nabla$, we have

$$
\begin{aligned}
L_1^2 \psi(\mathbf{r}_1, \mathbf{r}_2) &= -\hbar^2 (\mathbf{r}_1 \times \nabla_1) \cdot (\mathbf{r}_1 \times \nabla_1) \{ f(r_1^2) g(r_2^2) [\alpha(\mathbf{a} \cdot \mathbf{r}_1)(\mathbf{b} \cdot \mathbf{r}_2) \\
&\quad + \beta(\mathbf{b} \cdot \mathbf{r}_1)(\mathbf{a} \cdot \mathbf{r}_2) + \gamma(\mathbf{a} \cdot \mathbf{b})(\mathbf{r}_1 \cdot \mathbf{r}_2)] \} \\
&= -\hbar^2 f(r_1^2) g(r_2^2) (\mathbf{r}_1 \times \nabla_1) \cdot (\mathbf{r}_1 \times \nabla_1) \cdot [\alpha(\mathbf{a} \cdot \mathbf{r}_1)(\mathbf{b} \cdot \mathbf{r}_2) \\
&\quad + \beta(\mathbf{b} \cdot \mathbf{r}_1)(\mathbf{a} \cdot \mathbf{r}_2) + \gamma(\mathbf{a} \cdot \mathbf{b})(\mathbf{r}_1 \cdot \mathbf{r}_2)] \\
&= -\hbar^2 f(r_1^2) g(r_2^2) (\mathbf{r}_1 \times \nabla_1) \cdot [\alpha(\mathbf{r}_1 \times \mathbf{a})(\mathbf{b} \cdot \mathbf{r}_2) \\
&\quad + \beta(\mathbf{r}_1 \times \mathbf{b})(\mathbf{a} \cdot \mathbf{r}_2) + \gamma(\mathbf{a} \cdot \mathbf{b})(\mathbf{r}_1 \times \mathbf{r}_2)] \\
&= 2\hbar^2 f(r_1^2) g(r_2^2) [\alpha(\mathbf{a} \cdot \mathbf{r}_1)(\mathbf{b} \cdot \mathbf{r}_2) + \beta(\mathbf{b} \cdot \mathbf{r}_1)(\mathbf{a} \cdot \mathbf{r}_2) \\
&\quad + \gamma(\mathbf{a} \cdot \mathbf{b})(\mathbf{r}_1 \cdot \mathbf{r}_2)] \\
&= 1(1+1)\hbar^2 \psi(\mathbf{r}_1, \mathbf{r}_2),
\end{aligned}
$$

and similarly

$$
\begin{aligned}
L_2^2 \psi(\mathbf{r}_1 \cdot \mathbf{r}_2) &= -\hbar^2 (\mathbf{r}_2 \times \nabla_2) \cdot (\mathbf{r}_2 \times \nabla_2) \psi(\mathbf{r}_1 \cdot \mathbf{r}_2) \\
&= 1(1+1)\hbar^2 \psi(\mathbf{r}_1, \mathbf{r}_2).
\end{aligned}
$$

Hence the eigenvalues of L_1^2 and L_2^2 are each equal to $2\hbar^2$, and so each particle has the quantum number $l = 1$

b. We further note that

$$(\mathbf{r} \times \nabla) \cdot (\mathbf{a} \cdot \mathbf{r})\mathbf{e} = (\mathbf{r} \times \mathbf{a}) \cdot \mathbf{c},$$
$$(\mathbf{a} \times \mathbf{b}) \cdot (\mathbf{c} \times \mathbf{d}) = (\mathbf{a} \cdot \mathbf{c})(\mathbf{b} \cdot \mathbf{d}) - (\mathbf{a} \cdot \mathbf{d})(\mathbf{b} \cdot \mathbf{c}).$$

Thus we require

$$
\begin{aligned}
\mathbf{L}_1 \cdot \mathbf{L}_2 \psi(\mathbf{r}_1, \mathbf{r}_2) &= -\hbar^2 (\mathbf{r}_1 \times \nabla_1) \cdot (\mathbf{r}_2 \times \nabla_2) \{ f(r_1^2) g(r_2^2) [\alpha(\mathbf{a} \cdot \mathbf{r}_1) \\
&\quad \times (\mathbf{b} \cdot \mathbf{r}_2) + \beta(\mathbf{b} \cdot \mathbf{r}_1)(\mathbf{a} \cdot \mathbf{r}_2) + \gamma(\mathbf{a} \cdot \mathbf{b})(\mathbf{r}_1 \cdot \mathbf{r}_2)] \} \\
&= -\hbar^2 f(r_1^2) g(r_2^2)(\mathbf{r}_1 \times \nabla_1) \cdot (\mathbf{r}_2 \times \nabla_2)[\alpha(\mathbf{a} \cdot \mathbf{r}_1)(\mathbf{b} \cdot \mathbf{r}_2) \\
&\quad + \beta(\mathbf{b} \cdot \mathbf{r}_1)(\mathbf{a} \cdot \mathbf{r}_2) + \gamma(\mathbf{a} \cdot \mathbf{b})(\mathbf{r}_1 \cdot \mathbf{r}_2)] \\
&= -\hbar^2 f(r_1^2) g(r_2^2)(\mathbf{r}_1 \times \nabla_1) \cdot [\alpha(\mathbf{a} \cdot \mathbf{r}_1)(\mathbf{r}_2 \times \mathbf{b}) \\
&\quad + \beta(\mathbf{b} \cdot \mathbf{r}_1)(\mathbf{r}_2 \times \mathbf{a}) + \gamma(\mathbf{a} \cdot \mathbf{b})(\mathbf{r}_2 \times \mathbf{r}_1)] \\
&= -\hbar^2 f(r_1^2) g(r_2^2)[\alpha(\mathbf{r}_1 \times \mathbf{a}) \cdot (\mathbf{r}_2 \times \mathbf{b}) \\
&\quad + \beta(\mathbf{r}_1 \times \mathbf{b}) \cdot (\mathbf{r}_2 \times \mathbf{a}) + \gamma(\mathbf{a} \cdot \mathbf{b})(2\mathbf{r}_1 \cdot \mathbf{r}_2)] \\
&= -\hbar^2 f(r_1^2) g(r_2^2)[-\beta(\mathbf{a} \cdot \mathbf{r}_1)(\mathbf{b} \cdot \mathbf{r}_2) - \alpha(\mathbf{b} \cdot \mathbf{r}_1)(\mathbf{a} \cdot \mathbf{r}_2) \\
&\quad + (\alpha + \beta + 2\gamma)(\mathbf{a} \cdot \mathbf{b})(\mathbf{r}_1 \cdot \mathbf{r}_2)] \\
&= -\hbar^2 \lambda \psi(\mathbf{r}_1, \mathbf{r}_2)
\end{aligned}
$$

for $\psi(\mathbf{r}_1, \mathbf{r}_2)$ to be an eigenfunction of $\mathbf{L}_1 \cdot \mathbf{L}_2$. This demands that

$$-\beta = \lambda\alpha, -\alpha = \lambda\beta, \alpha + \beta + 2\gamma = \lambda\gamma,$$

which give three possible values of λ:

$$\lambda = -1, \quad \alpha = \beta = -\frac{3}{2}\gamma;$$

$$\lambda = +1, \quad \alpha = -\beta, \quad \gamma = 0;$$

$$\lambda = 2, \quad \alpha = \beta = 0.$$

Therefore the possible values of the total angular momentum squared, $\mathbf{J}^2 = \mathbf{L}_1^2 + \mathbf{L}_2^2 + 2\mathbf{L}_1 \cdot \mathbf{L}_2$, and the corresponding values of α, β and γ are

$$
\mathbf{J}^2 = 2\hbar^2 + 2\hbar^2 - 2\hbar^2 \lambda =
\begin{cases}
2(2+1)\hbar^2, & (\alpha = \beta = -\frac{3}{2}\gamma) \\
1(1+1)\hbar^2, & (\alpha = -\beta, \gamma = 0) \\
0. & (\alpha = \beta = 0)
\end{cases}
$$

3013

A quantum-mechanical state of a particle, with Cartesian coordinates x, y and z, is described by the normalized wave function

$$\psi(x, y, z) = \frac{\alpha^{5/2}}{\sqrt{\pi}} z \exp\left[-\alpha(x^2 + y^2 + z^2)^{1/2}\right].$$

Show that the system is in a state of definite angular momentum and give the values of L^2 and L_z associated with the state.

(Wisconsin)

ol: Transforming to spherical coordinates by

$$x = r\sin\theta\cos\varphi, \quad y = r\sin\theta\sin\varphi, \quad z = r\cos\theta,$$

we have

$$\psi(r,\ \theta,\ \varphi) = \frac{\alpha^{5/2}}{\sqrt{\pi}} r\cos\theta e^{-\alpha r} = f(r)Y_{10}\,.$$

Hence the particle is in a state of definite angular momentum. For this state, $l=1$, $L^2 = l(l+1)\hbar^2 = 2\hbar^2$, $L_z = 0$.

3014

free atom of carbon has four paired electrons in s-states and two more elec-
rons with p-wave orbital wave functions.

a. How many states are permitted by the Pauli exclusion principle for the
latter pair of electrons in this configuration?

b. Under the assumption of L-S coupling what are the "good" quantum
numbers? Give sets of values of these for the configuration of the two
p-wave electrons.

c. Add up the degeneracies of the terms found in (b), and show that it is
the same as the number of terms found in (a).

(Buffalo)

ol:

a. Each electron can occupy one of the $(2l+1)(2s+1) = 3\times2 = 6$ states, but it is
not permitted that two electrons occupy the same state. So the number of
permitted states is $C_2^6 = 15$.

The "good" quantum numbers are L^2, S^2, J^2 and J_z. Under the assumption of L-S
coupling, the total spin quantum numbers for two electrons, $S = s_1 + s_2$,
are $S = 0$, 1 and the total orbital quantum numbers, $L = l_1 + l_2$, are $L = 0$, 1, 2.
Considering the symmetry of exchange, for the singlet $S = 0$, L should be
even: $L = 0$, 2, corresponding to $^1S_0$, $^1D_2$ respectively; for the triplet $S = 1$,
L odd: $L = 1$, corresponding $^3P_{0,1,2}$.

b. The degeneracy equals to $2J+1$. For $J = 0$, 2 and 0, 1, 2 in (b), the total
number of degeneracies is $1+5+1+3+5 = 15$.

3015

a. Determine the energy levels of a particle bound by the isotropic potential $V(r) = kr^2/2$, where k is a positive constant.

b. Derive a formula for the degeneracy of the Nth excited state.

c. Identify the angular momenta and parities of the Nth excited state.

(Columbia)

Sol:

a. The Hamiltonian is

$$\hat{H} = -(\hbar^2/2m)\nabla^2 + kr^2/2.$$

Choose $(\hat{H}, \hat{l}^2, \hat{l}_z)$ to have common eigenstates. The energy levels of bound states are given by

$$E = (2n_r + l + 3/2)\hbar\omega_0 = (N + 3/2)\hbar\omega_0, \qquad \omega_0 = \sqrt{k}/m.$$

b. The degeneracy of the states is determined by n_r and l. As $N = 2n_r + l$, the odd-even feature of N is the same as that of l. For N even (i.e., l even), the degeneracy is

$$f = \sum_{\substack{l=0 \ (l \text{ even})}}^{N} (2l+1) = 4 \sum_{\substack{l=0 \ (l \text{ even})}}^{N} \left(\frac{l}{2} + \frac{1}{4}\right) = 4\sum_{l'=0}^{N/2}\left(l' + \frac{1}{4}\right)$$

$$= \frac{1}{2}(N+1)(N+2).$$

For N odd (i.e., l odd),

$$f = \sum_{\substack{l=1 \ (l \text{ odd})}}^{N} (2l+1) = \sum_{\substack{l=1 \ (l \text{ odd})}}^{N} \left[2(l-1)+3\right] = 4 \sum_{l'=0}^{(N-1)/2}\left(l' + \frac{3}{4}\right)$$

$$= \frac{1}{2}(N+1)(N+2).$$

Hence the degeneracy of the Nth excited state is $f = (N+1)(N+2)/2$.

c. In the common eigenstates of $(\hat{H}, l^2, l_z)$, the wave function of the system is

$$\psi_{n,l}(r, \theta, \varphi) = R(r)Y_{lm}(\theta, \varphi),$$

and the eigenenergy is

$$E_{n,l} = (2n_r + l + 3/2)\hbar\omega_0.$$

As $N = 2n_r + l$, the angular momentum l of the Nth excited state has $\frac{N}{2}+1$ values $0, 2, 4, 6 \ldots, N$ for N even, or $\frac{1}{2}(N+1)$ values $1, 3, 5, \ldots, N$ for N odd. Furthermore, the parity is

$$P = (-1)^l = (-1)^N.$$

3016

The ground state of the realistic helium atom is of course nondegenerate. However, consider a hypothetical helium atom in which the two electrons are replaced by two identical, spin-one particles of negative charge. Neglect spin-dependent forces. For this hypothetical atom, what is the degeneracy of the ground state? Give your reasoning.

(CUS)

Sol: The two new particles are Bosons; thus the wave function must be symmetrical. In the ground state, the two particles must stay in Is orbit. Then the space wave function is symmetrical, and consequently the spin wave function is symmetrical too. As $s_1 = 1$ and $s_2 = 1$, the total S has three possible values:

$S = 2$, the spin wave function is symmetric and its degeneracy is $2S+1 = 5$.

$S = 1$, the spin wave function is antisymmetric and its degeneracy is $2S+1 = 3$.

$S = 0$, the spin wave function is symmetric and its degeneracy is $2S+1 = 1$.

If the spin-dependent forces are neglected, the degeneracy of the ground state is $5+3+1 = 9$.

3017

Consider an electron in the quantum state given by

$$\Psi(r,\theta,\varphi) = \frac{1}{\sqrt{4\pi}}(2i\sin\varphi\sin\theta + \cos\theta)g(r)$$

Where $\int_0^\infty |g(r)|^2 r^2\, dr = 1$

a. Find the possible values for the z-component of the angular momentum L_z.

b. Find the probabilities for each of the possible values of L_z.

c. Find the expectation value of L_z.

Sol:

a. The expression for Ψ can be rewritten using the expressions for spherical harmonics $Y_{1,0} = \sqrt{\frac{3}{4\pi}}\cos\theta$, $Y_{1,1} = -\sqrt{\frac{3}{8\pi}}\sin\theta e^{i\varphi}$ and $Y_{1,-1} = \sqrt{\frac{3}{8\pi}}\sin\theta e^{-i\varphi}$ as

$$\Psi(r,\theta,\varphi) = \frac{1}{\sqrt{4\pi}}((e^{i\varphi} - e^{-i\varphi})\sin\theta + \cos\theta)g(r)$$

$$= \frac{1}{\sqrt{3}}Y_{1,0} - \sqrt{\frac{2}{3}}Y_{1,1} - \sqrt{\frac{2}{3}}Y_{1,-1} \tag{1}$$

From (1), the possible values of L_z are $+\hbar, 0 \ and -\hbar$

b. For a normalized wavefunction, the sum of the squares of coefficients should be equal to 1, therefore normalization constant is $\frac{3}{5}$ and

$$\psi_{normalized} = \sqrt{\frac{3}{5}}\left(\frac{1}{\sqrt{3}}Y_{1,0} - \sqrt{\frac{2}{3}}Y_{1,1} - \sqrt{\frac{2}{3}}Y_{1,-1} \right)$$

The probability of $L_z = 0$ is $\frac{1}{5}$, $L_z = +\hbar$ is $\frac{2}{5}$ and $L_z = -\hbar$ is $\frac{2}{5}$

c. Expectation value of L_z

$$L_z = \int \left[\sqrt{\frac{3}{5}}\left(\frac{1}{\sqrt{3}}Y_{1,0} - \sqrt{\frac{2}{3}}Y_{1,1} - \sqrt{\frac{2}{3}}Y_{1,-1} \right) \right]^* \times L_z \left[\sqrt{\frac{3}{5}}\left(\frac{1}{\sqrt{3}}Y_{1,0} - \sqrt{\frac{2}{3}}Y_{1,1} - \sqrt{\frac{2}{3}}Y_{1,-1} \right) \right]$$

$$\times |g(r)|^2 r^2 dr \sin\theta d\theta d\varphi$$

$$L_z = 0$$

3018

a. Consider a system of spin 1/2. What are the eigenvalues and normalized eigenvector of the operator $A\hat{s}_y + B\hat{s}_z$, where $\hat{s}_y, \hat{s}_z$ are the angular momentum operators, and A and B are real constants.

b. Assume that the system is in a state corresponding to the upper eigenvalue. What is the probability that a measurement of $\hat{s}_y$ will yield the value $\hbar/2$? The Pauli matrices are

$$\sigma_x = \begin{pmatrix} 0 & 1 \\ 1 & 0 \end{pmatrix}, \ \sigma_y = \begin{pmatrix} 0 & -i \\ i & 0 \end{pmatrix}, \ \sigma_z = \begin{pmatrix} 1 & 0 \\ 0 & -1 \end{pmatrix}.$$

(CUS)

Sol:

a. Using the definition of angular momentum operators let

$$\hat{T} = A\hat{s}_y + B\hat{s}_z = A\frac{1}{2}\hbar\sigma_y + B\frac{1}{2}\hbar\sigma_z.$$

Then

$$(\hat{T})^2 = \frac{1}{4}\hbar^2(A^2 + B^2 + AB\{\sigma_y, \sigma_z\}) = \frac{1}{4}\hbar^2(A^2 + B^2),$$

as

$$\sigma_i^2 = \begin{pmatrix} 1 & 0 \\ 0 & 1 \end{pmatrix} = I, \quad i = 1, 2, 3,$$

and

$$\{\sigma_i, \sigma_j\} \equiv \sigma_i\sigma_j + \sigma_j\sigma_i = 2\delta_{ij}.$$

Hence the two eigenvalues of $\hat{T}$ are

$$T_1 = \frac{1}{2}\hbar\sqrt{A^2 + B^2}, \quad T_2 = -\frac{1}{2}\hbar\sqrt{A^2 + B^2}.$$

In the representation of $\hat{s}^2$ and $\hat{s}_z$,

$$\hat{T} = \frac{\hbar}{2}(A\sigma_y + B\sigma_z) = \frac{\hbar}{2}\begin{pmatrix} B & -iA \\ iA & -B \end{pmatrix}.$$

Let the eigenvector be $\begin{pmatrix} a \\ b \end{pmatrix}$. Then the eigenequation is

$$\frac{\hbar}{2}\begin{pmatrix} B & -iA \\ iA & -B \end{pmatrix}\begin{pmatrix} a \\ b \end{pmatrix} = T\begin{pmatrix} a \\ b \end{pmatrix},$$

where

$$T = \frac{\hbar}{2}\begin{pmatrix} \pm\sqrt{A^2 + B^2} & 0 \\ 0 & \mp\sqrt{A^2 + B^2} \end{pmatrix},$$

or

$$\begin{pmatrix} B \mp \sqrt{A^2 + B^2} & -iA \\ iA & -B \mp \sqrt{A^2 + B^2} \end{pmatrix}\begin{pmatrix} a \\ b \end{pmatrix} = 0.$$

Hence

$$a : b = iA : B \mp \sqrt{A^2 + B^2},$$

and so the normalized eigenvector is

$$\begin{pmatrix} a \\ b \end{pmatrix} = \left[\frac{1}{A^2 + (B \mp \sqrt{A^2 + B^2})}\right]^{1/2}\begin{pmatrix} +iA \\ B \mp \sqrt{A^2 + B^2} \end{pmatrix}.$$

b.   In the representation of $\hat{s}^2$ and $\hat{s}_z$, the eigenvector of $\hat{s}_y$ is

$$\left|s_y=\frac{\hbar}{2}\right\rangle=\frac{1}{\sqrt{2}}\begin{pmatrix}-i\\1\end{pmatrix}.$$

Hence the probability of finding $s_y=\hbar/2$ is

$$P_{\mp}=\left|\frac{1}{\sqrt{2}}(i1)\begin{pmatrix}a\\b\end{pmatrix}\right|^2=\left|\frac{1}{\sqrt{2}}(ia+b)^2\right|^2=\frac{(B\mp\sqrt{A^2+B^2}-A)^2}{2[(B\mp\sqrt{A^2+B^2})^2+A^2]}.$$

Note that P_- is the probability corresponding to the system in the state of eigenvalue $T=\hbar\sqrt{A^2+B^2}/2$, and $P+$ is that corresponding to the state of $T=-\hbar\sqrt{A^2+B^2}/2.$

3019

A system of three (non-identical) spin one-half particles, whose spin operators are s_1, s_2 and s_3, is governed by the Hamiltonian

$$H=As_1\cdot s_2/\hbar^2+B(s_1+s_2)\cdot s_3/\hbar^2.$$

Find the energy levels and their degeneracies.

(Princeton)

Sol: Using a complete set of dynamical variables of the system (H, s_{12}^2, s^2, s_3), where $s_{12}=s_1+s_2$, $s=s_{12}+s_3=s_1+s_2+s_3$, the eigenfunction is $|s_{12}s_3sm_s\rangle$, and the stationary state equation is

$$\hat{H}|s_{12}s_3sm_s\rangle=E|s_{12}s_3sm_s\rangle.$$

As

$$s_1^2=s_2^2=s_3^2=\frac{3}{4}\hbar^2\begin{pmatrix}1&0\\0&1\end{pmatrix},$$

$$s_1\cdot s_2=\frac{1}{2}(s_{12}^2-s_1^2-s_2^2),$$

$$(s_1+s_2)\cdot s_3=\frac{1}{2}(s^2-s_{12}^2-s_3^2),$$

we have

$$\hat{H}=\frac{A}{\hbar^2}s_1\cdot s_2+\frac{B}{\hbar^2}(s_1+s_2)\cdot s_3$$

$$=\frac{A}{2}\left[\frac{1}{\hbar^2}s_{12}^2-\frac{3}{4}-\frac{3}{4}\right]$$

$$+\frac{B}{2}\left[\frac{1}{\hbar^2}s^2-\frac{1}{\hbar^2}s_{12}^2-\frac{3}{4}\right].$$

Now as the expectation value of s is $s(s+1)$, etc., we have

$$\hat{H}|s_{12}s_3sm_s\rangle=\left\{\frac{A}{2}\left[s_{12}(s_{12}+1)-\frac{3}{2}\right]\right.$$
$$\left.+\frac{B}{2}\left[s(s+1)-s_{12}(s_{12}+1)-\frac{3}{4}\right]\right\}|s_{12}s_3sm_s\rangle,$$

and hence

$$E=\frac{A}{2}\left[s_{12}(s_{12}+1)-\frac{3}{2}\right]$$
$$+\frac{B}{2}\left[s(s+1)-s_{12}(s_{12}+1)-\frac{3}{4}\right].$$

It follows that for $s_{12}=0$, $s=1/2$: $E=-3A/4$, the degeneracy of the energy level, $2s+1$, is 2; for $s_{12}=1$, $s=1/2$: $E=A/4-B$, the degeneracy of the energy level, is 2; for $s_{12}=1, s=3/2: E=A/4+B/2$, the degeneracy of the energy level is 4.

3020

A particle of spin one is subject to the Hamiltonian $H=As_z+Bs_x^2$, where A and B are constants. Calculate the energy levels of this system. If at time zero the spin is in an eigenstate of s with $s_z=+\hbar$, calculate the expectation value of the spin at time t.

(Princeton)

Sol: We first find the stationary energy levels of the system. The stationary Schrödinger equation is

$$E\psi=\hat{H}\psi=(As_z+Bs_x^2)\psi,$$

where

$$\psi=\begin{pmatrix}u_1\\u_2\\u_3\end{pmatrix}$$

is a vector in the spin space. As

$$s_x=\begin{pmatrix}0&0&0\\0&0&-i\\0&i&0\end{pmatrix}\hbar,$$

$$s_z=\begin{pmatrix}0&-i&0\\i&0&0\\0&0&0\end{pmatrix}\hbar,$$

$$s_x^2 = \begin{pmatrix} 0 & 0 & 0 \\ 0 & 1 & 0 \\ 0 & 0 & 1 \end{pmatrix} \hbar^2,$$

we have

$$\hat{H} = As_z + Bs_x^2 = \begin{pmatrix} 0 & -iA' & 0 \\ iA' & B' & 0 \\ 0 & 0 & B' \end{pmatrix},$$

where $A' = A\hbar$, $B' = B\hbar^2$. The energy levels are given by the eigenvalues of the above matrix, which are roots of the equation

$$\mathrm{Det} \begin{pmatrix} -E & -iA' & 0 \\ iA' & B'-E & 0 \\ 0 & 0 & B'-E \end{pmatrix} = 0,$$

i.e.,

$$(E - B')(E^2 - B'E - A') = 0.$$

Thus the energy levels are

$$E_0 = B', \ E_{\pm} = (B'' \pm \omega)\hbar / 2,$$

where $\omega = \sqrt{B''^2 + 4A'^2} / \hbar$, $B'' = B'/\hbar = B\hbar$. The corresponding eigenfunctions are

$$E_0 = B' : \ \varphi_{S_0} = \begin{pmatrix} 0 \\ 0 \\ 1 \end{pmatrix},$$

$$E_+ = \frac{B'}{2} + \frac{\sqrt{B'^2 + 4A'^2}}{2} : \ \varphi_{s+} = \sqrt{\frac{\omega - B''}{2\omega}} \begin{pmatrix} 1 \\ i\sqrt{\dfrac{\omega + B''}{\omega - B''}} \\ 0 \end{pmatrix},$$

$$E_- = \frac{B'}{2} + \frac{\sqrt{B'^2 + 4A'^2}}{2} : \ \varphi_{S-} = \sqrt{\frac{\omega + B''}{2\omega}} \begin{pmatrix} 1 \\ -i\sqrt{\dfrac{\omega - B''}{\omega + B''}} \\ 0 \end{pmatrix}.$$

The general wave function of the system is therefore

$$\varphi_S(t) = C_1 \varphi s_0 \exp\left[-\frac{iB'}{\hbar}t\right] + C_2 \varphi_{S+} \exp\left[-i\frac{E+}{\hbar}t\right]$$
$$+ C_3 \varphi s_- \exp\left[-i\frac{E_-}{\hbar}t\right].$$

Initially,

$$s_z\varphi_s(0)=\hbar\varphi_s(0).$$

Let

$$\varphi_s(0)=\begin{pmatrix}\alpha\\\beta\\\gamma\end{pmatrix}.$$

The above requires

$$\begin{pmatrix}0 & -i & 0\\i & 0 & 0\\0 & 0 & 0\end{pmatrix}\begin{pmatrix}\alpha\\\beta\\\gamma\end{pmatrix}=\begin{pmatrix}\alpha\\\beta\\\gamma\end{pmatrix},$$

i.e.,

$$\beta=i\alpha,\quad\gamma=0.$$

Thus we can take the initial wave function (normalized) as

$$\varphi s(0)=\frac{1}{\sqrt{2}}\begin{pmatrix}1\\i\\0\end{pmatrix}.$$

Equating $\varphi_s(0)$ with $C_1\varphi_{S0}+C_2\varphi_{S+}+C_3\varphi_{S-}$ gives

$$C_1=0,\ C_2=\frac{(\omega+B'')^{1/2}+(\omega-B'')^{1/2}}{2\omega^{1/2}},$$

$$C_3=\frac{(\omega+B'')^{1/2}-(\omega-B'')^{1/2}}{2\omega^{1/2}}.$$

We can now find the expectation value of the spin:

$$\langle s_x\rangle=\varphi_S^+(t)s_x\varphi_S(t)=0,$$

where we have used the orthogonality of φ_{S0}, φ_{S+} and φ_{S-}. Similarly,

$$\langle s_y\rangle=0,$$

$$\langle s_z\rangle=\varphi_S^+(t)s_z\varphi s(t)=\left[1-\frac{2B^2\hbar^2}{\omega^2}\sin^2(\omega t/2)\right]\hbar.$$

3021

A system of two particles each with spin 1/2 is described by an effective Hamiltonian

$$H=A(s_{1z}+s_{2z})+Bs_1\cdot s_2,$$

where s_1 and s_2 are the two spins, s_{1z} and s_{2z} are their z-components, and A and B are constants. Find all the energy levels of this Hamiltonian.

(Wisconsin)

Sol: We choose χ_{SM_S} as the common eigenstate of $\mathbf{S}^2 = (\mathbf{s}_1 + \mathbf{s}_2)^2$ and $S_z = s_{1z} + s_{2z}$. For $S = 1$, $M_S = 0, \pm 1$, it is a triplet and is symmetric when the two electrons are exchanged. For $S = 0$, $M_S = 0$, it is a singlet and is antisymmetric. For stationary states we use the time-independent Schrödinger equation

$$H\chi_{SM_S} = E\chi_{SM_S}.$$

As

$$\mathbf{S}^2 \chi_{1M_S} = S(S+1)\hbar^2 \chi_{1M_S} = 2\hbar^2 \chi_{1M_S}, \quad \mathbf{S}^2 \chi_{00} = 0,$$

$$\mathbf{S}^2 = (\mathbf{s}_1 + \mathbf{s}_2)^2 = \mathbf{s}_1^2 + \mathbf{s}_2^2 + 2\mathbf{s}_1 \cdot \mathbf{s}_2$$

$$= \frac{3\hbar^2}{4} + \frac{3}{4}\hbar^2 + 2\mathbf{s}_1 \cdot \mathbf{s}_2,$$

we have

$$\mathbf{s}_1 \cdot \mathbf{s}_2 \chi_{1M_S} = \left(\frac{\mathbf{S}^2}{2} - \frac{3}{4}\hbar^2\right)\chi_{1M_S} = \left(\hbar^2 - \frac{3}{4}\hbar^2\right)\chi_{1M_S} = \frac{\hbar^2}{4}\chi_{1M_S},$$

$$\mathbf{s}_1 \cdot \mathbf{s}_2 \chi_{00} = \left(0 - \frac{3}{4}\hbar^2\right)\chi_{00} = -\frac{3}{4}\hbar^2 \chi_{00},$$

and

$$S_z \chi_{1M_S} = (s_{1z} + s_{2z})\chi_{1M_S} = M_S \hbar \chi_{1M_S},$$

$$s_z \chi_{00} = 0.$$

Hence for the triplet state, the energy levels are

$$E = M_S \hbar A + \frac{\hbar^2}{4}B, \quad \text{with} \quad M_S = 0, \pm 1,$$

comprising three lines

$$E_1 = \hbar A + \frac{\hbar^2}{4}B, \quad E_2 = \frac{\hbar^2}{4}B, \quad E_3 = -\hbar A + \frac{\hbar^2}{4}B.$$

For the singlet state, the energy level consists of only one line

$$E_0 = -\frac{3}{4}\hbar^2 B.$$

3022

Suppose an atom is initially in an excited $^1S_0$ state (Fig. 3.1) and subsequently decays into a lower, short-lived $^1P_1$ state with emission of a photon γ_1

Fig. 3.2). Soon after, it decays into the $^1S_0$ ground state by emitting a second photon γ_2 (Fig. 3.3). Let θ be the angle between the two emitted photons.

a. What is the relative probability of θ in this process?

b. What is the ratio of finding both photons with the same circular polarization to that of finding the photons with opposite circular polarizations?

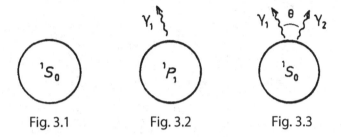

Fig. 3.1   Fig. 3.2   Fig. 3.3

It may be of some help to know the rotation matrices $d_{m'm}$, which relate one angular momentum representation in one coordinate system to another angular momentum representation in a rotated coordinate system, given below:

$$
d_{m'm} = \begin{array}{c} \\ m'=1 \\ \\ m'=0 \\ \\ m'=-1 \end{array}
\begin{array}{ccc}
m=1 & m=0 & m=-1 \\
\end{array}
\left(
\begin{array}{ccc}
\dfrac{1+\cos\alpha}{2} & -\dfrac{1}{\sqrt{2}}\sin\alpha & \dfrac{1-\cos\alpha}{2} \\[3mm]
\dfrac{1}{\sqrt{2}}\sin\alpha & \cos\alpha & -\dfrac{1}{\sqrt{2}}\sin\alpha \\[3mm]
\dfrac{1-\cos\alpha}{2} & \dfrac{1}{\sqrt{2}}\sin\alpha & \dfrac{1+\cos\alpha}{2}
\end{array}
\right)
$$

where α is the angle between the z-axis of one system and the z'-axis of the other.

(Columbia)

Sol: The atom is initially in the excited state $^1S_0$. Thus the projection of the atomic angular momentum on an arbitrary z direction is $L_z = 0$. We can take the direction of the first photon emission as the z direction. After the emission of the photon and the atom goes into the $^1P_1$ state, if the angular momentum of that photon is $L_z = \pm\hbar$, correspondingly the angular momentum of the atomic state

$^1P_1$ is $L_z = \mp\hbar$, i.e., $m_z = \mp 1$. If we let the direction of emission of the second photon be the z'-axis the projection of the eigenstate of the z-component of angular momentum on the z' direction is equivalent to multiplying the initial state with the $d_{m'm}$ matrix. Only atoms that are in states $m'_z = \pm 1$ can emit photon (as $L'_z = \pm\hbar$ must be satisfied) in z' direction and make the atom decay into the $^1S_0$ state $(m'_z = 0)$. Then the transitions are from $m = \pm 1$ to $m' = \pm 1$, and we have

$$C_1 = \langle m_{z'} = -1 \,|\, d_{m'm} \,|\, m_z = +1 \rangle$$

$$= (0,\ 0,\ 1) \begin{pmatrix} \dfrac{1+\cos\theta}{2} & -\dfrac{1}{\sqrt{2}}\sin\theta & \dfrac{1-\cos\theta}{2} \\[2mm] \dfrac{1}{\sqrt{2}}\sin\theta & \cos\theta & -\dfrac{1}{\sqrt{2}}\sin\theta \\[2mm] \dfrac{1-\cos\theta}{2} & \dfrac{1}{\sqrt{2}}\sin\theta & \dfrac{1+\cos\theta}{2} \end{pmatrix} \begin{pmatrix} 1 \\ 0 \\ 0 \end{pmatrix}$$

$$= \frac{1-\cos\theta}{2},$$

$$C_2 = \langle m_{z'} = +1 \,|\, d_{m'm} \,|\, m_z = +1 \rangle$$

$$= (1,\ 0,\ 0) \begin{pmatrix} \dfrac{1+\cos\theta}{2} & -\dfrac{1}{\sqrt{2}}\sin\theta & \dfrac{1-\cos\theta}{2} \\[2mm] \dfrac{1}{\sqrt{2}}\sin\theta & \cos\theta & -\dfrac{1}{\sqrt{2}}\sin\theta \\[2mm] \dfrac{1-\cos\theta}{2} & \dfrac{1}{\sqrt{2}}\sin\theta & \dfrac{1+\cos\theta}{2} \end{pmatrix} \begin{pmatrix} 1 \\ 0 \\ 0 \end{pmatrix}$$

$$= \frac{1+\cos\theta}{2},$$

$$C_3 = \langle m_{z'} = +1 \,|\, d_{m'm} \,|\, m_z = -1 \rangle = \frac{1-\cos\theta}{2},$$

$$C_4 = \langle m_{z'} = -1 \,|\, d_{m'm} \,|\, m_z = -1 \rangle = \frac{1+\cos\theta}{2}.$$

a. The relative probability of θ is

$$P(\theta) \propto |C_1|^2 + |C_2|^2 + |C_3|^2 + |C_4|^2 = 1 + \cos^2\theta.$$

b. The ratio of the probability of finding both photons with the same circular polarization to that of finding the photons with opposite circular polarizations is

$$(|C_2|^2 + |C_4|^2)/(|C_1|^2 + |C_3|^2) = (1+\cos\theta)^2 / (1-\cos\theta)^2.$$

3023

Consider an electron in a uniform magnetic field in the positive z direction. The result of a measurement has shown that the electron spin is along the positive x direction at $t=0$. Use Ehrenfest's theorem to compute the probability for $t>0$ that the electron is in the state (a) $s_x =1/2$, (b) $s_x =-1/2$, (c) $s_y =1/2$, (d) $s_y =-1/2$, (e) $s_z =1/2$, (f) $s_z =-1/2$.

Ehrenfest's theorem states that the expectation values of a quantum mechanical operator obey the classical equation of motion.

Hint: Recall the connection between expectation values and probability considerations].

(Wisconsin)

Sol: In the classical picture, an electron spinning with angular momentum **s** in a magnetic field **B** will, if the directions of **s** and **B** do not coincide, precess about the direction of **B** with an angular velocity **w** given by

$$\frac{ds}{dt}=s\times\omega,$$

where $\omega=\frac{e}{mc}\mathbf{B}$, m being the electron mass. Ehrenfest's theorem then states that in quantum mechanics we have

$$\frac{d}{dt}\langle s\rangle=\frac{e}{mc}\langle s\rangle\times\mathbf{B}.$$

This can be derived directly as follows.

An electron with spin angular momentum **s** has a magnetic moment $\mu=\frac{e}{mc}s$ and consequently a Hamiltonian

$$H=-\mu\cdot\mathbf{B}=-\frac{e}{mc}Bs_z,$$

taking the z axis along the direction of B. Then

$$\frac{d\langle s\rangle}{dt}=\frac{1}{i\hbar}\overline{[s,\hat{H}]}=-\frac{eB}{i\hbar mc}\overline{[s_x\hat{X}+s_y\hat{Y}+s_z\hat{Z},s_z]}$$

$$=-\frac{eB}{i\hbar mc}\{\overline{[s_x,s_z]}\hat{x}+\overline{[s_y,s_z]}\hat{y}\}$$

$$=-\frac{e}{mc}B\left(-\langle s_y\rangle\hat{x}+\langle s_x\rangle\hat{y}\right)$$

$$=\frac{e}{mc}\langle s\rangle\times\mathbf{B},$$

in agreement with the above. Note that use has been made of the commutation relations $[s_x, s_y] = i\hbar s_z$, etc.

Initially $\langle s_x \rangle = 1/2$, $\langle s_y \rangle = \langle s_z \rangle = 0$. and so we can write for $t > 0$, $\langle s_x \rangle = (\cos\omega t)/2$, $\langle s_y \rangle = (\sin\omega t)/2$, $\langle s_z \rangle = 0$.

Let the probability for $t > 0$ of the electron being in the state $s_x = 1/2$ be P and being in the state $s_x = -1/2$ be $1 - P$ since these are the only two states of s_x. Then

$$P\left(\frac{1}{2}\right) + (1-P)\left(-\frac{1}{2}\right) = \frac{1}{2}\cos \omega t,$$

giving

$$P = \cos^2(\omega t/2), \quad 1 - P = \sin^2(\omega t/2).$$

Similarly, let the probabilities for the electron being in the states $s_y = 1/2$, $s_y = -1/2$ be P and $1-P$ respectively. Then

$$P\left(\frac{1}{2}\right) + (1-P)\left(-\frac{1}{2}\right) = \frac{1}{2}\sin\omega t = \frac{1}{2}\cos\left(\frac{\pi}{2} - \omega t\right),$$

or

$$P = \frac{1}{2}\left[1 + \cos\left(\frac{\pi}{2} - \omega t\right)\right] = \cos^2\left(\frac{\omega t}{2} - \frac{\pi}{4}\right),$$

and hence $1 - P = \sin^2\left(\frac{\omega t}{2} - \frac{\pi}{4}\right)$. Lastly for (e) and (f), we have

$$P - \frac{1}{2} = 0, \quad \text{giving} \quad P = \frac{1}{2}, \ 1 - P = \frac{1}{2}.$$

3024

A particle with magnetic moment $\mu = \mu_0 \mathbf{s}$ and spin $\mathbf{s}$, with magnitude $1/2$, is placed in a constant magnetic field pointing along the x-axis. At $t = 0$, the particle is found to have $s_z = +1/2$. Find the probabilities at any later time of finding the particle with $s_y = \pm 1/2$.

(Columbia)

Sol: The Hamiltonian (spin part) of the system is

$$\hat{H} = -\mu_0 \hat{s} \cdot \mathbf{B} = -\frac{1}{2}\hat{\sigma}_x B\mu_0,$$

as $s = \frac{1}{2}\hbar\sigma_x$, being in the x direction. In the σ_x representation, the Schrödinger equation that the spin wave function $\begin{pmatrix} a_1 \\ a_2 \end{pmatrix}$ satisfies is

$$i\hbar\frac{d}{dt}\begin{pmatrix} a_1 \\ a_2 \end{pmatrix} + \frac{1}{2}\mu_0 B\begin{pmatrix} 0 & 1 \\ 1 & 0 \end{pmatrix}\begin{pmatrix} a_1 \\ a_2 \end{pmatrix} = 0,$$

or

$$\begin{cases} i\hbar\dfrac{d}{dt}a_1 + \dfrac{1}{2}\mu_0 Ba_2 = 0, \\[2mm] i\hbar\dfrac{d}{dt}a_2 + \dfrac{1}{2}\mu_0 Ba_1 = 0. \end{cases}$$

Elimination of a_1 or a_2 gives

$$\frac{d^2}{dt^2}a_{1,2} + \frac{\mu_0^2 B^2}{4\hbar^2}a_{1,2} = 0,$$

which have solutions

$$a_{1,2} = A_{1,2}e^{i\omega t} + B_{1,2}e^{-i\omega t},$$

where

$$\omega = \frac{\mu_0 B}{2h}$$

and $A_{1,2}$, $B_{1,2}$ are constants. As

$$\hat{s}_z = \frac{1}{2}\sigma_z = \frac{1}{2}\begin{pmatrix} 1 & 0 \\ 0 & -1 \end{pmatrix}$$

and

$$\hat{s}_z\begin{pmatrix} 1 \\ 0 \end{pmatrix} = \frac{1}{2}\begin{pmatrix} 1 \\ 0 \end{pmatrix},$$

the initial spin wave function is $\begin{pmatrix} 1 \\ 0 \end{pmatrix}$, i.e. $a_1(0) = 1$, $a_2(0) = 0$. The Schrödinger equation then gives

$$\frac{da_1(0)}{dt} = 0, \quad \frac{da_2(0)}{dt} = i\frac{\mu_0 B}{2\hbar} = i\omega .$$

These four initial conditions give

$$\begin{cases} A_1 + B_1 = 1, & A_2 + B_2 = 0, \\ \omega(A_1 - B_1) = 0, & \omega(A_2 - B_2) = \omega, \end{cases}$$

with the solution $A_1 = A_2 = B_1 = -B_2 = 1/2$. Substitution in the expressions for $a_{1,2}$ results in

$$\begin{pmatrix} a_1(t) \\ a_2(t) \end{pmatrix} = \begin{pmatrix} \cos \omega t \\ i \sin \omega t \end{pmatrix}.$$

As the eigenstate of $s_y = +1/2$ is

$$|s_y(+)\rangle = \frac{1}{\sqrt{2}} \begin{pmatrix} 1 \\ -i \end{pmatrix}$$

and that of $s_y = -1/2$ is

$$|s_y(+)\rangle = \frac{1}{\sqrt{2}} \begin{pmatrix} 1 \\ -i \end{pmatrix},$$

the probability of finding $s_y = +1/2$ is

$$P(+) = \left| \langle s_y(+) \, \psi(t) \rangle \right|^2$$

$$= \left| \frac{1}{\sqrt{2}} (1-i) \begin{pmatrix} \cos \omega t \\ i \sin \omega t \end{pmatrix} \right|^2$$

$$= \frac{1}{2} (1 + \sin 2\omega t).$$

Similarly the probability of finding $s_y = -1/2$ is

$$P(-) = |\langle s_y(-)|\psi(t)\rangle|^2 = \frac{1}{2}(1 - \sin 2\omega t).$$

3025

The Hamiltonian for a spin $-\frac{1}{2}$ particle with charge $+e$ in an external magnetic field is

$$H = -\frac{ge}{2mc} \mathbf{s} \cdot \mathbf{B}.$$

Calculate the operator ds/dt if $\mathbf{B} = B\hat{y}$. What is $s_z(t)$ in matrix form?

(Wisconsin)

Sol: In the Heisenberg picture,

$$\frac{d\mathbf{s}}{dt} = \frac{1}{i\hbar}[\mathbf{s}, H] = -\frac{ge}{i2m\hbar c}[\mathbf{s}, \mathbf{s} \cdot \mathbf{B}].$$

As

$$[\mathbf{s}, \mathbf{s} \cdot \mathbf{B}] = [s_x, \mathbf{s} \cdot \mathbf{B}]\hat{x} + [s_y, \mathbf{s} \cdot \mathbf{B}]\hat{y} + [s_z, \mathbf{s} \cdot \mathbf{B}]\hat{z}$$

and

$$[s_x, \mathbf{s} \cdot \mathbf{B}] = [s_x, s_z]B_x + [s_x, s_y]B_y + [s_x, s_z]B_z$$
$$= i\hbar(s_z B_y - s_y B_z)$$
$$= i\hbar(\mathbf{B} \times \mathbf{s})_x, \text{ etc}$$

we have

$$[\mathbf{s}, \mathbf{s} \cdot \mathbf{B}] = -i\hbar \mathbf{s} \times \mathbf{B},$$

and hence

$$\frac{d\mathbf{s}}{dt} = +\frac{ge}{2mc} \mathbf{s} \times \mathbf{B}.$$

If $\mathbf{B} = B\hat{y}$, the above gives

$$\frac{ds_x(t)}{dt} = -\frac{geB}{2mc} s_z(t),$$

$$\frac{ds_z(t)}{dt} = \frac{geB}{2mc} s_x(t),$$

and so

$$\frac{d^2 s_z(t)}{dt^2} + \left(\frac{geB}{2mc}\right)^2 s_z(t) = 0,$$

with the solution

$$s_z(t) = c_1 \cos(g\omega t) + c_2 \sin(g\omega t),$$

where $\omega = eB/2mc$. At $t = 0$ we have

$$s_z(0) = c_1, \quad s_z'(0) = c_2 g\omega = g\omega s_x(0),$$

and hence

$$s_z(t) = s_z(0) \cos\left(\frac{geB}{2mc} t\right) + s_x(0) \sin\left(\frac{geB}{2mc} t\right).$$

3026

Two electrons are tightly bound to different neighboring sites in a certain solid. They are, therefore, distinguishable particles which can be described in terms of their respective Pauli spin matrices $\sigma^{(1)}$ and $\sigma^{(2)}$. The Hamiltonian of these electrons takes the form

$$H = -J(\sigma_x^{(1)}\sigma_x^{(2)} + \sigma_y^{(1)}\sigma_y^{(2)}),$$

where *J* is a constant.

a. How many energy levels does the system have? What are their energies? What is the degeneracy of the different levels?

b. Now add a magnetic field in the *z* direction. What are the new energy levels? Draw an energy level diagram as a function of B_z

(Chicago)

Sol:

a. The Hamiltonian of the system is

$$H = -J[\sigma_x^{(1)}\sigma_x^{(2)} + \sigma_y^{(1)}\sigma_y^{(2)}]$$

$$= -J\left[\frac{(\sigma^{(1)}+\sigma^{(2)})^2 - \sigma^{(1)^2} - \sigma^{(2)^2}}{2}\right.$$

$$\left. - \frac{(\sigma_z^{(1)}+\sigma_z^{(2)})^2 - \sigma_z^{(1)^2} - \sigma_z^{(2)^2}}{2}\right]$$

$$= -J\left[\frac{(\sigma^{(1)}+\sigma^{(2)})^2 - 3 - 3}{2}\right.$$

$$\left. - \frac{(\sigma_z^{(1)}+\sigma_z^{(2)})^2 - 1 - 1}{2}\right]$$

$$= -J\left[\frac{(\sigma^{(1)}+\sigma^{(2)})^2 - (\sigma_z^{(1)}+\sigma_z^{(2)})^2}{2} - 2\right]$$

$$= -\frac{2J}{\hbar^2}(s^2 - s_z^2 - \hbar^2),$$

where

$$s = s^{(1)} + s^{(2)} = \frac{\hbar}{2}(\sigma^{(1)} + \sigma^{(2)})$$

is the total spin of the system and

$$s_z = s_z^{(1)} + s_z^{(2)} = \frac{\hbar}{2}(\sigma_z^{(1)} + \sigma_z^{(2)})$$

is its total *z*-component. s^2, s_z and *H* are commutable. Using the above and the coupling theory of angular momentum, we have (noting the eigen value of s^2 is $s(s+1)\hbar^2$)

s	s_z	number of states	energy
1	1	1	0
	0	2	$-2J$
	-1	1	0
0	0	2	$2J$

a. As seen from the table, the system has three energy levels $-2J, 0, 2J$, each with a degeneracy of 2. Note that if the electrons are indistinguishable, the second and fourth rows of the table would be different from the above.

b. In the presence of a magnetic field $|\mathbf{B}| = B_z$,

$$H = -J[\sigma_x^{(1)}\sigma_x^{(2)} + \sigma_y^{(1)}\sigma_y^{(2)}] - \mu \cdot \mathbf{B},$$

where

$$\mu = -\frac{e}{mc}s,$$

$-e$ and m being the electron charge and mass respectively. Thus

$$H = -J[\sigma_x^{(1)}\sigma_x^{(2)} + \sigma_y^{(1)}\sigma_y^{(2)}] + \frac{e}{mc}s_z B_z$$

$$= -2J\left[s(s+1) - \frac{s_z^2}{\hbar^2} - 1\right] + \frac{eB_z}{mc}s_z.$$

s^2, s_z and H are still commutable, so the new energy levels are: $2J$, $eB_z\hbar/mc$, $-eB_z\hbar/mc$, $-2J$. The energy level diagram is shown in Fig. 3.4 as a function of B_z (lines 1, 2, 3 and 4 for the above levels respectively).

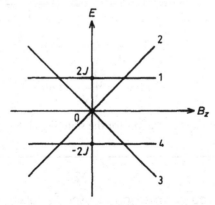

Fig. 3.4

3027

Consider a one electron atom in $n = 2$ state.

   a. Find out the possible spectroscopic states for this atom.

   b. What is the angle between $\vec{l}$ and $\vec{s}$ for the state $^2P_{3/2}$.

Sol:

   a. For $n = 2$ state, the orbital quantum number l can take values of 0 and 1.

       The value of spin quantum number S for electron is ½. So, the multiplicity is $2S + 1 = 2$.

       The possible values of the total quantum number j is given by

$$j = l \pm S$$
$$\text{For } l = 0, j = \frac{1}{2}$$

       and

$$\text{For } l = 1, j = \frac{3}{2}, \frac{1}{2}$$

       Hence, the possible spectroscopic states are

$$^2S_{1/2}, \, ^2P_{3/2} \text{ and } ^2P_{1/2}$$

   b. For the state $^2P_{3/2}$ we have,

$$l = 1, s = \frac{1}{2} \text{ and } j = \frac{3}{2}$$

       The angle between $\vec{l}$ and $\vec{s}$ is given by

$$\cos(\vec{l}.\vec{s}) = \frac{j(j+1) - l(l+1) - s(s+1)}{2\sqrt{l(l+1)}\sqrt{s(s+1)}}$$

$$= \frac{1}{\sqrt{6}}$$

$$\therefore \vec{l}.\vec{s} = \cos^{-1}\left(\frac{1}{\sqrt{6}}\right) = 65.9°$$

       The angle between $\vec{l}$ and $\vec{s}$ for the state $^2P_{3/2}$ is $65.9°$.

3028

A negatively charged π^- meason (a pseudoscalar particle: zero spin, odd parity) is initially bound in the lowest-energy Coulomb wave function around a deuteron. It is captured by the deuteron (a proton and a neutron in a $^3S_1$ state), which is converted into a pair of neutrons:

$$\pi^- + d \rightarrow n + n.$$

a. What is the orbital angular momentum of the neutron pair?

b. What is their total spin angular momentum?

c. What is the probability for finding both neutron spins directed opposite to the spin of the deuteron?

d. If the deuteron's spin is initially 100% polarized in the $\hat{\mathbf{R}}$ direction, what is the angular distribution of the neutron emission probability (per unit solid angle) for a neutron whose spin is opposite to that of the initial deuteron?

You may find some of the first few (not normalized) spherical harmonics useful:

$$Y_0^0 = 1, \qquad Y_1^{\pm 1} = \mp \sin\theta\, e^{\pm i\phi},$$
$$Y_1^0 = \cos\theta, \qquad Y_2^{\pm 1} = \mp \sin 2\theta\, e^{\pm i\phi}.$$

(CUS)

Sol:

a, b. Because of the conservation of parity in strong interactions, we have

$$p(\pi^-)p(d)(-1)^{L_1} = p(n)p(n)(-1)^{L_2},$$

where L_1, L_2 are the orbital angular momenta of $\pi^- + d$ and $n+n$ respectively. As the π^-, being captured, is in the lowest energy state of the Coulomb potential before the reaction, $L_1 = 0$. Since $p(\pi^-) = -1$, $p(d) = 1$, $p(n)p(n) = 1$, we have

$$(-1)^{L_2} = -1,$$

and so

$$L_2 = 2m+1, m = 0, 1, 2, \ldots$$

The deutron has $J = 1$ and π^- has zero spin, so that $J = 1$ before the reaction takes place. The conservation of angular momentum requires that after the reaction, $\mathbf{L}_2 + \mathbf{S} = \mathbf{J}$. The identity of n and n demands that the total wave function be antisymmetric. Then since the spacial wave function is antisymmetric, the spin wave function must be symmetric, i.e. $S = 1$ and so $L_2 = 2$, 1 or 0. As L_2 is odd, we must have $L_2 = 1, S = 1$.

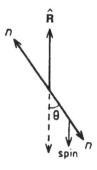

Fig. 3.5

The total orbital angular momentum and the total spin angular momentum are both $\sqrt{1(1+1)}\hbar = \sqrt{2}\hbar$.

c. Assume that the deuteron spin is in the direction $J_z = 1\hbar$ before the reaction. If both neutrons had spins in the reversed direction, we would have $S_z = -1\hbar$, $L_z = 2\hbar$, which is impossible since $L_z = 1$. Hence the probability is zero.

d. Take the z-axis along the $\hat{\mathbf{R}}$ direction. Then the initial state is $|J, J_z\rangle = |1, 1\rangle$. In the noncoupling representation, the state is $|L, L_z, S, S_z\rangle$, with $L = 1, S = 1$. Thus

$$|1, 1\rangle = \frac{\sqrt{2}}{2}|1, 0, 1, 1\rangle - \frac{\sqrt{2}}{2}|1, 1, 1, 0\rangle.$$

The state

$$|1, 1, 1, 0\rangle = Y_1^1(\theta, \phi)|1, 0\rangle = -\sqrt{3/8\pi}\,\sin\theta\, e^{i\phi}|1, 0\rangle$$

has $S_z = 0$ and so there must be one neutron with $s_z = -\hbar/2$. Hence the probability distribution required is

$$dP(\theta, \phi)/d\Omega = \frac{1}{2} \cdot \frac{3}{8\pi}\sin^2\theta = \frac{3}{16\pi}\sin^2\theta.$$

3029

An Ω^- hyperon (spin 3/2, mass 1672 MeV/c^2, intrinsic parity +) can decay via the weak interaction into a Λ hyperon (spin 1/2, mass 1116 MeV/c^2, intrinsic parity +) and a K^- meson (spin 0, mass 494 MeV/c^2, intrinsic parity –), i.e., $\Omega^- \rightarrow \Lambda + K^-$.

a. What is the most general form of the angular distribution of the K^- mesons relative to the spin direction of the Ω^- for the case when the Ω^- has a maximum possible component of angular momentum along the z-axis, i.e., the initial state $|\Omega_j^m\rangle = |\Omega_{3/2}^{+3/2}\rangle$. (Assume that the Ω^- is at rest).

b. What restrictions, if any, would be imposed on the form of the angular distribution if parity were conserved in the decay process?

(Berkeley)

Sol:

a. The initial state of the system is $|3/2, 3/2\rangle$, where the values are the orbital and spin momenta of the Ω^-. The spin part of the final state is $|1/2, s_z\rangle|0, 0\rangle = |1/2, s_z\rangle$, and the orbital part is $Y_{lm}(\theta, \varphi) = |l, m\rangle$. Thus the total final state of this system is

$$|l, m\rangle|1/2, s_z\rangle.$$

By conservation of angular momentum $l = 1, 2; m = 3/2 - s_z$. Thus the final state is a p wave if $l = 1$, the state being $|1, 1\rangle|1/2, 1/2\rangle$; a d wave if $l = 2$, the state being a combination of $|2, 2\rangle|1/2, -1/2\rangle$ and $|2, 1\rangle|1/2, 1/2\rangle$. Hence the wave functions are

$$\psi_p = Y_{11}(\theta, \varphi)\begin{pmatrix}1\\0\end{pmatrix},$$

$$\psi_d = \left[\sqrt{\frac{4}{5}}|2, 2\rangle\left|\frac{1}{2}, -\frac{1}{2}\right\rangle - \sqrt{\frac{1}{5}}|2, 1\rangle\left|\frac{1}{2}, -\frac{1}{2}\right\rangle\right]$$

$$= \begin{pmatrix}-\sqrt{\frac{1}{5}}Y_{21}(\theta, \varphi)\\-\sqrt{\frac{4}{5}}Y_{22}(\theta, \varphi)\end{pmatrix},$$

and

$$\psi = a_d\psi_d + a_p\psi_p$$

$$= \begin{pmatrix}a_pY_{11}(\theta, \varphi) - a_d\sqrt{\frac{1}{5}}Y_{21}(\theta, \varphi)\\\sqrt{\frac{4}{5}}a_dY_{22}(\theta, \varphi)\end{pmatrix}.$$

Therefore

$$\psi^* \psi = \frac{3}{8\pi} \sin^2\theta \, [|a_p|^2 + |a_d|^2 - 2Re a_p^* a_d \, \cos\theta],$$

i.e., the intensity of the emitted particles is

$$I \propto \sin^2\theta (1 + \alpha\cos\theta),$$

where

$$\alpha = -2Re \, a_p^* a_d \, / \, (|a_p|^2 + |a_d|^2).$$

This is the most general form of the angular distribution of the K^- mesons.

b. If parity were conserved in the decay process, the final state would have positive parity, i.e.

$$(-1)^l P_K P_\Lambda = +1.$$

Since

$$P_K P_\Lambda = (-1)(+1) = -1,$$

we get $l = 1$. It would follow that

$$\psi_f = Y_{11}(\theta, \varphi) \begin{pmatrix} 1 \\ 0 \end{pmatrix},$$

and

$$\psi_f^* \psi_f = \frac{3}{8\pi} \sin^2\theta = \frac{3}{8\pi}(1 - \cos^2\theta).$$

Hence parity conservation would impose an angular distribution of the form

$$I \propto (1 - \cos^2\theta).$$

3030

Given two angular momenta $\mathbf{J}_1$ and $\mathbf{J}_2$ (for example, $\mathbf{L}$ and $\mathbf{S}$) and the corresponding wave functions, where $j_1 = 1$ and $j_2 = 1/2$. Compute the Clebsch–Gordan coefficients for the states with $\mathbf{J} = \mathbf{J}_1 + \mathbf{J}_2$, $m = m_1 + m_2$, where:

a. $j = 3/2, \, m = 3/2,$

b. $j = 3/2, \, m = 1/2.$

Consider the reactions

$$K^- p \to \Sigma^- \pi^+,$$
$$\to \Sigma^+ \pi^-,$$
$$\to \Sigma^0 \pi^0,$$
$$K^- n \to \Sigma^- \pi^0,$$
$$\to \Sigma^0 \pi^-.$$

Assume they proceed through a resonance and hence a pure *I*-spin state. Find the relative rates based on *I*-spin conservation:

  c. for an $I=1$ resonance state,

  d. for an $I=0$ resonance state.

Use the Clebsch–Gordan cofficients supplied. The *I*-spins for $K, n, \Sigma,$ and π are 1/2, 1/2, 1, and 1 respectively.

(Berkeley)

Sol:

  a. As $j_1 = 1$, $j_2 = \frac{1}{2}$, we have

$$|3/2, 3/2\rangle = |1, 1\rangle |1/2, 1/2\rangle.$$

  b. Defining the operator $J_- = J_{1-} + J_{2-}$, we have

$$J_-|3/2, 3/2\rangle = (J_{1-} + J_{2-})|1, 1\rangle |1/2, 1/2\rangle,$$

or, using the properties of J_- (**Problem 3008**),

$$\hbar\sqrt{3}|3/2, 1/2\rangle = \hbar\sqrt{2}|1, 0\rangle |1/2, 1/2\rangle + \hbar|1, 1\rangle |1/2, -1/2\rangle,$$

and hence

$$|3/2, 1/2\rangle = \sqrt{2/3}|1, 0\rangle |1/2, 1/2\rangle + \sqrt{1/3}|1, 1\rangle |1/2, -1/2\rangle.$$

To calculate the relative reaction cross sections, we use the coupling representation to describe the initial and final *I*-spin states:

$$|K^- p\rangle = |1/2, -1/2\rangle |1/2, 1/2\rangle = \sqrt{1/2}|1, 0\rangle - \sqrt{1/2}|0, 0\rangle,$$
$$|\Sigma^- \pi^+\rangle = |1, -1\rangle |1, 1\rangle = \sqrt{1/6}|2, 0\rangle - \sqrt{1/2}|1, 0\rangle + \sqrt{1/3}|0, 0\rangle,$$
$$|\Sigma^+ \pi^-\rangle = |1, 1\rangle |1, -1\rangle = \sqrt{1/6}|2, 0\rangle + \sqrt{1/2}|1, 0\rangle + \sqrt{1/3}|0, 0\rangle,$$
$$|\Sigma^0 \pi^0\rangle = |1, 0\rangle |1, 0\rangle = \sqrt{2/3}|2, 0\rangle - \sqrt{1/3}|0, 0\rangle,$$
$$|K^- n\rangle = |1/2, -1/2\rangle |1/2, -1/2\rangle = |1, -1\rangle,$$

$$\left|\Sigma^{-}\pi^{0}\right\rangle = \left|1,-1\right\rangle\left|1,0\right\rangle = \sqrt{1/2}\left|2,-1\right\rangle - \sqrt{1/2}\left|1,-1\right\rangle,$$

$$\left|\Sigma^{0}\pi^{-}\right\rangle = \left|1,0\right\rangle\left|1,-1\right\rangle = \sqrt{1/2}\left|2,-1\right\rangle + \sqrt{1/2}\left|1,-1\right\rangle.$$

To $K^{-}p$ reactions going through the resonance state $I=1$, the final states $\left|\Sigma^{-}\pi^{+}\right\rangle$ contributes $\sqrt{\frac{1}{2}}\left|1,0\right\rangle$, $\left|\Sigma^{+}\pi^{-}\right\rangle$ contributes $\sqrt{\frac{1}{2}}\left|1,0\right\rangle$, while $\left|\Sigma^{0}\pi^{0}\right\rangle$ does not contribute. Hence

$$\sigma(\Sigma^{-}\pi^{0}):\sigma(\Sigma^{+}\pi^{-}):\sigma(\Sigma^{0}\pi^{0})=1:1:0.$$

Similarly for $K^{-}n$ reactions we have

$$\sigma(\Sigma^{-}\pi^{0}):\sigma(\Sigma^{0}\pi^{-})=1:1.$$

Only the $K^{-}p$ reactions go through the $I=0$ resonance state. A similar consideration gives the following reaction cross section ratios:

$$\sigma(\Sigma^{-}\pi^{+}):\sigma(\Sigma^{+}\pi^{-}):\sigma(\Sigma^{0}\pi^{0})=1:1:1.$$

3031

a. Compute the Clebsch-Gordan coefficients for the states with $\mathbf{J}=\mathbf{J}_{1}+\mathbf{J}_{2}$, $M=m_{1}+m_{2}$, where $j_{1}=1$ and $j_{2}=1/2$, and $j=3/2$, $M=1/2$ for the various possible m_{1} and m_{2} values.

b. Consider the reactions:

$$\pi^{+}p \rightarrow \pi^{+}p \tag{i}$$

$$\pi^{-}p \rightarrow \pi^{-}p \tag{ii}$$

$$\pi^{-}p \rightarrow \pi^{0}n \tag{iii}$$

These reactions, which conserve isospin, can occur in the isospin $I=3/2$ state (Δ resonance) or the $I=1/2$ state (N^{*} resonance). Calculate the ratios of these cross-sections, $\sigma_{i}:\sigma_{ii}:\sigma_{iii}$, for an energy corresponding to a Δ resonance and an N^{*} resonance respectively. At a resonance energy you can neglect the effect due to the other isospin states. Note that the pion is an isospin $I=1$ state and the nucleon an isospin $I=1/2$ state.

(Berkeley)

Sol:

a. As $M=m_{1}+m_{2}=1/2$, (m_{1}, m_{2}) can only be $(1,-1/2)$ or $(0,1/2)$. Consider

$$\left|3/2,3/2\right\rangle = \left|1,1\right\rangle\left|1/2,1/2\right\rangle.$$

As

$$M_{-}|3/2, 3/2\rangle = \sqrt{3}|3/2, 1/2\rangle,$$

and

$$M_{-}|3/2, 3/2\rangle = (M_{1_{-}} + M_{2_{-}})|1, 1\rangle|1/2, 1/2\rangle$$
$$= \sqrt{2}|1, 0\rangle|1/2, 1/2\rangle + |1, 1\rangle|1/2, -1/2\rangle,$$

we have

$$\langle 1, 1, 1/2, -1/2|3/2, 1/2\rangle = 1/\sqrt{3},$$
$$\langle 1, 0, 1/2, 1/2|3/2, 1/2\rangle = \sqrt{2/3}.$$

b. As

$$\pi^{+} = |1, 1,\rangle, \ \pi^{0} = |1, 0\rangle, \ \pi^{-} = |1, -1\rangle,$$
$$p = |1/2, 1/2\rangle, n = |1/2, -1/2\rangle,$$

we have

$$|\pi^{+}p\rangle = |1, 1\rangle|1/2, 1/2\rangle = |3/2, 3/2\rangle,$$
$$|\pi^{-}p\rangle = |1, -1\rangle|1/2, 1/2\rangle = a|3/2, -1/2\rangle + b|1/2, -1/2\rangle,$$
$$|\pi^{0}n\rangle = |1, 0\rangle|1/2, -1/2\rangle = c|3/2, -1/2\rangle + d|1/2, -1/2\rangle.$$

From a table of Clebsch-Gordan coefficients, we find $a = \sqrt{1/3}$, $b = -\sqrt{2/3}$, $c = \sqrt{2/3}$, $d = \sqrt{1/3}$. For the Δ resonance state, $I = 3/2$ and the ratios of the cross sections are

$$\sigma_{\mathrm{i}} : \sigma_{\mathrm{ii}} : \sigma_{\mathrm{iii}} = 1 : |a|^{4} : |ac|^{2} = 1 : \frac{1}{9} : \frac{2}{9}.$$

For the N^{*} resonance state, $I = 1/2$, and the ratios are

$$\sigma_{\mathrm{i}} : \sigma_{\mathrm{ii}} : \sigma_{\mathrm{iii}} = 0 : |b|^{4} : |bd|^{2} = 0 : \frac{4}{9} : \frac{2}{9}.$$

3032

Consider an electron in a uniform magnetic field along the z direction. Let the result of a measurement be that the electron spin is along the positive y direction at $t = 0$. Find the Schrödinger state vector for the spin, and the average polarization (expectation value of s_{x}) along the x direction for $t > 0$.

(Wisconsin)

Sol:

As we are only interested in the spin state and the magnetic field is uniform in space, we can leave out the space part of the wave function. Then the Hamiltonian can be taken to be

$$H = -\boldsymbol{\mu} \cdot \mathbf{B} = \mu_e \boldsymbol{\sigma} \cdot \mathbf{B} = \hbar \omega \sigma_z,$$

where $\boldsymbol{\mu} = -\mu_e \boldsymbol{\sigma}$, $\omega = \mu_e B / \hbar = eB / 2mc$, μ_e being the Bohr magneton $\dfrac{e\hbar}{2mc}$. As the electron is initially along the y direction, the initial spin wave function is

$$\psi(t = 0) = \frac{1}{\sqrt{2}} \begin{pmatrix} 1 \\ i \end{pmatrix}.$$

Let the spin wave function at a later time t be $\begin{pmatrix} \alpha \\ \beta \end{pmatrix}$. The Schrödinger equation

$$i\hbar \, d\psi/dt = H\psi$$

then gives

$$i\hbar \begin{pmatrix} \dot{\alpha} \\ \dot{\beta} \end{pmatrix} = \hbar \omega \begin{pmatrix} 1 & 0 \\ 0 & -1 \end{pmatrix} \begin{pmatrix} \alpha \\ \beta \end{pmatrix},$$

or

$$\dot{\alpha} = -i\omega\alpha, \quad \dot{\beta} = i\omega\beta,$$

with the solution

$$\psi(t) = \begin{pmatrix} \alpha(t) \\ \beta(t) \end{pmatrix} = \begin{pmatrix} e^{-i\omega t}\alpha_0 \\ e^{i\omega t}\beta_0 \end{pmatrix} = \frac{1}{\sqrt{2}} \begin{pmatrix} e^{-i\omega t} \\ ie^{-i\omega t} \end{pmatrix}.$$

Hence

$$\begin{aligned} \langle S_x \rangle &= \langle \psi(t) | \hat{s}_x | \psi(t) \rangle \\ &= \frac{\hbar}{2} \langle \psi | \sigma_x | \psi \rangle \\ &= \frac{\hbar}{2} \cdot \frac{1}{2} (e^{i\omega t} - ie^{-i\omega t}) \begin{pmatrix} 1 & 0 \\ 0 & -1 \end{pmatrix} \begin{pmatrix} e^{-i\omega t} \\ ie^{-i\omega t} \end{pmatrix} \\ &= \frac{i\hbar}{4} (e^{2i\omega t} - e^{-2i\omega t}) \\ &= -\frac{\hbar}{2} \sin(2\omega t). \end{aligned}$$

3033

Consider an electron in a uniform magnetic field pointing along the z direction. The electron spin is measured (at time t_0) to be pointing along the positive y-axis. What is the polarization along the x and z directions (i.e. the expectation values of $2s_x$ and $2s_z$) for $t > t_0$?

(Wisconsin)

Sol: The Schrödinger equation for the spin state vector is

$$i\hbar \frac{\partial}{\partial t}\begin{pmatrix} a(t) \\ b(t) \end{pmatrix} = -\mu \cdot \mathbf{B}\begin{pmatrix} a(t) \\ b(t) \end{pmatrix} = \mu_e \sigma \cdot \mathbf{B}\begin{pmatrix} a(t) \\ b(t) \end{pmatrix},$$

where $\mu_e = |e|\hbar/2m_e c$ is the magnitude of the magnetic moment of an electron. As $\mathbf{B}$ is along the z direction, the above becomes

$$i\hbar \partial_t \begin{pmatrix} a(t) \\ b(t) \end{pmatrix} = \mu_e B\begin{pmatrix} 1 & 0 \\ 0 & -1 \end{pmatrix}\begin{pmatrix} a(t) \\ b(t) \end{pmatrix},$$

or

$$\begin{cases} i\hbar \partial_t a(t) = \mu_e B a(t), \\ i\hbar \partial_t b(t) = -\mu_e B b(t). \end{cases}$$

The solutions are

$$a(t) = a(t_0)e^{-\frac{i}{\hbar}\mu_e B(t-t_0)},$$

$$b(t) = b(t_0)e^{\frac{i}{\hbar}\mu_e B(t-t_0)}.$$

At time t_0, the electron spin is in the positive y direction. Thus

$$s_y \begin{pmatrix} a(t_0) \\ b(t_0) \end{pmatrix} = \frac{\hbar}{2}\begin{pmatrix} 0 & -i \\ i & 0 \end{pmatrix}\begin{pmatrix} a(t_0) \\ b(t_0) \end{pmatrix} = \frac{\hbar}{2}\begin{pmatrix} a(t_0) \\ b(t_0) \end{pmatrix},$$

or

$$\begin{cases} -ib(t_0) = a(t_0), \\ ia(t_0) = b(t_0). \end{cases}$$

The normalization condition

$$|a(t_0)|^2 + |b(t_0)|^2 = 1,$$

then gives

$$|a(t_0)|^2 = |b(t_0)|^2 = \frac{1}{2}.$$

As $\dfrac{b(t_0)}{a(t_0)} = i$, we can take

$$a(t_0) = 1/\sqrt{2}, \quad b(t_0) = i/\sqrt{2}.$$

Hence for time $t > t_0$, the polarizations along x and z directions are respectively

$$\langle 2s_x \rangle = \hbar(a^*, b^*)\begin{pmatrix} 0 & 1 \\ 1 & 0 \end{pmatrix}\begin{pmatrix} a \\ b \end{pmatrix}$$

$$= \hbar(a^*b + b^*a) = -\hbar\sin\left[\frac{2\mu_e}{\hbar}B(t - t_0)\right];$$

$$\langle 2s_z \rangle = \hbar(a^*, b^*)\begin{pmatrix} 0 & 1 \\ 1 & 0 \end{pmatrix}\begin{pmatrix} a \\ b \end{pmatrix}$$

$$= \hbar(a^*a - b^*b) = 0.$$

3034

Two spin $-\dfrac{1}{2}$ particles form a composite system. Spin A is in the eigenstate $S_z = +1/2$ and spin B in the eigenstate $S_x = +1/2$. What is the probability that a measurement of the total spin will give the value zero?

(CUS)

Sol:

In the uncoupling representation, the state in which the total spin is zero can be written as

$$|0\rangle = \frac{1}{\sqrt{2}}\left(\left|S_{Az} = \frac{1}{2}\right\rangle\left|S_{Bz} = -\frac{1}{2}\right\rangle - \left|S_{Az} = -\frac{1}{2}\right\rangle\left|S_{Bz} = \frac{1}{2}\right\rangle\right),$$

where S_{Az} and S_{Bz} denote the z-components of the spins of A and B respectively. As these two spin $-\frac{1}{2}$ particles are now in the state

$$|Q\rangle = |S_{Az} = +1/2\rangle|S_{Bx} = +1/2\rangle,$$

the probability of finding the total spin to be zero is

$$P = |\langle 0|Q\rangle|^2.$$

In the representation of $\hat{\mathbf{S}}^2$ and $\hat{S}_z$, the spin angular momentum operator $\hat{S}_x$ is defined as

$$\hat{S}_x = \frac{\hbar}{2}\sigma_x = \frac{\hbar}{2}\begin{pmatrix} 0 & 1 \\ 1 & 0 \end{pmatrix}.$$

Solving the eigenequation of $\hat{S}_x$, we find that its eigenfunction $|S_x=+1/2\rangle$ can be expressed in the representation of $\mathbf{S}^2$ and S_z as

$$|S_x=+1/2\rangle=\frac{1}{\sqrt{2}}(|S_z=1/2\rangle+|S_z=-1/2\rangle).$$

Thus

$$\left\langle S_z=-\frac{1}{2}\middle|S_x=+\frac{1}{2}\right\rangle=\frac{1}{\sqrt{2}}\left(\left\langle S_z=-\frac{1}{2}\middle|S_z=\frac{1}{2}\right\rangle\right.$$

$$\left.+\left\langle S_z=-\frac{1}{2}\middle|S_z=-\frac{1}{2}\right\rangle\right)=\frac{1}{\sqrt{2}},$$

and hence

$$\langle 0|Q\rangle=\frac{1}{\sqrt{2}}\left(\left\langle S_{Az}=\frac{1}{2}\middle|S_{Az}=+\frac{1}{2}\right\rangle\left\langle S_{Bz}=-\frac{1}{2}\middle|S_{Bx}=+\frac{1}{2}\right\rangle\right.$$

$$\left.-\left\langle S_{Az}=-\frac{1}{2}\middle|S_{Az}=+\frac{1}{2}\right\rangle\left\langle S_{Bz}=\frac{1}{2}\middle|S_{Bx}=+\frac{1}{2}\right\rangle\right)$$

$$=\frac{1}{\sqrt{2}}\left\langle S_{Bz}=-\frac{1}{2}\middle|S_{Bx}=+\frac{1}{2}\right\rangle=\frac{1}{2}.$$

Therefore

$$P=|\langle 0|Q\rangle|^2=\frac{1}{4}=25\%.$$

3035

a. An electron has been observed to have its spin in the direction of the z-axis of a rectangular coordinate system. What is the probability that a second observation will show the spin to be directed in $x-z$ plane at an angle θ with respect to the z-axis?

b. The total spin of the neutron and proton in a deuteron is a triplet state. The resultant spin has been observed to be parallel to the z-axis of a rectangular coordinate system. What is the probability that a second observation will show the proton spin to be parallel to the z-axis?

(Berkeley)

Sol:

a. The initial spin state of the electron is

$$|\psi_0\rangle = \begin{pmatrix} 1 \\ 0 \end{pmatrix}.$$

The state whose spin is directed in $x-z$ plane at an angle θ with respect to the z-axis is

$$|\psi\rangle = \begin{pmatrix} \cos\dfrac{\theta}{2} \\ \sin\dfrac{\theta}{2} \end{pmatrix}.$$

Thus the probability that a second observation will show the spin to be directed in $x-z$ plane at an angle θ with respect to the z-axis is

$$P(\theta) = |\langle\psi|\psi_0\rangle|^2 = \left|\left(\cos\frac{\theta}{2}, \sin\frac{\theta}{2}\right)\begin{pmatrix} 1 \\ 0 \end{pmatrix}\right|^2$$

$$= \cos^2\left(\frac{\theta}{2}\right).$$

b. The initial spin state of the neutron-proton system is

$$|\psi_0\rangle = |1,1\rangle = |1/2, 1/2\rangle_n |1/2, 1/2\rangle_p.$$

Suppose a second observation shows the proton spin to be parallel to the z-axis. Since the neutron spin is parallel to the proton spin in the deuteron, the final state remains, as before,

$$|\psi_f\rangle = |1/2, 1/2\rangle_n |1/2, 1/2\rangle_p.$$

Hence the probability that a second observation will show the proton spin to be parallel to the z-axis is 1.

3036

The deuteron is a bound state of a proton and a neutron of total angular momentum $J=1$. It is known to be principally an $S(L=0)$ state with a small admixture of a $D(L=2)$ state.

a. Explain why a P state cannot contribute.
b. Explain why a G state cannot contribute.
c. Calculate the magnetic moment of the pure D state $n-p$ system with $J=1$. Assume that the n and p spins are to be coupled to make the total

spin S which is then coupled to the orbital angular momentum L to give the total angular momentum J. Express your result in nuclear magnetons. The proton and neutron magnetic moments are 2.79 and −1.91 nuclear magnetons respectively.

(CUS)

Sol:

a. The parities of the S and D states are positive, while the parity of the P state is negative. Because of the conservation of parity in strong interaction, a quantum state that is initially an S state cannot have a P state component at any later moment.

b. The possible spin values for a system composed of a proton and a neutron are 1 and 0. We are given $\mathbf{J}=\mathbf{L}+\mathbf{S}$ and $J=1$. If $S=0$, $L=1$, the system would be in a P state, which must be excluded as we have seen in (a). The allowed values are then $S=1$, $L=2,1,0$. Therefore a G state $(L=4)$ cannot contribute.

c. The total spin is $\mathbf{S}=\mathbf{s}_p+\mathbf{s}_n$. For a pure D state with $J=1$, the orbital angular momentum (relative to the center of mass of the n and p) is $L=2$ and the total spin must be $S=1$. The total magnetic moment arises from the coupling of the magnetic moment of the total spin, μ, with that of the orbital angular momentum, μ_L, where $\mu=\mu_p+\mu_n$, μ_p, μ_n being the spin magnetic moments of p and n respectively.

The average value of the component of μ in the direction of the total spin $\mathbf{S}$ is

$$\mu_s = \frac{\left(g_p\mu_N\mathbf{s}_p+g_n\mu_N\mathbf{s}_n\right)\cdot\mathbf{S}}{\mathbf{S}^2}\mathbf{S} = \frac{1}{2}(g_p+g_n)\mu_N\mathbf{S},$$

where

$$\mu_N = \frac{e\hbar}{2m_pc}, \quad g_p=5.58, \quad g_n=-3.82,$$

as $\mathbf{s}_p=\mathbf{s}_n=\frac{1}{2}\mathbf{S}$.

The motion of the proton relative to the center of mass gives rise to a magnetic moment, while the motion of the neutron does not as it is uncharged. Thus

$$\mu_L = \mu_N\mathbf{L}_p,$$

where $\mathbf{L}_p$ is the angular momentum of the proton relative to the center of mass. As $\mathbf{L}_p+\mathbf{L}_n=\mathbf{L}$ and we may assume $\mathbf{L}_p=\mathbf{L}_n$, we have $\mathbf{L}_p=\mathbf{L}/2$ (the center of mass is at the mid-point of the connecting line, taking $m_p \approx m_n$). Consequently, $\mu_L = \mu_N \mathbf{L} / 2$.

The total coupled magnetic moment along the direction of $\mathbf{J}$ is then

$$\mu_T = \frac{\left[\frac{1}{2}\mu_N \mathbf{L} \cdot \mathbf{J} + \frac{1}{2}(g_p + g_n)\mu_N \mathbf{S} \cdot \mathbf{J}\right]\mathbf{J}}{J(J+1)}.$$

Since $\mathbf{J} = \mathbf{L} + \mathbf{S}$, $\mathbf{S} \cdot \mathbf{L} = \frac{1}{2}\left(J^2 - L^2 - S^2\right)$. With $J = 1$, $L = 2$, $S = 1$ and so $J^2 = 2$, $L^2 = 6$, $S^2 = 2$, we have $\mathbf{S} \cdot \mathbf{L} = -3$ and thus $\mathbf{L} \cdot \mathbf{J} = 3$, $\mathbf{S} \cdot \mathbf{J} = -1$. Hence

$$\mu_T = \left[\frac{1}{2}\mu_N \cdot 3 + \frac{1}{2}(g_p + g_n)\mu_N(-1)\right]\mathbf{J} / 2$$

$$= \left[1.5 - \frac{1}{2}(g_p + g_n)\right]\frac{1}{2}\mu_N \mathbf{J} = 0.31\mu_N \mathbf{J}.$$

Taking the direction of $\mathbf{J}$ as the z-axis and letting J_z take the maximum value $J_z = 1$, we have $\mu_T = 0.31\mu_N$.

3037

A preparatory Stern-Gerlach experiment has established that the z-component of the spin of an electron is $-\hbar / 2$. A uniform magnetic field in the x-direction of magnitude B (use cgs units) is then switched on at time $t = 0$.

a. Predict the result of a single measurement of the z component of the spin after elapse of time T.

b. If, instead of measuring the z-component of the spin, the x-component is measured, predict the result of such a single measurement after elapse of time T.

(Berkeley)

Sol:

Method 1

The spin wave function $\begin{pmatrix} a \\ b \end{pmatrix}$ satisfies

$$i\hbar\partial_t\begin{pmatrix} a \\ b \end{pmatrix} = \frac{e\hbar B}{2mc}\sigma_x\begin{pmatrix} a \\ b \end{pmatrix} = \hbar\omega\begin{pmatrix} 0 & 1 \\ 1 & 0 \end{pmatrix}\begin{pmatrix} a \\ b \end{pmatrix} = \hbar\omega\begin{pmatrix} a \\ b \end{pmatrix},$$

where $\omega = eB / 2mc$, or

$$\begin{cases} i\dot{a} = \omega b, \\ i\dot{b} = \omega a. \end{cases}$$

Thus

$$\ddot{a} = -i\omega\dot{b} = -\omega^2 a,$$

the solution being

$$a = A e^{i\omega t} + C e^{-i\omega t},$$

$$b = \frac{i}{\omega}\dot{a} = -(A e^{i\omega t} - C e^{-i\omega t}),$$

where A and C are arbitrary constants. From the initial condition $a(0)=0$ and $b(0)=1$ as the initial spin is in the $-z$ direction, we get $A=-1/2$, $C=1/2$. Hence

$$\begin{cases} a = \dfrac{1}{2}(e^{-i\omega t} - e^{i\omega t}) = -i\sin\omega t, \\ b = \dfrac{1}{2}(e^{i\omega t} + e^{-i\omega t}) = \cos\omega t, \end{cases}$$

and the wave function at time t is

$$\psi(t) = \begin{pmatrix} -i\sin\omega t \\ \cos\omega t \end{pmatrix}$$

a. At $t = T$,

$$\psi(T) = \begin{pmatrix} -i\sin\omega T \\ \cos\omega t \end{pmatrix} = -i\sin\omega T \begin{pmatrix} 1 \\ 0 \end{pmatrix} + \cos\omega T \begin{pmatrix} 0 \\ 1 \end{pmatrix}.$$

As $\begin{pmatrix} 0 \\ 1 \end{pmatrix}$ and $\begin{pmatrix} 0 \\ 1 \end{pmatrix}$ are the eigenvectors for σ_z with eigenvalues $+1$ and -1 respectively the probability that the measured z-component of the spin is positive is $\sin^2\omega T$; the probability that it is negative is $\cos^2\omega T$.

b. In the diagonal representation of σ_z, the eigenvectors of σ_x are

$$\psi(\sigma_x = 1) = \frac{1}{\sqrt{2}}\begin{pmatrix} 1 \\ 1 \end{pmatrix}, \quad \psi(\sigma_x = -1) = \frac{1}{\sqrt{2}}\begin{pmatrix} -1 \\ 1 \end{pmatrix}.$$

As we can write

$$\psi(T) = \begin{pmatrix} -i\sin\omega T \\ \cos\omega T \end{pmatrix} = \frac{1}{\sqrt{2}} e^{-i\omega T} \cdot \frac{1}{\sqrt{2}}\begin{pmatrix} 1 \\ 1 \end{pmatrix} + \frac{1}{\sqrt{2}} e^{i\omega T} \cdot \frac{1}{\sqrt{2}}\begin{pmatrix} -1 \\ 1 \end{pmatrix}$$

$$= \frac{1}{\sqrt{2}} e^{-i\omega T} \psi(\sigma_x = 1) + \frac{1}{\sqrt{2}} e^{i\omega T} \psi(\sigma_x = -1),$$

the probabilities that the measured x-component of the spin is positive and is negative are equal, being

$$\left|\frac{e^{-i\omega T}}{\sqrt{2}}\right|^2 = \left|\frac{e^{i\omega T}}{\sqrt{2}}\right|^2 = \frac{1}{2}.$$

Method 2

The Hamiltonian for spin energy is

$$H = -\mu \cdot B = eB\hbar\sigma_x/2mc.$$

The eigenstates of σ_x are

$$\frac{1}{\sqrt{2}}\binom{1}{1}, \quad \frac{1}{\sqrt{2}}\binom{1}{-1}.$$

We can write the initial wave function as

$$\psi(t=0) = \binom{0}{1} = \frac{1}{\sqrt{2}}\left[\frac{1}{\sqrt{2}}\binom{1}{1} - \frac{1}{\sqrt{2}}\binom{1}{-1}\right].$$

The Hamiltonian then gives

$$\psi(t) = \frac{1}{2}\binom{1}{1}e^{-i\omega t} - \frac{1}{2}\binom{1}{-1}e^{i\omega t}, \quad \omega = \frac{eB}{2mc}.$$

a. As

$$\psi(t) = \binom{-i\sin\omega t}{\cos\omega t},$$

the probabilities at $t = T$ are

$$P_{z\uparrow} = \sin^2\omega T, \quad P_{z\downarrow} = \cos^2\omega T.$$

b. As in method 1 above,

$$\begin{cases} P_{x\uparrow} = \left|\frac{1}{\sqrt{2}}e^{-i\omega t}\right|^2 = \frac{1}{2}, \\ P_{x\downarrow} = \frac{1}{2}. \end{cases}$$

3038

An alkali atom in its ground state passes through a Stern–Gerlach apparatus adjusted so as to transmit atoms that have their spins in the $+z$ direction. The atom then spends time τ in a magnetic field H in the x direction. At the end of this time what is the probability that the atom would pass through a

Stern–Gerlach selector for spins in the $-z$ direction? Can this probability be made equal to unity? if so, how?

Sol: The Hamiltonian

$$\hat{H}=-\mu \cdot \mathbf{H}=\frac{|e|\hbar H}{2mc}\sigma_x \equiv \hbar\omega\sigma_x, \quad \omega=\frac{|e|H}{2mc},$$

gives the equation of motion

$$-\frac{\hbar}{i}\frac{d}{dt}\begin{pmatrix} \psi_1 \\ \psi_2 \end{pmatrix}=\hbar\omega\begin{pmatrix} 0 & 1 \\ 1 & 0 \end{pmatrix}\begin{pmatrix} \psi_1 \\ \psi_2 \end{pmatrix}=\hbar\omega\begin{pmatrix} \psi_2 \\ \psi_1 \end{pmatrix},$$

or

$$\begin{cases} i\dot{\psi}_1=\omega\psi_2, \\ i\dot{\psi}_2=\omega\psi_1. \end{cases}$$

and hence

$$\ddot{\psi}_1+\omega^2\psi_1=0.$$

The solution is $\psi=\begin{pmatrix} \psi_1 \\ \psi_2 \end{pmatrix}$, with

$$\begin{cases} \psi_1(t)=ae^{i\omega t}+be^{-i\omega t}, \\ \psi_2(t)=\dfrac{i}{\omega}\dot{\psi}_1=-ae^{i\omega t}+be^{-i\omega t}. \end{cases}$$

The initial condition

$$\psi(t=0)=\begin{pmatrix} 1 \\ 0 \end{pmatrix}$$

then gives $a=b=1/2$. Thus

$$\psi=\begin{pmatrix} \psi_1 \\ \psi_2 \end{pmatrix}=\frac{1}{2}\begin{pmatrix} e^{i\omega t}+e^{-ivt} \\ -e^{i\omega t}+e^{-i\omega t} \end{pmatrix}=\begin{pmatrix} \cos\omega t \\ -i\sin\omega t \end{pmatrix}$$

$$=\cos\omega t\begin{pmatrix} 1 \\ 0 \end{pmatrix}-i\sin\omega t\begin{pmatrix} 0 \\ 1 \end{pmatrix}.$$

In the above $\begin{pmatrix} 0 \\ 1 \end{pmatrix}$ is the eigenvector σ_z for eigenvalue -1. Hence the probability that the spins are in the $-z$ direction at time τ after the atom passes through the Stern–Gerlach selector is

$$\left| (0 \quad 1) \begin{pmatrix} \cos\omega\tau \\ -i\sin\omega\tau \end{pmatrix} \right|^2 = \sin^2\omega\tau = \frac{1-\cos 2\omega\tau}{2}.$$

The probability equals 1, if

$$1-\cos 2\omega\tau = 2,$$

or

$$\cos 2\omega\tau = -1,$$

i.e. at time

$$\tau = \frac{(2n+1)\pi}{2\omega} = (2n+1)\frac{mc\pi}{|e|H}.$$

Hence the probability will become unity at times $\tau = (2n+1)mc\pi/|e|H$.

3039

A beam of particles of spin 1/2 is sent through a Stern–Gerlach apparatus, which divides the incident beam into two spatially separated components depending on the quantum number m of the particles. One of the resulting beams is removed and the other beam is sent through another similar apparatus, the magnetic field of which has an inclination α with respect to that of the first apparatus (see Fig. 3.6). What are the relative numbers of particles that appear in the two beams leaving the second apparatus? Derive the result using the Pauli spin formalism.

(Berkeley)

Sol:

For an arbitrary direction in space $n = (\sin\theta\cos\varphi, \sin\theta\sin\varphi, \cos\theta)$ the spin operator is

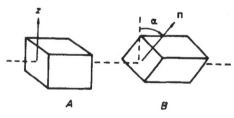

Fig. 3.6

$$\sigma \cdot \mathbf{n} = \sigma_x \sin\theta\cos\varphi + \sigma_y \sin\theta\sin\varphi + \sigma_z \cos\theta$$

$$= \begin{pmatrix} \cos\theta & \sin\theta e^{-i\varphi} \\ \sin\theta e^{i\varphi} & -\cos\theta \end{pmatrix},$$

where

$$\sigma_x = \begin{pmatrix} 0 & 1 \\ 1 & 0 \end{pmatrix}, \quad \sigma_y = \begin{pmatrix} 0 & -i \\ i & 0 \end{pmatrix}, \quad \sigma_z = \begin{pmatrix} 1 & 0 \\ 0 & -1 \end{pmatrix},$$

are Pauli's spin matrices. Let its eigenfunction and eigenvalue be $\begin{pmatrix} a \\ b \end{pmatrix}$ and λ respectively. Then

$$\sigma \cdot \mathbf{n} \begin{pmatrix} a \\ b \end{pmatrix} = \lambda \begin{pmatrix} a \\ b \end{pmatrix},$$

or

$$a(\cos\theta - \lambda) + be^{-i\varphi}\sin\theta = 0,$$
$$ae^{i\varphi}\sin\theta - b(\lambda + \cos\theta) = 0.$$

For a, b not to vanish identically,

$$\begin{vmatrix} \cos\theta - \lambda & e^{-i\varphi}\sin\theta \\ e^{i\varphi}\sin\theta & -(\lambda + \cos\theta) \end{vmatrix} = \lambda^2 - \cos^2\theta - \sin^2\theta = 0,$$

or $\lambda^2 = 1$, i.e., $\lambda = \pm 1$, corresponding to spin angular momenta $\pm\frac{1}{2}\hbar$. Also, normalization requires

$$(a^*b^*)\begin{pmatrix} a \\ b \end{pmatrix} = |a|^2 + |b|^2 = 1.$$

For $\lambda = +1$, we have

$$\frac{b}{a} = \frac{1-\cos\theta}{e^{-i\varphi}\sin\theta} = \frac{\sin\frac{\theta}{2}}{\cos\frac{\theta}{2}}e^{i\varphi},$$

and, with normalization, the eigenvector

$$|\uparrow n\rangle = \begin{pmatrix} \cos\dfrac{\theta}{2} \\ e^{i\varphi}\sin\dfrac{\theta}{2} \end{pmatrix}.$$

For $\lambda = -1$, we have

$$\frac{b}{a} = -\frac{\cos\theta + 1}{e^{-i\varphi}\sin\theta} = -\frac{\cos\frac{\theta}{2}}{e^{-i\varphi}\sin\frac{\theta}{2}}$$

and, with normalization, the eigenvector

$$|\downarrow n\rangle = \begin{pmatrix} -e^{-i\varphi} \sin\dfrac{\theta}{2} \\ \cos\dfrac{\theta}{2} \end{pmatrix}.$$

For the first Stern–Gerlach apparatus take the direction of the magnetic field as the z direction. Take $\mathbf{n}$ along the magnetic field in the second Stern–Gerlach apparatus. Then $\varphi = 0$, $\theta = \alpha$ in the above.

If the particles which are sent into the second S–G apparatus have spin up, we have

$$\left|\uparrow z\right\rangle = c\left|\uparrow n\right\rangle + d\left|\downarrow n\right\rangle,$$

$$c = \left\langle \uparrow n | \uparrow z \right\rangle = (\cos(\alpha/2),\ \sin(\alpha/2)) \begin{pmatrix} 1 \\ 0 \end{pmatrix} = \cos(\alpha/2),$$

$$d = \left\langle \downarrow n | \uparrow z \right\rangle = (-\sin(\alpha/2),\ \cos(\alpha/2)) \begin{pmatrix} 1 \\ 0 \end{pmatrix} = -\sin(\alpha/2).$$

Therefore, after they leave the second S–G apparatus, the ratio of the numbers of the two beams of particles is

$$\frac{|c|^2}{|d|^2} = \frac{\cos^2(\alpha/2)}{\sin^2(\alpha/2)} = \cot^2\frac{\alpha}{2}.$$

If the particles which are sent into the second S–G apparatus have spin down, we have

$$\left|\downarrow z\right\rangle = c\left|\uparrow n\right\rangle + d\left|\downarrow n\right\rangle,$$

$$c = \left\langle \uparrow n | \downarrow z \right\rangle = (\cos(\alpha/2),\ \sin(\alpha/2)) \begin{pmatrix} 0 \\ 1 \end{pmatrix} = \sin(\alpha/2),$$

$$d = \left\langle \downarrow n | \downarrow z \right\rangle = (-\sin(\alpha/2),\ \cos(\alpha/2)) \begin{pmatrix} 0 \\ 1 \end{pmatrix} = \cos(\alpha/2),$$

and the ratio of the numbers of the two beams of the particles is

$$\frac{\sin^2(\alpha/2)}{\cos^2(\alpha/2)} = \tan^2\frac{\alpha}{2}.$$

3040

The magnetic moment of a silver atom is essentially equal to the magnetic moment of its unpaired valence electron which is $\mu = -\gamma \mathbf{S}$, where $\gamma = e/mc$ and $\mathbf{s}$ is the electron's spin.

Suppose that a beam of silver atoms having velocity V is passed through a Stern–Gerlach apparatus having its field gradient in the z direction, and that only the beam with $m_s = \hbar/2$ is to be considered in what follows. This beam then enters a region of length L having a constant magnetic field B_0 directly along the axis of the beam (y-axis). It next enters another Stern–Gerlach apparatus identical to the first as shown in Fig. 3.7. Describe clearly what is seen when the beam exits from the second Stern–Gerlach apparatus. Express the intensities of the resulting beams in terms of V, L, B_0 and the constants of the problem.

Use quantum mechanical equations of motion to derive your result.

(Berkeley)

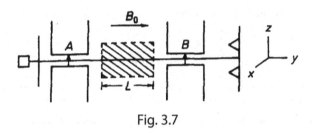

Fig. 3.7

Sol:

If we took a picture at the exit of the second S–G apparatus, we would see two black lines arising from the deposition of the two kinds of silver atoms with $m_s = \hbar/2$ and $m_s = -\hbar/2$.

Denote the state of the system in the region L by $|t\rangle$. If we consider only atoms of $m_s = \hbar/2$ in the beam that enters the region L at $t = 0$, then

$$|t=0\rangle = \begin{pmatrix} 1 \\ 0 \end{pmatrix}.$$

The Hamiltonian of the system in the region of length L is

$$H = -\mu \cdot B = \frac{\gamma \hbar B_0}{2} \sigma_y,$$

and so

$$|t\rangle = \exp\left(-\frac{i}{\hbar}Ht\right)|t=0\rangle$$

$$= \exp\left(-\frac{i\gamma B_0 t}{2}\sigma_y\right)\begin{pmatrix}1\\0\end{pmatrix}$$

$$= \left(\cos\frac{\gamma B_0 t}{2} - i\sigma_y\sin\frac{\gamma B_0 t}{2}\right)\begin{pmatrix}1\\0\end{pmatrix}$$

$$= \cos\frac{\gamma B_0 t}{2}\begin{pmatrix}1\\0\end{pmatrix} + \sin\frac{\gamma B_0 t}{2}\begin{pmatrix}0\\1\end{pmatrix}.$$

Hence at the exit of the region, the intensities of beams with $m_s = \frac{\hbar}{2}$ and $m_s = -\frac{\hbar}{2}$ are respectively

$$I_+ = I_0\cos^2\left(\frac{\gamma B_0 t}{2}\right), \quad I_- = I_0\sin^2\left(\frac{\gamma B_0 t}{2}\right),$$

where I_0 is the intensity of the beam that enters the region.

When the beam leaves the region $L, t = L/V$. So the ratio of intensities is

$$\cot^2(\gamma B_0 L / 2V).$$

The splitting of the beam is seen when it exists from the second Stern–Gerlach apparatus.

3041

Two oppositely charged spin$-\frac{1}{2}$ particles (spins s_1 and s_2) are coupled in a system with a spin-spin interaction energy ΔE.

The system is placed in a uniform magnetic field $\mathbf{H}=H\hat{z}$. The Hamiltonian for the spin interaction is

$$\hat{H}=(\Delta E/4)(\sigma_1\cdot\sigma_2)-(\mu_1+\mu_2)\cdot\mathbf{H}$$

where $\sigma_i = g_i\mu_0 s_i$ is the magnetic moment of the ith particle.

The spin wave functions for the 4 states of the system, in terms of the eigenstates of the z-component of the operators $\sigma_i = 2s_i$, are

$$\psi_1 = \alpha_1\alpha_2, \quad \psi_2 = s\beta_1\alpha_2 + c\alpha_1\beta_2, \quad \psi_3 = c\beta_1\alpha_2 - s\alpha_1\beta_2, \quad \psi_4 = \beta_1\beta_2,$$

where

$$(\sigma_z)_i\alpha_i = \alpha_i, \quad (\sigma_z)_i\beta_i = -\beta_i, \quad s = (1/\sqrt{2})\cdot(1-x/\sqrt{1+x^2})^{1/2},$$

$$c = (1/\sqrt{2})(1 + x/\sqrt{1+x^2})^{1/2},$$
$$x = \mu_0 H(g_2 - g_1)/\Delta E.$$

a. Find the energy eigenvalues associated with each state ψ_i. Discuss the limiting cases $\mu_0 H/\Delta E \gg 1$ and $\mu_0 H/\Delta E \ll 1$.

b. Assume that an initial state $\psi(0)$ is prepared in which particle 1 is polarized along the field direction z, but particle 2 is unpolarized. Find the time dependence of the polarization of particle 1:

$$P_{1z}(t) = \langle \psi(t) | \sigma_{1z} | \psi(t) \rangle.$$

Again discuss the limiting cases $\mu_0 H/\Delta E \ll 1$ and $\mu_0 H/\Delta E \gg 1$.

(Columbia)

Sol:

a. σ, α, β may be represented by matrices

$$\sigma_x = \begin{pmatrix} 0 & 1 \\ 1 & 0 \end{pmatrix}, \quad \sigma_y = \begin{pmatrix} 0 & -i \\ i & 0 \end{pmatrix}, \quad \sigma_z = \begin{pmatrix} 1 & 0 \\ 0 & -1 \end{pmatrix}$$

$$\alpha = \begin{pmatrix} 1 \\ 0 \end{pmatrix}, \quad \beta = \begin{pmatrix} 0 \\ 1 \end{pmatrix},$$

for which the following relations hold:

$$\sigma_{ix}\alpha_i = \beta_i, \qquad \sigma_{ix}\beta_i = \alpha_i,$$
$$\sigma_{iy}\alpha_i = i\beta_i, \qquad \sigma_{iy}\beta_i = -i\alpha_i,$$
$$\sigma_{iz}\alpha_i = \alpha_i, \qquad \sigma_{iz}\beta_i = -\beta_i.$$

Then as

$$\hat{H} = (\Delta E/4)(\sigma_1 \cdot \sigma_2) - (\mu_1 + \mu_2) \cdot \mathbf{H}$$
$$= (\Delta E/4)(\sigma_{1x}\sigma_{2x} + \sigma_{1y}\sigma_{2y} + \sigma_{1z}\sigma_{2z})$$
$$- \frac{1}{2}\mu_0 H(g_1\sigma_{1z} + g_2\sigma_{2z}),$$

where we have used $\mu = g\mu_0 s = \frac{1}{2}g\mu_0 s$, we have

$$\hat{H}\psi_1 = \hat{H}\alpha_1\alpha_2 = (\Delta E/4)(\beta_1\beta_2 - \beta_1\beta_2 + \alpha_1\alpha_2) - \frac{1}{2}\mu_0 H(g_1 + g_2)\alpha_1\alpha_2$$
$$= \left(\Delta E/4 - \frac{g_1 + g_2}{2}\mu_0 H\right)\alpha_1\alpha_2 = \left(\Delta E/4 - \frac{g_1 + g_2}{2} \cdot \mu_0 H\right)\psi_1,$$

and hence

$$E_1 = \Delta E / 4 - \frac{1}{2}(g_1 + g_2)\mu_0 H.$$

Similarly,

$$\hat{H}\psi_2 = \frac{\Delta E}{4}[s(\alpha_1\beta_2 + \alpha_1\beta_2 - \beta_1\alpha_2) + c(\beta_1\alpha_2 + \beta_1\alpha_2 - \alpha_1\beta_2)]$$

$$+\frac{1}{2}(g_1 - g_2)\mu_0 H(s\beta_1\alpha_2 - c\alpha_1\beta_2)$$

$$=[(\Delta E/4)(2c - s) - (\Delta E/2)xs]\beta_1\alpha_2$$

$$+[(\Delta E/4)(2s - c) + (\Delta E/2)xc]\alpha_1\beta_2$$

$$=(\Delta E/4)(2c/s - 2x - 1)s\beta_1\alpha_2 + (\Delta E/4)(2s/c + 2x - 1)c\alpha_1\beta_2.$$

Then as

$$2c/s - 2x = 2\frac{(1 + x/\sqrt{1+x^2})^{1/2}}{(1 - x/\sqrt{1+x^2})^{1/2}} - 2x = 2\sqrt{1+x^2}$$

$$= 2(\sqrt{1+x^2} - x) + 2x = 2s/c + 2x,$$

we have

$$\hat{H}\psi_2 = (\Delta E/4)(2\sqrt{1+x^2} - 1)\psi_2,$$

or

$$E_2 = (\Delta E/4)(2\sqrt{1+x^2} - 1).$$

By the same procedure we obtain

$$E_3 = (-\Delta E/4)(2\sqrt{1+x^2} + 1),$$

$$E_4 = (\Delta E/4) + \frac{g_1 + g_2}{2}\mu_0 H.$$

b.   As particle 2 is unpolarized and can be considered as in a mixed state, its state ξ_2 can be expanded in terms of α_2 and β_2:

$$\xi_2 = a\alpha_2 + b\beta_2,$$

where $|a|^2 = |b|^2 = 1/2$. Then the initial total wave function is

$$\psi(0) = \alpha_1\xi_2 = \alpha_1(a\alpha_2 + b\beta_2)$$

$$= a\psi_1 + b\frac{c}{c^2 + s^2}\psi_2 - b\frac{s}{c^2 + s^2}\psi_3$$

$$= a\psi_1 + bc\psi_2 - bs\psi_3,$$

as

$$c^2 + s^2 = \frac{1}{2}\left(1 + \frac{x}{\sqrt{1+x^2}}\right) + \frac{1}{2}\left(1 - \frac{x}{\sqrt{1+x^2}}\right) = 1.$$

Hence

$$\psi(t) = a\psi_1 \exp\left(-i\frac{E_1}{\hbar}t\right) + bc\psi_2 \exp\left(-i\frac{E_2}{\hbar}t\right)$$
$$- bs\psi_3 \exp\left(-i\frac{E_3}{\hbar}t\right).$$

Then using the relations

$$\sigma_{1z}\psi_1 = \sigma_{1z}\alpha_1\alpha_2 = \alpha_1\alpha_2,$$
$$\sigma_{1z}\psi_2 = \sigma_{1z}(s\beta_1\alpha_2 + c\alpha_1\beta_2) = -s\beta_1\alpha_2 + c\alpha_1\beta_2,$$
$$\sigma_{1z}\psi_3 = \sigma_{1z}(c\beta_1\alpha_2 - s\alpha_1\beta_2) = -(c\beta_1\alpha_2 + s\alpha_1\beta_2),$$

we obtain

$$P_{1z}(t) = \langle \psi(t)|\sigma_{1z}|\psi(t)\rangle$$
$$= |a|^2 + |b|^2 \langle \exp(-iE_2 t/\hbar)\, c(s\beta_1\alpha_2 + c\alpha_1\beta_2)$$
$$- \exp(-iE_3 t/\hbar)\, s(c\beta_1\alpha_2 - s\alpha_1\beta_2)|\exp(-iE_2 t/\hbar)$$
$$\times c(-s\beta_1\alpha_2 + c\alpha_1\beta_2) + \exp(-iE_3 / \hbar)\, s(c\beta_1\alpha_2 + s\alpha_1\beta_2)\rangle$$
$$= \frac{1}{2} + \frac{1}{2}[(s^2 - c^2)^2 + 4s^2c^2 \cos(E_2 - E_3)t/\hbar]$$
$$= \frac{1}{2} + \frac{1}{2(1+x^2)}[x^2 + \cos(\sqrt{1+x^2}\,\Delta Et/\hbar)]$$
$$= 1 - \frac{1}{1+x^2}\sin^2(\sqrt{1+x^2}\,\Delta Et/2\hbar).$$

c. In the limit $\mu_0 H/\Delta E \gg 1$, i.e., $x \gg 1$, we have

$$E_1 = \frac{\Delta E}{4} - \frac{g_1 + g_2}{2}\mu_0 H \approx -\frac{1}{2}(g_1 + g_2)\mu_0 H,$$

$$E_2 = \frac{\Delta E}{4}(2\sqrt{1+x^2} - 1) \approx (\Delta E/4)\times 2x = \frac{1}{2}(g_2 - g_1)\mu_0 H,$$

$$E_3 = -\frac{\Delta E}{4}(1 + 2\sqrt{1+x^2}) \approx -\frac{1}{2}(g_2 - g_1)\mu_0 H,$$

$$E_4 = \frac{\Delta E}{4} + \frac{g_1 + g_2}{2}\mu_0 H \approx \frac{1}{2}(g_1 + g_2)\mu_0 H,$$

$$P_{1z}(t) \approx 1.$$

When $\mu_0 H/\Delta E \ll 1$, i.e., $x \ll 1$, we have

$$E_1 \approx E_4 \approx \Delta E/4,$$

$$E_2 = \frac{\Delta E}{4}(2\sqrt{1+x^2} - 1) \approx \frac{\Delta E}{4},$$

$$E_3 \approx -\frac{3\Delta E}{4},$$

$$P_{1z}(t) \approx 1 - \sin^2(\Delta Et / 2\hbar).$$

3042

A hydrogen atom is in a $^2P_{1/2}$ state with total angular momentum up along the z-axis. In all parts of this problem show your computations and reasoning carefully.

a. With what probability will the electron be found with spin down?

b. Compute the probability per unit solid angle $P(\theta, \varphi)$ that the electron will be found at spherical angles θ, φ (independent of radial distance and spin).

c. An experimenter applies a weak magnetic field along the positive z-axis. What is the effective magnetic moment of the atom in this field?

d. Starting from the original state, the experimenter slowly raises the magnetic field until it overpowers the fine structure. What are the orbital and spin quantum numbers of the final state?
[Assume the Hamiltonian is linear in the magnetic field.]

e. What is the effective magnetic moment of this final state?

(Berkeley)

Sol:

a. For the state $^2P_{1/2}$, $l=1$, $s=1/2$, $J=1/2$, $J_z=1/2$. Transforming the coupling representation into the uncoupling representation, we have

$$|J, J_z\rangle = |1/2, 1/2\rangle = \sqrt{2/3}|1, 1\rangle|1/2, -1/2\rangle$$
$$-\sqrt{1/3}|1, 0\rangle|1/2, 1/2\rangle.$$

Therefore $P_\downarrow = 2/3$.

b. As

$$|J, J_z\rangle = \sqrt{\frac{1}{3}}\begin{pmatrix} \sqrt{2}Y_{11} \\ Y_{10} \end{pmatrix},$$

we have

$$P(\theta, \varphi)d\Omega = \frac{1}{3}(2Y_{11}^* Y_{11} + Y_{10}^* Y_{10})d\Omega.$$

Hence the probability per unit solid angle is

$$P(\theta, \varphi) = \frac{1}{3}\left(2 \times \frac{3}{8\pi}\sin^2\theta + \frac{3}{4\pi}\cos^2\theta\right) = \frac{1}{4\pi}.$$

c. In the weak magnetic field, J and J_z are "good" quantum numbers and the state remains unchanged. The effective magnetic moment is

$$\mu = \left\langle\frac{1}{2},\frac{1}{2}\bigg|\mu_z\bigg|\frac{1}{2},\frac{1}{2}\right\rangle = g\frac{e}{2mc}\left\langle\frac{1}{2},\frac{1}{2}\bigg|J_z\bigg|\frac{1}{2},\frac{1}{2}\right\rangle = g\cdot\frac{e\hbar}{4mc},$$

Where m is the electron mass and

$$g = 1 + \frac{J(J+1)-l(l+1)+s(s+1)}{2J(J+1)} = \frac{2}{3}.$$

Hence $\mu = e\hbar/6mc$.

d. In a strong magnetic field, the interaction of the magnetic moment with the field is much stronger than the coupling interaction of spin and orbit, so that the latter can be neglected. Here l and s are good quantum numbers. The Hamiltonian related to the magnetic field is

$$W = -\mu_l\cdot B - \mu_s\cdot B = eB\hat{l}_z/2mc + eB\hat{s}_z/mc.$$

When the magnetic field is increased slowly from zero, the state remains at the lowest energy. From the expression of W, we see that when the magnetic field becomes strong, only if $l_z = -\hbar$, $s_z = -\hbar/2$ can the state remain at the lowest energy. Thus the quantum numbers of the final state are $l=1$, $l_z=-1$, $s=1/2$, $s_z=-1/2$.

e. the effective magnetic moment of the final state is

$$\mu = \bar{\mu}_{l_z} + \bar{\mu}_{s_z} = -e\hbar/2mc - e\hbar/2mc = -e\hbar/mc.$$

3043

Consider a neutral particle with intrinsic angular momentum $\sqrt{s(s+1)}$, where $s = \hbar/2$, i.e., a spin-1/2 particle.

Assume the particle has a magnetic moment $\mathbf{M} = \gamma s$, where γ is a constant. The quantum-mechanical state of the particle can be described in a spin space spanned by the eigenvectors $|+\rangle$ and $|-\rangle$ representing alignments parallel and antiparallel to the z-axis:

$$\hat{s}_z|+\rangle = \frac{\hbar}{2}|+\rangle, \quad \hat{s}_z|-\rangle = -\frac{\hbar}{2}|-\rangle.$$

At time $t=0$ the state of the system is $|\psi(t=0)\rangle=|+\rangle$. The particle moves along the y-axis through a uniform magnetic field $B=B\hat{y}$ oriented along the y-axis.

    a. As expressed in the $|+\rangle$, $|-\rangle$ basis, what is $|\psi(t)\rangle$?

    b. What would be the expectation values for measurements of the observables s_x, s_y, s_z as functions of time?

(CUS)

Sol:

    a. The Hamiltonian of the particle is

$$\hat{H}=-\mathbf{M}\cdot\mathbf{B}=-\gamma\hat{s}_y B .$$

In the representation of $\hat{s}^2$, $\hat{s}_z$,

$$\hat{s}_y=\frac{\hbar}{2}\begin{pmatrix}0 & -i \\ i & 0\end{pmatrix},$$

and so the two eigenstates of $\hat{s}_y$ are

$$\left|s_y=\hbar/2\right\rangle=\frac{1}{\sqrt{2}}\begin{pmatrix}-i \\ 1\end{pmatrix}, \quad \left|s_y=-\hbar/2\right\rangle=\frac{1}{\sqrt{2}}\begin{pmatrix}i \\ 1\end{pmatrix}.$$

As

$$\hat{H}\left|s_y=\frac{1}{2}\hbar\right\rangle=-\gamma B\frac{1}{2}\hbar\left|s_y=\frac{1}{2}\hbar\right\rangle,$$

$$\hat{H}\left|s_y=-\frac{1}{2}\hbar\right\rangle=\gamma B\frac{1}{2}\hbar\left|s_y=-\frac{1}{2}\hbar\right\rangle,$$

any state of the particle can be expressed as

$$|\psi(t)\rangle=c_1\left|s_y=\frac{1}{2}\hbar\right\rangle\exp\left(+i\frac{1}{2}\gamma Bt\right)$$

$$+c_2\left|s_y=-\frac{1}{2}\hbar\right\rangle\exp\left(-i\frac{1}{2}\gamma Bt\right).$$

Then the initial condition

$$|\psi(t=0)\rangle=|s_z=\hbar/2\rangle$$

gives

$$\left| s_z = \frac{1}{2}\hbar \right\rangle = c_1 \left| s_y = \frac{1}{2}\hbar \right\rangle + c_2 \left| s_y = -\frac{1}{2}\hbar \right\rangle,$$

and so

$$c_1 = \left\langle \frac{1}{2}\hbar = s_y \middle| s_z = \frac{1}{2}\hbar \right\rangle = \frac{1}{\sqrt{2}} \begin{pmatrix} -i \\ 1 \end{pmatrix}^{\dagger} \begin{pmatrix} 1 \\ 0 \end{pmatrix}$$

$$= \frac{1}{\sqrt{2}} (i \ 1) \begin{pmatrix} 1 \\ 0 \end{pmatrix} = \frac{1}{\sqrt{2}} i,$$

$$c_2 = \left\langle -\frac{1}{2}\hbar = s_y \middle| s_z = \frac{1}{2}\hbar \right\rangle = \frac{1}{\sqrt{2}} \begin{pmatrix} i \\ 1 \end{pmatrix}^{\dagger} \begin{pmatrix} 1 \\ 0 \end{pmatrix}$$

$$= \frac{1}{\sqrt{2}} (-i \ 1) \begin{pmatrix} 1 \\ 0 \end{pmatrix} = -\frac{1}{\sqrt{2}} i.$$

Therefore

$$\left| \psi(t) \right\rangle = \frac{1}{\sqrt{2}} i \frac{1}{\sqrt{2}} \begin{pmatrix} -i \\ 1 \end{pmatrix} \exp\left(i\frac{1}{2}\gamma Bt \right)$$

$$- \frac{1}{\sqrt{2}} i \frac{1}{\sqrt{2}} \begin{pmatrix} i \\ 1 \end{pmatrix} \exp\left(-i\frac{1}{2}\gamma Bt \right)$$

$$= \cos\left(\frac{1}{2}\gamma Bt \right) \left| s_z = \frac{1}{2}\hbar \right\rangle - \sin\left(\frac{1}{2}\gamma Bt \right) \left| s_z = -\frac{1}{2}\hbar \right\rangle.$$

b.

$$\langle s_z \rangle = \left(\cos\left(\frac{1}{2}\gamma Bt \right) \quad -\sin\left(\frac{1}{2}\gamma Bt \right) \right) \frac{1}{2}\hbar \begin{pmatrix} 1 & 0 \\ 0 & -1 \end{pmatrix}$$

$$\times \begin{pmatrix} \cos\left(\frac{1}{2}\gamma Bt \right) \\ -\sin\left(\frac{1}{2}\gamma Bt \right) \end{pmatrix} = \frac{1}{2}\hbar \left(\left(\cos\left(\frac{1}{2}\gamma Bt \right)\right) \sin\left(\frac{1}{2}\gamma Bt \right) \right)$$

$$\times \begin{pmatrix} \cos\left(\frac{1}{2}\gamma Bt \right) \\ -\sin\left(\frac{1}{2}\gamma Bt \right) \end{pmatrix} = \frac{1}{2}\hbar \cos(\gamma Bt).$$

$\langle s_y \rangle = 0$, because $\langle s_y \rangle = 0$ at $t = 0$ and s_y is conserved.

$$\langle s_x \rangle = \left(\cos\left(\frac{1}{2}\gamma Bt\right) - \sin\left(\frac{1}{2}\gamma Bt\right) \right) \frac{1}{2}\hbar \begin{pmatrix} 0 & 1 \\ 1 & 0 \end{pmatrix}$$

$$\times \begin{pmatrix} \cos\left(\frac{1}{2}\gamma Bt\right) \\ -\sin\left(\frac{1}{2}\gamma Bt\right) \end{pmatrix} = \left(-\sin\left(\frac{1}{2}\gamma Bt\right) \cos\left(\frac{1}{2}\gamma Bt\right) \right)$$

$$\times \frac{1}{2}\hbar \begin{pmatrix} \cos\left(\frac{1}{2}\gamma Bt\right) \\ -\sin\left(\frac{1}{2}\gamma Bt\right) \end{pmatrix} = -\frac{1}{2}\hbar \sin(\gamma Bt).$$

3044

A particle of spin 1/2 and magnetic moment μ is placed in a magnetic field

$$B = B_0\hat{\mathbf{z}} + B_1\cos\omega t\,\hat{\mathbf{x}} - B_1\sin\omega t\,\hat{\mathbf{y}},$$

which is often employed in magnetic resonance experiments. Assume that the particle has spin up along the $+z$ -axis at $t = 0\,(m_s = +1/2)$. Derive the probability to find the particle with spin down $(m_s = -1/2)$ at time $t > 0$.

(Berkeley)

Sol:

The Hamiltonian of the system is

$$H = -\mu\sigma \cdot \mathbf{B}.$$

Letting

$$\omega_0 = \mu B_0 / \hbar, \quad \omega_1 = \mu B_1 / \hbar,$$

we have

$$H = -\mu(B_0\sigma_z + B_1\sigma_x\cos\omega t - B_1\sigma_y\sin\omega t)$$

$$= -\hbar\omega_0\sigma_z - \hbar\omega_1 \begin{pmatrix} 0 & e^{i\omega t} \\ e^{-i\omega t} & 0 \end{pmatrix},$$

where

$$\sigma_x = \begin{pmatrix} 0 & 1 \\ 1 & 0 \end{pmatrix}, \quad \sigma_y = \begin{pmatrix} 0 & -i \\ i & 0 \end{pmatrix}, \quad \sigma_z = \begin{pmatrix} 1 & 0 \\ 0 & -1 \end{pmatrix}.$$

Let the wave function of the system be

$$|t\rangle = \begin{pmatrix} a(t) \\ b(t) \end{pmatrix}.$$

The Schrödinger equation $i\hbar\partial_t|t\rangle = H|t\rangle$, or

$$i\begin{pmatrix} \dot{a} \\ \dot{b} \end{pmatrix} = -\omega_0\begin{pmatrix} 1 & 0 \\ 0 & -1 \end{pmatrix}\begin{pmatrix} a \\ b \end{pmatrix} - \omega_1\begin{pmatrix} 0 & e^{i\omega t} \\ e^{i\omega t} & 0 \end{pmatrix}\begin{pmatrix} a \\ b \end{pmatrix}.$$

gives

$$\begin{cases} \dot{a} = i\omega_0 a + i\omega_1 e^{i\omega t}b, \\ \dot{b} = -i\omega_0 b + i\omega_1 e^{-i\omega t}a. \end{cases}$$

Try a solution of the type

$$\begin{cases} a = \alpha\exp(i\omega_0 t), \\ b = \beta\exp(-i\omega_0 t). \end{cases}$$

Substitution in the above equations gives

$$\begin{cases} \dot{\alpha} = i\omega_1\exp[i(-2\omega_0 + w)t]\beta, \\ \dot{\beta} = i\omega_1\exp[-i(-2\omega_0 + \omega)t]\alpha. \end{cases}$$

Assume that α and β have the forms

$$\begin{cases} \alpha = A_1\exp[i(-2\omega_0 + \omega + \Omega)t], \\ \beta = A_2 e^{i\Omega t}, \end{cases}$$

Where A_1, A_2, and Ω are constants. Substitution gives

$$\begin{cases} (-2\omega_0 + \omega + \Omega)A_1 - \omega_1 A_2 = 0, \\ -\omega_1 A_1 + \Omega A_2 = 0. \end{cases}$$

For this set of equations to have nontrivial solutions the determinant of the coefficients of A_1, A_2 must be zero, i.e.,

$$(-2\omega_0 + \omega + \Omega)\Omega - \omega_1^2 = 0,$$

giving

$$\Omega_\pm = -\left(-\omega_0 + \frac{\omega}{2}\right) \pm \sqrt{(-\omega_0 + \omega/2)^2 + \omega_1^2}.$$

Therefore the general form of β is

$$\beta = A_{2+}\exp(i\Omega_+ t) + A_{2-}\exp(i\Omega_- t),$$

and that of α is

$$\alpha = \frac{\beta \exp[i(-2\omega_0 + \omega)t]}{i\omega_1}$$

$$= \frac{1}{\omega_1} \exp[i(-2\omega_0 + \omega)t][\Omega_+ A_{2+} \exp(i\Omega_+ t)$$

$$+ \Omega_- A_{2-} \exp(i\Omega_- t)].$$

Initially the spin is up along the z-axis, so

$$|t = 0\rangle = \begin{pmatrix} 1 \\ 0 \end{pmatrix} = \begin{pmatrix} a(0) \\ b(0) \end{pmatrix} = \begin{pmatrix} \alpha(0) \\ \beta(0) \end{pmatrix},$$

giving

$$\frac{1}{\omega_1}(\Omega_+ A_{2+} + \Omega_- A_{2-}) = 1,$$

$$A_{2+} + A_{2-} = 0.$$

The solution is

$$A_{2+} = -A_{2-} = \omega_1 / (\Omega_+ - \Omega_-)$$

$$= \frac{\omega_1}{2\sqrt{(\omega_0 - \omega/2)^2 + \omega_1^2}}.$$

Hence

$$b(t) = \exp(-i\omega_0 t)\beta(t)$$

$$= \exp(-i\omega_0 t)A_{2+}$$

$$\times [\exp(i\Omega_+ t) - \exp(i\Omega_- t)]$$

$$= \exp(-i\omega t/2)2iA_{2+}$$

$$\times \sin(\sqrt{(\omega_0 - \omega/2)^2 + \omega_1^2}\, t)$$

$$= \frac{i\omega_1 \exp[-i(\omega/2)t\,|}{\sqrt{(\omega_0 - \omega/2)^2 + \omega_1^2}}$$

$$\times \sin(\sqrt{(\omega_0 - \omega/2)^2 + \omega_1^2}\, t).$$

The probability that the particle has spin down along the z-axis at time t is

$$P = |\langle z \downarrow| t \rangle|^2$$

$$= |b(t)|^2 = \frac{\omega_1^2 \sin^2(\sqrt{\omega_0 - \omega/2)^2 + \omega_1^2}\, t)}{(\omega_0 - \omega/2)^2 + \omega_1^2}.$$

3045

A spin-$\frac{1}{2}$ system with magnetic moment $\mu = \mu_0 \sigma$ is located in a uniform time-independent magnetic field B_0 in the positive z direction. For the time interval $0 < t < T$ an additional uniform time-independent field B_1 is applied in the positive x direction. During this interval, the system is again in a uniform constant magnetic field, but of different magnitude and direction z' from the initial one. At and before $t = 0$, the system is in the $m = 1/2$ state with respect to the z-axis.

   a. At $t = 0+$, what are the amplitudes for finding the system with spin projections $m' = \pm 1/2$ with respect to the z' direction?

   b. What is the time development of the energy eigenstates with respect to the z' direction, during the time interval $0 < t < T$?

   c. What is the probability amplitude at $t = T$ of observing the system in the spin state $m = -1/2$ along the original z-axis?

   [Express your answers in terms of the angle θ between the z and z' axes and the frequency $\omega_0 = \mu_0 B_0 / \hbar$.]

(Berkeley)

Sol:

   a. In the representation of s_z, the eigenvectors of $s_{z'}$ are

$$\begin{pmatrix} \cos\dfrac{\theta}{2} \\[2mm] \sin\dfrac{\theta}{2} \end{pmatrix}, \quad \begin{pmatrix} -\sin\dfrac{\theta}{2} \\[2mm] \cos\dfrac{\theta}{2} \end{pmatrix},$$

corresponding to the eigenvalues $s_{z'} = 1/2$ and $-1/2$ respectively. Then the probability amplitudes for $m' = \pm 1/2$ are respectively

$$C_+ = \begin{pmatrix} \cos\dfrac{\theta}{2} & \sin\dfrac{\theta}{2} \end{pmatrix} \begin{pmatrix} 1 \\ 0 \end{pmatrix} = \cos\dfrac{\theta}{2},$$

$$C_- = \begin{pmatrix} -\sin\dfrac{\theta}{2} & \cos\dfrac{\theta}{2} \end{pmatrix} \begin{pmatrix} 1 \\ 0 \end{pmatrix} = -\sin\dfrac{\theta}{2}.$$

b. The Hamiltonian in the interval $0 < t < T$ is

$$H = -\mu \cdot B = -\mu_0 (B_0 \sigma_z + B_1 \sigma_x)$$

$$= -\mu_0 \begin{pmatrix} B_0 & B_1 \\ B_1 & -B_0 \end{pmatrix}.$$

The initial eigenfunctions are

$$\chi_+(0) = \begin{pmatrix} \cos\dfrac{\theta}{2} \\ \sin\dfrac{\theta}{2} \end{pmatrix}, \quad \chi_-(0) = \begin{pmatrix} -\sin\dfrac{\theta}{2} \\ \cos\dfrac{\theta}{2} \end{pmatrix},$$

where

$$\theta = \tan^{-1}\left(\frac{B_1}{B_0}\right).$$

Substitution in the Schrödinger equation $H\chi\pm(0) = \pm E\chi\pm(0)$ gives

$$E = -\mu_0 B_0 / \cos\theta = -\mu_0 B,$$

where

$$B = \sqrt{B_0^2 + B_1^2}\ .$$

At a later time t in $0 < t < T$, the eigenstates are

$$\chi\pm(t) = \exp(\mp iEt/\hbar)\chi\pm(0) = \exp(\pm i\mu_0 Bt/\hbar)\chi\pm(0).$$

c. The probability amplitude at $t = T$ is

$$C_-(T) = (0\ 1)\exp(-iHT/\hbar)\begin{pmatrix} 1 \\ 0 \end{pmatrix}$$

$$= (iB_1 / \sqrt{B_0^2 + B_1^2})\sin(\mu_0 T\sqrt{B_0^2 + B_1^2}\ / \hbar)$$

$$= i\sin\theta\sin(\mu_0 BT / \hbar).$$

An alternative way is to make use of

$$\psi(0) = \begin{pmatrix} 1 \\ 0 \end{pmatrix} = \cos\frac{\theta}{2}\chi_+(0) - \sin\frac{\theta}{2}\chi_-(0),$$

and so

$$\psi(t) = \chi_+(0)\cos\frac{\theta}{2}\exp(i\mu_0 Bt/\hbar)$$

$$- \chi_-(0)\sin\frac{\theta}{2}\exp(-i\mu_0 Bt/\hbar),$$

to get

$$C_-(T)=\beta^+\psi(T)=\cos\frac{\theta}{2}\sin\frac{\theta}{2}$$
$$\times\{\exp(i\mu_0 BT/\hbar)-\exp(-i\mu_0 BT/\hbar)\},$$
$$=i\sin\theta\sin\frac{\mu_0 BT}{\hbar},\quad\text{where }\beta=\begin{pmatrix}0\\1\end{pmatrix}.$$

3046

A spin-$\frac{1}{2}$ system of magnetic moment μ is placed in a dc magnetic field $H_0 e_z$ in which the energy of the spin state $|+1/2\rangle$ is $\hbar\omega_0$, that of $|-1/2\rangle$ being taken as 0. The system is in the state $|-1/2\rangle$ when at $t=0$, a magnetic field $H(e_x\cos\omega_0 t+e_y\sin\omega_0 t)$ is suddenly turned on. Ignoring relaxation find the energy of the spin system as a function of ω_0, H, c and t, where

$$c=\langle+1/2|\mu_x+i\mu_y|-1/2\rangle.$$

Why is the energy of the spin system not conserved?

(Columbia)

Sol: The Hamiltonian is

$$\hat{H}=-\mu(H+H_0)=-\mu\sigma\cdot(H+H_0)$$
$$=-\mu(H\sigma_x\cos\omega_0 t+H\sigma_y\sin\omega_0 t+H_0\sigma_z)$$
$$=-\mu\begin{pmatrix}H_0 & H\exp(-i\omega_0 t)\\ H\exp(i\omega_0 t) & -H_0\end{pmatrix}.$$

In the Schrödinger equation

$$i\hbar\partial\psi/\partial t=\hat{H}\psi,$$

setting

$$\psi=\begin{pmatrix}a(t)\\b(t)\end{pmatrix},$$

we get

$$\begin{cases}\dfrac{da}{dt}=\dfrac{\mu i}{\hbar}[H_0 a+H\exp(-i\omega_0 t)b],\\[2mm]\dfrac{db}{dt}=\dfrac{\mu i}{\hbar}[H\exp(i\omega_0 t)a-H_0 b].\end{cases}$$

230 *Problems and Solutions in Quantum Mechanics*

Try a solution of the type

$$a = A\exp\left[-i\left(\Omega + \frac{1}{2}\omega_0\right)t\right],$$

$$b = B\exp\left[-i\left(\Omega - \frac{1}{2}\omega_0\right)t\right],$$

where A, B and Ω are constants. Substitution gives

$$\begin{cases} \left(\Omega + \frac{1}{2}\omega_0 + \omega\right)A + \omega'B = 0, \\ \left(\Omega - \frac{1}{2}\omega_0 - \omega\right)B + \omega'A = 0, \end{cases}$$

where

$$\omega = \frac{\mu H_0}{\hbar},$$

$$\omega' = \frac{\mu H}{\hbar}.$$

For nontrivial solutions we require

$$\begin{vmatrix} \Omega + \left(\frac{1}{2}\omega_0 + \omega\right) & \omega' \\ \omega' & \Omega - \left(\frac{1}{2}\omega_0 + \omega\right) \end{vmatrix} = 0,$$

giving

$$\Omega = \pm\sqrt{\omega'^2 + \left(\frac{1}{2}\omega_0 + \omega\right)^2} = \pm Q,$$

where $Q = \left|\sqrt{\omega'^2 + \left(\frac{1}{2}\omega_0 + \omega\right)^2}\right|$. Hence

$$\psi(t) = (A_1 e^{-iQt} + A_2 e^{iQt})\exp\left(-i\frac{\omega_0}{2}t\right)\alpha$$

$$+ (B_1 e^{-iQt} + B_2 e^{iQt})\exp\left(i\frac{\omega_0}{2}t\right)\beta,$$

where

$$B_{1,2} = -A_{1,2}\frac{\Omega_{1,2} + \left(\omega + \frac{1}{2}\omega_0\right)}{\omega'},$$

the subscripts 1, 2 corresponding to the values of Ω with $+$, $-$ signs respectively, and

$$\alpha = \begin{pmatrix} 1 \\ 0 \end{pmatrix}, \quad \beta = \begin{pmatrix} 0 \\ 1 \end{pmatrix}.$$

At $t=0$ the system is in the $\left|-\frac{1}{2}\right\rangle$ state and $\psi = \begin{pmatrix} 0 \\ 1 \end{pmatrix}$. Thus $B_1 + B_2 = 1$, $A_1 + A_2 = 0$.

Then as

$$\left(Q + \frac{1}{2}\omega_0 + \omega\right) A_1 + \omega' B_1 = 0,$$

$$\left(-Q + \frac{1}{2}\omega_0 + \omega\right) A_2 + \omega' B_2 = 0,$$

we have

$$Q(A_1 - A_2) + \omega' = 0,$$

$$\left(\frac{1}{2}\omega_0 + \omega\right)(A_1 - A_2) + \omega'(B_1 - B_2) = 0,$$

giving

$$A_1 = -\frac{\omega'}{2Q}, \quad A_2 = \frac{\omega'}{2Q},$$

$$B_1 = \frac{Q + \left(\omega + \frac{1}{2}\omega_0\right)}{2Q},$$

$$B_2 = \frac{Q - \left(\omega + \frac{1}{2}\omega_0\right)}{2Q}.$$

Therefore the wave function of the system is

$$\psi(t) = \frac{\omega'}{Q} i \sin(Qt) \exp\left(-i\frac{\omega_0}{2}t\right)\alpha$$

$$+ \left[\cos Qt - i\frac{\left(\omega + \frac{1}{2}\omega_0\right)}{Q}\sin Qt\right] \exp\left(i\frac{\omega_0}{2}t\right)\beta,$$

and the energy of the system is

$$E = \langle \psi | H | \psi \rangle = -\mu\left[-H_0 \cos^2 Qt\right.$$

$$\left. + \frac{\omega'^2 H_0 - \left(\omega + \frac{1}{2}\omega_0\right)^2 H_0 - 2\omega' H\left(\omega + \frac{\omega_0}{2}\right)}{Q^2}\sin^2 Qt\right].$$

Note that as

$$c = \left\langle +\frac{1}{2} \middle| \mu_x + i\mu_y \middle| -\frac{1}{2} \right\rangle$$

$$= \mu \left\langle +\frac{1}{2} \middle| \sigma_x + i\sigma_y \middle| -\frac{1}{2} \right\rangle$$

$$= \mu (1 0) \begin{pmatrix} 0 & 2 \\ 0 & 0 \end{pmatrix} \begin{pmatrix} 0 \\ 1 \end{pmatrix} = 2\mu,$$

$$\mu = \frac{c}{2}.$$

As the energy of the system changes with time t, it is not conserved. This is because with regard to spin it is not an isolated system.

3047

A beam of neutrons of velocity v passes from a region (I) (where the magnetic field is $\mathbf{B} = B_1 \mathbf{e}_z$) to a region (II) (where the field is $\mathbf{B} = B_2 \mathbf{e}_x$). In region (I) the beam is completely polarized in the $+z$ direction.

a. Assuming that a given particle passes from (I) to (II) at time $t = 0$, what is the spin wave function of that particle for $t > 0$?

b. As a function of time, what fraction of the particles would be observed to have spins in the $+x$ direction; the $+y$ direction; the $+z$ direction?

c. As a practical matter, how abrupt must the transition between (I) and (II) be to have the above description valid?

(Wisconsin)

Sol:

a. Considering only the spin wave function, the Schrödinger equation is

$$i\hbar \partial |\chi\rangle / \partial t = \hat{H} |\chi\rangle,$$

where

$$\hat{H} = -\boldsymbol{\mu} \cdot \mathbf{B} = -\mu_n B_2 \sigma_x,$$

with $\mu_n = -1.9103 \mu_N$ being the anomalous magnetic moment of the neutron, $\mu_N = e\hbar/2m_p c$ the nuclear magneton, m_p the mass of a proton. Thus

$$\frac{d}{dt} |\chi\rangle = \frac{i}{\hbar} \mu_n B_2 \sigma_x |\chi\rangle \equiv -i\omega_2 \sigma_x |\chi\rangle,$$

where $\omega_2 = \frac{\mu_n|B_2}{\hbar}$. Let $|\chi\rangle = \begin{pmatrix} a \\ b \end{pmatrix}$. The above gives

$$\begin{cases} \dot{a} = -i\omega_2 b, \\ \dot{b} = -i\omega_2 a. \end{cases}$$

The initial condition is $a(0)=1$, $b(0)=0$ as the beam is initially polarized in the $+z$ direction. Solving for a and b and using the initial condition we obtain for $t > 0$

$$|\chi\rangle = \begin{pmatrix} \cos\omega_2 t \\ -i\sin\omega_2 t \end{pmatrix},$$

b. The mean value of the spin in the state $|\chi\rangle$, i.e., the polarized vector for neutron, is

$$\mathbf{p} = \langle\chi|\sigma|\chi\rangle = \langle\chi|\sigma_x e_x + \sigma_y e_y + \sigma_z e_z|\chi\rangle$$
$$= (0, -\sin 2\omega_2 t, \cos 2\omega_2 t)$$

Thus in the region (**II**), the neutron spin is in the yz plane and precesses about the x direction with angular velocity $2\omega_2$.

c. For the descriptions in (a) and (b) to be valid, the time of the transition between (I) and (II) must satisfy

$$t \ll \frac{2\pi}{w_2} = \frac{h}{|\mu_n|B_2}.$$

For example if $B_2 \sim 10^3$ Gs, then $t \ll 0.7\mu$ s.

If the kinetic energy of the incident neutrons is given, we can calculate the upper limit of the width of the transition region.

3048

The Hamiltonian for a $(\mu^+ e^-)$ atom in the $n=1$, $l=0$ state in an external magnetic field **B** is

$$H = as_\mu \cdot s_e + \frac{|e|}{m_e c} s_e \cdot \mathbf{B} - \frac{|e|}{m_\mu c} s_\mu \cdot \mathbf{B}.$$

a. What is the physical significance of each term? Which term dominates in the interaction with the external field?

b. Choosing the z-axis along **B** and using the notation (**F**, **M**), where $\mathbf{F} = s_\mu + s_e$, show that $(1, +1)$ is an eigenstate of H and give its eigenvalue.

c. An *rf* field can be applied to cause transitions to the state $(0, 0)$. Describe qualitatively how an observation of the decay $\mu^+ \to e^+ v_e v_\mu$ could be used to detect the occurrence of this transition.

(Wisconsin)

Sol:

a. The first term in $\hat{H}$ is due to the magnetic interaction between μ and e, the second and third terms respectively account for the magnetic interactions of μ and e with the external field **B**. Of the latter, the term $|e| s_e \cdot \mathbf{B} / m_e c$ is dominant as $m_e \approx m_\mu / 200$.

b. As

$$\mathbf{F} = \mathbf{s}_\mu + \mathbf{s}_e$$

$$\hat{H} = \frac{1}{2} a[\mathbf{F}^2 - \mathbf{s}_\mu^2 - \mathbf{s}_e^2] + \frac{eB}{m_e c} s_{ez} - \frac{eB}{m_\mu c} s_{\mu z}.$$

Consider the state

$$(1, +1) = \begin{pmatrix} 1 \\ 0 \end{pmatrix}_e \begin{pmatrix} 1 \\ 0 \end{pmatrix}_\mu.$$

As the eigenvalues of $F^2, s_\mu^2, s_e^2, s_{\mu z}, s_{ez}$ are $1 \ (1+1)\hbar^2, \frac{1}{2}(\frac{1}{2}+1)\hbar^2, \frac{1}{2}(\frac{1}{2}+1)\hbar^2,$ $\frac{1}{2}\hbar, \frac{1}{2}\hbar$ respectively, we have

$$\hat{H}(1, +1) = \left\{ \frac{1}{2} a\hbar^2 \left[2 - 2 \cdot \frac{3}{4} \right] + \frac{e\hbar}{2m_e c} B - \frac{e\hbar}{2m_\mu c} B \right\}(1, +1)$$

$$= \left(\frac{1}{4} a\hbar^2 + \frac{e\hbar}{2m_e c} B - \frac{e\hbar}{2m_\mu c} B \right)(1, +1).$$

Thus $(1,+1)$ is an eigenstate of $\hat{H}$, with the eigenvalue

$$a\hbar^2/4 + e\hbar B/2m_e c - e\hbar B/2m_\mu c.$$

c. The decay $\mu^+ \to e^+ v_e v_\mu$ can be detected through the observation of the annihilation of the positronium $e^+ e^- \to 2\gamma$. For the state $(1,+1)$, the total angular momentum of the $e^+ e^-$ system is 1, and so $e^+ e^-$ cannot decay into 2γ whose total angular momentum is 0. For the state $(0, 0)$, the total angular momentum of the e^+e^- system is 0 and so it can decay into 2γ. Hence, detection of $e^+e^- \to 2\gamma$ implies the decay $\mu^+ \to e^+ \nu_e \nu_\mu$ of the Coulomb potential before the reaction, $(\mu^+ e^-)$ system in the state $(0, 0)$, as well as the transition $(1, +1) \to (0, 0)$.

Part IV

Motion in Electromagnetic Field

4001

We may generalize the semi-classical Bohr-Sommerfeld relation

$$\oint P \cdot dr = (n + 1/2)h,$$

(where the integral is along a closed orbit) to apply to the case where electro-magnetic field is present by replacing **P** with $\mathbf{p} - e\mathbf{A}/c$, where e is the charge of the particle. Use this and the equation of motion for the linear momentum **p** to derive a quantization condition on the magnetic flux of a semi-classical electron which is in a magnetic field **B** in an arbitrary orbit. For electrons in solids this condition can be restated in terms of the size S of the orbit in **k**-space. Obtain the quantization condition on S in terms of B (Ignore spin effects).

(*Chicago*)

Sol: In the presence of an electromagnetic field, the mechanical momentum **P** is
$$\mathbf{P} = \mathbf{p} - e\mathbf{A}/c,$$

where **p** is the canonical momentum, e is the charge of particle. The generalized Bohr-Sommerfeld relation becomes

$$\oint \mathbf{P} \cdot dr = \oint \left(\mathbf{p} - \frac{e}{c}\mathbf{A} \right) \cdot d\mathbf{r} = (n + 1/2)h,$$

or

$$\oint \mathbf{p} \cdot d\mathbf{r} - \frac{e}{c}\phi = (n + 1/2)h,$$

where

$$\phi = \int_S \mathbf{B} \cdot ds = \int_S (\nabla \times \mathbf{A}) \cdot ds = \oint \mathbf{A} \cdot d\mathbf{r},$$

using Stokes theorem. The classical equation of the motion of an electron in a constant magnetic field **B**,

$$\frac{d\mathbf{p}}{dt} = -\frac{e}{c}\frac{d\mathbf{r}}{dt}\times\mathbf{B},$$

gives $\mathbf{p} = -e\mathbf{r}\times\mathbf{B}/c$ and

$$\oint\mathbf{p}\cdot d\mathbf{r} = -\oint\frac{e}{c}(\mathbf{r}\times\mathbf{B})\cdot d\mathbf{r} = \int_S\frac{e}{c}\nabla\times(\mathbf{B}\times\mathbf{r})\cdot d\mathbf{s}$$

$$= \frac{e}{c}\int_S 2\mathbf{B}\cdot d\mathbf{s} = 2e\phi/c,$$

Hence

$$\phi = (n+1/2)\phi_0,$$

where $\phi_0 = hc/e$.

Defining **k** by $\mathbf{p} = \hbar\mathbf{k} = -e\mathbf{r}\times\mathbf{B}/c$, we have, assuming **r** is perpendicular to **B**,

$$\hbar\Delta k = -Be\Delta r/c,$$

or

$$\Delta r = -\hbar c\Delta k/Be.$$

Therefore, if the orbit occupies an area S_n in **k**-space and an area A_n in **r**-space, we have the relation

$$A_n = (\hbar c/Be)^2 S_n.$$

As

$$\phi = \int BdA_n = \left(\frac{\hbar c}{Be}\right)^2\int BdS_n = \left(\frac{\hbar c}{Be}\right)^2 BS_n = (n+1/2)hc/e,$$

we have

$$S_n = 2\pi Be(n+1/2)/\hbar c.$$

4002

A particle of charge q and mass m is subject to a uniform electrostatic field **E**.

a. Write down the time-dependent Schrödinger equation for this system.

b. Show that the expectation value of the position operator obeys Newton's second law of motion when the particle is in an arbitrary state $\psi(\mathbf{r}, t)$.

c. It can be shown that this result is valid if there is also a uniform magnetostatic field present. Do these results have any practical application to the design of instruments such as mass spectrometers, particle accelerators, etc.? Explain.

(Buffalo)

Sol:

a. The Schrödinger equation for the particle is

$$i\hbar \frac{\partial \psi}{\partial t} = -\frac{\hbar^2}{2m} \nabla^2 \psi - q\mathbf{E} \cdot \mathbf{r}\psi.$$

b. The Hamiltonian of the particle is

$$H = \frac{\mathbf{p}^2}{2m} - q\mathbf{E} \cdot \mathbf{r}.$$

Then

$$\frac{dx}{dt} = \frac{1}{i\hbar}[x, H] = \frac{1}{i\hbar}\left[x, \frac{p_x^2}{2m}\right] = \frac{p_x}{m}\frac{1}{i\hbar}[x, p_x] = \frac{p_x}{m},$$

$$\frac{dp_x}{dt} = \frac{1}{i\hbar}[p_x, H] = \frac{1}{i\hbar}[p_x, -E_x x] = -\frac{qE_x}{i\hbar}[p_x, x] = qE_x,$$

and hence

$$\frac{d\langle \mathbf{r} \rangle}{dt} = \frac{\langle \mathbf{p} \rangle}{m},$$

$$\frac{d\langle \mathbf{p} \rangle}{dt} = q\mathbf{E}.$$

Thus

$$\frac{d^2}{dt^2}\langle \mathbf{r} \rangle = \frac{1}{m}\frac{d\langle \mathbf{p} \rangle}{dt},$$

or

$$m\frac{d^2}{dt^2}\langle \mathbf{r} \rangle = q\mathbf{E},$$

which is just Newton's second law of motion.

c. These results show that we could use classical mechanics directly when we are computing the trajectory of a charged particle in instruments such as mass spectrometers, particle accelerators, etc.

4003

The Hamiltonian for a spinless charged particle in a magnetic field $\mathbf{B} = \nabla \times \mathbf{A}$ is

$$H = \frac{1}{2m}\left(\mathbf{p} - \frac{e}{c}\mathbf{A}(\mathbf{r})\right)^2 ,$$

where e is the charge of particle, $\mathbf{p} = (p_x, p_y, p_z)$ is the momentum conjugate to the particle's position $\mathbf{r}$. Let $A = -B_0 y \mathbf{e}_x$, corresponding to a constant magnetic field $\mathbf{B} = B_0 \mathbf{e}_z$.

a. Prove that p_x and p_z are constants of motion.

b. Find the (quantum) energy levels of this system.

(MIT)

Sol: The Hamiltonian for the particle can be written as

$$H = \frac{1}{2m}\left(p_x + \frac{eB_0}{c}y\right)^2 + \frac{1}{2m}p_y^2 + \frac{1}{2m}p_z^2 .$$

a.   As H does not depend on x and z explicitly, the basic commutation relations in quantum mechanics

$$[x_i, p_j] = i\hbar\delta_{ij}, \; [p_i, p_j] = 0,$$

require

$$[p_x, H] = 0, [p_z, H] = 0,$$

which show that p_x, p_z are constants of the motion.

b.   In view of (a) we can choose $\{H, p_x, p_z\}$ as a complete set of mechanical variables. The corresponding eigenfunction is

$$\psi(x, y, z) = e^{i(xp_x + zp_z)/\hbar}\phi(y),$$

where p_x, p_z are no longer operators but are now constants. The Schrödinger equation

$$H\psi(x, y, z) = E\psi(x, y, z)$$

then gives

$$\frac{1}{2m}\left[\left(p_x + \frac{eB_0}{c}y\right)^2 - \hbar^2\frac{d^2}{dy^2} + p_z^2\right]\phi(y) = E\phi(y),$$

or

$$-\frac{\hbar^2}{2m}\frac{d^2\phi}{dy^2}+\frac{m}{2}\left(\frac{eB_0}{mc}\right)^2\left(y+\frac{cp_x}{eB_0}\right)^2\phi=\left(E-\frac{p_z^2}{2m}\right)\phi.$$

Setting

$$\omega=\frac{|e|B_0}{mc},\ y'=y+\frac{cp_x}{eB_0},\ E'=E-\frac{p_z^2}{2m},$$

we can write the equation as

$$-\frac{\hbar^2}{2m}\frac{d^2\phi}{dy'^2}+\frac{m}{2}\omega^2 y'^2\phi=E'\phi,$$

which is the energy eigenequation for a one-dimensional harmonic oscillator. The energy eigenvalues are therefore

$$E'=E-p_z^2/2m=(n+1/2)\hbar\omega,\ n=0,1,2,....$$

Hence the energy levels for the system are

$$E_n=p_z^2/2m+(n+1/2)\hbar\omega,\ n=0,1,2,....$$

4004

An electron of mass m and charge $-e$ moves in a region where a uniform magnetic field $\mathbf{B}=\nabla\times\mathbf{A}$ exists in z direction.

  a. Set up the Schrödinger equation in rectangular coordinates.

  b. Solve the equation for all energy levels.

  c. Discuss the motion of the electron.

(*Buffalo*)

Sol:

  a. The Hamiltonian is

$$\hat{H}=\frac{1}{2m}\left(\hat{\mathbf{P}}+\frac{e}{c}\mathbf{A}\right)^2.$$

  As

$$\frac{\partial A_z}{\partial y}-\frac{\partial A_y}{\partial z}=0,$$

$$\frac{\partial A_x}{\partial z}-\frac{\partial A_z}{\partial x}=0,$$

$$\frac{\partial A_y}{\partial x}-\frac{\partial A_x}{\partial y}=B,$$

we can take $A_x = A_z = 0$, $A_y = Bx$, i.e. $\mathbf{A} = Bx\hat{y}$, and write the Schrödinger equation as

$$\hat{H}\psi = \frac{1}{2m}\left[\hat{P}_x^2 + \left(\hat{P}_y + \frac{eBx}{c}\right)^2 + \hat{P}_z^2\right]\psi = E\psi.$$

b. As $[\hat{P}_y, \hat{H}] = [\hat{P}_z, \hat{H}] = 0$, P_y and P_z are conserved. Choose H, P_y, P_z as a complete set of mechanical variables and write the Schrödinger equation as

$$\left[\frac{1}{2m}\hat{P}_x^2 + \frac{1}{2m}\left(\hat{P}_y + \frac{eBx}{c}\right)^2\right]\psi = \left(E - \frac{P_z^2}{2m}\right)\psi.$$

Let $\xi = x + cP_y/eB$, $\hat{P}_\xi = \hat{P}_x$. Then $[\xi, \hat{P}_\xi] = i\hbar$ and

$$\hat{H} = \frac{1}{2m}\hat{P}_\xi^2 + \frac{m}{2}\left(\frac{eB}{mc}\right)^2 \xi^2 + \frac{1}{2m}\hat{P}_z^2.$$

The above shows that $\hat{H} - \frac{\hat{P}_z^2}{2m}$ is the Hamiltonian of a one-dimensional harmonic oscillator of angular frequency $\omega = \frac{eB}{mc}$. Hence the energy levels of the system are

$$E = (n + 1/2)\hbar\omega + P_z^2/2m, \quad n = 0, 1, 2, \ldots.$$

Because the expression of E does not contain P_y explicitly, the degeneracies of the energy levels are infinite.

c. In the coordinate frame chosen, the energy eigenstates correspond to free motion in the z direction and circular motion in the $x - y$ plane, i.e. a helical motion. In the z direction, the mechanical momentum $mv_z = P_z$ is conserved, describing a uniform linear motion. In the x direction there is a simple harmonic oscillation round the equilibrium point $x = -cP_y / eB$. In the y direction, the mechanical momentum is $mv_y = P_y + eBx/c = eB\xi/c = m\omega\xi$ and so there is a simple harmonic oscillation with the same amplitude and frequency.

4005

Write down the Hamiltonian for a spinless charged particle in a magnetic field. Show that the gauge transformation $\mathbf{A}(\mathbf{r}) \to \mathbf{A}(\mathbf{r}) + \nabla f(\mathbf{r})$ is equivalent to multiplying the wave function by the factor $\exp[ief(\mathbf{r})/\hbar c]$. What is the significance of this result? Consider the case of a uniform field B directed along the z-axis. Show that the energy levels can be written as

$$E = (n + 1/2)\frac{|e|\hbar}{mc}B + \frac{\hbar^2 k_z^2}{2m} .$$

Discuss the qualitative features of the wave functions.

Hint: use the gauge where $A_x = -By$, $A_y = A_z = 0$.

(Wisconsin)

Sol: The Hamiltonian for the particle is

$$\hat{H} = \frac{1}{2m}\left(\hat{\mathbf{p}} - \frac{e}{c}\mathbf{A}\right)^2 ,$$

where **A** is related to the magnetic field by

$$\mathbf{B} = \nabla \times \mathbf{A}.$$

The Schrödinger equation is then

$$\frac{1}{2m}\left(\hat{\mathbf{p}} - \frac{e}{c}\mathbf{A}\right)^2 \psi(\mathbf{r}) = E\psi(\mathbf{r}).$$

Suppose we make the transformation

$$\mathbf{A}(\mathbf{r}) \rightarrow \mathbf{A}'(\mathbf{r}) = \mathbf{A}(\mathbf{r}) + \nabla f(\mathbf{r}),$$

$$\psi(\mathbf{r}) \rightarrow \psi'(\mathbf{r}) = \psi(\mathbf{r})\exp\left\{\frac{ie}{\hbar c}f(\mathbf{r})\right\},$$

and consider

$$\left(\hat{\mathbf{p}} - \frac{e}{c}\mathbf{A}'\right)\psi'(\mathbf{r}) = \hat{\mathbf{p}}\,\psi'(\mathbf{r}) - \left[\frac{e}{c}\mathbf{A} + \frac{e}{c}\nabla f(\mathbf{r})\right]\exp\left[\frac{ie}{\hbar c}f(\mathbf{r})\right]\psi(\mathbf{r})$$

$$= \exp\left[\frac{ie}{\hbar c}f(\mathbf{r})\right]\left(\hat{\mathbf{p}} - \frac{e}{c}\mathbf{A}\right)\psi(\mathbf{r}),$$

$$\left(\hat{\mathbf{p}} - \frac{e}{c}\mathbf{A}'\right)^2 \psi'(\mathbf{r}) = \exp\left[\frac{ie}{\hbar c}f(\mathbf{r})\right]\left(\hat{\mathbf{p}} - \frac{e}{c}\mathbf{A}\right)^2 \psi(\mathbf{r}),$$

where we have used

$$\hat{\mathbf{p}}\,\psi'(\mathbf{r}) = \frac{\hbar}{i}\nabla\left\{\exp\left[\frac{ie}{\hbar c}f(\mathbf{r})\right]\psi(\mathbf{r})\right\} = \exp\left[\frac{ie}{\hbar c}f(\mathbf{r})\right]\left[\frac{e}{c}\nabla f(\mathbf{r}) + \hat{\mathbf{p}}\right]\psi(\mathbf{r}).$$

Substitution in Schrödinger's equation gives

$$\frac{1}{2m}\left(\hat{\mathbf{p}} - \frac{e}{c}\mathbf{A}'\right)^2 \psi'(\mathbf{r}) = E\psi'(\mathbf{r}).$$

This shows that under the gauge transformation $\mathbf{A}' = \mathbf{A} + \nabla f$, the Schrödinger equation remains the same and that there is only a phase difference between the original and the new wave functions. Thus the system has gauge invariance.

Now consider the case of a uniform field $\mathbf{B} = \nabla \times \mathbf{A} = B e_z$, for which we have

$$A_x = -By, \qquad A_y = A_z = 0.$$

The Hamiltonian can be written as

$$\hat{H} = \frac{1}{2m}\left[\left(\hat{p}_x + \frac{eB}{c}y\right)^2 + \hat{p}_y^2 + \hat{p}_z^2\right].$$

Since $[\hat{p}_x, \hat{H}] = [\hat{p}_z, \hat{H}] = 0$ as H does not depend on x, z explicitly, we may choose the complete set of mechanical variables $(\hat{p}_x, \hat{p}_z, \hat{H})$. The corresponding eigenstate is

$$\psi(x, y, z) = e^{i(p_x x + p_z z)/\hbar}\chi(y).$$

Substituting it into the Schrödinger equation, we have

$$\frac{1}{2m}\left[\left(p_x + \frac{eB}{c}y\right)^2 - \hbar^2\frac{\partial^2}{\partial y^2} + p_z^2\right]\chi(y) = E\chi(y).$$

Let $cp_x/eB = -y_0$. Then the above equation becomes

$$-\frac{\hbar^2}{2m}\chi'' + \frac{m}{2}\left(\frac{eB}{mc}\right)^2 (y - y_0)^2 \chi = (E - p_z^2/2m)\chi,$$

which is the equation of motion of a harmonic oscillator. Hence the energy levels are

$$E = \frac{\hbar^2}{2m}k_z^2 + \left(n + \frac{1}{2}\right)\hbar\frac{|e|B}{mc}, \quad n = 0, 1, 2, \ldots,$$

where $k_z = p_z/h$, and the wave functions are

$$\psi_{p_x p_z n}(x, y, z) = e^{i(p_x x + p_z z)/\hbar}\chi_n(y - y_0),$$

where

$$\chi_n(y - y_0) \sim \exp\left[-\frac{|e|B}{2\hbar c}(y - y_0)^2\right]H_n\left(\sqrt{\frac{|e|B}{\hbar c}}(y - y_0)\right),$$

H_n being Hermite polynomials. As the expressions for energy does not depend on p_x and p_z explicitly, there are infinite degeneracies with respect to p_x and p_z.

4006

A point particle of mass m and charge q moves in spatially constant crossed magnetic and electric fields $\mathbf{B} = B_0\hat{z}$, $\mathbf{E} = E_0\hat{x}$.

a. Solve for the complete energy spectrum.

b. Evaluate the expectation value of the velocity $\mathbf{v}$ in a state of zero momentum.

(*Princeton*)

Sol:

a. Choose a gauge $A = B_0 x\hat{y}$, $\varphi = -E_0 x$ so that $\nabla \times A = B_0\hat{z}$, $-\nabla \cdot \varphi = E_0$. Then

$$H = \frac{1}{2m}\left(\mathbf{p} - \frac{q}{c}\mathbf{A}\right)^2 + q\varphi = \frac{1}{2m}\left[p_x^2 + \left(p_y - \frac{q}{c}B_0 x\right)^2 + p_z^2\right] - qE_0 x.$$

As H does not depend on y and z explicitly, p_y and p_z each commutes with H, so that p_y and p_z are conserved. Thus they can be replaced by their eigenvalues directly. Hence

$$H = \frac{1}{2m}p_x^2 + \frac{q^2 B_0^2}{2mc^2}\left(x - \frac{cp_y}{qB_0} - \frac{c^2 mE_0}{qB_0^2}\right)^2$$

$$+ \frac{1}{2m}p_z^2 - \frac{mc^2 E_0^2}{2B_0^2} - \frac{cp_y E_0}{B_0} = \frac{1}{2m}p_\xi^2$$

$$+ \frac{m}{2}\omega^2 \xi^2 + \frac{1}{2m}p_z^2 - \frac{mc^2 E_0^2}{2B_0^2} - \frac{cp_y E_0}{B_0},$$

where

$$p_\xi = p_x, \quad \xi = x - \frac{cp_y}{qB_0} - \frac{mc^2 E_0}{qB_0^2}$$

are a new pair of conjugate variables. Let $\omega = |q|B_0/mc$. By comparing the expression of H with that for a one-dimensional harmonic oscillator, we get the eigenvalues of H:

$$E_n = (n+1/2)\hbar\omega + p_z^2/2m - mc^2 E_0^2/2B_0^2 - cp_y E_0/B_0, \quad n = 0, 1, 2, \dots.$$

The fact that only p_y and p_z, but not y and z, appear in the expression for energy indicates an infinite degeneracy exists with respect to p_y and to p_z.

b. A state of zero momentum signifies one in which the eigenvalues of p_y and p_z as well as the expectation value of p_x are all zero. As velocity is defined as

$$\mathbf{v} = \frac{1}{m}\mathbf{p}_{\text{mec}} = \frac{1}{m}\left(\mathbf{p} - \frac{q}{c}\mathbf{A}\right),$$

its expectation value is

$$\langle\mathbf{v}\rangle = \frac{1}{m}\left\langle\mathbf{p} - \frac{q}{c}\mathbf{A}\right\rangle = -\frac{q}{mc}\langle\mathbf{A}\rangle = -\frac{qB_0}{mc}\langle x\rangle\hat{\mathbf{y}}.$$

Then as

$$\langle x\rangle = \langle\xi\rangle + \frac{cp_y}{qB_0} + \frac{mc^2 E_0}{qB_0^2} = \frac{mc^2 E_0}{qB_0^2},$$

since $\langle\xi\rangle = 0$ for a harmonic oscillator and $p_y = 0$, we have

$$\langle\mathbf{v}\rangle = -\frac{cE_0}{B_0}\hat{\mathbf{y}}.$$

4007

Determine the energy levels, their degeneracy and the corresponding eigenfunctions of an electron contained in a cube of essentially infinite volume L^3. The electron is in an electromagnetic field characterized by the vector potential

$$\mathbf{A} = H_0 x\hat{\mathbf{e}}_y \ (|\hat{\mathbf{e}}_y| = 1).$$

(Chicago)

Sol: As $\mathbf{A} = H_0 x\hat{\mathbf{e}}_y$, we have the Schrödinger equation

$$\hat{H}\psi = \frac{1}{2m}[\hat{p}_x^2 + \hat{p}_z^2 + (\hat{p}_y - H_0 xe/c)^2]\psi = E\psi,$$

where e is the electron charge $(e < 0)$.

As $[\hat{H}, \hat{p}_y] = [\hat{H}, \hat{p}_z] = 0, [\hat{p}_y, \hat{p}_z] = 0$, we can choose $\hat{H}, \hat{p}_y, \hat{p}_z$ as a complete set of mechanical variables, the corresponding eigenfunction being

$$\psi = e^{i(p_y y + p_z z)/\hbar}\psi_0(x),$$

where p_y, p_z are arbitrary real numbers. Substitution of ψ in the Schrödinger equation gives

$$\frac{1}{2m}[\hat{p}_x^2 + (eH_0/c)^2(x - cp_y/eH_0)^2]\psi_0 = E_0\psi_0,$$

where $E_0 = E - p_z^2/2m$, or

$$-\frac{\hbar^2}{2m}d^2\psi_0/dx^2 + \frac{m}{2}(H_0 e/cm)^2(x-x_0)^2\psi_0 = E_0\psi_0,$$

where $x_0 = cp_y/eH_0$.

The last equation is the energy eigenequation of a one-dimensional oscillator of natural frequency $\omega_0 = -H_0 e/mc$ and equilibrium position $x = x_0$, the energy eigenvalues being

$$E_0 = (n+1/2)\hbar\omega_0, \quad n = 0, 1, 2, \ldots,$$

or

$$E = p_z^2/2m - (n+1/2)H_0 e\hbar/mc, \quad n = 0, 1, 2, \ldots.$$

The corresponding eigenfunctions are

$$\psi_{0n} \sim \exp\left[\frac{eH_0}{2\hbar c}(x-x_0)^2\right]H_n\left(-\frac{eH_0}{\hbar c}(x-x_0)\right),$$

where H_n are Hermite polynomials.

As no p_y terms occur in the expression for energy levels and p_y can be any arbitrary real number, the degeneracies of energy levels are infinite.

The eigenfunctions for the original system are therefore

$$\psi(x) = C_n \exp\left[\frac{i(p_y y + p_z z)}{\hbar} + \frac{eH_0}{2\hbar c}(x-x_0)^2\right]$$
$$\times H_n\left(-\frac{eH_0}{\hbar c}(x-x_0)\right),$$

where C_n is the normalization constant.

4008

Consider a hydrogen atom placed in a uniform constant magnetic field B. Find the energy spectrum of this atom.

Sol:

The Hamiltonian of a particle in a magnetic field is given by

$$H = \frac{1}{2\mu}\left(\vec{p} + \frac{e}{c}\vec{A}\right)^2 \tag{1}$$

The Schrodinger wave equation is

$$H\psi = E\psi$$

$$\frac{1}{2\mu}\left(\vec{p}+\frac{e}{c}\vec{A}\right)^2 \psi = E\psi$$

$$\frac{1}{2\mu}\left(\vec{p}+\frac{e}{c}\vec{A}\right)\left(\vec{p}\psi+\frac{e}{c}\vec{A}\psi\right) = E\psi$$

$$\frac{1}{2\mu}\left(-i\hbar\nabla+\frac{e}{c}\vec{A}\right)\left(-i\hbar\nabla\psi+\frac{e}{c}\vec{A}\psi\right) = E\psi$$

$$\frac{-\hbar^2}{2\mu}\nabla^2\psi - \frac{ie\hbar}{\mu c}\vec{A}.\nabla\psi - \frac{ie\hbar}{\mu c}(\nabla.\vec{A})\psi + \frac{e^2}{2mc^2}A^2\psi = E\psi$$

$$\frac{-\hbar^2}{2\mu}\nabla^2\psi - \frac{ie\hbar}{\mu c}\vec{A}.\nabla\psi + \frac{e^2}{2mc^2}A^2\psi = E\psi \qquad (\because \nabla.\vec{A}=0) \qquad (2)$$

For a constant magnetic field B, we can choose the direction along z axis and

$$\vec{A}=\frac{1}{2}\vec{r}\times\vec{B}$$

To evaluate equation (2), we use

$$\vec{r}\times\vec{B}.\nabla\psi = -\vec{B}.\vec{r}\times\nabla\psi = -\frac{i}{\hbar}\vec{B}.\vec{L}\psi$$

$$(\vec{r}\times\vec{B})^2 = r^2B^2 - (\vec{r}.\vec{B})^2$$

and write

$$\frac{-\hbar^2}{2\mu}\nabla^2\psi - \frac{e}{2\mu c}\vec{B}.\vec{L}\psi + \frac{e^2}{8mc^2}[r^2B^2 - (\vec{r}.\vec{B})^2]\psi = E\psi \qquad (3)$$

For small values of r, the third term in equation (3) is very small and can be neglected.

Therefore, the Schrodinger wave equation for a particle in a uniform constant magnetic field becomes

$$\frac{-\hbar^2}{2\mu}\nabla^2\psi - \frac{e}{2\mu c}\vec{B}.\vec{L}\psi = E\psi$$

In the above equation, the additional term $\dfrac{e}{2\mu c}\vec{B}.\vec{L} = \dfrac{e}{2\mu c}B_z.L_z$ is called the Zeeman term.

The Hamiltonian of the Hydrogen atom in a magnetic field is the sum of the Hamiltonian of a free Hydrogen atom and Zeeman term,

$$(H_{\text{Hydrogen}} + H_{\text{Zeeman}})\psi_{nlm} = (E_n + m\mu_B B)\psi_{nlm} \qquad (4)$$

where μ_B is the Bohr magneton and m is the magnetic Quantum number.
The energy spectrum of the system is discrete and given by equation (4).

4009

a. Assuming that nonrelativistic quantum mechanics is invariant under time reversal, derive the time reversed form of the Schrödinger wave function.

b. What is the quantum mechanical Hamiltonian for a free electron with magnetic moment μ in the external constant magnetic field H_z in the z-direction, in the reference frame of the electron?

c. Suppose that an extra constant magnetic field H_y is imposed in the y-direction. Determine the form of the quantum mechanical operator for the time rate of change of μ in this case.

(Buffalo)

Sol:

a. Consider the Schrödinger equation

$$i\hbar\frac{\partial}{\partial t}\psi(t)=\hat{H}\psi(t).$$

Making the time reversal transformation $t\to-t$, we obtain

$$-i\hbar\frac{\partial}{\partial t}\psi(-t)=\hat{H}(-t)\psi(-t),$$

or

$$i\hbar\frac{\partial}{\partial t}\psi^*(-t)=\hat{H}^*(-t)\psi^*(-t).$$

If $\hat{H}^*(-t)=\hat{H}(t)$, then the Schrödinger equation is covariant under time reversal and the time reversed form of the wave function is $\psi^*(-t)$.

b. Let $-e$ be the charge of the electron. Then $\mu=-\frac{e\hbar}{2mc}\sigma$ and in the reference frame of the electron,

$$\hat{H}=-\mu\cdot\mathbf{H}=-\mu_z H_z=\frac{e\hbar}{2mc}\sigma_z H_z.$$

c. The magnetic field is now $H_y\hat{\mathbf{y}}+H_z\hat{\mathbf{z}}$, and so

$$\hat{H}=\frac{e\hbar}{2mc}(\sigma_z H_z+\sigma_y H_y),$$

so

$$\frac{d\mu}{dt}=\frac{1}{ih}[\mu,\hat{H}]=\frac{1}{ih}\left(\frac{e\hbar}{2mc}\right)^2[-\sigma_x\hat{x}-\sigma_y\hat{y}-\sigma_z\hat{z},$$

$$\sigma_zH_z+\sigma_yH_y]=\frac{2}{\hbar}\left(\frac{e\hbar}{2mc}\right)^2[(\sigma_yH_z-\sigma_zH_y)\hat{x}-\sigma_xH_z\hat{y}$$

$$+i\sigma_xH_y\hat{z}]=\frac{2}{\hbar}\left(\frac{e\hbar}{2mc}\right)^2\sigma\times\mathbf{H}$$

$$=\frac{e}{mc}\mathbf{H}\times\mu,$$

where use has been made of the relations $\sigma_x\sigma_y=i\sigma_z$, $\sigma_y\sigma_z=i\sigma_x$, $\sigma_z\sigma_x=i\sigma_y$.

4010

An electron is subject to a static uniform magnetic field $B=B_0\,\mathbf{z}$ and occupies the spin eigenstate $|\uparrow\rangle$. At a given moment $t=0$, an additional time-dependent, spatially uniform magnetic field $B_1(t)=B_1(\cos\omega t\,x+\sin\omega t\,y)$ is turned on. Calculate the probability of finding the electron with its spin along the negative z-axis at time $t>0$. Ignore spatial degrees of freedom.

Sol:

The Schrodinger equation for the system is

$$i\hbar\frac{d}{dt}|\psi(t)\rangle=-\frac{e}{m_e}(B_0S_z+B_1\cos\omega tS_x+B_1\sin\omega tS_y)|\psi(t)\rangle$$

Setting

$$\psi(t)=(a(t),b(t))^T$$

we obtain

$$\frac{da}{dt}=i\frac{e}{2m_e}(B_0a+B_1e^{-i\omega t}b)$$

$$\frac{db}{dt}=i\frac{e}{2m_e}(-B_0b+B_1e^{-i\omega t}a)$$

Introducing

$$\omega_0=\frac{|e|B_0}{2m_e},\quad \omega_1=\frac{|e|B_1}{2m_e}$$

we obtain the above equations in the form

$$\frac{da}{dt} = -i\omega_0 a - i\omega_1 e^{-i\omega t} b$$

$$\frac{db}{dt} = i\omega_0 b - i\omega_1 e^{i\omega t} a$$

At this point, let us substitute the trial solutions

$$a(t) = \exp(-i\omega t/2 + i\Omega t A), b(t) = \exp(i\omega t/2 + i\Omega t B)$$

We immediately obtain

$$\Omega = \pm\frac{1}{2}\sqrt{(\omega - 2\omega_0)^2 + 4\omega_1^2} = \pm\frac{1}{2}\gamma$$

and

$$B = -\frac{\pm\gamma - (\omega - 2\omega_0)}{2\omega_1} A$$

Finally, we obtain

$$a(t) = e^{-i\omega t/2}(A_+ e^{i\gamma t/2} + A_- e^{-i\gamma t/2})$$

$$b(t) = e^{i\omega t/2}(-\frac{\gamma - (\omega - 2\omega_0)}{2\omega_1} A_+ e^{i\gamma t/2} + \frac{\gamma + (\omega - 2\omega_0)}{2\omega_1} A_- e^{-i\gamma t/2})$$

Applying the initial condition $a(0) = 1$, $b(0) = 0$, we arrive at

$$A_\pm = \frac{\gamma \pm (\omega - 2\omega_0)}{2\gamma}$$

and

$$\psi(t) = \left(e^{-i\omega t/2}(\cos\frac{\gamma t}{2} + i\frac{\omega - 2\omega_0}{\gamma}\sin\frac{\gamma t}{2}), -2ie^{-i\omega t/2}\frac{\omega_1}{\gamma}\sin\frac{\gamma t}{2}\right)^T$$

The probability of finding the spin of the electron pointing along $-z$ is

$$P_\downarrow = \frac{4\omega_1^2}{\omega - 2\omega_0 + 4\omega_1^2}\sin^2\frac{\gamma t}{2}$$

Note that for $\omega = 2\omega_0$, this probability exhibits the resonance phenomenon. For this choice, the probability of spin flip is $P_{\downarrow\omega=2\omega_0} = \sin^2\omega_1 t$ and becomes unity at $t = (2n+1)\pi/2\omega_1$.

4011

In a recent classic table-top experiment, a monochromatic neutron beam ($\lambda = 1.445\,\overset{\circ}{A}$) was split by Bragg reflection at point A of an interferometer into two beams which were recombined (after another reflection) at point D (see Fig. 4.1). One beam passes through a region of transverse magnetic field of strength B for a distance l. Assume that the two paths from A to D are identical except for the region of the field.

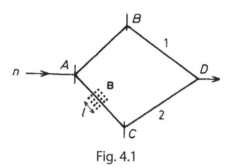

Fig. 4.1

Find the explicit expressions for the dependence of the intensity at point D on B, l and the neutron wavelength, with the neutron polarized either parallel or anti-parallel to the magnetic field.

(Chicago)

Sol: This is a problem on spinor interference. Consider a neutron in the beam. There is a magnetic field $\mathbf{B}$ in the region where the Schrödinger equation for the (uncharged) neutron is

$$\left(-\frac{\hbar^2}{2m}\nabla^2 - \mu\,\boldsymbol{\sigma}\cdot\mathbf{B} \right)\psi = E\psi .$$

Supposing $\mathbf{B}$ to be constant and uniform, we have

$$\psi(t_1) = \exp[-i\,\hat{H}(t_1 - t_0)/\hbar]\psi(t_0),$$

where t_0, t_1 are respectively the instants when the neutron enters and leaves the magnetic field.

Write $\psi(t) = \psi(\mathbf{r},t)\psi(\mathbf{s},t)$, where $\psi(\mathbf{r},t)$ and $\psi(\mathbf{s},t)$ are respectively the space and spin parts of ψ. Then

$$\psi(\mathbf{r}, t_1) = \exp\left[-\frac{i}{\hbar}\left(-\frac{\hbar^2}{2m}\nabla^2\right)(t_1 - t_0)\right]\psi(\mathbf{r}, t_0),$$

which is the same as the wave function of a free particle, and

$$\psi(\mathbf{s}, t_1) = \exp\left[\frac{i}{\hbar}\mu\boldsymbol{\sigma}\cdot\mathbf{B}(t_1 - t_0)\right]\psi(\mathbf{s}, t_0).$$

The interference arises from the action of $\mathbf{B}$ on the spin wave function. As $\psi(\mathbf{r}, t)$ is the wave function of a free particle, we have $t_1 - t_0 = l/v = ml/\hbar k$ and

$$\psi(\mathbf{s}, t_1) = \exp[i2\pi\mu ml\lambda\boldsymbol{\sigma}\cdot\mathbf{B}/h^2]\psi(s, t_0),$$

where $k = \frac{2\pi}{\lambda} = \frac{mv}{\hbar}$ is the wave number of the neutron. The intensity of the interference of the two beams at D is then proportional to

$$|\psi_D^{(1)}(\mathbf{r}, t)\,\psi_D^{(1)}(\mathbf{s}, t) + \psi_D^{(2)}(\mathbf{r}, t)\psi_D^{(2)}(\mathbf{s}, t)|^2$$

$$\propto |\psi_D^{(1)}(\mathbf{s}, t) + \psi_D^{(2)}(\mathbf{s}, t)|^2 = |\psi^{(2)}(\mathbf{s}, t_0) + \psi^{(2)}(\mathbf{s}, t_1)|^2.$$

As

$$\exp\left(i\frac{2\pi\mu ml\lambda}{h^2}\boldsymbol{\sigma}\cdot\mathbf{B}\right) = \cos\frac{2\pi\mu ml\lambda B}{h^2} + i\boldsymbol{\sigma}\cdot\frac{\mathbf{B}}{B}$$

$$\times \sin\frac{2\pi\mu ml\lambda B}{h^2},$$

and $\boldsymbol{\sigma}\cdot\mathbf{B} = \pm\sigma B$ depending on whether $\boldsymbol{\sigma}$ is parallel or anti-parallel to $\mathbf{B}$, we have

$$\left|\psi^{(2)}(\mathbf{s}, t_0) + \psi^{(2)}(\mathbf{s}, t_1)\right|^2 = \left|1 + \exp\left(i\frac{2\pi\mu ml\lambda}{h^2}\boldsymbol{\sigma}\cdot\mathbf{B}\right)\right|^2 \left|\psi(\mathbf{s}, t_0)\right|^2$$

$$= \left|1 + \cos\frac{2\pi\mu ml\lambda B}{h^2} \pm i\sigma\sin\frac{2\pi\mu ml\lambda B}{h^2}\right|^2$$

$$= \left(1 + \cos\frac{2\pi\mu ml\lambda B}{h^2}\right)^2$$

$$+ \sin^2\frac{2\pi\mu ml\lambda B}{h^2} = 4\cos^2\frac{\pi\mu ml\lambda B}{h^2}.$$

Therefore, the interference intensity at $D \propto \cos^2(\pi\mu ml\lambda B/h^2)$, where μ is the intrinsic magnetic moment of the neutron ($\mu < 0$).

4012

A neutron interferometer beam splitter plus mirrors as shown in Fig. 4.2 has been built out of a single crystal.

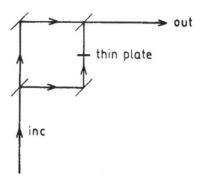

Fig. 4.2

a. By varying the thickness of a thin plastic sheet placed in the beam in one arm of the interferometer one can vary the relative phase and hence shift the fringes. Give a brief qualitative explanation of the origin of the phase shift.

b. By inserting in one arm a magnetic field which is normal to the beam, time independent and very nearly uniform so that the force on the neutrons can be neglected, and by choosing the field so each neutron spin vector precesses through just one rotation, one finds the relative phase of the two beams is shifted by π radians, or one-half cycle. Explain, with appropriate equations, why this is so.

(Princeton)

Sol:

a. When a neutron passes through the thin plastic sheet, it is under the action of an additional potential, and so its momentum changes together with its de Broglie wavelength. The phase change of the neutron when it passes through the plastic sheet is different from that when it passes through a vacuum of the same thickness. If the thickness of the plastic sheet is varied, the relative phase of the two beams (originating from the same beam) also changes, causing a shift in the fringes.

b. The neutron possesses an anomalous magnetic moment $\mu_n = -\mu_n \sigma$ and its Schrödinger equation is

$$(\mathbf{p}^2/2m_n + \mu_n \sigma \cdot \mathbf{B})\psi = E\psi.$$

We may neglect the reflection that occurs when a neutron wave is incident on the "surface" of the sheet-like magnetic field as the action of the field on the neutron is rather weak. Under such an approximation we may show (by solving the above two-spin-component Schrödinger equation for a one-dimensional square well): The wave function ψ_{inc} for a neutron incident normally on the sheet-like magnetic field is related to the transmitted wave function ψ_{out} out of the field by a unitary transformation

$$\psi_{out} = \exp(-i\sigma \cdot \rho/2)\psi_{inc},$$

where $\rho = \omega_L \tau e_B$, with $\omega_L = 2\mu_n B/\hbar$ being the Larmor frequency, $\tau = Lm_n/\hbar k$ the time taken for the neutron to pass through the magnetic field of thickness L, e_B the unit vector in the direction of $\mathbf{B}$, k the wave number of the incident neutron.

If a neutron is polarized in the (θ, φ) direction before entering the field, i.e., its polarized vector is

$$\langle \psi_{inc} | \sigma | \psi_{inc} \rangle = \{\sin\theta\cos\varphi, \sin\theta\sin\varphi, \cos\theta\},$$

then we can take

$$\psi_{inc} = \begin{pmatrix} e^{-i\varphi/2} & \cos\dfrac{\theta}{2} \\ e^{i\varphi/2} & \sin\dfrac{\theta}{2} \end{pmatrix} e^{i\mathbf{k}\cdot\mathbf{x}},$$

where θ is the angle the polarized vector makes with the direction of the magnetic field. Taking the latter as the z direction, we have $\rho \cdot \sigma = \rho\sigma_z$. Then as

$$\exp\left(-i\frac{\rho}{2}\sigma_z\right) = \cos\frac{\rho}{2} - i\sigma_z\sin\frac{\rho}{2} = \cos\frac{\rho}{2}\begin{pmatrix} 1 & 0 \\ 0 & 1 \end{pmatrix}$$

$$-i\sin\frac{\rho}{2}\begin{pmatrix} 1 & 0 \\ 0 & -1 \end{pmatrix} = \begin{pmatrix} e^{-i\rho/2} & 0 \\ 0 & e^{i\rho/2} \end{pmatrix},$$

we have

$$\psi_{out} = \begin{pmatrix} e^{-i\rho/2} & 0 \\ 0 & e^{i\rho/2} \end{pmatrix} \psi_{inc} = \begin{pmatrix} e^{-i(\varphi+\rho)/2}\cos\dfrac{\theta}{2} \\ e^{i(\varphi+\rho)/2}\sin\dfrac{\theta}{2} \end{pmatrix} e^{i\mathbf{k}\cdot\mathbf{x}}.$$

By adjusting $\mathbf{B}$ (or L) so that $\rho = 2\pi$, we make the polarized vector of a neutron precess through one rotation as it traverses the region of the magnetic field. Then

$$\psi_{\text{out}} = \begin{pmatrix} e^{-i\varphi/2} \cos\dfrac{\theta}{2} \\ e^{i\varphi/2} \sin\dfrac{\theta}{2} \end{pmatrix} e^{i\mathbf{k}\cdot\mathbf{x}+\pi},$$

i.e., the phase of the transmitted wave increases by π. Hence, compared with the wave traversing the other arm (without magnetic field), the relative phase of the beam changes by a half-cycle.

4013

a. A hydrogen atom is in its 2P state, in a state of $L_x = +\hbar$. At time $t = 0$ a strong magnetic field of strength $|\mathbf{B}|$ pointing in the z direction is switched on. Assuming that the effects of electron spin can be neglected, calculate the time dependence of the expectation value of L_x.

b. How strong must the magnetic field in part (a) be so that the effects of electron spin can actually be neglected? The answer should be expressed in standard macroscopic units.

c. Suppose that, instead, the magnetic field is very weak. Suppose, further, that at $t = 0$ the atom has $L_x = +\hbar$ and $s_x = \frac{1}{2}\hbar$, and the magnetic field is still oriented in the z direction. Sketch how you would calculate the time dependence of the expectation value of L_x in this case. You need not do the full calculation, but explain clearly what the main steps would be.

Note: All effects of nuclear spin are to be ignored in this problem.

(Princeton)

Sol:

a. The initial wave function of the atom is

$$\psi(\mathbf{r}, t=0) = R_{21}(r)\Theta(\theta, \varphi),$$

where

$$\Theta(\theta, \varphi) = \frac{1}{2}(Y_{11} + Y_{1-1} + \sqrt{2}\,Y_{10}),$$

is the eigenstate of $L_x = \hbar$.

At $t = 0$ a strong magnetic field Be_z is switched on. Then for $t \geq 0$ the Hamiltonian of the system is

$$\hat{H} = \frac{\mathbf{p}^2}{2m_e} + \frac{eBl_z}{2m_e c} + \frac{e^2 B^2 (x^2 + y^2)}{8m_e c^2} - \frac{e^2}{r}.$$

For a not too strong magnetic field $B \sim 10^5$ Gs, we can neglect the B^2 term and take as the Hamiltonian

$$\hat{H} = \frac{\mathbf{p}^2}{2m_e} + \frac{eBl_z}{2m_e c} - \frac{e^2}{r}.$$

The Schrödinger equation

$$i\hbar \partial \psi / \partial t = \hat{H}\psi$$

then gives the eigenstate solutions

$$\psi_n(\mathbf{r}, t) = R_{nl}(r) Y_{lm}(\theta, \varphi) e^{-iE_{nlm} t/\hbar},$$

where

$$E_{nlm} = E_{nl} + \frac{eB}{2m_e c} m\hbar.$$

Thus the general solution is

$$\psi(\mathbf{r}, t) = \sum_{n,l,m} a_n \psi_{nlm}(\mathbf{r}) \exp\left(-i\frac{E_{nlm}}{\hbar} t\right).$$

For $t = 0$, we then have

$$\sum_{n,l,m} a_n \psi_{nlm}(\mathbf{r}) = R_{21}(r)\left(\frac{1}{2}Y_{11} + \frac{1}{2}Y_{1-1} + \frac{1}{\sqrt{2}}Y_{10}\right),$$

or

$$a_2 \psi_{211}(\mathbf{r}) = \frac{1}{2}R_{21}(r)Y_{11}, \text{ etc.}$$

Hence

$$\psi(\mathbf{r}, t) = R_{21}(r)\left[\frac{1}{2}Y_{11}\exp\left(-i\frac{E_{211}}{\hbar}t\right)\right.$$

$$+ \frac{1}{2}Y_{1-1}\exp\left(-i\frac{E_{21-1}}{\hbar}t\right)$$

$$+ \left.\frac{1}{\sqrt{2}}Y_{10}\exp\left(-i\frac{E_{210}}{\hbar}t\right)\right].$$

The expectation value of L_x is given by $\langle \psi(\mathbf{r},t)| L_x | \psi(\mathbf{r},t) \rangle$. As $L_x = (L_+ + L_-)/2$,

$$L_+ Y_{lm} = \hbar \sqrt{(l+m+1)(l-m)} Y_{l,m+1},$$
$$L_- Y_{lm} = \hbar \sqrt{(l-m+1)(l+m)} Y_{l,m-1},$$

we have

$$L_x Y_{11} = \frac{\hbar}{\sqrt{2}} Y_{10}, \qquad L_x Y_{1-1} = \frac{\hbar}{\sqrt{2}} Y_{10},$$

$$L_x Y_{10} = \frac{\hbar}{\sqrt{2}} (Y_{11} + Y_{1-1}),$$

and hence

$$\overline{L_x(t)} = \langle \psi(\mathbf{r},t)| L_x | \psi(\mathbf{r},t) \rangle = \hbar \cos \frac{eBt}{2m_e c}.$$

b. The effects of electron spin can be neglected if the additional energy due to the strong magnetic field is much greater than the coupling energy for spin-orbit interaction, i.e.,

$$\frac{e\hbar B}{2m_e c} \gg \Delta E_{\text{spin-orbit}} \approx 10^{-3} \,\text{eV},$$

or

$$B \geq 10^6 \,\text{Gs}.$$

Thus when the magnetic field B is greater than 10^6 Gs, the effects of electron spin can be neglected.

c. If the magnetic field is very weak, the effects of electron spin must be taken into consideration. To calculate the time dependence of the expectation value of L_x, follow the steps outlined below.

   i. The Hamiltonian is now

$$\hat{H} = \frac{\mathbf{p}^2}{2m_e} - \frac{e^2}{r} + \frac{e^2}{2m_e^2 c^2 r^3}(\hat{\mathbf{s}} \cdot \hat{\mathbf{L}}) + \frac{eB}{2m_e c}\hat{j}_z + \frac{eB}{2m_e c}\hat{s}_z,$$

   which is the Hamiltonian for anomalous Zeeman effect and we can use the coupling representation. When calculating the additional energy due to the $\hat{s}_x$ term, we can regard $\hat{s}_x$ as approximately diagonal in this representation.

   ii. Write down the time-dependent wave function which satisfies the initial condition $L_x = +\hbar$ and $s_x = \frac{1}{2}\hbar$. At time $t = 0$ the wave function is

$$\psi_0(\mathbf{r}, s_x) = R_{21}(r)\Theta(\theta, \varphi)\phi_s,$$

where Θ and ϕ_s are the eigenfunctions of $L_x = \hbar$ and $s_x = \hbar/2$ in the representations (l^2, l_z), (s^2, s_z) respectively. Explicitly,

$$\psi_0(\mathbf{r}, s_x) = R_{21}(r)\frac{1}{2}[Y_{11} + Y_{1-1} + \sqrt{2}Y_{10}]\frac{1}{\sqrt{2}}(\alpha + \beta)$$

$$= \frac{R_{12}(r)}{2\sqrt{2}}(Y_{11}\alpha + Y_{11}\beta + Y_{1-1}\alpha + Y_{1-1}\beta$$

$$+ \sqrt{2}Y_{10}\alpha + \sqrt{2}Y_{10}\beta).$$

As

$$\phi_j = \frac{3}{2}, \quad m_j = \frac{3}{2} = Y_{11}\alpha, \quad \phi_{\frac{3}{2}\frac{1}{2}} = \sqrt{\frac{1}{3}}Y_{11}\beta + \sqrt{\frac{2}{3}}Y_{10}\alpha,$$

$$\phi_{\frac{3}{2}-\frac{1}{2}} = \sqrt{\frac{1}{3}}Y_{1-1}\alpha + \sqrt{\frac{2}{3}}Y_{10}\beta, \quad \phi_{\frac{3}{2}-\frac{3}{2}} = Y_{1-1}\beta,$$

ψ_0 can be written in the coupling representation as

$$\psi_0(\mathbf{r}, s_x) = \frac{1}{2\sqrt{2}}R_{21}(r)\left(\phi_{\frac{3}{2}\frac{3}{2}} + \sqrt{3}\phi_{\frac{3}{2}\frac{1}{2}}\right.$$

$$\left. + \sqrt{3}\phi_{\frac{3}{2}-\frac{1}{2}} + \phi_{\frac{3}{2}-\frac{3}{2}}\right),$$

where ϕ_{jm_j} is the eigenfunction of (j^2, j_z) for the energy level E_{nljm_j}.

Therefore, the time-dependent wave function for the system is

$$\psi(\mathbf{r}, s, t) = \frac{1}{2\sqrt{2}}R_{21}(r) \cdot \left[\phi_{\frac{3}{2}\frac{3}{2}}\exp\left(-i\frac{E_{21\frac{3}{2}\frac{3}{2}}}{\hbar}t\right)\right.$$

$$+ \sqrt{.3}\phi_{\frac{3}{2}\frac{1}{2}}\exp\left(-i\frac{E_{21\frac{3}{2}\frac{1}{2}}}{\hbar}t\right)$$

$$+ \sqrt{.3}\phi_{\frac{3}{2}-\frac{1}{2}}\exp\left(-i\frac{E_{21\frac{3}{2}-\frac{1}{2}}}{\hbar}t\right)$$

$$\left. + \phi_{\frac{3}{2}-\frac{3}{2}}\exp\left(-i\frac{E_{21\frac{3}{2}-\frac{3}{2}}}{\hbar}t\right)\right].$$

iii. Calculate the expectation value of L_x in the usual manner:

$$\langle\psi(\mathbf{r}, s, t) | L_x | \psi(\mathbf{r}, s, t)\rangle.$$

4014

Consider the one-dimensional motion of an uncharged particle of spin 1/2 and magnetic moment $\mu = -2\mu_0 s/h$. The particle is confined in an infinite square well extending from $x = -L$ to $x = L$. In region I ($x < 0$) there is a uniform magnetic field in the z direction $\mathbf{B} = B_0\mathbf{e}_z$; in region II ($x > 0$) there is a uniform field of the same magnitude but pointing in the x direction $\mathbf{B} = B_0\mathbf{e}_x$. Here $\mathbf{e}_x$ and $\mathbf{e}_z$ are unit vectors in the x and z directions.

a. Use perturbation theory to find the ground state energy and ground state wave function (both space and spin parts) in the weak field limit $B_0 \ll (\hbar/L)^2/2m\mu_0$.

b. Now consider fields with B_0 of arbitrary strength. Find the general form of the energy eigenfunction ψ_I (both space and spin parts) in region I which satisfies the left-hand boundary condition. Find also the form ψ_{II} that the eigenfunction has in region II which satisfies the right-hand boundary condition (Fig. 4.3).

c. Obtain an explicit determinantal equation whose solutions would give the energy eigenvalues E.

(MIT)

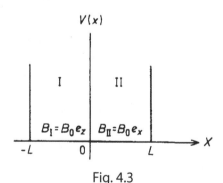

Fig. 4.3

Sol:

a. In the absence of magnetic field, $H = H_0$ and the energy eigenfunctions (space part) and eigenvalues are respectively

$$\psi_n = \sqrt{1/L}\sin\frac{n\pi(x+L)}{2L},$$

$$E_n = \frac{\pi^2\hbar^2 n^2}{8mL^2}, \quad n=1, 2, 3, \ldots.$$

As for the spin part, we know that each energy level has a degeneracy of 2. When a magnetic field is present, $H = H_0 + H'$, where

$$H' = -\boldsymbol{\mu}\cdot\mathbf{B} = \frac{2\mu_0\mathbf{s}\cdot\mathbf{B}}{\hbar} = \mu_0\boldsymbol{\sigma}\cdot\mathbf{B} = \begin{cases} \mu_0 B_0\sigma_z, & -L\leq x\leq 0, \\ \mu_0 B_0\sigma_x, & 0\leq x\leq L, \\ 0 & \text{elsewhere.} \end{cases}$$

If the field is weak, let $u_1 = \psi_1(x)\begin{pmatrix}1\\0\end{pmatrix}$, $u_2 = \psi_1(x)\begin{pmatrix}0\\1\end{pmatrix}$ be the base vectors.

Then

$$H'_{11} = \langle u_1 | H' | u_1\rangle = \mu_0 B_0^*(1\ 0)\begin{pmatrix}1 & 0\\0 & -1\end{pmatrix}\begin{pmatrix}1\\0\end{pmatrix}$$

$$\times \int_{-L}^0 \psi_1^*(x)\psi_1(x)dx = \frac{\mu_0 B_0}{2},$$

$$H'_{21} = H'_{12} = \langle u_1 | H' | u_2\rangle$$

$$= \mu_0 B_0(1\ 0)\begin{pmatrix}0 & 1\\1 & 0\end{pmatrix}\begin{pmatrix}0\\1\end{pmatrix}\int_0^L \psi_1^*(x)\psi_1(x)dx = \frac{\mu_0 B_0}{2},$$

$$H'_{22} = \langle u_2 | H' | u_2\rangle$$

$$= -\mu_0 B_0(0\ 1)\begin{pmatrix}1 & 0\\0 & -1\end{pmatrix}\begin{pmatrix}0\\1\end{pmatrix}\int_{-L}^0 \psi_1^*(x)\psi_1(x)dx = -\frac{\mu_0 B_0}{2},$$

and from $\det(H' - E^{(1)}I) = 0$ we get

$$\left(\frac{\mu_0 B_0}{2} - E^{(1)}\right)\left(-\frac{\mu_0 B_0}{2} - E^{(1)}\right) - \frac{\mu_0^2 B_0^2}{4} = 0,$$

or

$$E^{(1)} = \pm\frac{1}{\sqrt{2}}\mu_0 B_0.$$

The ground state energy level is therefore

$$E_0 = \frac{\pi^2\hbar^2}{8mL^2} - \frac{1}{\sqrt{2}}\mu_0 B_0.$$

From

$$(H' - E^{(1)}I)\begin{pmatrix}a\\b\end{pmatrix} = \begin{pmatrix}0\\0\end{pmatrix},$$

we get the ground state wave function

$$\varphi_0 = au_1 + bu_2 = \psi_1(x)\begin{pmatrix} 1-\sqrt{2} \\ 1 \end{pmatrix}. \quad \text{(unnormalized)}$$

b.   The space part of the wave function in region I is

$$\psi_{1k_1} = \begin{cases} A\sin k_1(x+L) + B\cos k_1(x+L), & -L \leq x \leq 0, \\ 0, & x < -L. \end{cases}$$

The continuity condition of the wave function gives $B=0$. In region I, the spin is aligned to the z direction, the eigenvectors being $\begin{pmatrix} 0 \\ 1 \end{pmatrix}$ for $z\downarrow$ and $\begin{pmatrix} 1 \\ 0 \end{pmatrix}$ for $z\uparrow$. Hence

$$\begin{cases} \psi_{1k_1 z\downarrow} = \sin k_1(x+L)\begin{pmatrix} 0 \\ 1 \end{pmatrix}, & E = \dfrac{\hbar^2 k_1^2}{2m} - \mu_0 B_0; \\[2mm] \psi_{1k_1 z\uparrow} = \sin k_1(x+L)\begin{pmatrix} 1 \\ 0 \end{pmatrix}, & E = \dfrac{\hbar^2 k_1^2}{2m} + \mu_0 B_0. \end{cases}$$

In a similar way we obtain the eigenfunctions for region II $(0 \leq x \leq L)$

$$\begin{cases} \psi_{\mathrm{II} k_2 x\downarrow} = \sin k_2(x-L)\begin{pmatrix} 1 \\ -1 \end{pmatrix}, & E = \dfrac{\hbar^2 k_2^2}{2m} - \mu_0 B_0; \\[2mm] \psi_{\mathrm{II} k_2 x\uparrow} = \sin k_2(x-L)\begin{pmatrix} 1 \\ 1 \end{pmatrix}, & E = \dfrac{\hbar^2 k_2^2}{2m} + \mu_0 B_0. \end{cases}$$

c.   Considering the whole space the energy eigenfunction is

$$\psi_E = \begin{cases} A\psi_{1k_1 z\downarrow} + B\psi_{1k_1' z\uparrow}, & -L \leq x \leq 0, \\ C\psi_{\mathrm{II} k_2 x\downarrow} + D\psi_{\mathrm{II} k_2' x\uparrow}, & 0 \leq x \leq L, \\ 0, & \text{elsewhere.} \end{cases}$$

Thus

$$H\psi_E =$$

$$\begin{cases} \left(\dfrac{\hbar^2 k_1^2}{2m} - \mu_0 B_0\right) A\psi_{1k_1 z\downarrow} + \left(\dfrac{\hbar^2 k_1'^2}{2m} + \mu_0 B_0\right) B\psi_{1k_1' z\uparrow}, & -L \leq x \leq 0, \\[2mm] \left(\dfrac{\hbar^2 k_2^2}{2m} - \mu_0 B_0\right) C\psi_{\mathrm{II} k_2 x\downarrow} + \left(\dfrac{\hbar^2 k_2'^2}{2m} + \mu_0 B_0\right) D\psi_{\mathrm{II} k_2' x\uparrow}, & 0 \leq x \leq L, \\[2mm] & \text{elsewhere.} \\ 0. \end{cases}$$

From $H\psi_E = E\psi_E$ for each region we have

$$E = \frac{\hbar^2 k_1^2}{2m} - \mu_0 B_0 = \frac{\hbar^2 k_1'^2}{2m} + \mu_0 B_0 = \frac{\hbar^2 k_2^2}{2m} - \mu_0 B_0$$
$$= \frac{\hbar^2 k_2'^2}{2m} + \mu_0 B_0,$$

and so $k_1 = k_2 = k$, $k_1' = k_2' = k'$.

Then the continuity of the wave function at $x = 0$ gives

$$B \sin k'L = -C \sin kL - D \sin k'L,$$
$$A \sin kL = C \sin kL - D \sin k'L,$$

and the continuity of the derivative of the wave function at $x = 0$ gives

$$Bk' \cos k'L = Ck \cos kL + Dk' \cos k'L,$$
$$Ak \cos kL = -Ck \cos kL + Dk' \cos k'L.$$

To solve for A, B, C, D, for nonzero solutions we require

$$\begin{vmatrix} 0 & \sin k'L & \sin kL & \sin k'L \\ \sin kL & 0 & -\sin kL & \sin k'L \\ 0 & k' \cos k'L & -k \cos kL & -k' \cos k'L \\ k \cos kL & 0 & k \cos kL & -k' \cos k'L \end{vmatrix} = 0,$$

i.e.,

$$k \sin kL \cos k'L - k' \sin k'L \cos kL = 0.$$

This and

$$E = \frac{\hbar^2 k^2}{2m} - \mu_0 B_0 = \frac{\hbar^2 k'^2}{2m} + \mu_0 B_0$$

determine the eigenvalues E.

4015

Consider an infinitely long solenoid which carries a current I so that there is a constant magnetic field inside the solenoid. Suppose in the region outside the solenoid the motion of a particle with charge e and mass m is described by the Schrödinger equation. Assume that for $I = 0$, the solution of the equation is given by

$$\psi_0(\mathbf{x}, t) = e^{iE_0 t} \psi_0(\mathbf{x}). \qquad (\hbar = 1)$$

a. Write down and solve the Schrödinger equation in the region outside the solenoid for the case $I \neq 0$.

b. Consider a two-slit diffraction experiment for the particles described above (see Fig. 4.4). Assume that the distance d between the two splits is large compared to the diameter of the solenoid. Compute the shift ΔS of the diffraction pattern on the screen due to the presence of the solenoid with $I \neq 0$. Assume $I \gg \Delta S$.

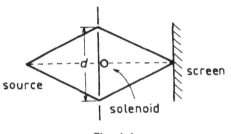

Fig. 4.4

Hint: Let

$$\psi(\mathbf{x},t) = \psi_0(\mathbf{x},t)\psi_A(\mathbf{x}),$$

where

$$\left(\nabla - i\frac{e}{c}\mathbf{A}(\mathbf{x})\right)\psi_A(\mathbf{x}) = 0. \quad (\hbar = 1).$$

(*Chicago*)

Sol:

a. In the presence of a vector potential $\mathbf{A}$, $\mathbf{p} \rightarrow \mathbf{p} - e\mathbf{A}/c$. In the absence of electromagnetic field the Schrödinger equation is

$$i\frac{\partial}{\partial t}\psi_0(\mathbf{x},t) = \left[\frac{1}{2m}\mathbf{p}^2 + V(\mathbf{x})\right]\psi_0(\mathbf{x},t),$$

where, as below, we shall use units such that $\hbar = 1$. The Schrödinger equation in the presence of an electromagnetic field can thus be obtained (using the minimum electromagnetic coupling theory) as

$$i\frac{\partial}{\partial t}\psi(\mathbf{x},t) = \left[\frac{1}{2m}\left(-i\nabla - \frac{e}{c}\mathbf{A}\right)^2 + V(\mathbf{x})\right]\psi(\mathbf{x},t),$$

where A is given by $\nabla \times \mathbf{A} = \mathbf{B}$. Let

$$\psi(\mathbf{x}, t) = \psi_1(\mathbf{x}, t)\exp\left(i\int^x \frac{e}{c}\mathbf{A} \cdot d\mathbf{x}\right).$$

Then the above becomes

$$i\frac{\partial}{\partial t}\psi_1(\mathbf{x}, t) = \left[\frac{1}{2m}\mathbf{p}^2 + V(\mathbf{x})\right]\psi_1(\mathbf{x}, t),$$

which is the Schrödinger equation for zero magnetic field. Hence

$$\psi_1(\mathbf{x}, t) = \psi_0(\mathbf{x}, t) = e^{iE_0 t}\psi_0(\mathbf{x}),$$

and so

$$\psi(\mathbf{x}, t) = e^{iE_0 t}\psi_0(\mathbf{x})\exp\left(i\int^x \frac{e}{c}\mathbf{A} \cdot d\mathbf{x}\right).$$

b. This is a problem on the Aharonov-Bohm effect. When $I = 0$, for any point on the screen the probability amplitude f is $f = f_+ + f_-$, where $f+$ and f_- represent the contributions of the upper and lower slits respectively. When the current is on, i.e., $I \neq 0$, we have the probability amplitude $f' = f'_+ + f'_-$ with

$$f'_+ = \exp\left(i\int_{c+}^x \frac{e}{c}\mathbf{A} \cdot d\mathbf{x}\right)f_+,$$

$$f'_- = \exp\left(i\int_{c-}^x \frac{e}{c}\mathbf{A} \cdot d\mathbf{x}\right)f_-,$$

where c_+ and c_- denote integral paths above and below the solenoid respectively. Thus

$$f' = f'_+ + f'_- = \exp\left(i\int_{c+}^x \frac{e}{c}\mathbf{A} \cdot d\mathbf{x}\right)f_+ + \exp\left(i\int_{c-}^x \mathbf{A} \cdot d\mathbf{x}\right)f_-$$

$$\sim \exp\left(i\oint \frac{e}{c}\mathbf{A} \cdot d\mathbf{x}\right)f_- + f_+,$$

on dividing the two contributions by a common phase factor $\exp(i\int_{c+}^x \frac{e}{c}\mathbf{A} \cdot d\mathbf{x})$, which does not affect the interference pattern. The closed line integral, to be taken counterclockwise along an arbitrary closed path around the solenoid, gives

$$\oint \frac{e}{c}\mathbf{A} \cdot d\mathbf{x} = \frac{e}{c}\int \nabla \times \mathbf{A} \cdot d\mathbf{s} = \frac{e}{c}\int \mathbf{B} \cdot d\mathbf{s} = \frac{e\phi}{c},$$

where ϕ is the magnetic flux through the solenoid.

Thus the introduction of the solenoid gives a phase factor $e\phi/c$ to the probability amplitude at points on the screen contributed by the lower slit.

Using a method analogous to the treatment of Young's interference in optics, we see that the interference pattern is shifted by ΔS. Assuming $l \gg d$ and $l \gg \Delta S$, we have

$$\Delta S \cdot \frac{d}{l} \cdot k = \frac{e}{c}\phi,$$

k being the wave number of the particles, and so

$$\Delta S = \frac{e l \phi}{c d k} = \frac{e l \phi}{c d \sqrt{2mE_0}}.$$

Note the treatment is only valid nonrelativistically.

4016

a. What are the energies and energy eigenfunctions for a nonrelativistic particle of mass m moving on a ring of radius R as shown in the Fig. 4.5.?

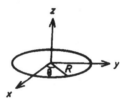

Fig. 4.5

b. What are the energies and energy eigenfunctions if the ring is doubled (each loop still has radius R) as shown in Fig. 4.6?

Fig. 4.6

c. If the particle has charge q, what are the energies and energy eigenfunctions if a very long solenoid containing a magnetic flux passes the rings in (a) as shown in the Fig. 4.7.? and in (b)? Assume the system does not radiate electromagnetically.

(Columbia)

Fig. 4.7

Sol:

a. As

$$\hat{H} = \frac{\hat{\mathbf{P}}^2}{2mR^2} = -\frac{\hbar^2}{2mR^2}\frac{d^2}{d\theta^2},$$

we have the Schrödinger equation

$$-\frac{\hbar^2}{2I}\frac{d^2}{d\theta^2}\Psi(\theta) = E\Psi(\theta),$$

where

$$I = mR^2,$$

or

$$\frac{d^2\Psi(\theta)}{d\theta^2} + n^2\Psi(\theta) = 0,$$

with

$$n^2 = \frac{2IE}{\hbar^2}$$

Thus the solutions are

$$\Psi_n(\theta) = Ae^{in\theta}.$$

For single-valuedness we require

$$\Psi(\theta + 2\pi) = \Psi(\theta),$$

i.e.,

$$n = 0, \pm 1, \pm 2, \ldots.$$

Normalization requires

$$A^*A = 1, \quad \text{or} \quad A = \frac{1}{\sqrt{2\pi}}.$$

Hence the eigenfunctions are

$$\Psi_n(\theta) = \frac{1}{\sqrt{2\pi}}e^{in\theta}, \quad n = 0, \pm 1, \pm 2, \ldots,$$

and the energy eigenvalues are

$$E_n(\theta) = \frac{n^2 \hbar^2}{2I}.$$

b. The same Hamiltonian applies, and so we still have the same Schrödinger equation

$$-\frac{\hbar^2}{2I}\frac{d^2}{d\theta^2}\Psi(\theta) = E\Psi(\theta).$$

However, the single valuedness of the solutions now requires

$$\Psi(\theta + 4\pi) = \Psi(\theta).$$

Hence the normalized eigenfunctions and the energy eigenvalues are now

$$\Psi_n(\theta) = \frac{1}{\sqrt{4\pi}} e^{i\frac{n}{2}\theta}, \quad n = 0, \pm 1, \pm 2, \ldots,$$

and

$$E_n(\theta) = \frac{n^2 \hbar^2}{8I}.$$

c. The Hamiltonian in the presence of a magnetic field is

$$\hat{H} = \frac{1}{2m}(\mathbf{P} - q\mathbf{A}(\mathbf{x}))^2 = -\frac{\hbar^2}{2m}\left(\nabla - \frac{iq}{\hbar}\mathbf{A}(\mathbf{x})\right)^2.$$

In the region where the particle moves, $\mathbf{B} = \nabla \times \mathbf{A} = 0$ and we can choose $\mathbf{A} = \nabla\varphi$. From the symmetry, we have $\mathbf{A} = A_\theta \mathbf{e}_\theta$, $A_\theta = $ constant. Then

$$\oint \mathbf{A} \cdot d\mathbf{l} = \int_0^{2\pi} A_\theta R \, d\theta = 2\pi R A_\theta = \phi, \quad \text{say.}$$

Thus

$$\mathbf{A} = \frac{\phi \mathbf{e}_\theta}{2\pi R} = \nabla(\phi\theta / 2\pi),$$

and we can take $\varphi = \phi\theta/2\pi$, neglecting possibly a constant phase factor in the wave functions. The Schrödinger equation is

$$\hat{H}\Psi = -\frac{\hbar^2}{2m}\left(\nabla - \frac{iq\phi}{2\pi\hbar}\nabla\theta\right)^2 \Psi$$

$$= -\frac{\hbar^2}{2m}\exp\left(i\frac{q\phi}{2\pi h}\theta\right)\nabla^2\left[\exp\left(-\frac{iq\phi}{2\pi\hbar}\theta\right)\Psi\right] = E\Psi.$$

On writing

$$\psi'(\theta) = \exp\left(-i\frac{q\phi}{2\pi h}\theta\right)\Psi(\theta),$$

it becomes

$$-\frac{\hbar^2}{2I}\frac{d^2}{d\theta^2}\psi'(\theta) = E\psi'(\theta),$$

with solutions

$$\psi'(\theta) = \exp\left(\pm i \sqrt{\frac{2IE}{\hbar^2}} \theta \right).$$

Hence

$$\psi(\theta) = c \exp\ (i\alpha\theta)\psi'(\theta)$$
$$= c \exp\ [i(\alpha \pm \beta)\theta],$$

where

$$\alpha = \frac{q\phi}{2\pi\hbar}, \qquad \beta = \sqrt{\frac{2IE}{\hbar^2}}, \qquad c = a \text{ constant.}$$

For the ring of (a), the single-valuedness condition

$$\Psi(\theta + 2\pi) = \Psi(\theta),$$

requires

$$\alpha \pm \beta = n, \qquad\qquad n = 0, \pm 1, \pm 2, \ldots$$

i.e.,

$$\frac{q\phi}{2\pi\hbar} \pm \sqrt{2I\, E/\hbar^2} = n.$$

Hence

$$E_n = \frac{\hbar^2}{2I}\left(n - \frac{q\phi}{2\pi\hbar} \right)^2,$$

and

$$\psi_n(\theta) = \frac{1}{\sqrt{2\pi}} e^{in\theta},$$

where

$$n = 0, \pm 1, \pm 2, \ldots.$$

Similarly for the ring of (b), we have

$$E_n = \frac{\hbar^2}{2I}\left(\frac{n}{2} - \frac{q\phi}{2\pi\hbar} \right)^2 = \frac{\hbar^2}{8I}\left(n - \frac{q\phi}{\pi\hbar} \right)^2,$$

and

$$\psi_n(\theta) = \frac{1}{\sqrt{4\pi}} \exp\left(i\frac{n}{2}\theta \right),$$

where

$$n = 0, \pm 1, \pm 2, \ldots.$$

Part V
Perturbation Theory

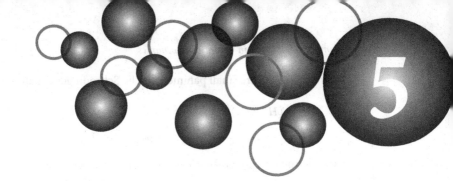

5001

a. Show that in the usual stationary state perturbation theory, if the Hamiltonian can be written $H = H_0 + H'$ with $H_0 \phi_0 = E_0 \phi_0$, then the correction ΔE_0 is

$$\Delta E_0 \approx \langle \phi_0 | H' | \phi_0 \rangle.$$

b. For a spherical nucleus, the nucleons may be assumed to be in a spherical potential well of radius R given by $V_{sp} = \begin{cases} 0, & r < R, \\ \infty, & r > R. \end{cases}$

For a slightly deformed nucleus, it may be correspondingly assumed that the nucleons are in an elliptical well, again with infinite wall height, that is:

$$V_{el} = \begin{cases} 0, & \text{inside the ellipsoid } \dfrac{x^2 + y^2}{b^2} + \dfrac{z^2}{a^2} = 1, \\ \infty, & \text{otherwise,} \end{cases}$$

where $a \cong R(1 + 2\beta/3)$, $b \cong R(1 - \beta/3)$, and $\beta \ll 1$.

Calculate the approximate change in the ground state energy E_0 due to the ellipticity of the non-spherical nucleus by finding an appropriate H' and using the result obtained in (a). HINT: Try to find a transformation of variables that will make the well look spherical.

(Buffalo)

Sol:

a. Assuming that H' is very small compared with H_0 so that the wave function Ψ can be expanded as

$$\Psi = |\phi_0\rangle + \lambda_1 |\phi_1\rangle + \cdots + \lambda_n |\phi_n\rangle + \cdots,$$

275

where $\lambda_1 \cdots \lambda_n \cdots$ are small parameters. The Schrödinger equation is then

$$(H' + H_0)\big(|\phi_0\rangle + \lambda_1|\phi_1\rangle + \cdots + \lambda_n|\phi_n\rangle + \cdots\big)$$
$$= (E_0 + \Delta E_0)\big(|\phi_0\rangle + \lambda_1|\phi_1\rangle + \cdots + \lambda_n|\phi_n\rangle + \cdots\big).$$

Considering only the first order correction, we have

$$H'|\phi_0\rangle + H_0\big(\lambda_1|\phi_1\rangle + \cdots + \lambda_n|\phi_n\rangle + \cdots\big)$$
$$= \Delta E_0|\phi_0\rangle + E_0\big(\lambda_1|\phi_1\rangle + \cdots + \lambda_n|\phi_n\rangle + \cdots\big).$$

Multiplying both sides of the equation by $\langle\phi_0|$ and noting the orthonormality of the eigenfunctions we get

$$\Delta E_0 = \langle\phi_0|H'|\phi_0\rangle.$$

b. For the stationary state,

$$\widehat{H} = -\frac{\hbar^2}{2m}\nabla^2 + V,$$

where

$$V = \begin{cases} 0 & \text{inside the ellipsoid } \dfrac{x^2 + y^2}{b^2} + \dfrac{z^2}{a^2} = 1, \\[2mm] \infty & \text{otherwise.} \end{cases}$$

Replacing the variables x, y, z by $\frac{b}{R}\xi$, $\frac{b}{R}\eta$, $\frac{a}{R}\zeta$ respectively, we can write the equation of the ellipsoid as $\xi^2 + \eta^2 + \zeta^2 = R^2$ and

$$H \equiv -\frac{\hbar^2}{2m}\left(\frac{\partial^2}{\partial x^2} + \frac{\partial^2}{\partial y^2} + \frac{\partial^2}{\partial z^2}\right)$$
$$= -\frac{\hbar^2}{2m}\left(\frac{R^2}{b^2}\frac{\partial^2}{\partial \xi^2} + \frac{R^2}{b^2}\frac{\partial^2}{\partial \eta^2} + \frac{R^2}{a^2}\frac{\partial^2}{\partial \zeta^2}\right)$$
$$\approx -\frac{\hbar^2}{2m}\left(\frac{\partial^2}{\partial \xi^2} + \frac{\partial^2}{\partial \eta^2} + \frac{\partial^2}{\partial \zeta^2}\right)$$
$$\quad -\frac{\hbar^2\beta}{3m}\left(\frac{\partial^2}{\partial \xi^2} + \frac{\partial^2}{\partial \eta^2} - 2\frac{\partial^2}{\partial \zeta^2}\right)$$
$$= -\frac{\hbar^2}{2m}\nabla'^2 - \frac{\hbar^2\beta}{3m}\left(\frac{\partial^2}{\partial \xi^2} + \frac{\partial^2}{\partial \eta^2} - 2\frac{\partial^2}{\partial \zeta^2}\right).$$

The second term in H can be considered a perturbation as $\beta \ll 1$. Thus

$$\Delta E_0 = \langle\phi_0|H'|\phi_0\rangle = \langle\phi_0| - \frac{\hbar^2\beta}{3m}\left(\frac{\partial^2}{\partial \xi^2} + \frac{\partial^2}{\partial \eta^2} - 2\frac{\partial^2}{\partial \zeta^2}\right)|\phi_0\rangle,$$

where ϕ_0 is the ground state wave function for the spherical potential well,

$$\phi_0 = \sqrt{\frac{2}{R}} \frac{\sin \frac{\pi r}{R}}{r}, \quad \gamma^2 = \xi^2 + \eta^2 + \zeta^2.$$

As ϕ_0 is spherically symmetric,

$$\left\langle \phi_0 \left| \frac{\partial^2}{\partial \xi^2} \right| \phi_0 \right\rangle = \left\langle \phi_0 \left| \frac{\partial^2}{\partial \eta^2} \right| \phi_0 \right\rangle = \left\langle \phi_0 \left| \frac{\partial^2}{\partial \zeta^2} \right| \phi_0 \right\rangle,$$

and so $\Delta E_0 = 0$.

5002

Calculate the first-order energy correction to the ground state and first excited state of an infinite square well of width "a" whose portion has been sliced off as shown in the figure, where $V_0 x/a$ is the perturbed potential.

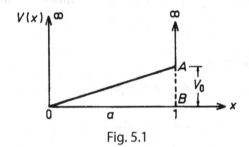

Fig. 5.1

Sol:

The wave function of a particle inside an infinite potential well of width a is given by

$$\psi = \sqrt{\frac{2}{a}} \sin(n\pi x / a) \tag{1}$$

The correction to the energy in first order is given by $\int dx \, \psi \times H\psi$

$$2 / a \int_0^a dx \sin^2(n\pi x / a) V_0 x / a$$

For ground state $n = 1$, so the integral evaluates to $2V_0 / a^2 \int_0^a x \, dx \sin^2(\pi x / a)$

Using the identity $\sin^2 x = [1 - \cos(2\pi x)]/2$

So the integral becomes

$$V_0 / a^2 \int_0^a x \, dx [1 - \cos(2\pi x / a] = V_0 / a^2 \left\{ \int_0^a x \, dx - \int_0^a \cos(2\pi x / a) dx \right\}$$

where the second term in the integral vanishes and the correction to the energy is given by

$$E_1^1 = V_0/2$$

In a similar manner, the correction to the first-order energy is given by

$$E_2^1 = V_0/2$$

5003

A particle of mass m moves one-dimensionally in the oscillator potential $V(x) = \frac{1}{2}m\omega^2 x^2$. In the nonrelativistic limit, where the kinetic energy T and momentum p are related by $T = p^2/2m$, the ground state energy is well known to be $\frac{1}{2}\hbar\omega$.

Allow for relativistic corrections in the relation between T and p and compute the ground state level shift ΔE to order $\frac{1}{c^2}$ (c = speed of light).

(Buffalo)

Sol: In relativistic motion, the kinetic energy T is

$$T \equiv E - mc^2 = \sqrt{m^2c^4 + p^2c^2} - mc^2$$

$$= mc^2\left(1 + \frac{p^2}{m^2c^2}\right)^{\frac{1}{2}} - mc^2$$

$$\approx mc^2\left(1 + \frac{p^2}{2m^2c^2} - \frac{p^4}{8m^4c^4}\right) - mc^2$$

$$= \frac{p^2}{2m} - \frac{p^4}{8m^3c^2}$$

to order $\frac{1}{c^2}$. The $\frac{-p^4}{8m^3c^2}$ term may be considered as a perturbation. Then the energy shift of the ground state is

$$\Delta E = \left\langle -\frac{p^4}{8m^3c^2}\right\rangle = \int_{-\infty}^{\infty}\phi_0^*\left(\frac{-\hat{p}^4}{8m^3c^2}\right)\phi_0\,dx$$

$$= \int_{-\infty}^{\infty}\left(\frac{m\omega}{\pi\hbar}\right)^{1/4}\exp\left[-\frac{m\omega}{2\hbar}x^2\right]$$

$$\times \left(-\frac{\hbar^4}{8m^3c^2}\frac{\partial^4}{\partial x^4}\right)\left(\frac{m\omega}{\pi\hbar}\right)^{\frac{1}{4}}\exp\left[-\frac{m\omega}{2\hbar}x^2\right]dx$$

$$= -\frac{15}{32}\frac{(\hbar\omega)^2}{mc^2}.$$

5004

An electron moves in a Coulomb field centered at the origin of coordinates. With neglect of spin and relativistic corrections the first excited level $(n = 2)$ is well known to be 4-fold degenerate: $l = 0$, $m_l = 0$; $l = 1$, $m_l = 1, 0, -1$. Consider what happens to this level in the presence of an additional non-central potential V_{pert} : $V_{pert} = f(r)xy$, where $f(r)$ is some central function, well-behaved but not otherwise specified (it falls off rapidly enough as $r \to \infty$). This perturbation is to be treated to first order. To this order the originally degenerate $n = 2$ level splits into several levels of different energies, each characterized by an energy shift ΔE and by a degeneracy (perhaps singly degenerate, i.e., nondegenerate; perhaps multiply degenerate).

   a. How many distinct energy levels are there?

   b. What is the degeneracy of each?

   c. Given the energy shift, call it A $(A > 0)$, for one of the levels, what are the values of the shifts for all the others?

(Princeton)

Sol: With $V = f(r)xy = f(r)r^2 \sin^2 \theta \sin\varphi \cos\varphi$ treated as perturbation, the unperturbed wave functions for energy level $n = 2$ are

$$
\begin{aligned}
l = 0, \quad & m_l = 0, \quad && R_{20}(r)Y_{00}, \\
l = 1, \quad & m_l = 1, \quad && R_{21}(r)Y_{11}, \\
l = 1, \quad & m_l = 0, \quad && R_{21}(r)Y_{10}, \\
l = 1, \quad & m_l = -1, \quad && R_{21}(r)Y_{1,-1}.
\end{aligned}
$$

As they all correspond to the same energy, i.e., degeneracy occurs, we have first to calculate

$$
\begin{aligned}
H'_{l'm'lm} &= \langle l'm' | V | lm \rangle \\
&= \int R_{2l'}(r)R_{2l}(r)r^2 f(r)Y^*_{l'm'} \sin^2 \theta \sin\varphi \cos\varphi Y_{lm} \, dV.
\end{aligned}
$$

The required spherical harmonics are

$$
Y_{00} = \left(\frac{1}{4\pi}\right)^{\frac{1}{2}}, \qquad Y_{11} = \left(\frac{3}{8\pi}\right)^{\frac{1}{2}} \sin\theta e^{i\varphi},
$$

$$
Y_{10} = \left(\frac{3}{4\pi}\right)^{\frac{1}{2}} \cos\theta, \qquad Y_{1,-1} = \left(\frac{3}{8\pi}\right)^{\frac{1}{2}} \sin\theta e^{-i\varphi}.
$$

Considering the factor involving φ in the matrix elements $H_{l'm'lm}$ we note that all such elements have one of the following factors:

$$\int_0^{2\pi} \sin\varphi\cos\varphi\, d\varphi = 0, \qquad \int_0^{2\pi} e^{\pm i2\varphi}\sin\varphi\cos\varphi\, d\varphi = 0,$$

except $H'_{1,-1,1,1}$ and $H'_{1,1,1,-1}$, which have nonzero values

$$H'_{1,-1,1,1} = \frac{3}{8\pi}\int [R_{21}(r)]^2 r^4 f(r)dr \int_0^{\pi}\sin^5\theta d\theta$$

$$\times \int_0^{2\pi}\sin\varphi\cos\ \varphi e^{-2i\varphi}d\varphi = iA,$$

$$H'_{1,1,1,-1} = -iA, \quad \text{with } A = \frac{1}{5}\int [R(r)]^2 r^4 f(r)dr.$$

We then calculate the secular equation

$$\left\| \begin{pmatrix} 0 & 0 & 0 & 0 \\ 0 & 0 & 0 & iA \\ 0 & 0 & 0 & 0 \\ 0 & -iA & 0 & 0 \end{pmatrix} - \Delta EI \right\| = \begin{vmatrix} \Delta E & 0 & 0 & 0 \\ 0 & \Delta E & 0 & iA \\ 0 & 0 & \Delta E & 0 \\ 0 & -iA & 0 & \Delta E \end{vmatrix} = 0,$$

whose solutions are $\Delta E = 0$, $\Delta E = 0$, $\Delta E = A$, $\Delta E = -A$.

Thus with the perturbation there are three distinct energy levels with $n = 2$. The energy shifts and degeneracies are as follow.

$$\Delta E = \begin{cases} A, & \text{one-fold degeneracy,} \\ -A, & \text{one-fold degeneracy,} \\ 0, & \text{two-fold degeneracy.} \end{cases}$$

Thus there are three distinct energy levels with $n = 2$.

5005

A particle moves in a one-dimensional box with a small potential dip (Fig. 5.2):

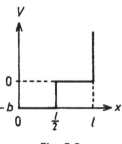

Fig. 5.2

$$V = \infty \text{ for } x < 0 \text{ and } x > l$$
$$V = -b \text{ for } 0 < x < (1/2)l,$$
$$V = 0 \text{ for } (1/2)l < x < l.$$

Treat the potential dip as a perturbation to a "regular" rigid box ($V = \infty$ for $x < 0$ and $x > l$, $V = 0$ for $0 < x < l$). Find the first order energy of the ground state.

(Wisconsin)

Sol: For the regular rigid box, the energy and wave function of the ground state are respectively

$$E^{(0)} = \frac{\pi^2 \hbar^2}{2ml^2}, \quad \psi^{(0)}(x) = \sqrt{\frac{2}{l}} \sin\frac{\pi x}{l}.$$

The perturbation is $H^{(1)} = -b$, $0 \le x \le \frac{l}{2}$. Hence, the energy correction of first order perturbation is

$$E^{(1)} = \int_0^{\frac{l}{2}} \phi^{(0)^*}(x)(-b)\varphi^{(0)}(x)dx$$

$$= \int_0^{\frac{l}{2}} \frac{2}{l} \sin^2\left(\frac{\pi x}{l}\right)(-b)dx$$

$$= -\frac{b}{l}\int_0^{\frac{l}{2}}\left(1 - \cos\frac{2\pi x}{l}\right)dx = -\frac{b}{2}.$$

Thus the energy of the ground state with first order perturbation correction is

$$E = E^{(0)} + E^{(1)} = \frac{\hbar^2 \pi^2}{2ml^2} - \frac{b}{2}.$$

5006

An infinitely deep one-dimensional square well has walls at $x = 0$ and $x = L$. Two small perturbing potentials of width a and height V are located at $x = L/4$, $x = (3/4)L$, where a is small ($a \ll L/100$, say) as shown in Fig. 5.3. Using perturbation methods, estimate the difference in the energy shifts between the $n = 2$ and $n = 4$ energy levels due to this perturbation.

(Wisconsin)

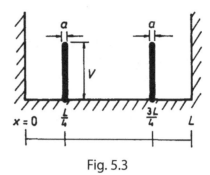

Fig. 5.3

Sol: The energy levels and wave functions for a one-dimensional infinite potential well are respectively

$$E_n^{(0)} = \frac{\pi^2 \hbar^2}{2\mu L^2} n^2,$$

$$\psi_n(x) = \sqrt{\frac{2}{L}} \sin \frac{n\pi}{L} x, \quad n = 1, 2, \dots .$$

The shift of the energy level n, $E_n^{(1)} = H'_{nn}$, according to first order perturbation is given by

$$H'_{nn} = \int_{L/4-a/2}^{L/4+a/2} V \cdot \frac{2}{L} \sin^2\left(\frac{\pi n}{L} x\right) dx$$

$$+ \int_{3L/4-a/2}^{3L/4+a/2} V \cdot \frac{2}{L} \sin^2\left(\frac{\pi n}{L} x\right) dx.$$

As $a \ll L/100$, we can apply the mean value theorem to the integrals and obtain

$$H'_{nn} = \frac{2Va}{L}\left[\sin^2\left(\frac{\pi n}{L} \cdot \frac{L}{4}\right) + \sin^2\left(\frac{\pi n}{L} \cdot \frac{3L}{4}\right)\right]$$

$$= \frac{2Va}{L}\left(\sin^2\frac{\pi n}{4} + \sin^2\frac{3\pi n}{4}\right).$$

Therefore, the change of energy difference between energy levels $n = 2$ and $n = 4$ is

$$E_2^{(1)} - E_4^{(1)} = \frac{2Va}{L}\left(\sin^2\frac{\pi}{2} + \sin^2\frac{3\pi}{2} - \sin^2\pi - \sin^2 3\pi\right)$$

$$= \frac{4Va}{L}.$$

5007

A particle of mass m moves in a one-dimensional potential box

$$V(x) = \begin{cases} \infty & \text{for } |x| > 3|a|, \\ 0 & \text{for } a < x < 3a, \\ 0 & \text{for } -3a < x < -a, \\ V_0 & \text{for } -a < x < a, \end{cases}$$

as shown in Fig. 5.4.

Consider the V_0 part as a perturbation on a flat box ($V = 0$ for $-3a < x < 3a$, $V = \infty$ for $|x| > 3|a|$) of length 6a. Use the first order perturbation method to calculate the energy of the ground state.

(Wisconsin)

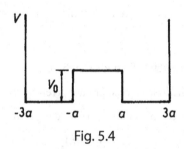

Fig. 5.4

Sol: The energies and wave functions of a particle in a flat box of length 6a are respectively

$$E^{(0)} = \frac{\pi^2 \hbar^2 n^2}{72ma^2}, \quad n = 1, 2, \ldots,$$

$$\psi^{(0)}(x) = \sqrt{\frac{1}{3a}} \cos\frac{n\pi x}{6a}, \quad n = \text{odd integer},$$

$$\psi^{(0)}(x) = \sqrt{\frac{1}{3a}} \sin\frac{n\pi x}{6a}, \quad n = \text{even integer}.$$

Particularly for the ground state, we have

$$\psi_1^{(0)}(x) = \sqrt{\frac{1}{3a}} \cos\frac{\pi x}{6a}$$

$$E_1^{(0)} = \frac{\pi^2 \hbar^2}{72ma^2}.$$

The energy correction of first order perturbation is given by

$$E^{(1)} = (\psi_1^{(0)}(x), \quad \hat{V}\psi_1^{(0)}(x)),$$

where $\hat{V} = V_0$ for $-a \le x \le a$. Thus

$$E^{(1)} = \int_{-0}^{a} \frac{V_0}{3a} \cos^2\left(\frac{\pi x}{6a}\right) dx = V_0\left(\frac{1}{3} + \frac{\sqrt{3}}{2\pi}\right).$$

Hence, the energy of ground state given by first order perturbation is

$$E = E^{(0)} + E^{(1)} = \frac{\pi^2\hbar^2}{72ma^2} + V_0\left(\frac{1}{3} + \frac{\sqrt{3}}{2\pi}\right).$$

5008

Consider a free particle confined to a one-dimensional infinite potential box of length L. A uniform electric field is applied along the x direction. Find the first-order correction to the energy to the first three states.

Sol:

The X component of electric field is given by $E\hat{\imath}$.

The corrections to the first three states are given by E_1^1, E_2^1, and E_3^1.

We know that the force on a charged particle is given by

$$F = qE\hat{\imath} \tag{1}$$

The work done is stored in terms of potential energy, which acts as a perturbation. Work is defined as

$$W = F.X$$

which gives $W = qE\hat{\imath} . x\hat{\imath} = qEx$

Now, the first-order correction is given as

$$<H'> = <qEx>$$

where q and E are constants, so $<H'> = qE<x>$

The $<X>$ is given by $\int \psi \times X \psi \, dx$ with the system limits $0 \to L$ and the wave function is given by $\psi = \sqrt{2/L} \sin(n\pi x / L)$ where n takes values 1, 2, and 3.

For the ground state $n = 1$, $<X> = 2/L \int_0^L dx \sin^2(\pi x / L)x$

Using the identity $\sin^2 x = [1 - \cos(2\pi x)] / 2$

$$<X>=1/L\left\{\int_0^L dx\ x+\int_0^L dx\cos(2\pi x/L)\right\}$$

The second integral is zero after integrating and applying the limits, only the first part survives and $<X> = L/2$.

So the energy correction is given by $<H'> = qEL/2$

In a similar manner, the energy corrections for the second and third states are given by $qEL/2$.

The energies of the first three levels after corrections are given by

$E_1 + qEL/2$ for the ground state.

$4E_1 + qEL/2$ for the first excited state.

$9E_1 + qEL/2$ for the third excited state, where $E_1 = \pi^2\hbar^2/2mL^2$.

5009

A perfectly elastic ball is bouncing between two parallel walls.

   a. Using classical mechanics, calculate the change in energy per unit time of the ball as the walls are slowly and uniformly moved closer together.

   b. Show that this change in energy is the same as the quantum mechanical result if the ball's quantum number does not change.

   c. If the ball is in the quantum state with $n = 1$, under what conditions of wall motion will it remain in that state?

(Chicago)

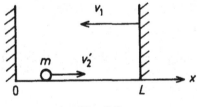

Fig. 5.5

Sol:

   a. In classical mechanics, the energy of the ball is

$$E = \frac{p^2}{2m},$$

so
$$\frac{dE}{dt} = \frac{p}{m}\frac{dp}{dt}.$$

At a certain instant t, the walls are separated by L and the ball moves to the right with speed v_2'. Because the collision is perfectly elastic, the speed of the ball relative to the right wall before and after bouncing from it remains the same:

$$v_2' + v_1 = v_2 + (-v_1),$$

where v_2 is the speed of the ball after bouncing. Thus,

$$v_2' - v_2 = -2v_1,$$

$$\Delta p = m(v_2' - v_2) = -2mv_1,$$

$$dE/dt = \frac{p}{m}\frac{dp}{dt} \approx \frac{p}{m}\frac{\Delta p}{\Delta t} = -\frac{p}{m}2mv_1 \cdot \frac{v_2}{2L} = -\frac{pv_1v_2}{L},$$

where $\Delta t = \dfrac{2L}{v_2}$ is the time interval between two successive collisions. As the right wall moves very slowly,

$$dE/dt = -\frac{pv_2}{L}v_1 = -\frac{p}{L}\cdot\frac{p}{m}\cdot\frac{dL}{dt}$$

$$= -\frac{2}{L}\cdot\frac{p^2}{2m}\frac{dL}{dt} = -\frac{2E}{L}\frac{dL}{dt},$$

which is the rate of change of the energy of the ball according to classical mechanics.

b. As the motion of the right wall is very slow, the problem can be treated as one of perturbation. If the wall motion can be neglected, we have

$$E_n = \frac{n\pi^2\hbar^2}{2mL^2}.$$

If n does not change,

$$dE_n/dt = \frac{n^2\pi^2\hbar^2}{2m}(-2)\frac{1}{L^3}\frac{dL}{dt} = -\frac{2E_n}{L}dL/dt,$$

same as the classical result.

c. If the energy change during one collision is much smaller than $E_2 - E_1$, the ball can remain in the state $n = 1$ (analogous to adiabatic processes in thermodynamics). More precisely, as

$$E = \frac{p^2}{2m},$$

we have

$$\Delta E = \frac{p}{m} \Delta p = -\sqrt{\frac{2E}{m}} \cdot 2m\upsilon_1 = -2\sqrt{2mE}\upsilon_1.$$

The condition

$$E_2 - E_1 \gg |\Delta E|$$

then gives

$$\frac{\pi^2\hbar^2}{2mL^2}(2^2 - 1^2) \gg 2\sqrt{2m\frac{\pi^2\hbar^2}{2mL^2}}|\upsilon_1|,$$

or

$$|\upsilon_1| \ll \frac{3\pi\hbar}{4mL}.$$

This means that the speed of the right wall should be much smaller than $\frac{3\pi\hbar}{4mL}$.

5010

Consider a free particle confined to a one-dimensional box of length a (from $x=0$ to $x=a$), if it is subjected to a perturbation $V(x)=V_0 \cos(\pi x / a)$ in the range $0 < x < a/2$. Find the first-order correction to the energy of the ground state.

Sol:

$\psi_{gs}(x) = \sqrt{\frac{2}{a}} \sin(\pi x / a)$ and the first-order correction to the energy is given by

$<\psi | \hat{H} | \psi >$

$$E_1^1 = 2V_0 / a \int_0^{a/2} \cos(\pi x / a)\sin^2(\pi x / a)\,dx$$

The limits are taken only up to the perturbed part.

This integral can be done by substitution. Let $\sin(\pi x / a) = t, \cos(\pi x / a) = (a / \pi)dt$ where the limits change from

$$\text{At } x = 0 \rightarrow t = 0$$
$$\text{At } x = a/2 \rightarrow t = 1$$

So the integral changes to $(a / \pi)\int_0^1 dt\, t^2$, which gives $a/3\pi$.

Now, $E_1^1 = 2V_0/a \times (a/3\pi) = 2V_0/3\pi.$

5011

A charged particle is bound in a harmonic oscillator potential $V = \frac{1}{2}kx^2$. The system is placed in an external electric field E that is constant in space and time. Calculate the shift of the energy of the ground state to order E^2.

(Columbia)

Sol: Take the direction of electric field as the x-direction. The Hamiltonian of the system is

$$\hat{H} = -\frac{\hbar^2}{2m}\frac{d^2}{dx^2} + \frac{1}{2}kx^2 - qEx = \hat{H}_0 + \hat{H}',$$

where $\hat{H}' = -qEx$ is to be treated as a perturbation.

The wave function of the ground state of a harmonic oscillator is

$$\psi(x) \equiv \langle x|0 \rangle = \sqrt{\frac{a}{\pi^{1/2}}} \exp\left(-\frac{1}{2}\alpha^2 x^2\right).$$

where

$$\alpha = \sqrt{\frac{m\omega}{\hbar}}, \quad \omega = \sqrt{\frac{k}{m}}.$$

As ψ_0 is an even function, the first order correction $\langle 0|H'|0 \rangle = 0$ and we have to go to the second order. For the harmonic oscillator we have

$$\langle n'|x|n \rangle = \frac{1}{\alpha}\left[\sqrt{\frac{n'}{2}}\delta_{n,n'-1} + \sqrt{\frac{n'+1}{2}}\delta_{n,n'+1}\right],$$

and hence

$$H'_{0,n} = -qE\langle 0|x|n \rangle = -(qE/\sqrt{2}\alpha)\delta_{n,1}.$$

Thus the energy correction for the ground state to order E^2 is

$$\Delta E_0^{(2)} = {\sum_n}' \frac{|H'_{0,n}|^2}{E_0^{(0)} - E_n^{(0)}} = {\sum_n}' \frac{\frac{q^2E^2}{2\alpha^2}\delta_{n,1}}{-n\hbar\omega}$$

$$= -\frac{q^2E^2}{2\hbar\omega\alpha^2} = -\frac{q^2E^2}{2m\omega^2},$$

where the partial sum ${\sum_n}'$ excludes $n = 0$.

5012

For a one-dimensional harmonic oscillator, introduction of the dimensionless coordinate and energy variables $y = x(m\omega_0/\hbar)^{1/2}$ and $\varepsilon_n = 2E_n/\hbar\omega_0$ gives a

Schrödinger equation with kinetic energy operator $T = -\dfrac{d^2}{dy^2}$ and potential energy $V = y^2$.

a. Using the fact that the only non-vanishing dipole matrix element is $\langle n+1|y|n\rangle = \sqrt{\dfrac{n+1}{2}}$ (and its Hermitean conjugate), find values for all the non-vanishing matrix elements of y^3 that connect to the ground state $|0\rangle$.

b. The oscillator is perturbed by an harmonic potential $V' = \alpha y^3$. Find the correction to the ground state energy in the lowest non-vanishing order. (If you did not get complete answers in part(a), leave your result in terms of clearly defined matrix elements, etc.)

(Berkeley)

Sol:

a. As

$$\langle m|y|n\rangle = \sqrt{\frac{m}{2}}\delta_{n,m-1} + \sqrt{\frac{n}{2}}\delta_{m,n-1},$$

the non-vanishing matrix elements connected to $|0\rangle$,

$$\langle m|y^3|0\rangle = \sum_{k,l}\langle m|y|k\rangle\,\langle k|y|l\rangle\,\langle l|y|0\rangle,$$

are those with $m = 3$, $k = 2$, $l = 1$, and with $m = 1$, and $k = 0, l = 1$, or $k = 2, l = 1$, namely

$$\frac{\sqrt{3}}{2}\delta_{m,3} \quad \text{and} \quad \frac{3}{2\sqrt{2}}\delta_{m,1}.$$

b. Because ψ_0 is an even function, $\langle 0|y^3|0\rangle = 0$, or the first order energy correction $\langle 0|\alpha y^3|0\rangle$ is zero, and we have to calculate the second order energy correction:

$$\Delta E_2 = \sum_{n\neq 0}\frac{|\langle 0|\alpha y^3|n\rangle|^2}{1 - \varepsilon_n} = |\alpha|^2\left(\frac{|\langle 0|y^3|1\rangle|^2}{1 - \varepsilon_1} + \frac{|\langle 0|y^3|3\rangle|}{1 - \varepsilon_3}\right).$$

As

$$\varepsilon_n = \frac{2E_n}{\hbar\omega_0} = 2n+1,$$

$$\Delta E = |\alpha|^2 \left(\frac{|\langle 0| y^3|1\rangle|^2}{-2} + \frac{|\langle 0| y^3|3\rangle|^2}{-6} \right) = -\frac{11}{16}|\alpha|^2.$$

5013

Consider a one-dimensional harmonic oscillator of frequency ω_0. Denote the energy eigenstates by n, starting with $n = 0$ for the lowest. To the original harmonic oscillator potential a time-independent perturbation $\mathcal{H} = V(x)$ is added. Instead of giving the form of the perturbation $V(x)$, we shall give explicitly its matrix elements, calculated in the representation of the unperturbed eigenstates. The matrix elements $\mathcal{H}$ are zero unless m and n are even. A portion of the matrix is given below, where ε is a small dimensionless constant. [Note that the indices on this matrix run from $n = 0$ to 4.]

$$\varepsilon\hbar\omega_0 \begin{pmatrix} 1 & 0 & -\sqrt{1/2} & 0 & \sqrt{3/8} \\ 0 & 0 & 0 & 0 & 0 \\ -\sqrt{1/2} & 0 & 1/2 & 0 & -\sqrt{3/16} \\ 0 & 0 & 0 & 0 & 0 \\ \sqrt{3/8} & 0 & -\sqrt{3/16} & 0 & 3/8 \end{pmatrix}$$

a. Find the new energies for the first five energy levels to *first order* in perturbation theory.

b. Find the new energies for $n = 0$ and 1 to *second order* in perturbation theory.

Sol:

a. The energy levels to first order in perturbatiosswn theory are

$$E'_n = E_n + H'_{nn}.$$

where $E_n = \left(n + \frac{1}{2}\right)\hbar\omega_0$, $H'_{nn} = \langle n|\mathcal{H}|n\rangle$. Thus the energy for the first five energy levels are

$$E'_0 = \frac{1}{2}\hbar\omega_0 + \varepsilon\hbar\omega_0 = \left(\frac{1}{2} + \varepsilon\right)\hbar\omega_0,$$

$$E'_1 = \frac{3}{2}\hbar\omega_0,$$

$$E'_2 = \left(\frac{5}{2} + \frac{1}{2}\varepsilon\right)\hbar\omega_0,$$

$$E'_3 = \frac{7}{2}\hbar\omega_0,$$

$$E'_4 = \left(\frac{9}{2} + \frac{3}{8}\varepsilon\right)\hbar\omega_0.$$

b. The energies for $n = 0$ and 1 to the second order in perturbation theory are

$$E''_0 = E_0 + H'_{00} + \sum_{k \neq 0} \frac{|H_{k0}|^2}{E_0 - E_k}$$

$$= \frac{1}{2}\hbar\omega_0 + \varepsilon\hbar\omega_0 + \sum_{k \neq 0} \frac{1}{-k\hbar\omega_0}|H'_{k0}|^2$$

$$= \hbar\omega_0\left[\frac{1}{2} + \varepsilon - \varepsilon^2\left(\frac{1}{4} + \frac{3}{32} + \cdots\right)\right],$$

$$E''_1 = E_1 + H'_{11} + \sum_{k \neq 0} \frac{|H'_{k1}|^2}{E_1 - E_k}$$

$$= \frac{3}{2}\hbar\omega_0 + 0 + \sum_{k \neq 1} \frac{1}{(1-k)\hbar\omega_0}|H'_{k1}|^2$$

$$= \hbar\omega_0\left(\frac{3}{2} + 0 + 0\right) = \frac{3}{2}\hbar\omega_0.$$

5014

A mass m is attached by a massless rod of length l to a pivot P and swings in a vertical plane under the influence of gravity (see Fig. 5.6).

Fig. 5.6

a. In the small angle approximation find the energy levels of the system.

b. Find the lowest order correction to the ground state energy resulting from inaccuracy of the small angle approximation.

(Columbia)

Sol:

a. Take the equilibrium position of the point mass as the zero point of potential energy. For small angle approximation, the potential energy of the system is

$$V = mgl(1-\cos\theta) \approx \frac{1}{2}mgl\theta^2,$$

and the Hamiltonian is

$$H = \frac{1}{2}ml^2\dot{\theta}^2 + \frac{1}{2}mgl\theta^2.$$

By comparing it with the one-dimensional harmonic oscillator, we obtain the energy levels of the system

$$E_n = \left(n+\frac{1}{2}\right)\hbar\omega,$$

where $\omega = \sqrt{g/l}$.

b. The perturbation Hamiltonian is

$$H' = mgl(1-\cos\theta) - \frac{1}{2}mgl\theta^2$$

$$\approx -\frac{1}{24}mgl\theta^4 = -\frac{1}{24}\frac{mg}{l^3}x^4,$$

with $x = l\theta$. The ground state wave function for a harmonic oscillator is

$$\psi_0 = \sqrt{\frac{\alpha}{\pi^{1/2}}}\exp\left(-\frac{1}{2}\alpha^2x^2\right)$$

with $\alpha = \sqrt{\frac{m\omega}{\hbar}}$. The lowest order correction to the ground state energy resulting from inaccuracy of the small angle approximation is

$$E' = \langle 0|H'|0\rangle = -\frac{1}{24}\frac{mg}{l^3}\langle 0|x^4|0\rangle.$$

As

$$\langle 0|x^4|0\rangle = \frac{\alpha}{\sqrt{\pi}}\int_{-\infty}^{\infty}x^4\exp(-\alpha^2x^2)dx = \frac{3}{4\alpha^4},$$

$$E' = -\frac{\hbar^2}{32ml^2}.$$

5015

A quantum mechanical rigid rotor constrained to rotate in one plane has moment of inertia I about its axis of rotation and electric dipole moment μ (in the plane).

This rotor is placed in a weak uniform electric field ε, which is in the plane of rotation. Treating the electric field as a perturbation, find the first non-vanishing corrections to the energy levels of the rotor.

(Wisconsin)

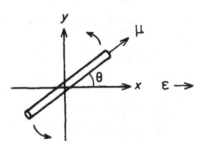

Fig. 5.7

Sol: Take the plane of rotation of the rotor as the xy plane with the x-axis parallel to ε as shown in Fig. 5.7. In the absence of external electric field the Hamiltonian of rotor is

$$H_0 = -\frac{\hbar^2}{2I}\frac{\partial^2}{\partial\theta^2},$$

and the eigenequation is

$$-\frac{\hbar^2}{2I}\frac{\partial^2\psi}{\partial\theta^2} = E\psi,$$

which has solutions

$$\psi = \frac{1}{\sqrt{2\pi}}e^{im\theta}, \quad m = 0, \pm1, \pm2, \ldots,$$

corresponding to energy levels

$$E_m^{(0)} = \frac{\hbar^2 m^2}{2I}.$$

When the external electric field acts on the system and may be treated as perturbation, the Hamiltonian of perturbation is

$$H' = -\mu\cdot\varepsilon = -\mu\varepsilon\cos\theta.$$

The first order energy correction is

$$E^{(1)} = \langle m | H' | m \rangle = -\frac{\mu\varepsilon}{2\pi} \int_0^{2\pi} \cos\theta d\theta = 0$$

The second order energy correction is

$$E^{(2)} = \sum_{m' \neq m}{}' \frac{\left| \langle m' | H' | m \rangle \right|^2}{E_m^{(0)} - E_{m'}^{(0)}}.$$

As

$$\langle m' | H' | m \rangle = -\frac{\mu\varepsilon}{2\pi} \int_0^{2\pi} e^{i(m-m')\theta} \cos\theta d\theta$$

$$= -\frac{\mu\varepsilon}{4\pi} \int_0^{2\pi} \left[e^{i(m-m'+1)\theta} + e^{i(m-m'-1)\theta} \right] d\theta$$

$$= -\frac{\mu\varepsilon}{2} \left(\delta_{m',m+1} + \delta_{m',m-1} \right),$$

we have

$$E^{(2)} = \frac{\mu^2\varepsilon^2}{4} \cdot \frac{2I}{\hbar^2} \left[\frac{1}{m^2 - (m-1)^2} + \frac{1}{m^2 - (m+1)^2} \right]$$

$$= \frac{\mu^2\varepsilon^2 I}{\hbar^2} \cdot \frac{1}{4m^2 - 1}.$$

5016

The polarization of a diatomic molecule in weak electric fields may be treated by considering a rigid rotator with moment of inertia I and electric dipole moment **d** in a weak electric field **E**.

a. Ignoring the motion of the center of mass write down the Hamiltonian H of the rigid rotator in the form of $H_0 + H'$.

b. Solve the unperturbed problem exactly. How are the levels degenerate?

c. Calculate the lowest order correction to all the energy levels by non-degenerate perturbation method.

d. Explain why is nondegenerate perturbation method applicable here and how are the levels degenerate.

The following relation may be useful:

$$\cos\theta Y_{lm} = \sqrt{\frac{(l+1-m)(l+1+m)}{(2l+1)(2l+1)}} Y_{l+1,m} + \sqrt{\frac{(l+m)(l-m)}{(2l+1)(2l-1)}} Y_{l-1,m}.$$

(Buffalo)

ol:

a. Take the z-axis in the direction of the field. The Hamiltonian of the rotator is

$$H = \frac{\mathbf{J}^2}{2I} - \mathbf{d}\cdot\mathbf{E} = \frac{\mathbf{J}^2}{2I} - dE\cos\theta.$$

Considering $-dE\cos\theta$ as a perturbation, we have

$$H_0 = \frac{\mathbf{J}^2}{2I}, \quad H' = -dE\cos\theta.$$

b. The eigenfunctions of the unperturbed system are

$$\psi_{jm} = Y_{jm}(\theta,\varphi),$$

where $m = -j, (-j+1), \equiv , (j-1), j,$ and the energy eigenvalues are

$E_{jm}^{(0)} = \frac{j(j+1)\hbar^2}{2I}$. The levels are $(2j+1)$-fold degenerate since $E_{jm}^{(0)}$ does not depend on m at all.

c. The first order energy correction is

$$\langle jm| - dE\cos\theta | jm\rangle = -dE\langle jm|\cos\theta| jm\rangle$$

$$= -dE\int_0^\pi\int_0^{2\pi}\cos\theta\sin\theta\, d\theta\, d\varphi = 0.$$

For the second order correction we calculate

$$E_n^{(2)} = \sum_i{}' \frac{|\langle n|H'|i\rangle|^2}{E_n^{(0)} - E_i^{(0)}},$$

Where n, i denote pairs of j, m and the prime signifies exclusion of $i = n$ in the summation. As the non-vanishing matrix elements are only

$$\langle j+1, m| - dE\cos\theta | jm\rangle = -dE\sqrt{\frac{(j+1-m)(j+1+m)}{(2j+1)(2j+3)}},$$

$$\langle j-1, m| - dE\cos\theta | jm\rangle = -dE\sqrt{\frac{(j+m)(j-m)}{(2j+1)(2j-1)}},$$

the lowest order energy correction is

$$E^{(2)} = \frac{2Id^2E^2}{\hbar^2}$$

$$\times\left\{\frac{(j+1-m)(j+1+m)}{(2j+1)(2j+3)[j(j+1)-(j+1)(j+2)]}\right.$$

$$\left. + \frac{(j+m)(j-m)}{(2j+1)(2j-1)[j(j+1)-j(j-1)]}\right\}$$

$$= \frac{Id^2E^2[j(j+1)-3m^2]}{\hbar^2 j(j+1)(2j-1)(2j+3)}.$$

d. Because H' is already diagonal in the j subspace, the nondegenerate perturbation theory is still applicable. However even with the perturbation, degeneracy does not completely disappear. In fact states with same j but opposite m are still degenerate.

5017

A rigid rotator with electric dipole moment **P** is confined to rotate in a plane. The rotator has moment of inertia I about the (fixed) rotation axis. A weak uniform electric field **E** lies in the rotation plane. What are the energies of the three lowest quantum states to order E^2?

(MIT)

Sol: The Hamiltonian for the free rotator is

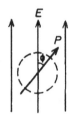

Fig. 5.8

$$H_0 = \frac{1}{2I}L_\phi^2 = -\frac{\hbar^2}{2I}\frac{d^2}{d\phi^2}.$$

A Schrödinger equation then gives the eigenvalues and eigenfunctions

$$E_m^{(0)} = \frac{\hbar^2 m^2}{2I}, \quad \psi_m^{(0)}(\phi) = \frac{1}{\sqrt{2\pi}}e^{im\phi}.$$

$$(m = 0, \pm 1, \pm 2, \ldots)$$

When a weak uniform electric field **E** is applied, the perturbation Hamiltonian is

$$H' = -\mathbf{E}\cdot\mathbf{P} = \lambda\cos\phi,$$

where $\lambda = -EP$. The matrix elements of the perturbation Hamiltonian is

$$\langle n|H'|m\rangle = \frac{\lambda}{2\pi}\int_0^{2\pi}e^{i(m-n)\phi}\cos\phi\,d\phi = \frac{\lambda}{2}\left(\delta_{n,m+1}+\delta_{n,m-1}\right).$$

Define $E_m^{(0)}$ and E by

$$H_0|m\rangle = \left|E_m^{(0)}|m\rangle\right., \quad (H_0+H')|\rangle = E|\rangle,$$

and expand $|\rangle$ in $|m\rangle \equiv \psi_m^{(0)}(\phi)$:

$$|\rangle = \sum_{m=-\infty}^{\infty} C_m |m\rangle.$$

Then

$$(H_0 + H') \sum_m C_m |m\rangle = \sum_m C_m E |m\rangle.$$

or

$$\sum_m C_m \left(E - E_m^{(0)}\right)|m\rangle - \sum_m C_m H' |m\rangle = 0.$$

Multiplying both sides by $\langle n|$ we have, on account of the orthonormality of $|m\rangle$ and the property of $\langle n|H'|m\rangle$ given above,

$$\sum_m C_m \left(E - E_m^{(0)}\right)\delta_{mn} - \frac{\lambda}{2}\sum_m C_m \left(\delta_{m,n-1} + \delta_{m,n+1}\right) = 0,$$

or

$$\left(E - E_n^{(0)}\right)C_n - \frac{\lambda}{2}C_{n-1} - \frac{\lambda}{2}C_{n+1} = 0.$$

Expanding E and C_n as power series in λ:

$$E = \sum_{\rho=0}^{\infty} E^{(\rho)}\lambda^\rho, \quad C_n = \sum_{\rho=0}^{\infty} C_n^{(\rho)}\lambda^\rho,$$

and substituting in the above, we obtain perturbation equations of different orders:

$$\lambda^0: \quad \left(E^{(0)} - E_n^{(0)}\right)C_n^{(0)} = 0,$$

$$\lambda^1: \quad \left(E^{(0)} - E_n^{(0)}\right)C_n^{(1)} + E^{(1)}C_n^{(0)} - \frac{1}{2}C_{n-1}^{(0)} - \frac{1}{2}C_{n+1}^{(0)} = 0,$$

$$\lambda^2: \quad \left(E^{(0)} - E_n^{(0)}\right)C_n^{(2)} + E^{(1)}C_n^{(1)} + E^{(2)}C_n^{(0)}$$

$$-\frac{1}{2}C_{n-1}^{(1)} - \frac{1}{2}C_{n+1}^{(1)} = 0,$$

$$\cdots$$

To find the energy level E_k^0, we first see that the zeroth order equation $(E_k^{(0)} - E_n^{(0)})C_n^{(0)} = 0$ requires $C_n^{(0)} = 0$ if $k \neq n$. Hence, we write

$$C_n^{(0)} = a_k\delta_{n,k} + a_{-k}\delta_{n,-k}. \tag{1}$$

Substitution in the first order equation gives

$$\left(E_k^{(0)} - E_n^{(0)}\right)C_n^{(1)} + E^{(1)}\left(a_k\delta_{n,k} + a_{-k}\delta_{n,-k}\right)$$

$$-\frac{1}{2}(a_k\delta_{n-1,k} + a_{-k}\delta_{n-1,-k} + a_k\delta_{n+1,k} + a_{-k}\delta_{n+1,-k}) = 0.$$

When $n = \pm k$, we have

$$E^{(1)} = 0.$$

When $n \neq \pm k$, we have

$$C_n^{(1)} = -\frac{1}{2} \frac{1}{E_n^{(0)} - E_k^{(0)}} \left(a_k \delta_{n-1,k} + a_{-k} \delta_{n-1,-k} + a_k \delta_{n+1,k} + a_{-k} \delta_{n+1,-k} \right)$$

$$= -\frac{1}{2} \frac{1}{E_n^{(0)} - E_k^{(0)}} \left(C_{n-1}^{(0)} + C_{n+1}^{(0)} \right).$$

Substituting $C_n^{(1)}$ in the second order equation gives for $n = k$

$$\left(E_k^{(0)} - E_n^{(0)} \right) C_n^{(2)} + E^{(2)} C_n^{(0)} - \frac{1}{2} C_{n-1}^{(1)} - \frac{1}{2} C_{n+1}^{(1)} = 0,$$

i.e.,

$$E^{(2)} C_k^{(0)} = \frac{1}{2} \left(C_{k-1}^{(1)} + C_{k+1}^{(1)} \right)$$

$$= \frac{1}{2} \left\{ -\frac{1}{2} \frac{1}{E_{k-1}^{(0)} - E_k^{(0)}} \left(C_{k-2}^{(0)} + C_k^{(0)} \right) \right.$$

$$\left. -\frac{1}{2} \frac{1}{E_{k+1}^{(0)} - E_k^{(0)}} \left(C_k^{(0)} + C_{k+2}^{(0)} \right) \right\}$$

$$= -\frac{1}{4} \left\{ \frac{1}{E_{k-1}^{(0)} - E_k^{(0)}} \left(C_{k-2}^{(0)} + C_k^{(0)} \right) \right.$$

$$\left. + \frac{1}{E_{k+1}^{(0)} - E_k^{(0)}} \left(C_k^{(0)} + C_{k+2}^{(0)} \right) \right\}. \qquad (2)$$

For the ground state $k = 0$, Eq. (1) gives $C_0^{(0)} \neq 0$, $C_{-2}^{(0)} = C_2^{(0)} = 0$, and so Eq. (2) becomes

$$E^{(2)} C_0^{(0)} = -\frac{1}{4} \left\{ \frac{1}{E_{-1}^{(0)} - E_0^{(0)}} + \frac{1}{E_1^{(0)} - E_0^{(0)}} \right\} C_0^{(0)}.$$

Hence,

$$E_0^{(2)} = -\frac{1}{2} \frac{1}{E_1^{(0)} - E_0^{(0)}} = -\frac{I}{\hbar^2}$$

For the first exited state $k = \pm 1$, Eq. (1) gives $C_{\pm 1}^{(0)} \neq 0$, and Eq. (2) becomes

$$E^{(2)} C_1^{(0)} = -\frac{1}{4} \left\{ \frac{1}{E_0^{(0)} - E_1^{(0)}} \left(C_{-1}^{(0)} + C_1^{(0)} \right) \right.$$

$$\left. + \frac{1}{E_2^{(0)} - E_1^{(0)}} \left(C_1^{(0)} + C_3^{(0)} \right) \right\}$$

$$= -\frac{1}{4}\left\{ \left(\frac{1}{E_2^{(0)} - E_1^{(0)}} - \frac{1}{E_1^{(0)}} \right) C_1^{(0)} - \frac{1}{E_1^{(0)}} C_{-1}^{(0)} \right\}$$

and

$$E^{(2)} C_{-1}^{(0)} = -\frac{1}{4}\left\{ -\frac{1}{E_{-1}^{(0)}} C_1^{(0)} + \left(\frac{1}{E_{-2}^{(0)} - E_{-1}^{(0)}} - \frac{1}{E_{-1}^{(0)}} \right) C_{-1}^{(0)} \right\}$$

$$= -\frac{1}{4}\left\{ -\frac{1}{E_1^{(0)}} C_1^{(0)} + \left(\frac{1}{E_2^{(0)} - E_1^{(0)}} - \frac{1}{E_1^{(0)}} \right) C_{-1}^{(0)} \right\},$$

as $E_0^{(0)} = 0$, $C_{\pm 3}^{(0)} = 0$, or

$$\left[-\frac{1}{4}\left(\frac{1}{E_2^{(0)} - E_1^{(0)}} - \frac{1}{E_1^{(0)}} \right) - E^{(2)} \right] C_1^{(0)} + \frac{1}{4}\frac{1}{E_1^{(0)}} C_{-1}^{(0)} = 0,$$

and

$$\frac{1}{4}\frac{1}{E_1^{(0)}} C_1^{(0)} + \left[-\frac{1}{4}\left(\frac{1}{E_2^{(0)} - E_1^{(0)}} - \frac{1}{E_1^{(0)}} \right) - E^{(2)} \right] C_{-1}^{(0)} = 0.$$

These are homogeneous equations in C_1^0 and C_{-1}^0. Solving the secular equation, we obtain for the first excited state two energies

$$E_{1+}^{(2)} = \frac{1}{4}\frac{1}{E_1^{(0)}} - \frac{1}{4}\left(\frac{1}{E_2^{(0)} - E_1^{(0)}} - \frac{1}{E_1^{(0)}} \right) = \frac{5I}{6\hbar^2},$$

$$E_{1-}^{(2)} = -\frac{1}{4}\frac{1}{E_1^{(0)}} - \frac{1}{4}\left(\frac{1}{E_2^{(0)} - E_1^{(0)}} - \frac{1}{E_1^{(0)}} \right) = -\frac{I}{6\hbar^2}.$$

For the second exited state $k = \pm 2$, Eq. (1) gives $C_{\pm 2}^{(0)} \neq 0$ and Eq. (2) becomes

$$E^{(2)} C_{\pm 2}^{(2)} = -\frac{1}{4}\left(\frac{1}{E_1^{(0)} - E_2^{(0)}} + \frac{1}{E_3^{(0)} - E_2^{(0)}} \right) C_{\pm 2}^{(0)}.$$

Thus

$$E_2^{(2)} = -\frac{1}{4}\left(\frac{1}{E_1^{(0)} - E_2^{(0)}} + \frac{1}{E_3^{(0)} - E_2^{(0)}} \right) = \frac{I}{15\hbar^2}.$$

Therefore, the energy correction to the second order perturbation for the ground state is

$$E_0 = \lambda^2 E_0^{(2)} = -\frac{I(Ep)^2}{\hbar^2},$$

for the first exited state is

$$E_{1+} = \frac{\hbar^2}{2I} + \frac{5}{6}\frac{I(Ep)^2}{\hbar^2}, \qquad E_{1-} = \frac{\hbar^2}{2I} - \frac{1}{6}\frac{I(Ep)^2}{\hbar^2},$$

and for the second exited state is

$$E_2 = \frac{2\hbar^2}{I} + \frac{1}{15}\frac{I(Ep)^2}{\hbar^2}.$$

5018

Consider a particle of mass *m* in a potential $V(X) = (m\omega^2X^2/2) + g\cos(kX)$. Calculate the change in the ground-state energy compared to the harmonic oscillator $V(X) = m\omega^2X^2/2$ to the first order in *g*.

Sol:

The perturbed Hamiltonian $\hat{H}' = g\cos(kX)$.

The first order correction to the ground state is given by $<n|\hat{H}'|n>$, where $|n>$ is the basis.

In integral form, $<\hat{H}> = \int_{-\infty}^{\infty} dx\, \psi_0^* \hat{H}' \psi_0$, where ψ_0 is given by $(\alpha^2/\pi)^{1/4}\exp(\alpha^2X^2/2)$.

So the correction term is given by

$$<\hat{H}'> = (\alpha^2/\pi)^{1/2} g \int_{-\infty}^{\infty} dx \exp(\alpha^2X^2)\cos(kX).$$

Now here is the standard integral $\int_{-\infty}^{\infty} dx\exp(-aX)^2\cos bX = (\pi/a)^{1/2}\exp(-b^2/4a)$. Using this standard integral, the above integral evaluates to $(\pi/\alpha^2)^{1/2}\exp(-k^2/4\alpha^2)$. So the correction to the first order energy is $<\hat{H}'> = g\exp(-k^2/4\alpha^2)$.

5019

a. State all the energy levels of a symmetric top with principal moments of inertia $I_1 = I_2 = I \neq I_3$.

b. A slightly asymmetric top has no two *I*'s exactly equal, but $I_1 - I_2 = \Delta \neq 0$, $I_1 + I_2 = 2I$, $(\Delta/2I) \ll 1$. Compute the $J = 0$ and $J = 1$ energies through $0(\Delta)$.

(Berkeley)

Sol:

a. Let (x, y, z) denote the rotating coordinates fixed in the top. The Hamiltonian of the system is

$$H = \frac{1}{2}\left(\frac{J_x^2}{I_1} + \frac{J_y^2}{I_2} + \frac{J_z^2}{I_3}\right)$$

$$= \frac{1}{2I}\left(J_x^2 + J_y^2\right) + \frac{1}{2I_3}J_z^2$$

$$= \frac{1}{2I}J^2 + \frac{1}{2}\left(\frac{1}{I_3} - \frac{1}{I}\right)J_z^2.$$

Hence, a state with quantum numbers J, m has energy

$$E = \frac{\hbar^2}{2I}J(J+1) + \frac{\hbar^2}{2}\left(\frac{1}{I_3} - \frac{1}{I}\right)m^2,$$

which gives the energy levels of the symmetric top.

b. For the slightly asymmetric top the Hamiltonian is

$$H = \frac{1}{2I}J^2 + \frac{1}{2}\left(\frac{1}{I_3} - \frac{1}{I}\right)J_z^2 + \frac{\Delta}{4I^2}(J_y^2 - J_x^2)$$

$$= H_0 + H',$$

where $H' = \frac{\Delta}{4I^2}(J_y^2 - J_x^2)$, to be considered as a perturbation.

Defining $J\pm = J_x \pm iJ_y$ we have

$$J_x^2 - J_y^2 = \frac{1}{2}(J_+^2 + J_-^2).$$

Noting that

$$J_+ \mid jm\rangle = \sqrt{(j+m+1)(j-m)}\mid j, m+1\rangle,$$

$$J_- \mid jm\rangle = \sqrt{(j-m+1)(j+m)}\mid j, m-1\rangle,$$

$$J_\pm^2 \mid jm\rangle = J_\pm\left(J_\pm \mid jm\rangle\right),$$

we have

$$J_+^2 \mid 00\rangle = J_-^2 \mid 00\rangle = 0,$$

$$J_+^2 \mid 10\rangle = J_-^2 \mid 10\rangle = 0,$$

$$J_+^2 \mid 1,1\rangle = 0, \quad J_+^2 \mid 1, -1\rangle = 2\hbar^2 \mid 11\rangle,$$

$$J_-^2 \mid 1,-1\rangle = 0, \quad J_-^2 \mid 11\rangle = 2\hbar^2 \mid 1,-1\rangle.$$

Hence, for the perturbed states:

i. $J = 0$, $m = 0$, (nondegenerate):

$$E_0' = E_0^{(0)} + \langle 00 \mid H' \mid 00\rangle = E_0^{(0)} = 0.$$

ii. $J = 1$, $m = 0$, (nondegenerate):

$$E_1' = E_1^{(0)} + \langle 10|H'|10\rangle = E_1^{(0)} = \frac{\hbar^2}{I},$$

iii. $J = 1$, $m = \pm 1$, (two-fold degenerate):

As degeneracy occurs, we first calculate

$$\langle 1, \, -1|H'|1, \, -1\rangle = \langle 1,1|H'|11\rangle = 0,$$

$$\langle 1,-1|H'|11\rangle = \langle 11|H|1,-1\rangle = \frac{\Delta\hbar^2}{4I^2}.$$

We then form the secular equation

$$\left| \lambda 1 - \begin{pmatrix} 0 & \dfrac{\Delta\hbar^2}{4I^2} \\ \dfrac{\Delta\hbar^2}{4I^2} & 0 \end{pmatrix} \right| = 0,$$

i.e.,

$$\begin{vmatrix} \lambda & -\dfrac{\Delta\hbar^2}{4I^2} \\ -\dfrac{\Delta\hbar^2}{4I^2} & \lambda \end{vmatrix} = 0.$$

This equation has two solutions

$$\lambda_\pm = \pm\frac{\Delta\hbar^2}{4I^2},$$

which means that the energy of the states $J = 1$, $m = \pm 1$, $E_{1,\pm 1}$, splits into two levels:

$$E_{1,\pm 1} = \frac{\hbar^2}{2I} + \frac{\hbar^2}{2I_3} \pm \frac{\Delta\hbar^2}{4I^2}.$$

5020

a. Using a simple hydrogenic wave function for each electron, calculate by perturbation theory the energy in the ground state of the He atom associated with the electron-electron Coulomb interaction (neglect exchange effects). Use this result to estimate the ionization energy of helium.

b. Calculate the ionization energy by using the variational method, with the effective charge Z in the hydrogenic wave function as the variational parameter. Compare the results of (a) and (b) with the experimental ionization energy of $1.807E_0$, where $E_0 = \alpha^2 mc^2 / 2$.

Note:

$$\psi_{1s}(r) = \sqrt{\frac{Z^3}{\pi a_0^3}} \exp(-Zr / a_0), \quad a_0 = \frac{\hbar^2}{me^2},$$

$$\int\int d^3 r_1 d^3 r_2 e^{-\alpha(r_1+r_2)} / |r_1 - r_2| = 20\pi^2 / \alpha^5.$$

(Columbia)

Sol:

a. The unperturbed Hamiltonian of the system is

$$H_0 = -\frac{\hbar^2}{2m}\left(\nabla_2^1 + \nabla_2^2\right) - \frac{Ze^2}{r_1} - \frac{Ze^2}{r_2}.$$

The wave functions have the form

$$\Phi = \phi(r_1, r_2)\chi_0(s_{1z}, s_{2z}).$$

For the ground state,

$$\phi(r_1, r_2) = \psi_{100}(r_1)\psi_{100}(r_2),$$

where

$$\psi_{100}(r) = \sqrt{\frac{Z^3}{\pi a_0^3}} \exp\{-Zr / a_0\}$$

with $a_0 = \hbar^2 / me^2$. Treating the electron-electron interaction $\dfrac{e^2}{|r_1 - r_2|}$ as a perturbation, the energy correction to first order in perturbation theory is

$$\Delta E = \langle H' \rangle = e^2 \int\int \frac{d^3 r_1 d^3 r_2}{|r_1 - r_2|} |\psi_{100}(r_1)|^2 |\psi_{100}(r_2)|^2$$

$$= e^2 \left(\frac{Z^3}{\pi a_0^3}\right)^2 \int\int \frac{d^3 r_1 d^3 r_2}{|r_1 - r_2|} \exp\left[-\frac{2Z}{a_0}(r_1 + r_2)\right]$$

$$= e^2 \left(\frac{Z^3}{\pi a_0^3}\right)^2 \cdot \frac{20\pi^2}{\left(\frac{2Z}{a_0}\right)^5} = \frac{5Ze^2}{8a_0}.$$

The energy levels of a hydrogen-like atom are given by

$$E_n = -\frac{e^2}{2a_0}\frac{Z^2}{n^2}.$$

and so the energy of the ground state of the system excluding the electro-nelectron Coulomb interaction is

$$E_0 = -2\frac{e^2 Z^2}{2a_0} = -\frac{e^2 Z^2}{a_0}.$$

Hence the corrected energy of the ground state of helium is

$$E = -\frac{e^2 Z^2}{a_0} + \frac{5Ze^2}{8a_0} = -\frac{11e^2}{4a_0},$$

with $Z = 2$ for helium nucleus.

The ionization energy is the energy required to remove the two electrons of helium atom to infinity. Thus for the ground state

$$I = -\frac{e^2 Z^2}{2a_0} - \left(-\frac{e^2 Z^2}{a_0} + \frac{5Ze^2}{8a_0} \right) = \frac{3e^2}{4a_0}$$

with $Z = 2$ for helium nucleus, i.e.,

$$I = 1.5\, E_0,$$

with

$$E_0 = \frac{e^2}{2a_0} = \frac{1}{2}\left(\frac{e^2}{\hbar c} \right)^2 mc^2 = \frac{1}{2}\alpha^2 mc^2,$$

α being the fine structure constant.

b. The Hamiltonian of He with electron-electron interaction is

$$H = -\frac{\hbar^2}{2m}(\nabla_1^2 + \nabla_2^2) - Ze^2/r_1 - Ze^2/r_2 + e^2/r_{12}$$

with $Z = 2$, $r_{12} = |r_1 - r_2|$. For the ground state use a trial wave function

$$\phi(r_1, r_2, \lambda) = \frac{\lambda^3}{\pi} e^{-\lambda(r_1 + r_2)}.$$

Let $u(r) = e^{-\lambda r}$. Then

$$\left(-\frac{1}{2}\nabla^2 - \frac{\lambda}{r} \right) u(r) = -\frac{\lambda^2}{2} u(r),$$

i. e.,

$$\left(-\frac{\hbar^2}{2m}\nabla^2 - \frac{\lambda\hbar^2}{mr} \right) u(r) = -\frac{\lambda^2\hbar^2}{2m} u(r).$$

Setting

$$Ze^2 - \frac{\lambda\hbar^2}{m} = \sigma,$$

we have, using the above results,

$$\bar{H} = \iint d^3 r_1 d^3 r_2 \Phi^* \left(-\frac{\hbar^2}{2m}\nabla_1^2 - \frac{\hbar^2}{2m}\nabla_2^2 - \frac{Ze^2}{r_1} - \frac{Ze^2}{r_2} + \frac{e^2}{r_{12}} \right)\Phi$$

$$= \iint d^3 r_1 d^3 r_2 \Phi^* \left(-\frac{\lambda^2\hbar^2}{m} - \frac{\sigma}{r_1} - \frac{\sigma}{r_2} + \frac{e^2}{r_{12}} \right)\Phi$$

$$= -\frac{\lambda^2\hbar^2}{m} - \frac{2\sigma\lambda^3}{\pi}\int \frac{e^{-2\lambda r_1}}{r_1}d^3 r_1 + \frac{e^2\lambda^6}{\pi^2}\iint d^3 r_1 d^3 r_2 \frac{e^{-2\lambda(r_1+r_2)}}{r_{12}}.$$

As

$$\int \frac{e^{-2\lambda r_1}}{r_1}d^3 r_1 = \int_0^\infty \frac{e^{-2\lambda r_1}}{r_1} 4\pi r_1^2\, dr_1 = \pi/\lambda^2,$$

$$\bar{H} = -\frac{\lambda^2\hbar^2}{m} - 2\sigma\lambda + \frac{\lambda^6}{\pi^2}\frac{20\pi^2}{(2\lambda)^5}$$

$$= \frac{\lambda^2\hbar^2}{m} - \left(2Z - \frac{5}{8}\right)e^2\lambda.$$

Letting $\dfrac{\partial \bar{H}}{\partial \lambda} = 0$, we get

$$\lambda = \frac{me^2}{2\hbar^2}\left(2Z - \frac{5}{8}\right).$$

Therefore, the energy of the ground state is

$$E = -\left(Z - \frac{5}{16}\right)^2 \frac{me^4}{\hbar^2} = -\left(\frac{27}{16}\right)^2 \frac{e^2}{a_0},$$

as $Z = 2\ a_0 = \dfrac{\hbar^2}{me^2}$, and the ground state ionization energy is

$$I = -\frac{Z^2 e^2}{2a_0} + \left(Z - \frac{5}{16}\right)^2 \frac{e^2}{a_0} = \left[\left(\frac{27}{16}\right)^2 - 2\right]\frac{e^2}{a_0} = 1.695 E_0$$

Thus the result from the variational method is in better agreement with experiment.

5021

A particle of mass m moves in one dimension in the periodic potential

$$V(x) = V_0 \cos\left(\frac{2\pi x}{\alpha}\right).$$

We know that the energy eigenstates can be divided into classes characterized by an angle θ with wave functions $\phi(x)$ that obey $\phi(x+a) = e^{i\theta}\phi(x)$ for all x. For the class $\theta = \pi$, this becomes $\phi(x+a) = -\phi(x)$ (antiperiodic over length a).

a. Even when $V_0 = 0$, we can still classify eigenstates by θ. For which values of k does the plane wave $\phi(x) = e^{ikx}$ satisfy the antiperiodic condition over length a? What is the energy spectrum of the class $\theta = \pi$ for $V_0 = 0$?

b. When V_0 is small (i.e. $V_0 \ll \hbar^2 / ma^2$), calculate the lowest two energy eigenvalues by first order perturbation theory.

(MIT)

Sol:

a. For the plane wave $\psi(x) = e^{ikx}$, we have

$$\psi(x+a) = e^{ik(x+a)} = e^{ika}\psi(x).$$

If k satisfies

$$ka = (2n+1)\pi, \quad (n = 0, \pm 1, \pm 2, \ldots)$$

the plane wave satisfies the antiperiodic condition

$$\psi(x+a) = -\psi(x).$$

The corresponding energy spectrum is

$$E_n = \frac{\hbar^2 \pi^2}{2ma^2}(2n+1)^2. \quad (n = 0, \pm 1, \pm 2, \ldots)$$

b. If $V_0 \ll \dfrac{\hbar^2}{ma^2}$, one can treat

$$H' = V_0 \cos\left(\frac{2\pi x}{a}\right)$$

as a perturbation imposed on the free motion of a particle. For the ground state, the eigenvalue and eigenfunction of the free particle are respectively $(n = 0, -1; \text{i.e., } ka = \pi, -\pi)$

$$E^{(0)}_{\binom{0}{-1}} = \frac{\hbar^2 \pi^2}{2ma^2},$$

$$\psi^{(0)}_0(x) = \frac{1}{\sqrt{a}}e^{i\pi x/a},$$

$$\psi^{(0)}_{-1}(x) = \frac{1}{\sqrt{a}}e^{-i\pi x/a}.$$

Let $\dfrac{2\pi}{a} = \beta$ =and consider $\langle m|H'|n\rangle$. We have

$$\langle -1|H'|-1\rangle = \langle 0|H'|0\rangle = \frac{V_0}{a}\int_0^a \cos\left(\frac{2\pi x}{a}\right)dx = 0,$$

$$\langle -1|H'|0\rangle = \langle 0|H'|-1\rangle = \frac{V_0}{2a}\int_0^a e^{\pm i\beta x}(e^{i\beta x}+e^{-i\beta x})dx$$

$$=\frac{V_0}{2}.$$

Hence, for ground state,

$$H' = \begin{pmatrix} 0 & \dfrac{V_0}{2} \\[2ex] \dfrac{V_0}{2} & 0 \end{pmatrix},$$

and the secular equation for first order perturbation is

$$\begin{vmatrix} E^{(1)} & \dfrac{V_0}{2} \\[2ex] \dfrac{V_0}{2} & E^{(1)} \end{vmatrix} = 0,$$

giving

$$E^{(1)} = \pm\frac{V_0}{2}.$$

Thus the ground state energy level splits into two levels

$$E_1 = \frac{\hbar^2\pi^2}{2ma^2}-\frac{V_0}{2}, \quad E_2 = \frac{\hbar^2\pi^2}{2ma^2}+\frac{V_0}{2}.$$

These are the lowest energy eigenvalues of the system.

5022

a. A rigid rotor in a plane is acted on by a perturbation $H' = (V_0/2)(3\cos^2\varphi - 1)$. Calculate the first-order correction to the energy of the ground state, where $\psi(\varphi) = e^{im\varphi}/\sqrt{2\pi}$ is the normalized wave function and m takes integer values.

Consider a free particle of mass m confined to move in a one-dimensional infinite box of length a. If the system is subjected to perturbation $V(x) = V_0\delta(x - a/2)$, find the first order correction to the energy of the ground state if $\psi n(x) = \sqrt{\dfrac{2}{a}}\sin\dfrac{n\pi x}{a}$ is the normalized wave function where n takes integer values.

Sol:

a. For the ground state, $m=0, \psi_0 = 1/\sqrt{2\pi}$.

So the correction is given by $E_0^1 = <\psi|\hat{H}'|\psi> = (V_0/4\pi)\int_0^{2\pi} d\varphi(3\cos^2\varphi - 1)$

$$E_0^1 = V_0/4\pi \left[\int_0^{2\pi} d\varphi \, 3\cos^2\varphi - \int_0^{2\pi} d\varphi\right] = V_0/4$$

where we have used the formula $\cos^2 x = (1 + \cos^2 x)/2$.

b. The perturbed Hamiltonian is given by $\hat{H}' = V_0\delta(x - a/2)$.

The correction to the energy is given by

$$<\hat{H}'> = (2V_0/a)\int_0^a dx \, \sin^2(\pi x/a)\, \delta(x - a/2).$$

Using the property of Dirac delta integral,

$$\int_a^b f(x)\delta(x - c)dx = f(c), \quad \text{where } [a < c < b]$$

$$<\hat{H}'> = (2V_0/a)\sin^2(\pi x/a)|_{x=a/2} = 2V_0/a.$$

5023

Consider the one-dimensional motion of an electron confined to a potential well $V(x) = \frac{1}{2}kx^2$ and subjected also to a perturbing electric field $\mathbf{F} = F\hat{\mathbf{x}}$.

a. Determine the shift in the energy levels of this system due to the electric field.

b. The dipole moment of this system in state n is defined as $P_n = -e\langle x \rangle_n$, where $\langle x \rangle_n$ is the expectation value of x in the state n. Find the dipole moment of the system in the presence of the electric field.

(Wisconsin)

Sol:

a. The Hamiltonian of the system is

$$H = -\frac{\hbar^2}{2m}\nabla^2 + \frac{1}{2}kx^2 - qFx$$

$$= -\frac{\hbar^2}{2m}\nabla^2 + \frac{1}{2}k\left(x - \frac{qF}{k}\right)^2 - \frac{q^2F^2}{2k}$$

$$= -\frac{\hbar^2}{2m}\nabla'^2 + \frac{1}{2}kx'^2 - \frac{q^2F^2}{2k},$$

where $x' = x - \dfrac{qF}{k}$.

Hence the energy shift due to the perturbing electric field $F\hat{x}$ is

$$E' = \frac{q^2 F^2}{2k} = \frac{e^2 F^2}{2k}.$$

b. The expectation value of x in state n is

$$\langle x \rangle_n = \left\langle x' + \frac{qF}{k} \right\rangle = \langle x' \rangle + \left\langle \frac{qF}{k} \right\rangle = \frac{qF}{k}.$$

Therefore the dipole moment of the system is

$$P_n = -e\frac{qF}{k} = e^2\frac{F}{k}, \quad (q=-e).$$

5024

If a very small uniform-density sphere of charge is in an electrostatic potential $V(r)$, its potential energy is $U(\mathbf{r}) = V(\mathbf{r}) + \frac{1}{6}r_0^2\nabla^2 V(\mathbf{r}) + \cdots$, where $\mathbf{r}$ is the position of the center of the charge and r_0 is its very small radius. The "Lamb shift" can be thought of as the small correction to the energy levels of the hydrogen atom because the physical electron does have this property. If the r_0^2 term of U is treated as a very small perturbation compared to the Coulomb interaction $V(r) = -e^2/r$, what are the Lamb shifts for the 1s and 2p levels of the hydrogen atom? Express your result in terms of r_0 and fundamental constants. The unperturbed wave functions are

$$\psi_{1s}(\mathbf{r}) = 2a_B^{-3/2} \cdot e^{-r/a_B} Y_0^0; \quad \psi_{2pm}(\mathbf{r}) = \frac{1}{\sqrt{24}} a_B^{-5/2} r e^{-r/2a_B} Y_1^m,$$

where $a_B = \hbar^2/m_e e^2$.

(CUS)

Sol: The state 1s is nondegenerate, so the energy correction is

$$\Delta E = \langle 1s | H' | 1s \rangle.$$

As

$$H' = \frac{r_0^2}{6}\nabla^2 V(\mathbf{r}) = \frac{r_0^2}{6}(-e^2)\nabla^2\frac{1}{r}$$

$$= \frac{r_0^2}{6}(-e^2)(-4\pi)\delta(\mathbf{r}) = \frac{2\pi}{3}r_0^2 e^2 \delta(\mathbf{r}),$$

$$Y_0^0 = \left(\frac{1}{4\pi}\right)^{\frac{1}{2}},$$

we have

$$\Delta E = \int \frac{2\pi}{3} r_0^2 e^2 \delta(\mathbf{r}) |\psi_{1s}(x)|^2 \, dx$$

$$= \frac{2\pi}{3} r_0^2 e^2 |\psi_{1s}(0)|^2 = \frac{2}{3} \frac{r_0^2 e^2}{a_B^3}.$$

The perturbation H', being a δ-function, has an effect only if $\psi(0) \neq 0$. As $\psi_{2pm}(0) = 0, H'$ has no effect on the energy, i.e., $\Delta E_{2pm} = 0$.

5025

Positronium is a hydrogen atom but with a positron as "nucleus", instead of a proton. In the nonrelativistic limit, the energy levels and wave functions are the same as for hydrogen, except for scale.

  a. From your knowledge of the hydrogen atom, write down the normalized wave function for the 1s ground state of positronium. Use spherical coordinates and the hydrogenic Bohr radius a_0 as a scale parameter.

  b. Evaluate the root-mean-square radius for the 1s state in units of a_0. Is this an estimate of the physical diameter or the radius of positronium?

  c. In the s states of positronium there is a contact hyperfine interaction

$$H_{int} = -\frac{8\pi}{3} \mu_e \cdot \mu_p \delta(\mathbf{r}),$$

where μ_e and μ_p are the electron and positron magnetic moments

$$\left(\mu = g \frac{e}{2mc} s \right).$$

For electrons and positrons, $|g| = 2$. Using first order perturbation theory compute the energy difference between the singlet and triplet ground states. Determine which state lies lowest. Express the energy splitting in GHz (i.e., energy divided by Planck's constant). Get a number!

(Berkeley)

Sol:

  a. By analogy with the hydrogen atom the normalized wave function for the 1s ground state of positronium is

$$\psi_{100}(\mathbf{r}) = \frac{1}{\pi}\left(\frac{1}{2a_0}\right)^{3/2} e^{-r/2a_0}$$

with $a_0 = \dfrac{\hbar}{me^2}$, m being the electron rest mass. Note that the factor 2 in front of a_0 is to account for the fact that the reduced mass is $\mu = \frac{1}{2}m$.

b. The mean-square radius for the 1s state is

$$\langle r^2 \rangle = \frac{1}{8\pi a_0^3}\int_0^\infty e^{-r/a_0} r^2 \cdot r^2 \, dr$$

$$= \frac{a_0^2}{8\pi}\int_0^\infty e^{-x} x^4 \, dx = \frac{3a_0^2}{\pi},$$

and the root-mean-square radius is

$$\sqrt{\langle r^2 \rangle} = \sqrt{\frac{3}{\pi}} a_0.$$

This can be considered a physical estimate of the radius of positronium.

c. Taking the spin into account a state of the system is to be described by $|n, l, m, S, S_z\rangle$, where S and S_z are respectively the total spin and the z-component of the spin. Thus

$$\langle 100 S' S_z' | H_{\text{int}} | 100 \, S \, S_z \rangle$$

$$= \int d^3 r \psi_{100}^*(\mathbf{r})\left(-\frac{8\pi}{3}\right)\delta(\mathbf{r})\psi_{100}(\mathbf{r})\chi_s^+, \, (S_z')\mu_e \cdot \mu_p \chi_s(S_z)$$

$$= \frac{8\pi}{3}\left(\frac{e}{mc}\right)^2 |\psi_{100}(0)|^2 \, \chi_s^+(S_z')s_e \cdot s_p \chi_s(S_z)$$

$$= \frac{1}{3}\left(\frac{e^2}{\hbar c}\right)^2 \cdot \frac{e^2}{a_0}\cdot\left[\frac{1}{2}S(S+1)-\frac{3}{4}\right]\delta_{SS'}\delta_{S_z S_z'},$$

where we have used

$$S = s_e + s_p$$

and so

$$s_e \cdot s_p = \frac{1}{2}\left[S^2 - (s_e^2 + s_p^2)\right]$$

$$= \frac{1}{2}\left[S(S+1) - 2\left(\frac{1}{2}\right)\left(\frac{1}{2}+1\right)\right].$$

For the singlet state, $S = 0$, $S_z = 0$,

$$\Delta E_0 = -\frac{1}{4}\left(\frac{e^2}{\hbar c}\right)^2 \frac{e^2}{a_0} < 0.$$

For the triplet state, $S = 1$, $S_z = 0, \pm 1$,

$$\Delta E_1 = \frac{1}{12}\left(\frac{e^2}{\hbar c}\right)^2 \frac{e^2}{a_0} > 0.$$

Thus the singlet ground state has the lowest energy and the energy splitting of the ground state is

$$\Delta E_1 - \Delta E_0 = \left(\frac{1}{12} + \frac{1}{4}\right)\left(\frac{e^2}{\hbar c}\right)^2 \frac{e^2}{a_0} = \frac{1}{3}\left(\frac{e^2}{\hbar c}\right)^4 mc^2$$

$$= \frac{1}{3}\left(\frac{1}{137}\right)^4 \cdot 0.51 \times 10^6 = 4.83 \times 10^{-4}\,\text{eV},$$

corresponding to

$$\nu = \Delta E / h = 1.17 \times 10^{11}\,\text{Hz} = 117\,\text{GHz}.$$

5026

Consider the proton to be a spherical shell of charge of radius R. Using first order perturbation theory calculate the change in the binding energy of hydrogen due to the non-point-like nature of the proton. Does the sign of your answer make sense physically? Explain.

Note: You may use the approximation $R \ll a_0$ throughout this problem, where a_0 is the Bohr radius.

(MIT)

Sol: If we consider the proton to be a spherical shell of radius R and charge e, the potential energy of the electron, of charge $-e$, is

$$V(r) = \begin{cases} -\dfrac{e^2}{R}, & 0 \leq r \leq R, \\[2mm] -\dfrac{e^2}{r}, & R \leq r \leq \infty. \end{cases}$$

Take the difference between the above $V(r)$ and the potential energy $-\dfrac{e^2}{r}$ due to a point-charge proton as perturbation:

$$H' = \begin{cases} \dfrac{e^2}{r} - \dfrac{e^2}{R}, & 0 \leq r \leq R, \\[2mm] 0, & R \leq r \leq \infty. \end{cases}$$

The energy correction given by first order perturbation is

$$\Delta E = \int_0^R \left(\frac{e^2}{r} - \frac{e^2}{R} \right) r^2 R_{10}^2 dr$$

$$= \frac{4}{a_0^3} \int_0^R \left(\frac{e^2}{r} - \frac{e^2}{R} \right) r^2 e^{-2r/a_0} \, dr$$

$$\approx \frac{4}{a_0^3} \int_0^R \left(\frac{e^2}{r} - \frac{e^2}{R} \right) r^2 \, dr = \frac{2e^2 R^2}{3a_0^3}. \quad (R \ll a_0)$$

As $\Delta E > 0$, the ground state·energy level of the hydrogen atom would increase due to the non-point-like nature of the proton, i.e., the binding energy of the hydrogen atom would decrease. Physically, comparing the point-like and shell-shape models of the proton nucleus, we see that in the latter model there is an additional repulsive action. As the hydrogen atom is held together by attractive force, any non-point-like nature of the proton would weaken the attractive interaction in the system, thus reducing the binding energy.

5027

Assume that the proton has a nonzero radius $r_p \approx 10^{-13}$ cm and that its charge is distributed uniformly over this size. Find the shift in the energy of the 1s and 2p states of hydrogen due to the difference between a point charge distribution and this extended charge.

(Columbia)

Sol: The Coulomb force an electron inside the sphere of the proton experiences is

$$F = -e^2 \left(\frac{r}{r_p} \right)^3 \frac{1}{r^2} e_r = -\frac{e^2}{r_p^3} r e_r.$$

The electrical potential energy of the electron is

$$V_1 = \frac{e^2}{2r_p^3} r^2 + C \qquad \text{for} \qquad r \le r_p,$$

$$V_2 = -\frac{e^2}{r} \qquad \text{for} \qquad r \ge r_p,$$

Continuity at r_p requires that $V_1(r_p) = V_2(r_p)$, giving $C = -\dfrac{3}{2}\dfrac{e^2}{r_p}$. Thus

$$
V(r) = \begin{cases}
-\dfrac{e^2}{r}, & r > r_p, \\[3mm]
\dfrac{e^2}{2r_p}\left[\left(\dfrac{r}{r_p}\right)^2 - 3\right], & r \le r_p.
\end{cases}
$$

The Hamiltonian of the system is

$$
\hat{H} = \hat{H}_0 + \hat{H}',
$$

where

$$
\hat{H}' = \begin{cases}
0, & r > r_p, \\[3mm]
\dfrac{e^2}{2r_p}\left[\left(\dfrac{r}{r_p}\right)^2 + \dfrac{2r_p}{r} - 3\right], & r \le r_p
\end{cases}
$$

$$
\hat{H}_0 = \dfrac{\hat{P}^2}{2m} - \dfrac{e^2}{r}.
$$

Hence

$$
\begin{aligned}
E'_{nl} &= \langle nlm | H' | nlm \rangle = \langle nl | H' | nl \rangle \\
&= \int_0^\infty R_{nl}^* R_{nl} H'(r) r^2 \, dr \\
&= \int_0^{r_p} R_{nl}^*(r) R_{nl}(r) \\
&\quad \times \dfrac{e^2}{2r_p}\left[\left(\dfrac{r}{r_p}\right)^2 + \dfrac{2r_p}{r} - 3\right] r^2 dr,
\end{aligned}
$$

where

$$
R_{10} = \dfrac{2}{a^{3/2}} e^{-r/a}, \qquad R_{21} = \dfrac{1}{2\sqrt{6}\,a^{3/2}} \times \dfrac{r}{a} e^{-r/2a}
$$

with $a = \frac{\hbar^2}{me^2}$.

As $r_p \ll a$ we can take $e^{-r/a} \approx 1$. Integrating the above gives the energy shifts of 1s and 2p states:

$$
E'_{10} = \langle 10 | H' | 10 \rangle \approx \dfrac{2e^2 r_p^2}{5a^3},
$$

$$
E'_{21} = \langle 21 | H' | 21 \rangle \approx \dfrac{e^2 r_p^4}{1120 a^5}.
$$

5028

Consider a free particle of mass m confined in a three-dimensional (3D) cubic box of length "a," where the potential $V(X) = 0$ if $0 < X, Y, Z < a$ and infinite elsewhere. If the system is subjected to a perturbation that is given by $H' = V_0\delta(x - a/2)\delta(y - a/4)$, evaluate the first order correction to the ground-state energy.

Sol:

The ground-state wave function for a particle in 3D box is given by

$$\psi_{gs} = (2/a)^{3/2} \sin(\pi x/a) \sin(\pi y/a) \sin(\pi z/a)$$

$$<\hat{H}'> = V_0(2/a)^3 \int_0^a dx \, \sin^2(\pi x/a)\delta(x-a/2)\int_0^a dy \, \sin^2(\pi y/a)\delta(x-a/4)$$

$$\int_0^a dz \, \sin^2(\pi z/a) .$$

These are three separate integrals, and using the property of Dirac delta integrals, i.e., $\int_a^b f(x)\delta(x-c)dx = f(c)$, where $[a \le c \le b]$ the first two integrals can be evaluated. The third integral can be evaluated by

Using the identity $\sin^2 x = [1 - \cos(2\pi x)]/2$

so, $\int_0^a dx \, \sin^2(\pi x/a)\delta(x-a/2) = \sin^2(\pi x/a)|_{x=a/2} = 1$

$\int_0^a dy \, \sin^2(\pi y/a)\delta(x-a/4) = \sin^2(\pi y/a)|_{x=a/4} = 1/2$

$\int_0^a dz \, \sin^2(\pi z/a) = (1/2) \int_0^a dz(1-\cos 2\pi z/a) = a/2$

So the correction to the energy of the ground state is given by
$$8/a_0^3 1 \times (1/2) \times (a/2) = 2V_0/a^2.$$

5029

A hydrogen atom subjected to a perturbation $V_{pert} = aL^2$, where L is the angular momentum. Evaluate the shift in the energy level of the 2P state when the effects of spin is neglected up to second order in a.

Sol:

The given perturbation is $\hat{H}' = aL^2$.

The first order correction to the energy is given by $<\hat{H}'> = a<L^2>$.

We know that $L^2|l,m> = \hbar^2 l(l+1)|l,m>$ which is an eigenvalue equation, where the eigenvalue turns out to be $\hbar^2 l(l+1)$.

Now, for a P state, $l = 1$ so, $L^2|l,m> = 2\hbar^2|l,m>$ and sandwiching $<l,m|$ to the left gives the expectation value of $<L^2> = 2\hbar^2$ as $<l,m|l,m> = 1$ from the definition of Kronecker delta.

So, the correction to the energy turns out to be $2a\hbar^2$.

5030

A proposal has been made to study the properties of an atom composed of a $\pi^+(m_{\pi^+} = 237.2m_e)$ and a $\mu^-(m_\mu{-} = 206.77m_e)$ in order to measure the charge radius of the pion. Assume that all of the pion charge is spread uniformly on a spherical shell at $R_0 = 10^{-13}$ cm and that the μ is a point charge. Express the potential as a Coulomb potential for a point charge plus a perturbation and use perturbation theory to calculate a numerical value for the percentage shift in the ls -2p energy difference Δ. Neglect spin orbit effects and Lamb shift. Given

$$a_0 = \frac{\hbar^2}{me^2}, \quad R_{10}(r) = \left(\frac{1}{a_0}\right)^{\frac{3}{2}} 2e^{-r/a_0}, \quad R_{21}(r) = \left(\frac{1}{2a_0}\right)^{\frac{3}{2}} \frac{r}{a_0}\frac{e^{-r/a_0}}{\sqrt{3}}.$$

(Wisconsin)

Sol: The Coulomb potential energy of the muon is

$$V = \begin{cases} -\dfrac{e^2}{r} & \text{for } r \geq R, \\ -\dfrac{e^2}{R} & \text{for } r \leq R. \end{cases}$$

It can be written in the form

$$V = V_0 + V' = -\frac{e^2}{r} + V',$$

where

$$V' = \begin{cases} 0 & \text{for } r \geq R, \\ \left(\dfrac{1}{r} - \dfrac{1}{R}\right)e^2 & \text{for } r \leq R, \end{cases}$$

is to be treated as a perturbation.

The energy levels and wave functions of the unperturbed system are

$$E_n = -\frac{e^2}{2a_0 n^2}, \quad \psi_n^{(0)} = R_{nl}(r) Y_{lm}(\theta, \varphi).$$

As spin orbit and Lamb effects are to be neglected, we need only consider R_{nl} in perturbation calculations. Thus

$$\Delta E_{1s} = e^2 \int_0^\infty R_{10}^2 V' \, r^2 \, dr = \frac{4e^2}{a_0^3} \int_0^R e^{-\frac{2r}{a_0}} \left(\frac{1}{r} - \frac{1}{R} \right) r^2 \, dr,$$

where $a_0 = \dfrac{\hbar^2}{me^2}$, m being the reduced mass of the system:

$$m = \frac{m_\pi m_\mu}{m_\pi + m_\mu} = \frac{\rho m_\mu}{\rho + 1}$$

with

$$\rho = \frac{m_\pi}{m_\mu} = 1.15.$$

Hence

$$a_0 = \frac{m_e}{m} \left(\frac{\hbar^2}{m_e e^2} \right) = \frac{1}{110.5} \times 0.53 \times 10^{-8} = 4.8 \times 10^{-11} \text{ cm}.$$

Thus $a_0 \gg R$ and the factor $\exp(-2r/a_0)$ in the integrand above may be neglected. Hence

$$\Delta E_{1s} \approx \frac{4e^2}{a_0^3} \int_0^R \left(\frac{1}{r} - \frac{1}{R} \right) r^2 \, dr = \frac{2}{3} \left(\frac{e^2}{a_0} \right) \left(\frac{R}{a_0} \right)^2,$$

$$\Delta E_{2p} \approx \frac{e^2}{24 a_0^5} \int_0^R \left(\frac{1}{r} - \frac{1}{R} \right) r^4 \, dr = \frac{1}{480} \left(\frac{e^2}{a_0} \right) \left(\frac{R}{a_0} \right)^2.$$

Therefore

$$\frac{\Delta E_{2p} - \Delta E_{1s}}{E_{2p} - E_{1s}} \approx \frac{-\dfrac{2}{3} \dfrac{e^2}{a_0} \left(\dfrac{R}{a_0} \right)^2}{\dfrac{e^2}{a_0} \left(-\dfrac{1}{8} + \dfrac{1}{2} \right)} = -\frac{16}{9} \left(\frac{R}{a_0} \right)^2$$

$$= -7.7 \times 10^{-6} = -7.7 \times 10^{-4} \%.$$

5031

Muonic atoms consist of mu mesons (mass $m_\mu = 206 m_e$) bound to atomic nuclei in hydrogenic orbits. The energies of the mu mesic levels are shifted relative to their values for a point nucleus because the nuclear charge is distributed over a region with radius R. The effective Coulomb potential can be approximated as

$$V(r) = \begin{cases} \dfrac{-Ze^2}{r}, & r \geq R, \\[3mm] \dfrac{-Ze^2}{R}\left(\dfrac{3}{2} - \dfrac{1}{2}\dfrac{r^2}{R^2}\right), & r \leq R. \end{cases}$$

a. State qualitatively how the energies of the 1s, 2s, 2p, 3s, 3p, 3d muonic levels will be shifted absolutely and relative to each other, and explain physically any differences in the shifts. Sketch the unperturbed and perturbed energy level diagrams for these states.

b. Give an expression for the first order change in energy of the 1s state associated with the fact that the nucleus is not point-like.

c. Estimate the 2s-2p energy shift under the assumption that $R/a_\mu \ll 1$, where a_μ is the "Bohr radius" for the muon and show that this shift gives a measure of R.

d. When is the method of part (b) likely to fail? Does this method underestimate or overestimate the energy shift? Explain your answer in physical terms. Useful information:

$$\psi_{1s} = 2N_0 e^{-r/a_\mu}\, Y_{00}(\theta, \phi),$$

$$\psi_{2s} = \frac{1}{\sqrt{8}} N_0 \left(2 - \frac{r}{a_\mu}\right) e^{-r/2a_\mu} Y_{00}(\theta, \phi),$$

$$\psi_{2p} = \frac{1}{\sqrt{24}} N_0 \frac{r}{a_\mu} e^{-r/2a_\mu} Y_{1m}(\theta, \phi),$$

where

$$N_0 = \left(\frac{1}{a_\mu}\right)^{3/2}.$$

(Wisconsin)

ol:

a. If the nucleus were a point particle of charge Ze, the Coulomb potential energy of the muon would be $V_0 = -\dfrac{Ze^2}{r}$. Let $H' = V - V_0$ and consider it as perturbation. Then the perturbation Hamiltonian of the system is

$$H' = \begin{cases} 0, & r \geq R, \\ Ze^2\left[\dfrac{1}{r} - \dfrac{1}{R}\left(\dfrac{3}{2} - \dfrac{1}{2}\dfrac{r^2}{R^2}\right)\right], & r \leq R \end{cases}$$

When $r < R$, $H' > 0$ and the energy levels shift up on account of the perturbation. The shifts of energy levels of s states are larger than those of p and d states because a muon in s state has a greater probability of staying in the $r \sim 0$ region than a muon in p and d states. Besides, the larger the quantum number l, the greater is the corresponding orbital angular momentum and the farther is the spread of μ cloud from the center, leading to less energy correction. In Fig. 5.9, the solid lines represent unperturbed energy levels, while the dotted lines represent perturbed energy levels. It is seen that the unperturbed energy level of d state almost overlaps the perturbed energy level.

b. The energy shift of 1s state to first order perturbation is given by

$$\Delta E_{1s} = \langle 1s | H' | 1s \rangle.$$

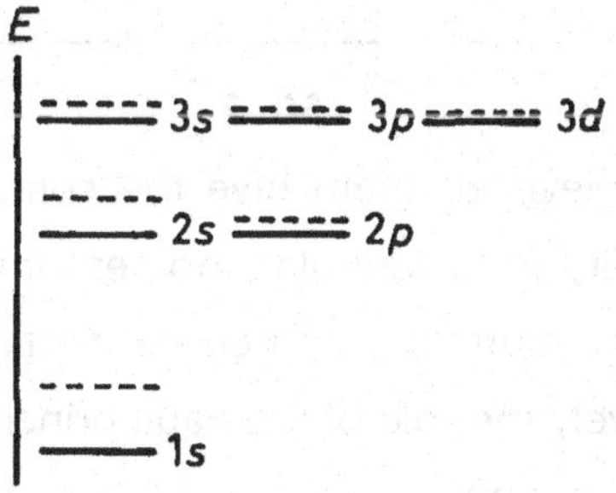

Fig. 5.9

As $R \ll a_\mu$, we can take $e^{-r/a_\mu} \approx 1$. Thus

$$\Delta E_{1s} \approx \frac{4N_0^2 Ze^2}{4\pi} \int_0^R \left[\frac{1}{r} - \frac{1}{R}\left(\frac{3}{2} - \frac{1}{2}\frac{r^2}{R^2}\right)\right] 4\pi r^2 \, dr$$

$$= \frac{2}{5}\left(\frac{R}{a_\mu}\right)^2 \frac{Ze^2}{a_\mu}.$$

c. By the same procedure,

$$\Delta E_{2s} \approx \frac{1}{20}\frac{Ze^2}{a_\mu}\left(\frac{R}{a_\mu}\right)^2,$$

$$\Delta E_{2p} \approx 0,$$

and so

$$\Delta E_{2s} - \Delta E_{2p} \approx \Delta E_{2s} = \frac{1}{20}\frac{Ze^2}{a_\mu}\left(\frac{R}{a_\mu}\right)^2,$$

$$= \frac{1}{20}\frac{Ze^2}{a_0}\left(\frac{R}{a_0}\right)^2\left(\frac{m_\mu}{m_e}\right)^3,$$

where a_0 is the Bohr radius. Thus by measuring the energy shift, we can deduce the value of R. Or, if we assume $\dot{R} = 10^{-13}$ cm, $Z = 5$, we get $\Delta E_{2s} - \Delta E_{2p} \approx 2\times10^{-2}$ eV.

d. In the calculation in (b) the approximation $R \ll a_\mu$ is used. If R is not much smaller than a_μ, the calculation is not correct. In such a case, the actual energy shifts of p and d states are larger than what we obtain in (b) while the actual energy shifts of s states are smaller than those given in (b). In fact the calculation in (b) overestimates the probability that the muon is located inside the nucleus $\left(\text{probability density} \propto |\psi_{1s}(0)|^2\right)$.

5032

a. Using an energy-level diagram give the complete set of electronic quantum numbers (total angular momentum, spin, parity) for the ground state and the first two excited states of a helium atom.

b. Explain qualitatively the role of the Pauli principle in determining the level order of these states.

c. Assuming Coulomb forces only and a knowledge of $Z = 2$ hydrogenic wave functions, denoted by $|1s\rangle$, $|2s\rangle$, $|2p\rangle$, etc., together with associated $Z = 2$ hydrogenic energy eigenvalues E_{1s}, E_{2s}, E_{2p},..., give perturbation formulas for the energies of these helium states. Do not evaluate integrals, but carefully explain the notation in which you express your result.

(Berkeley)

ol:

a. Figure 5.10 shows the ground and first two excited states of a helium atom in para (left) and ortho states with the quantum numbers (J, S, P).

b. Pauli's exclusion principle requires that a system of electrons must be described by an antisymmetric total wave function. For the two electrons of a helium atom, as the triplet states have symmetric spin wave functions the space wave functions must be antisymmetry. Likewise, the singlet states must have symmetric space wave functions. In the latter case, the overlap of the electron clouds is large and as the repulsive energy between the electrons is greater (because $|\mathbf{r}_1 - \mathbf{r}_2|$ is smaller). So, the corresponding energy levels are higher.

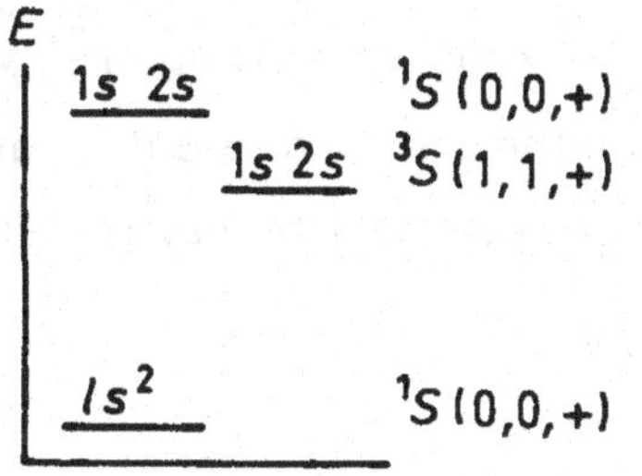

E

$\underline{1s\ 2s}$ $^1S\,(0,0,+)$

    $\underline{1s\ 2s}$ $^3S\,(1,1,+)$

$\underline{1s^2}$ $^1S\,(0,0,+)$

Fig. 5.10

c. The Hamiltonian of a helium atom is

$$\hat{H} = -\frac{\hbar^2}{2m}\nabla_1^2 - \frac{\hbar^2}{2m}\nabla_2^2 - \frac{2e^2}{r_1} - \frac{2e^2}{r_2} + \frac{e^2}{|\mathbf{r}_1 - \mathbf{r}_2|}.$$

Treating the last term as perturbation, the energy correction of $|1s1s\rangle$ state is

$$\Delta E_{1s} = \left\langle 1s1s \left| \frac{e^2}{|\mathbf{r}_1 - \mathbf{r}_2|} \right| 1s1s \right\rangle.$$

The perturbation energy correction of spin triplet states is

$$\Delta E_{nl}^{(3)} = \frac{1}{2}\left[\left(\langle 1snl| - \langle nl1s| \right) \frac{e^2}{|\mathbf{r}_1 - \mathbf{r}_2|} \left(|1snl\rangle - |nl1s\rangle \right) \right]$$

$$= \frac{1}{2}\left\langle 1snl \left| \frac{e^2}{|\mathbf{r}_1 - \mathbf{r}_2|} \right| 1snl \right\rangle - \frac{1}{2}\left\langle nl1s \left| \frac{e^2}{|\mathbf{r}_1 - \mathbf{r}_2|} \right| 1snl \right\rangle$$

$$- \frac{1}{2}\left\langle 1snl \left| \frac{e^2}{|\mathbf{r}_1 - \mathbf{r}_2|} \right| nl1s \right\rangle + \frac{1}{2}\left\langle nl1s \left| \frac{e^2}{|\mathbf{r}_1 - \mathbf{r}_2|} \right| nl1s \right\rangle$$

$$= \left\langle 1snt \left| \frac{e^2}{|\mathbf{r}_1 - \mathbf{r}_2|} \right| 1snl \right\rangle - \left\langle 1snl \left| \frac{e^2}{|\mathbf{r}_1 - \mathbf{r}_2|} \right| nl1s \right\rangle.$$

The perturbation energy correction of spin singlet states is

$$\Delta E_{nl}^{(1)} = \left\langle 1snl \left| \frac{e^2}{\|\mathbf{r}_1 - \mathbf{r}_2\|} \right| 1snl \right\rangle$$

$$+ \left\langle 1snl \left| \frac{e^2}{\|\mathbf{r}_1 - \mathbf{r}_2\|} \right| nl1s \right\rangle.$$

The first term of the above result is called direct integral and the second term, exchange integral.

5033

A particle of mass m is confined to a circle of radius a, but is otherwise free. A perturbing potential $H = A\sin\theta\cos\theta$ is applied, where θ is the angular position on the circle. Find the correct zero-order wave functions for the two lowest states of this system and calculate their perturbed energies to second order.

(Berkeley)

Sol: The unperturbed wave functions and energy levels of the system are respectively

$$\psi_n(\theta) = \frac{1}{\sqrt{2\pi}} e^{in\theta},$$

$$E_n = \frac{n^2\hbar^2}{2ma^2}, \quad n = \pm 1, \pm 2, \ldots$$

The two lowest states are given by $n = \pm 1$, which correspond to the same energy. To first order perturbation, we calculate for the two degenerate states $n = \pm 1$

$$\langle \pm 1 | H | \pm 1 \rangle = \frac{A}{2\pi} \int_0^{2\pi} \sin\theta\cos\theta \, d\theta = 0,$$

$$\langle +1 | H | -1 \rangle = \frac{A}{2\pi} \int_0^{2\pi} e^{-2i\theta} \sin\theta\cos\theta \, d\theta$$

$$= \frac{A}{4\pi} \int_0^{2\pi} (\cos 2\theta - i\sin 2\theta)\sin 2\theta \, d\theta$$

$$= \frac{-iA}{4},$$

$$\langle -1 | H | +1 \rangle = \frac{iA}{4}.$$

The perturbation matrix is thus

$$
\begin{pmatrix}
0 & -\dfrac{iA}{4} \\[2ex]
\dfrac{iA}{4} & 0
\end{pmatrix}.
$$

Diagonalizing, we obtain $\Delta E^{(1)} = \pm \dfrac{A}{4}$. Hence the two non-vanishing wave functions and the corresponding energy corrections are

$$\psi_1' = (|-1\rangle + i|1\rangle)/\sqrt{2}, \qquad\qquad \Delta E_1^{(1)} = \frac{A}{4}.$$

$$\psi_2' = (|1\rangle + i|-1\rangle)/\sqrt{2}, \qquad\qquad \Delta E_2^{(1)} = -\frac{A}{4}.$$

To second order perturbation, the energy correction is given by

$$\Delta E^{(2)} = {\sum_{n \neq k}}' \frac{\left| \langle \psi_k' | H' | n \rangle \right|^2}{E_k - E_n}.$$

As

$$
\begin{aligned}
H'|n\rangle &= \frac{A}{2} \sin 2\theta |n\rangle \\
&= \frac{A}{4i} \frac{1}{\sqrt{2\pi}} \left(e^{i2\theta} - e^{-i2\theta} \right) e^{in\theta} \\
&= \frac{A}{4i} \frac{1}{\sqrt{2\pi}} \left[e^{i(n+2)\theta} - e^{i(n-2)\theta} \right] \\
&= \frac{A}{4i}|n+2\rangle - \frac{A}{4i}|n-2\rangle,
\end{aligned}
$$

we have

$$
\begin{aligned}
\Delta E_1^{(2)} &= {\sum_{n \neq \pm 1}}' \frac{\left| \langle \psi_1' | n+2 \rangle - \langle \psi_1' | n-2 \rangle \right|^2}{\dfrac{\hbar^2}{2ma^2}(1 - n^2) + \dfrac{A}{4}} \times \left(\frac{A}{4} \right)^2 \\[2ex]
&= \frac{\dfrac{1}{2}}{\dfrac{\hbar^2}{2ma^2}\left[1 - (-3)^2 \right] + \dfrac{A}{4}} \times \left(\frac{A}{4} \right)^2 \\[2ex]
&\quad + \frac{\dfrac{1}{2}}{\dfrac{\hbar^2}{2ma^2}\left(1 - 3^2 \right) + \dfrac{A}{4}} \times \left(\frac{A}{4} \right)^2 \\[2ex]
&\approx -\frac{ma^2 A^2}{64\hbar^2},
\end{aligned}
$$

and similarly $\Delta E_2^{(2)} \approx -\dfrac{ma^2 A^2}{64\hbar^2}$.

Therefore

$$E_1 = \frac{\hbar^2}{2ma^2} + \frac{A}{4} - \frac{ma^2 A^2}{64\hbar^2},$$

$$E_2 = \frac{\hbar^2}{2ma^2} - \frac{A}{4} - \frac{ma^2 A^2}{64\hbar^2}.$$

5034

An electron at a distance x from a liquid helium surface feels a potential

$$V(x) = -\frac{K}{x}, \qquad x > 0, \quad K = \text{constant},$$

$$V(x) = \infty, \qquad x \leq 0.$$

a. Find the ground state energy level. Neglect spin.

b. Compute the Stark shift in the ground state using first order perturbation theory.

(Berkeley)

Sol:

a. At $x \leq 0$, the wave function is $\psi(x) = 0$. At $x > 0$, the Schrödinger equation is

$$\left(-\frac{\hbar^2}{2m} \frac{d^2}{dx^2} - \frac{K}{x} \right) \psi(x) = E\psi(x).$$

In the case of the hydrogen atom, the radial wave function $R(r)$ satisfies the equation

$$\left[-\frac{\hbar^2}{2m} \frac{1}{r^2} \frac{d}{dr}\left(r^2 \frac{d}{dr} \right) + \frac{l(l+1)\hbar^2}{2mr^2} + V(r) \right] R = ER.$$

Let $R(r) = \chi(r)/r$. For $l = 0$, the equation becomes

$$\left(-\frac{\hbar^2}{2m} \frac{d^2}{dr^2} - \frac{e^2}{r} \right) \chi(r) = E\chi(r).$$

This is mathematically identical with the Schrödinger equation above and both satisfy the same boundary condition, so the solutions must also be the same (with $r \leftrightarrow x$, $e^2 \leftrightarrow K$).

As the wave function and energy of the ground state of the hydrogen atom are respectively

$$E_{10} = -\frac{me^4}{2\hbar^2},$$

$$\chi_{10}(r) = \frac{2r}{a_0^{3/2}} e^{-r/a_0} \quad \text{with} \quad a_0 = \hbar^2/me^2,$$

the required wave function and energy are

$$E_1 = -\frac{mK^2}{2\hbar^2},$$

$$\psi_1(x) = \frac{2}{a^{3/2}} xe^{-x/a}, \quad a = \hbar^2/mK.$$

b. Suppose an electric field ε_e is applied in the x direction. Then the perturbation potential is $V' = e\varepsilon_e x$ and the energy correction to the ground state to first order in perturbation theory is

$$\Delta E_1 = \langle \psi_1 | V' | \psi_1 \rangle$$

$$= \int_0^\infty \frac{2}{a^{3/2}} xe^{-x/a} \cdot e\varepsilon_e x \cdot \frac{2}{a^{3/2}} xe^{-x/a} dx$$

$$= \frac{3}{2} e\varepsilon_e a = \frac{3\hbar^2 e\varepsilon_e}{2mK}.$$

5035

Discuss and compute the Stark effect for the ground state of the hydrogen atom.

(Berkeley)

Sol: Suppose the external electric field is along the z-axis, and consider its potential as a perturbation. The perturbation Hamiltonian of the system is

$$H' = e\varepsilon \cdot \mathbf{r} = e\varepsilon z.$$

As the ground state of the hydrogen atom is nondegenerate, we can employ the stationary perturbation theory. To first order perturbation, the energy correction is

$$E^{(1)} = \langle n=1, l=0, m=0 | e\varepsilon z | n=1, l=0, m=0 \rangle.$$

For the hydrogen atom the parity is $(-1)^l$, so the ground state $(l=0)$ has even parity. Then as z is an odd parity operator, $E^{(1)} = 0$.

To second order perturbation, the energy correction is given by

$$E^{(2)} = e^2 \varepsilon^2 \sum_{\substack{n \neq 1 \\ l,m}} \frac{|\langle 1,0,0 | z | n,l,m \rangle|^2}{E_1 - E_n}.$$

As $E_n = E_1/n^2$, where $E_1 = -\dfrac{e^2}{2a}$, $a = \dfrac{\hbar^2}{me^2}$, we have $E_1 - E_n < 0$, $(n \neq 1)$. Thus the energy correction $E^{(2)}$ is negative and has a magnitude proportional to ε^2. Hence increasing the electric field strength would lower the energy level of the ground state. We can easily perform the above summation, noting that only matrix elements with $l = \pm 1$ $m = 0$ are non-vanishing.

5036

Describe and calculate the Zeeman effect of the hydrogen 2p state.

(Berkeley)

Sol: The change in the energy levels of an atom caused by an external uniform magnetic field is called the Zeeman effect. We shall consider such change for a hydrogen atom to first order in the field strength H. We shall first neglect any interaction between the magnetic moment associated with the electron spin and the magnetic field. The effect of electron spin will be discussed later. A charge e in an external magnetic field H has Harmiltonian

$$\hat{H} = \frac{1}{2m}\left(\mathbf{P} - \frac{e}{c}\mathbf{A}\right)^2 + e\phi,$$

which gives the Schrödinger equation

$$i\hbar\frac{\partial\psi}{\partial t} = \left(-\frac{1}{2m}\nabla^2 + \frac{ie\hbar}{mc}\mathbf{A}\cdot\nabla + \frac{ie\hbar}{2mc}\nabla\cdot\mathbf{A} + \frac{e^2}{2mc}\mathbf{A}^2 + e\phi\right)\psi.$$

As $\mathbf{H}$ is uniform it can be represented by the vector potential

$$\mathbf{A} = \frac{1}{2}\mathbf{H}\times\mathbf{r}$$

since $\mathbf{H} = \nabla\times\mathbf{A}$. Then $\nabla\cdot\mathbf{A} = \dfrac{1}{2}(\mathbf{r}\cdot\nabla\times\mathbf{H} - \mathbf{H}\cdot\nabla\times\mathbf{r}) = 0$ and so the only terms involving $\mathbf{A}$ that appear in the Hamiltonian for an electron of charge $-e$ and reduced mass μ are

$$-\frac{ie\hbar}{\mu c}\mathbf{A}\cdot\nabla + \frac{e^2}{2\mu c^2}\mathbf{A}^2 = \frac{e}{2\mu c}(\mathbf{H}\times\mathbf{r})\cdot\mathbf{P} + \frac{e^2}{8\mu c^2}(\mathbf{H}\times\mathbf{r})\cdot(\mathbf{H}\times\mathbf{r})$$

$$= \frac{e}{2\mu c}\mathbf{H}\cdot\mathbf{L} + \frac{e^2}{8\mu c^2}H^2 r^2\sin^2\theta,$$

where $\mathbf{L} = \mathbf{r}\times\mathbf{P}$ and θ is the angle between $\mathbf{r}$ and $\mathbf{H}$.

To first order in **H**, we can take the perturbation Hamiltonian as

$$\hat{H}' = \frac{e}{2\mu c} \mathbf{H} \cdot \mathbf{L}.$$

Taking the direction of the magnetic field as the z direction we can choose for the energy eigenfunctions of the unperturbed hydrogen atom the eigenstates of L_z with eigenvalues $m\hbar$, where m is the magnetic quantum number. Then the energy correction from first order perturbation is

$$W_l' = \left\langle m \mid \hat{H}' \mid m \right\rangle = \frac{e}{2\mu c} H m \hbar.$$

Thus the degeneracy of the $2l+1$ states of given n and l is removed in the first order. In particular, for the $2p$ state, where $l=1$, the three-fold degeneracy is removed.

We shall now consider the effect of electron spin. The electron has an intrinsic magnetic moment in the direction of its spin, giving rise to a magnetic moment operator $-(e/mc)\mathbf{S}$.

For a weak field, we shall consider only the first order effects of **H**. The Hamiltonian is

$$\hat{H} = -\frac{\hbar^2}{2m}\nabla^2 + V(r) + \xi(r)\mathbf{L}\cdot\mathbf{S} + \varepsilon(l_z + 2s_z) \quad \text{with} \quad \varepsilon = \frac{eH}{2mc},$$

where the field is taken to be along the z-axis.

We choose the following eigenfunctions of $\mathbf{J}^2$ and J_z as the wave functions:

$$
{}^2P_{\frac{3}{2}}\begin{cases}
m = \dfrac{3}{2} & (+)Y_{1,1}, \\[2mm]
\dfrac{1}{2} & 3^{-\frac{1}{2}}\left[2^{\frac{1}{2}}(+)Y_{1,0} + (-)Y_{1,1}\right], \\[2mm]
-\dfrac{1}{2} & 3^{-\frac{1}{2}}\left[2^{\frac{1}{2}}(-)Y_{1,0} + (+)Y_{1,-1}\right], \\[2mm]
-\dfrac{3}{2} & (-)Y_{1,-1},
\end{cases}
$$

$$
{}^2P_{\frac{1}{2}}\begin{cases}
m = \dfrac{1}{2} & 3^{-\frac{1}{2}}\left[-(+)Y_{1,0} + 2^{\frac{1}{2}}(-)Y_{1,1}\right], \\[2mm]
-\dfrac{1}{2} & 3^{-\frac{1}{2}}\left[(-)Y_{1,0} + 2^{\frac{1}{2}}(+)Y_{1,-1}\right],
\end{cases}
$$

$$
{}^2S_{\frac{1}{2}}\begin{cases}
m = \dfrac{1}{2} & (+)Y_{0,0}, \\[2mm]
-\dfrac{1}{2} & (-)Y_{0,0},
\end{cases}
$$

where $Y_{0,0}, Y_{1,0}, Y_{1,1}$ and $Y_{1,-1}$ are spherical harmonic functions, (+) and (−) are spin wave functions. It can be shown that the magnetic energy $\varepsilon(l_z + 2s_z) = \varepsilon(J_z + s_z)$ has non-vanishing matrix elements between states of different j, but not between states of the same j and different m. We can neglect the former because of the relatively large energy separation between states of different j. Thus the magnetic energy is diagonal with respect to m for each j and shifts the energy of each state above by its expectation value for the state. In each case, J_z is diagonal, and so its expectation value is $m\hbar$. The expectation value of s_z for the $P_{3/2}$ state with $m = 1/2$, for example, is

$$\iint 3^{-\frac{1}{2}}\left[2^{\frac{1}{2}}(+)^\dagger Y_{1,0}^* + (-)^\dagger Y_{1,1}^*\right]\frac{1}{2}\hbar\sigma_z 3^{-\frac{1}{2}}\left[2^{\frac{1}{2}}(+)Y_{1,0} + (-)Y_{1,1}\right]\sin\theta\,d\theta\,d\phi$$

$$= \frac{\hbar}{6}\iint\left[2^{\frac{1}{2}}(+)^\dagger Y_{1,0}^* + (-)^\dagger Y_{1,1}^*\right]\left[2^{\frac{1}{2}}(+)Y_{1,0} - (-)Y_{1,1}\right]\sin\theta\,d\theta\,d\phi$$

$$= \frac{\hbar}{6}(2-1) = \frac{\hbar}{6}.$$

Hence, the magnetic energy of this state is $\varepsilon\hbar\left(\frac{1}{2} + \frac{1}{6}\right) = \frac{2}{3}\varepsilon\hbar$.

This and similar results for the other states can be expressed in terms of the Landé g^- factor as $\varepsilon m\hbar g$, with

$$g = \frac{4}{3} \text{ for } {}^2P_{3/2}, \quad g = \frac{2}{3} \text{ for } {}^2P_{1/2}, \quad g = 2 \text{ for } {}^2S_{1/2}.$$

5037

Explain why excited states of atomic hydrogen can show a linear Stark effect in an electric field, but the excited states of atomic sodium show only a quadratic one.

(MIT)

Sol: The potential energy of the electron of the atom in an external electric field **E** is
$$H' = e\mathbf{E}\cdot\mathbf{r}.$$

If we make the replacement $\mathbf{r} \to -\mathbf{r}$ in $\langle l'|H'|l\rangle$, as the value of the integral does not change we have

$$\langle l'|H'|l\rangle(\mathbf{r}) = \langle l'|H'|l\rangle(-\mathbf{r})$$
$$= (-1)^{l'+l+1}\langle l'|H'|l\rangle(\mathbf{r}).$$

This means that if the l' and l states have the same parity (i.e. l and l' are both even or both odd) we must have $\langle l'|H'|l\rangle = 0$.

If the electric field is not too small, we need not consider the fine structure of the energy spectrum caused by electron spin. In such cases, an exited state of the hydrogen atom is a superposition of different parity states, i.e. there is degeneracy with respect to l and the perturbation theory for degenerate states is to be used. Because of the existence of non-vanishing perturbation Hamiltonian matrix elements, exited states of the hydrogen atom can show a linear Stark effect.

For exited states of atomic sodium, each energy level corresponds to a definite parity, i.e., there is no degeneracy in l. When we treat it by nondegenerate perturbation theory, the first order energy correction $\langle l'|H'|l\rangle$ vanishes.

We then have to go to second order energy correction. Thus the exited states of atomic sodium show only quadratic Stark effect.

5038

The Stark effect. The energy levels of the $n = 2$ states of atomic hydrogen are illustrated in Fig. 5.11.

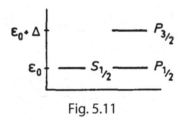

Fig. 5.11

The $S_{1/2}$ and $P_{1/2}$ levels are degenerate at an energy ε_0 and the $P_{3/2}$ level is degenerate at an energy $\varepsilon_0 + \Delta$.

A uniform static electric field E applied to the atom shifts the states to energies $\varepsilon_1, \varepsilon_2$ and ε_3. Assuming that all states other than these three are far enough away to be neglected, determine the energies $\varepsilon_1, \varepsilon_2$ and ε_3 to second order in the electric field E.

(Princeton)

Sol: Suppose the matrix elements of the perturbation Hamiltonian $H' = -e\mathbf{E}\cdot\mathbf{r}$ are

	$P_{3/2}$	$P_{1/2}$	$S_{1/2}$
$P_{3/2}$	0	0	b
$P_{1/2}$	0	0	a
$S_{1/2}$	b^*	a^*	0

since $\langle l'|H'|l\rangle = 0$ for l', l states of the same parity (**Problem 5037**). Then for energy level $P_{3/2}$, we have

$$E_{P3/2} = E_{P3/2}^{(0)} + \frac{\left|\langle P_{3/2}|H'|S_{1/2}\rangle\right|^2}{E_{P3/2}^{(0)} - E_{S1/2}^{(0)}}$$

$$= \varepsilon_0 + \Delta + \frac{|b|^2}{\Delta}.$$

For energy levels $P_{1/2}$ and $S_{1/2}$, we diagonalize the Hamiltonian in the corresponding subspace, i.e, solve

$$\begin{vmatrix} -\lambda & a \\ a^* & -\lambda \end{vmatrix} = 0.$$

The roots are $\lambda = \pm|a|$, which give the new wave functions

$$|1\rangle = \frac{1}{\sqrt{2}}\left(\frac{a}{|a|}|P_{1/2}\rangle + |S_{1/2}\rangle\right) = \frac{a|P_{1/2}\rangle + |a||S_{1/2}\rangle}{\sqrt{2}|a|},$$

$$|2\rangle = \frac{1}{\sqrt{2}}\left(-\frac{a}{|a|}|P_{1/2}\rangle + |S_{1/2}\rangle\right) = \frac{-a|P_{1/2}\rangle + |a||S_{1/2}\rangle}{\sqrt{2}|a|},$$

with energies

$$E_1 = E_1^{(0)} + |a| + \frac{\left|\langle 1|H'|P_{3/2}\rangle\right|^2}{E_1^{(0)} - E_{P3/2}^{(0)}}$$

$$= \varepsilon_0 + |a| + \frac{|b|^2}{2(-\Delta)}$$

$$= \varepsilon_0 + |a| - \frac{|b|^2}{2\Delta},$$

$$E_2 = E_2^{(0)} - |a| + \frac{\left|\langle 2|H'|P_{3/2}\rangle\right|^2}{E_2^{(0)} - E_{P3/2}^{(0)}}$$

$$= \varepsilon_0 - |a| - \frac{|b|^2}{2\Delta}.$$

5039

The Stark effect in atoms (shift of energy levels by a uniform electric field) is usually observed to be quadratic in the field strength. Explain why. But for some states of the hydrogen atom the Start effect is observed to be linear in the field strength. Explain why. Illustrate by making a perturbation calcula-tion of the Stark effect to lowest non-vanishing order for the ground and first excited states of the hydrogen atom.

To within an uninteresting overall constant, the wave functions are

$$\psi_{100} = 4\sqrt{2}a_0 e^{-r/a_0},$$

$$\psi_{200} = (2a_0 - r)e^{-r/2a_0},$$

$$\psi_{21\pm1} = \pm r e^{-r/2a_0}\sin\theta e^{\pm i\phi}/\sqrt{2},$$

$$\psi_{210} = r e^{-r/2a_0}\cos\theta.$$

(Wisconsin)

Sol: The electric dipole moment of an atomic system with several electrons is

$$d = -\sum_j e_i \mathbf{r}_i$$

In general, the energy levels do not have degeneracy other than with respect to l_z. The energy depends on the quantum numbers n and l. As the perturbation Hamiltonian $H' = -\mathbf{d}\cdot\mathbf{E}$ is an odd parity operator, only matrix elements between opposite parity states do not vanish. Thus its expectation value for any one given state is zero, i.e.,

$$\langle nlm|-\mathbf{d}\cdot\mathbf{E}|nlm'\rangle = 0.$$

This means that first order energy corrections are always zero and one needs to consider energy corrections of second order perturbation. Hence, the energy corrections are proportional to E^2.

As regards the hydrogen atom, degeneracy occurs for the same n but different l. And so not all the matrix elements $\langle nl'|H'|nl\rangle$ between such states are zero. So, shifts of energy levels under first order (degenerate) perturbation may be nonzero, and the Stark effect is a linear function of the electric field E. Write the perturbation Hamiltonian in spherical coordinates, taking the z-axis in the direction of **E**. As $H' = eEz = eEr\cos\theta$, the ground state, which is not degenerate, has wave function

$$\psi_{100} = 4\sqrt{2}a_0 e^{-r/a}0,$$

and so (**Problem 5037**)

$$V^{(1)} = \langle 100|H'|100\rangle = 0.$$

The second order energy correction is

$$V^{(2)} = \sum_{n \neq 1}' \frac{|H'_{n0}|^2}{E_1^{(0)} - E_n^{(0)}}$$

$$= e^2 E^2 \sum_{n=2}^{\infty} \frac{|\langle nl0 | z | 100 \rangle|^2}{\left(1 - \frac{1}{n^2}\right) \frac{e^2}{2a}}$$

$$= 2aE^2 \cdot \frac{9}{8} a^2 = \frac{9}{4} a^3 E^2.$$

Note that for $H'_{n0} \neq 0$ we require $\Delta l = \pm 1$.

The first exited state $n = 2$ is four-fold degenerate, the wave functions being

$$\psi_{200}, \ \psi_{210}, \ \psi_{21,\pm1}.$$

As

$$eE\langle n, l+1, m | z | n, l, m \rangle = -eE \sqrt{\frac{(l-m+1)(l+m+1)}{(2l+1)(2l+3)}} \cdot \frac{3a_0 n}{2} \sqrt{n^2 - l^2}$$

$$= -3eEa_0$$

are the only non-vanishing elements of H' for $n = 2$, we have the secular equation

$$H' - E^{(1)} I = \begin{vmatrix} -E^{(1)} & -3eEa_0 & 0 & 0 \\ -3eEa_0 & -E^{(1)} & 0 & 0 \\ 0 & 0 & -E^{(1)} & 0 \\ 0 & 0 & 0 & -E^{(1)} \end{vmatrix} = 0,$$

which gives the energy corrections

$$E^{(1)} = \pm 3eE_a, \ 0,0.$$

Therefore, the energy level for $n = 2$ splits into

$$-\frac{e^2}{2a_0} \cdot \frac{1}{2^2} + 3eEa_0,$$

$$-\frac{e^2}{8a_0} - 3eEa_0,$$

where a_0 is the Bohr radius $\frac{\hbar^2}{me^2}$. The splitting of the first excited state $n = 2$ is shown in Fig. 5.12. Note that the two other states are still degenerate.

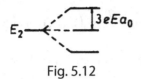

$$E_2 \underset{}{\overset{}{\longleftarrow}} \left. \begin{array}{c} \\ \\ \end{array} \right\} 3eEa_0$$

Fig. 5.12

5040

Consider an ionized atom (Z, A) with only a single electron remaining. Calculate the Zeeman splitting in the $n = 2$ state in a "weak" magnetic field

a. for an electron

b. for a hypothetical spin $= 0$ particle with electron mass.

c. Calculate the first-order Stark effect (energy levels and wave functions) for an electron in the $n = 2$ state.

(After you define the radial integrals you can express the term by a parameter; you need not evaluate them. The same holds for nonzero angular integrals.)

(Berkeley)

Sol:

a. Take the direction of the external magnetic field as the z direction. For an electron and a weak external magnetic field, in comparison with its effect the spin-orbit coupling cannot be neglected, which gives rise to anomalous Zeeman effect. The Hamiltonian of the system

$$\hat{H} = \frac{p^2}{2m_e} - \frac{Ze^2}{r} + \frac{eB}{2m_ec}(\hat{l}_z + 2\hat{s}_z) + \xi(r)\hat{s} \cdot \hat{l}\beta$$

can be written as

$$\hat{H} = H_0 + \frac{eB}{2m_ec}\hat{j}_z + \frac{eB}{2m_ec}\hat{s}_z,$$

with

$$H_0 \equiv \frac{p^2}{2m_e} - \frac{Ze^2}{r} + \xi(r)\hat{s} \cdot \hat{l}, \quad \hat{j}_z = \hat{l}_z + \hat{s}_z.$$

Before applying the magnetic field, we have

$$H_0 \psi_{nl_jm_j} = E_{nl_j} \psi_{nl_jm_j} \cdot \left(j = l \pm \frac{1}{2} \right)$$

If we neglect the term $\dfrac{eB}{2m_ec}\hat{s}_z$, $\left(\hat{L}^2,\hat{J}^2,\hat{j}_z\right)$ are still conserved quantities. Then

$\left\langle jm_j\,|\,\hat{j}_z\,|\,jm_j\right\rangle=m_j\hbar$ and the energy of the system is

$$E_{nl_j}+m_j\hbar\omega_L,$$

where

$$\omega_L=\frac{eB}{2m_ec}.$$

When the weak magnetic field is applied, the contribution of the term $\dfrac{eB}{2m_ec}\hat{s}_z$ is **(Problem 5057)**

$$\omega_L\overline{s}_z=\frac{\hbar\omega_L}{2}\left\langle jm_j\,|\,\hat{\sigma}_z\,|\,jm_j\right\rangle$$

$$=\begin{cases}\dfrac{m_j}{2j}\hbar\omega_L, & j=l+\dfrac{1}{2},\\[3mm]-\dfrac{m_j}{2j+2}\hbar\omega_L, & j=l-\dfrac{1}{2}.\end{cases}$$

Hence,

$$E_{nl_jm_j}=E_{nl_j}+\begin{cases}\left(1+\dfrac{1}{2j}\right)m_j\hbar\omega_L, & j=l+\dfrac{1}{2},\\[3mm]\left(1-\dfrac{1}{2j+2}\right)m_j\hbar\omega_L, & j=l-\dfrac{1}{2}.\end{cases}$$

For $n=2$, we have

$$\begin{cases}E_{20\frac{1}{2}m_j}=E_{20\frac{1}{2}}+2m_j\hbar\omega_L, & m_j=\pm\dfrac{1}{2},\\[3mm]E_{21\frac{3}{2}m_j}=E_{21\frac{3}{2}}+\dfrac{4}{3}m_j\hbar\omega_L, & m_j=\pm\dfrac{3}{2},\pm\dfrac{1}{2},\\[3mm]E_{21\frac{1}{2}m_j}=E_{21\frac{1}{2}}+\dfrac{2}{3}m_j\hbar\omega_L, & m_j=\pm\dfrac{1}{2}.\end{cases}$$

b. When spin $=0$, there is no spin-related effect so that

$$\hat{H}=\frac{\hat{P}^2}{2m_e}+V(r)+\frac{eB}{2m_ec}\hat{l}_z.$$

The eigenfunction is

$$\psi_{nlm}(r,\theta,\varphi)=R_{nlm}(r)Y_{lm}(\theta,\varphi),$$

and the energy eigenvalue is

$$E_{nlm} = E_{nl} + \frac{eB}{2m_e c} m\hbar.$$

For $n = 2$,

$$E_{200} = E_{20},$$
$$E_{210} = E_{21},$$
$$E_{21,\pm 1} = E_{21} \pm \frac{eB}{2m_e c}\hbar.$$

c. See the solution of **Problem 5042.**

5041

Stark showed experimentally that, by applying an external weak uniform electric field, the 4-fold degeneracy in the $n = 2$ level of atomic hydrogen could be removed. Investigate this effect by applying perturbation theory, neglecting spin and relativistic effects.

Specifically:

a. What are the expressions for the first order corrections to the energy level? (Do not attempt to evaluate the radial integrals).

b. Are there any remaining degeneracies?

c. Draw an energy level diagram for $n = 2$ which shows the levels before and after application of the electric field. Describe the spectral lines that originate from these levels which can be observed.

(Chicago)

Sol: Write the Hamiltonian of the system as $H = H_0 + H'$, where

$$H_0 = -\frac{e^2}{r} + \frac{\mathbf{L}^2}{2mr^2},$$
$$H' = eEz,$$

taking the direction of the electric field **E** as the z direction. For a weak field, $H' \ll H_0$ and we can treat H' as perturbation.

Let $(0,0)$, $(1,0)$, $(1,1)$ and $(1,-1)$ represent the four degenerate eigenfunctions (l,m) of the state $n = 2$ of the hydrogen atom.

The matrix representation of H' in the subspace is

$$H' = \begin{pmatrix} 0 & \langle 1,0|H'|1,0\rangle & 0 & 0 \\ \langle 1,0|H'|0,0\rangle & 0 & 0 & 0 \\ 0 & 0 & 0 & 0 \\ 0 & 0 & 0 & 0 \end{pmatrix},$$

where

$$\langle 1,0|H'|0,0\rangle = \langle 0,0|H'|1,0\rangle^*$$
$$= eE\int u_{210}^*(r)r\cos\theta u_{200}(r)d^3r$$
$$= -3eEa_0, \quad a_0 = \frac{\hbar^2}{me^2}$$

being the Bohr radius. Note that $\langle l'|H'|l\rangle = 0$ unless the l', l states have opposite parities.

Solving the secular equation

$$\begin{vmatrix} -w_1 & \langle 0,0|H'|1,0\rangle & 0 & 0 \\ \langle 1,0|H'|0,0\rangle & -w_1 & 0 & 0 \\ 0 & 0 & -w_1 & 0 \\ 0 & 0 & 0 & -w_1 \end{vmatrix} = 0,$$

we get four roots

$$w_1^{(1)} = 3eEa_0,$$
$$w_1^{(2)} = w_1^{(3)} = 0,$$
$$w_1^{(4)} = 3eEa_0.$$

a. The first order energy corrections are thus

$$\Delta E = w_1 = \begin{cases} 3eEa_0, \\ 0, \\ 0, \\ -3eEa_0. \end{cases}$$

b. As $\omega_1^{(2)} = \omega_1^{(3)} = 0$, there is still a two-fold degeneracy.

c. Figure 5.13 shows the $n = 2$ energy levels. The selection rules for electric dipole transitions are $\Delta l = \pm 1$, $\Delta m = 0, \pm 1$, which give rise to two spectral lines:

$$hv_1 = 3ea_0E, \qquad v_1 = 3ea_0E/h;$$
$$hv_2 = 2\times 3ea_0E, \qquad v_2 = 6ea_0E/h.$$

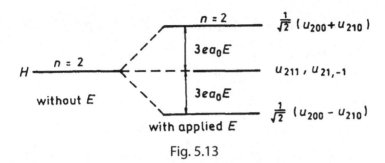

Fig. 5.13

5042

Consider the $n=2$ levels of a hydrogen-like atom. Suppose the spins of the orbiting particle and nucleus to be zero. Neglect all relativistic effects.

a. Calculate to lowest order the energy splittings in the presence of a uniform magnetic field.

b. Do the same for the case of a uniform electric field.

c. Do the same for both fields present simultaneously and at right angles to each other.

(Any integral over radial wave functions need not be evaluated; it can be replaced by a parameter for the rest of the calculation. The same may be done for any integral over angular wave function, once you have ascertained that it does not vanish.)

(Berkeley)

Sol:

a. Take the direction of the magnetic field as the z direction. Then the Hamiltonian of the system is

$$H = \frac{1}{2m_e}\mathbf{p}^2 + V(r) + \frac{eB}{2m_e c}\hat{l}_z,$$

where $V(r) = -\dfrac{e^2}{r}$. Considering $H' = \dfrac{eB}{2m_e c}\hat{l}_z$ as perturbation, the eigenfunctions for the unperturbed states are

$$\psi_{nlm}(r,\theta,\varphi) = R_{nl}(r)Y_{lm}(\theta,\varphi),$$

with $n = 1, 2, 3\ldots,$ $\quad l = 0, 1, 2, \ldots, n-1.$

$$m = -l, -l+1, \ldots l-1, l.$$

As (H, l^2, l_z) are still conserved quantities, $\langle nlm | \hat{l}_z | nlm \rangle = m\hbar$ and the energy splittings to first order for $n = 2$ are given by

$$E_{2lm} = E_{2l} + \frac{eB}{2m_ec} m\hbar$$

$$= \begin{cases} E_{20}, & l = 0; \\ E_{21} + \begin{cases} \dfrac{eB}{2m_ec}\hbar, \\ 0, \\ \dfrac{-eB}{2m_ec}\hbar, \end{cases} l = 1 \begin{cases} m = 1, \\ m = 0, \\ m = -1. \end{cases} \end{cases}$$

b. The energy level for $n = 2$ without considering spin is four-fold degenerate. The corresponding energy and states are respectively

$$E_2 = -\frac{Z^2 e^2}{2a_0} \frac{1}{2^2}, \quad \psi_{200}, \psi_{210}, \psi_{211}, \psi_{21-1}.$$

Suppose a uniform electric field is applied along the z-axis. Take as perturbation $H' = e\varepsilon z = E_0 V'$, where $E_0 = e\varepsilon a_0$, $V' = z / a_0 = r\cos\theta / a_0$, $a_0 = \dfrac{\hbar^2}{m_e e^2}$. Since

$$\cos\theta Y_{lm} = \sqrt{\frac{(l+1)^2 - m^2}{(2l+1)(2l+3)}} Y_{l+1,m}$$
$$+ \sqrt{\frac{l^2 - m^2}{(2l+1)(2l-1)}} Y_{l-1,m},$$

$H'_{nl'm',nlm} \neq 0$ for only $\Delta l = \pm 1$, $\Delta m = 0$. Hence, the non-vanishing elements of the perturbation matrix are

$$(H')_{200,210} = \int \psi_{200}^* H' \psi_{210} d^3x = \int \psi_{200} H' \psi_{210} d^3x,$$
$$(H')_{210,200} = \int \psi_{210}^* H' \psi_{200} d^3x = \int \psi_{210} H' \psi_{200} d^3x.$$

Let $(H')_{200,210} = (H')_{210,200} = E'$, i.e., $H_{01} = H_{10} = E'$, and solve the secular equation

$$\det | H_{\mu\nu} - E^{(1)} \delta_{\mu\nu} | = 0.$$

The roots are $E^{(1)} = \pm E', 0, 0$. Hence, the energy state $n = 2$ splits into three levels:

$$E_2 \pm E', E_2 \quad \text{(two-fold degeneracy for } E_2\text{).}$$

c. Assuming that the magnetic field is along the z-axis and the electric field is along the x-axis, the perturbation Hamiltonian of the system is

$$H' = \frac{eB}{2m_ec}\hat{l}_z + e\varepsilon x = \beta l_z / \hbar + \sqrt{2}\gamma x / 3a,$$

where

$$\beta = eB\hbar/2m_ec, \ \gamma = 3e\varepsilon a/\sqrt{2}, \ a = a_0 / Z.$$

The non-vanishing matrix elements of x are

$$(x)_{l,\,m-1}^{l-1,m} = (x)_{l-1,m}^{l,m-1}$$

$$= \frac{3}{4}n\sqrt{\frac{(n^2-l^2)(l-m+1)(l-m)}{(2l+1)(2l-1)}}a,$$

$$(x)_{l-1,m-1}^{l,m} = (x)_{l,m}^{l-1,m-1}$$

$$= -\frac{3}{4}n\sqrt{\frac{(n^2-l^2)(l+m-1)(l+m)}{(2l+1)(2l-1)}}a.$$

Thus, for $n = 2$,

$$x_{00}^{11} = -\frac{3}{\sqrt{2}}a = x_{11}^{00},$$

$$x_{1-1}^{00} = \frac{3}{\sqrt{2}}a = x_{00}^{1-1},$$

and the perturbation matrix is

$$
\begin{array}{l}
\left.\begin{array}{l} l=1 \\ m=1 \end{array}\right\} \\[1em]
\left.\begin{array}{l} l=1 \\ m=0 \end{array}\right\} \\[1em]
\left.\begin{array}{l} l=1 \\ m=-1 \end{array}\right\} \\[1em]
\left.\begin{array}{l} l=0 \\ m=0 \end{array}\right\}
\end{array}
\qquad
\begin{pmatrix}
\beta & 0 & 0 & -\gamma \\
0 & 0 & 0 & 0 \\
0 & 0 & -\beta & \gamma \\
-\gamma & 0 & \gamma & 0
\end{pmatrix}
$$

The secular equation

$$\det\begin{vmatrix}
\beta - E^{(1)} & 0 & -\gamma \\
0 & \beta - E^{(1)} & \gamma \\
-\gamma & \gamma & -E^{(1)}
\end{vmatrix} = 0$$

has roots

$$E_1^{(1)} = 0, \ E_{2,3}^{(1)} = \pm\sqrt{\beta^2 + 2\gamma^2}.$$

Hence, the energy state $n = 2$ splits into three levels, of energies

$$E_2, \ E_2 \pm \sqrt{\beta^2 + 2\gamma^2}.$$

5043

A nonrelativistic hydrogen atom, with a spinless electron, is placed in an electric field $\mathcal{E}$ in the z direction and a magnetic field $\mathcal{H}$ in the x direction. The effect of the two fields on the energy levels are comparable.

a. If the atom is in a state with n, the principal quantum number, equal to two, state which matrix elements in the first-order perturbation calculation of the energy shifts are zero.

b. Now obtain an equation for the energy shifts; once you have the determinantal equation you need not go through the algebra of evaluating the determinant. Do not insert the precise forms of the radial wave functions; express your results in terms of matrix elements of r^n (where n is an appropriate power) between radial wave functions.

$$(\ell_x \pm i\ell_y)|\ell,m\rangle = \sqrt{\{(\ell \mp m)(\ell \pm m + 1)\}}\,|\ell, m \pm 1\rangle.$$

(Berkeley)

Sol:

a. The perturbation Hamiltonian is

$$H' = \frac{eB}{2mc}\hat{l}_x + e\varepsilon\hat{z}.$$

Let the state vectors for $n = 2$ be $|200\rangle, |210\rangle, |211\rangle, |21,-1\rangle$. As

$$(\ell_x \pm i\ell_y)|\ell,m\rangle = \sqrt{\{(\ell \mp m)(\ell \pm m + 1)\}}\,|\ell, m \pm 1\rangle,$$

we have

$$\hat{l}_x|0,0\rangle = 0, \hat{l}_x|1,1\rangle = \hat{l}_x|1,-1\rangle = \frac{\sqrt{2}}{2}\hbar|1,0\rangle,$$

$$\hat{l}_x|1,0\rangle = \frac{\sqrt{2}}{2}\hbar\{|1,1\rangle + |1,-1\rangle\}.$$

As $z = r\cos\theta$, we have

$$\langle 210 \,|\, r\cos\theta \,|\, 200 \rangle = \langle 200 \,|\, r\cos\theta \,|\, 210 \rangle = \sqrt{\frac{1}{3}}\langle r \rangle$$

with $\langle r \rangle = \int_0^\infty r^3 R_{20} R_{21}\, dr$, other matrix elements of z being zero. Hence, the

perturbation matrix is

$$H' = \begin{pmatrix} 0 & \sqrt{\dfrac{1}{3}}e\varepsilon\langle r\rangle & 0 & 0 \\[2ex] \sqrt{\dfrac{1}{3}}e\varepsilon\langle r\rangle, & 0 & \dfrac{\sqrt{2}eB\hbar}{4mc} & \dfrac{\sqrt{2}eB\hbar}{4mc} \\[2ex] 0 & \dfrac{\sqrt{2}eB\hbar}{4mc} & 0 & 0 \\[2ex] 0 & \dfrac{\sqrt{2}eB\hbar}{4mc} & 0 & 0 \end{pmatrix}.$$

b. The secular equation $|H - \lambda \mathbf{I}| = 0$, i.e.,

$$\begin{vmatrix} -\lambda & \alpha & 0 & 0 \\ \alpha & -\lambda & \beta & \beta \\ 0 & \beta & -\lambda & 0 \\ 0 & \beta & 0 & -\lambda \end{vmatrix} = 0,$$

where $\alpha = \sqrt{\dfrac{1}{3}}e\varepsilon\langle r\rangle$, $\beta = \dfrac{\sqrt{2}eB\hbar}{4mc}$, has roots $\lambda = 0, 0, \pm\sqrt{2\beta^2 + \alpha^2}$, which are

the energy shifts. Note that a two-fold degeneracy still remains.

5044

Two non-identical particles, each of mass m, are confined in one dimension to an impenetrable box of length L. What are the wave functions and energies of the three lowest-energy states of the system (i.e., in which at most one particle is excited out of its ground state)? If an interaction potential of the form $V_{12} = \lambda\delta(x_1 - x_2)$ is added, calculate to first order in λ the energies of these three lowest states and their wave functions to zeroth order in λ.

(Wisconsin)

Sol: Both particles can stay in the ground state because they are not identical. The energy and wave function are respectively

$$E_{11} = \frac{\hbar^2 \pi^2}{mL^2}, \quad \psi_{11} = \frac{2}{L} \sin \frac{\pi x_1}{L} \sin \frac{\pi x_2}{L}.$$

If one particle is in the ground state, the other in the first exited state, the energies and corresponding wave functions are

$$E_{12} = \frac{5\hbar^2 \pi^2}{2mL^2}, \qquad \psi_{12} = \frac{2}{L} \sin \frac{\pi x_1}{L} \sin \frac{2\pi x_2}{L},$$

$$E_{21} = \frac{5\hbar^2 \pi^2}{2mL^2}, \qquad \psi_{21} = \frac{2}{L} \sin \frac{2\pi x_1}{L} \sin \frac{\pi x_2}{L}.$$

When both particles are in the single-particle ground state, i.e., the system is in the ground state, we have the energy correction

$$E^{(1)} = (\psi_{11}, V_{12}\psi_{11}) = \frac{4}{L^2} \lambda \int_0^L \sin^4 \left(\frac{\pi x_1}{L} \right) dx_1 = \frac{3\lambda}{2L},$$

and the wave function to zeroth order in λ

$$\phi_{11} = \psi_{11}.$$

When one particles is in the ground state and the other in the first excited state, the energy level is two-fold degenerate and we have to use the perturbation theory for degenerate states. We first calculate the elements of the perturbation Hamiltonian matrix:

$$\int\int \psi_{12}^* V_{12} \psi_{12} dx_1 dx_2 = \int\int \psi_{21}^* V_{12} \psi_{21} dx_1 dx_2$$

$$= \frac{4}{L^2} \lambda \int_0^L \sin^2 \frac{\pi x_1}{L} \sin^2 \frac{2\pi x_1}{L} dx_1 = \frac{\lambda}{L},$$

$$\int\int \psi_{12}^* V_{12} \psi_{21} dx_1 dx_2 = \int\int \psi_{21}^* V_{12} \psi_{12} dx_1 dx_2 = \frac{\lambda}{L}.$$

We then solve secular equation

$$\det \begin{vmatrix} \dfrac{\lambda}{L} - E^{(1)} & \dfrac{\lambda}{L} \\ \dfrac{\lambda}{L} & \dfrac{\lambda}{L} - E^{(1)} \end{vmatrix} = 0,$$

and obtain the roots

$$E_+^{(1)} = \frac{2\lambda}{L}, \qquad E_-^{(1)} = 0,$$

which are the energy corrections. The corresponding zeroth order wave functions are

$$\phi_{12} = \frac{1}{\sqrt{2}} (\psi_{12} \pm \psi_{21}).$$

5045

Consider a three-level system described by the Hermitian Hamiltonian

$$H = H_0 + \lambda H_1,$$

where λ is a real number. The eigenstates of H_0 are $|1\rangle, |2\rangle$ and $|3\rangle$, and

$$H_0|1\rangle = 0,$$
$$H_0|2\rangle = \Delta|2\rangle,$$
$$H_0|3\rangle = \Delta|3\rangle.$$

a. Write down the most general 3×3 matrix representation of H_1 in the $\{|1\rangle,|2\rangle,|3\rangle\}$ basis.

b. When the spectrum of H is computed using perturbation theory, it is found that the eigenstates of H to lowest order in λ are $|1\rangle, |\pm\rangle \equiv \frac{1}{\sqrt{2}}(|2\rangle\pm|3\rangle)$ and that the corresponding eigenvalues are

$$E_1 = -\frac{\lambda^2}{\Delta} + O(\lambda^3),$$

$$E_+ = \Delta + \lambda + \frac{\lambda^2}{\Delta} + O(\lambda^3),$$

$$E_- = \Delta - \lambda + O(\lambda^3).$$

Determine as many of the matrix elements of H_1 from part (a) as you can.

(Buffalo)

Sol:

a. Since λ is a real number, the Hermition perturbation Hamiltonian matrix has the form

$$H_1 = \begin{pmatrix} a & d & e \\ d^* & b & f \\ e^* & f^* & c \end{pmatrix},$$

where a, b, c are real numbers.

b.   To first order approximation, energy eigenvalue is the expectation value of the Hamiltonian with respect to the selected state vectors. Thus

$$E_+ = \langle + | H_0 + \lambda H_1 | + \rangle$$
$$= \langle + | H_0 | + \rangle + \lambda \langle + | H_1 | + \rangle.$$
$$= \Delta + \lambda + \frac{\lambda^2}{\Delta} + O(\lambda^3)$$

Comparing the coefficients of λ gives

$$\langle + | H_1 | + \rangle = 1.$$

Similarly

$$\langle - | H_1 | - \rangle = -1.$$

As the energy levels corresponding to $|2\rangle$ and $|3\rangle$ are degenerate, we transform to the following state vectors in whose representation the degeneracy disappears,

$$|\pm\rangle = \frac{1}{\sqrt{2}}(|2\rangle \pm |3\rangle).$$

Thus H_1 is transformed to a representation in basic vectors $|1\rangle, |+\rangle, |-\rangle$:

$$\begin{pmatrix} 1 & 0 & 0 \\ 0 & \frac{1}{\sqrt{2}} & \frac{1}{\sqrt{2}} \\ 0 & \frac{1}{\sqrt{2}} & \frac{-1}{\sqrt{2}} \end{pmatrix} \begin{pmatrix} a & d & e \\ d^* & b & f \\ e^* & f^* & c \end{pmatrix} \begin{pmatrix} 1 & 0 & 0 \\ 0 & \frac{1}{\sqrt{2}} & \frac{1}{\sqrt{2}} \\ 0 & \frac{1}{\sqrt{2}} & \frac{-1}{\sqrt{2}} \end{pmatrix}$$

$$= \begin{pmatrix} a & \frac{d+e}{\sqrt{2}} & \frac{d-e}{\sqrt{2}} \\ \frac{d^*+e^*}{\sqrt{2}} & 1 & 0 \\ \frac{d^*-e^*}{\sqrt{2}} & 0 & -1 \end{pmatrix}.$$

In the above we have used

$$\langle + | H_1 | + \rangle = \frac{1}{2}(b + f + f^* + c) = 1,$$
$$\langle - | H_1 | - \rangle = \frac{1}{2}(b - f - f^* + c) = -1$$

and chosen the solution

$$b = c = 0, \; f = f^* = 1.$$

Perturbation theory for nondegenerate states gives

$$E_m = E_m^{(0)} + \lambda H'_{mm} + \lambda^2 \sum_{n \neq m}{}' \frac{|H'_{mn}|^2}{E_m - E_n} + O(\lambda^3).$$

Thus

$$E_1 = 0 + \lambda a + \frac{\lambda^2 |d+e|^2}{2(0-\Delta)} + \frac{\lambda^2 |d-e|^2}{2(0-\Delta)} + O(\lambda^3)$$

$$= \lambda a - \lambda^2 \left(\frac{|d+e|^2}{2\Delta} + \frac{|d-e|^2}{2\Delta} \right) + O(\lambda^3),$$

$$E_2 = \Delta + \lambda + \frac{\lambda^2 |d+e|^2}{2\Delta} + O(\lambda^3),$$

$$E_3 = \Delta - \lambda + \frac{\lambda^2 |d-e|^2}{2\Delta} + O(\lambda^3).$$

Identifying E_1, E_2, E_3 with the given energies $E_1, E+, E_-$ and comparing the coefficients of λ and λ^2 give $a = 0$ and

$$|d+e|^2 + |d-e|^2 = 2,$$

$$|d+e|^2 = 2,$$

$$|d-e|^2 = 0,$$

or $d + e = \sqrt{2} e^{i\delta}$, $d - e = 0$, where δ is an arbitrary constant. Hence $a = 0$, $d = e = \dfrac{e^{i\delta}}{\sqrt{2}}$ and

$$H_1 = \begin{pmatrix} 0 & \frac{1}{\sqrt{2}} e^{i\delta} & \frac{1}{\sqrt{2}} e^{i\delta} \\ \frac{1}{\sqrt{2}} e^{-i\delta} & 0 & 1 \\ \frac{1}{\sqrt{2}} e^{-i\delta} & 1 & 0 \end{pmatrix}$$

is the representation in state vectors $|1\rangle$, $|2\rangle$, $|3\rangle$.

5046

Two identical spin $-\dfrac{1}{2}$ fermions are bound in a three-dimensional isotropic harmonic oscillator potential with classical frequency ω. There is, in addition, a weak short-range spin-independent interaction between the fermions.

a. Give the spectroscopic notation for the energy eigenstates up through energy $5\hbar\omega$ (measured from the bottom of the well).

b. Write the appropriate approximate (i.e., to lowest order in the interaction) wave functions of the system, expressed in terms of single-particle harmonic oscillator wave functions, for all states up through energy $4\hbar\omega$.

c. For a specific interparticle interaction $V_{12} = -\lambda\delta^3(\mathbf{r}_1 - \mathbf{r}_2)$, find the energies of the states of (b) correct to first order in λ. You may leave your result in the form of integrals.

(Wisconsin)

Sol:

a. For a three-dimensional harmonic oscillator,

$$E_n = \left(n + \frac{3}{2}\right)\hbar\omega, \quad n = 2n_r + l,$$

where n_r and l are integers not smaller than zero. For the system of two identical fermions in harmonic oscillator potential, we have, from the Hamiltonian,

$$E_N = \left(n_1 + \frac{3}{2}\right)\hbar\omega + \left(n_2 + \frac{3}{2}\right)\hbar\omega = (N+3)\hbar\omega, \quad N = n_1 + n_2.$$

Consequently, for

$$E_0 = 3\hbar\omega, \quad (l_1, l_2) = (0,0),$$

there is only one state ${}^1S_0$; for

$$E_1 = 4\hbar\omega, \quad (l_1, l_2) = (0,0) \text{ or } (0,1), (s_1, s_2) = \left(\frac{1}{2}, \frac{1}{2}\right),$$

there are two states ${}^1P_1$ and ${}^3P_{210}$; for

$E_2 = 5\hbar\omega$, and

1. $(n_1, n_2) = (2, 0)$ or $(0, 2)$,

$$\begin{cases} (l_1, l_2) = (0,0), & \text{there are two states } {}^1S_0, {}^3S_1; \\ (l_1, l_2) = (2,0) \text{ or } (0,2), & \text{there are two states } {}^1D_2, {}^3D_{321}; \end{cases}$$

2. $(n_1, n_2) = (1,1)$, $(l_1, l_2) = (1,1)$, there are three states ${}^1S_0, {}^1D_2, {}^3P_{210}$.

b. Let ψ_0 be the ground state and ψ_{1m} the first exited state of a single-particle system, where $m = 0, \pm 1$, and χ_0 and χ_{1m} be spin singlet and triplet states. With the states labeled as $\left| NLL_z SS_z \right\rangle$, the wave functions required are

$$\left| 00000 \right\rangle = \chi_0 \psi_0(1)\psi_0(2), \text{ state } {}^1S_0;$$

$$\left| 11m00 \right\rangle = \chi_0 \frac{1}{\sqrt{2}}(1+\hat{P}_{12})\psi_0(1)\psi_{1m}(2), \text{ state } {}^1P_1;$$

$$\left| 11m1M \right\rangle = \chi_{1M} \frac{1}{\sqrt{2}}(1-\hat{P}_{12})\psi_0(1)\psi_{1m}(2), \text{ state } {}^3P_{210};$$

where $m, M = 0, \pm 1 \left(L_z = m, S_z = M \right)$.

c. For the ground state ${}^1S_0$, the energy correction to first order in λ is

$$\left\langle {}^1S_0 \left| V_{12} \right| {}^1S_0 \right\rangle \approx -\lambda \int d\mathbf{r}_1 \, d\mathbf{r}_2 \delta(\mathbf{r}_1 - \mathbf{r}_2)[\psi_0(\mathbf{r}_1)\psi_0(\mathbf{r}_2)]^2$$

$$= -\lambda \int d\mathbf{r}\psi_0^4(\mathbf{r}) = -\lambda\left(\frac{\alpha}{\sqrt{2\pi}} \right)^3$$

with $\alpha = \sqrt{m\omega} / \hbar$. Hence the ground state energy is

$$E\left(\left| {}^1S_0 \right\rangle \right) = 3\hbar\omega - \lambda\left(\frac{m\omega}{2\pi\hbar} \right)^{3/2}.$$

The first exited state consists of 12 degenerate states (but as there is no spin in V_{12}, $\left\langle {}^1P_1 \left| V_{12} \right| {}^3P_1 \right\rangle = 0$).

As the spatial wave function is antisymmetric when $S = 1$, the expectation value of $-\lambda\delta^3(\mathbf{r}_1 - \mathbf{r}_2)$ equals to zero, i.e., $\left\langle 11m'1M' \left| V_{12} \right| 11m1M \right\rangle = \delta_{M'M}$
$\left\langle 1m' \left| V_{12} \right| 1m \right\rangle = 0$. As

$$E\left(\left| {}^3P_{210} \right\rangle \right) = 4\hbar\omega,$$

$$\left\langle 11m'00 \left| V_{12} \right| 11m00 \right\rangle = -\lambda \int \frac{4}{2} d\mathbf{r}\psi_0^2(r)\psi_{1m'}^*(r)\psi_{1m}(r)$$

$$= -2\lambda \int dr \left| \psi_0(r)\psi_{1m}(\mathbf{r}) \right|^2 \delta_{m'm}$$

$$= -\lambda\left(\frac{\alpha}{\sqrt{2\pi}} \right)^3 \delta_{m'm},$$

we have

$$E\left(\left| {}^1P_{1m} \right\rangle \right) = 4\hbar\omega - \lambda\left(\frac{m\omega}{2\pi\hbar} \right)^{3/2},$$

Where m is the eigenvalue of L_z.

5047

The Hamiltonian for an isotropic harmonic oscillator in two dimensions is

$$H = \omega(n_1 + n_2 + 1),$$

where $n_i = a_i^\dagger a_i$, with $[a_i, a_j^\dagger] = \delta_{ij}$ and $[a_i, a_j] = 0$.

a. Work out the commutation relations of the set of operators $\{H, J_1, J_2, J_3\}$ where

$$J_1 = \frac{1}{2}(a_2^\dagger a_1 + a_1^\dagger a_2), \quad J_2 = \frac{i}{2}(a_2^\dagger a_1 - a_1^\dagger a_2),$$

$$J_3 = \frac{1}{2}(a_1^\dagger a_1 - a_2^\dagger a_2).$$

b. Show that $\mathbf{J}^2 \equiv J_1^2 + J_2^2 + J_3^2$ and J_3 form a complete commuting set and write down their orthonormalized eigenvectors and eigenvalues.

c. Discuss the degeneracy of the spectrum and its splitting due to a small perturbation $\mathbf{V} \cdot \mathbf{J}$ where $\mathbf{V}$ is a constant three-component vector.

(Buffalo)

Sol:

a. The system can be considered a system of bosons, which has two single-particle states. The operators $a_i^\dagger$ and a_i are respectively creation and destruction operators. As among their commulators only $[a_i, a_i^\dagger]$ is not zero, we can use the relation

$$[ab, cd] = a[b, c]d + ac[b, d] + [a, c]bd + c[a, d]b$$

to obtain

$$[a_1^\dagger a_1, a_1^\dagger a_1] = [a_2^\dagger a_2, a_2^\dagger a_2] = [a_1^\dagger a_1, a_2^\dagger a_2] = 0,$$
$$[a_1^\dagger a_1, a_2^\dagger a_1] = -[a_2^\dagger a_2, a_2^\dagger a_1] = -a_2^\dagger a_1,$$
$$[a_1^\dagger a_1, a_1^\dagger a_2] = -[a_2^\dagger a_2, a_1^\dagger a_2] = a_1^\dagger a_2,$$
$$[a_2^\dagger a_1, a_1^\dagger a_2] = a_2^\dagger a_2 - a_1^\dagger a_1.$$

Hence,

$$[H, J_1] = [H, J_2] = [H, J_3] = 0,$$
$$[J_1, J_2] = iJ_3, \quad [J_2, J_3] = iJ_1, \quad [J_3, J_1] = iJ_2.$$

b. The above commutation relations show that J_1, J_2, J_3 have the same properties as the components of the angular momentum **L**. Hence $\hat{J}^2$ and J_3 commute and form a complete set of dynamical variables of the two-dimensional system.

The commutation relations of a, a^+,

$$[a_i, a_j^+] = \delta_{ij}, \quad [a_i, a_j] = 0,$$

can be satisfied if we define

$$a_1 |n_1, n_2\rangle = \sqrt{n_1} |n_1 - 1, n_2\rangle, \qquad a_2 |n_1, n_2\rangle = \sqrt{n_2} |n_1, n_2 - 1\rangle$$

$$a_1^+ |n_1, n_2\rangle = \sqrt{n_1 + 1} |n_1 + 1, n_2\rangle, \qquad a_2^+ |n_1, n_2\rangle = \sqrt{n_2 + 1} |n_1, n_2 + 1\rangle$$

and thus

$$|n_1, n_2\rangle = (n_1! n_2!)^{-\frac{1}{2}} (a_1^+)^{n_1} (a_2^+)^{n_2} |0, 0\rangle.$$

These can be taken as the common normalized eigenvectors of the complete set of dynamical variables $\hat{J}^2$ and J_3. As

$$J^2 = \frac{1}{4} \left\{ (a_2^+ a_1 + a_1^+ a_2)^2 - (a_2^+ a_1 - a_1^+ a_2)^2 + (a_1^+ a_1 - a_2^+ a_2)^2 \right\}$$

$$= \frac{1}{4} \{ 2 a_2^+ a_1 a_1^+ a_2 + 2 a_1^+ a_2 a_2^+ a_1 + a_1^+ a_1 a_1^+ a_1$$

$$+ a_2^+ a_2 a_2^+ a_2 - a_1^+ a_1 a_2^+ a_2 - a_2^+ a_2 a_1^+ a_1 \}$$

$$= \frac{1}{4} \left\{ a_2^+ a_1 a_1^+ a_2 + a_1^+ a_2 a_2^+ a_1 + a_2^+ a_2 [a_1, a_1^+] \right.$$

$$\left. + a_1^+ a_1 [a_2, a_2^+] + a_1^+ a_1 a_1^+ a_1 + a_2^+ a_2 a_2^+ a_2 \right\}$$

$$= \frac{1}{4} \{ a_1^+ a_2 a_1^+ a_2 + a_2^+ a_1 a_2^+ a_1 + a_2^+ a_2 + a_1^+ a_1 + a_1^+ a_1 a_1^+ a_1 + a_2^+ a_2 a_2^+ a_2 \},$$

where use has been made of

$$a_2^+ a_1 a_1^+ a_2 = a_2^+ a_1 a_2 a_1^+ = a_2^+ a_2 a_1 a_1^+, \text{ etc.,}$$

and

$$a_1^+ a_2 |n_1, n_2\rangle = \sqrt{(n_1 + 1) n_2} |n_1, n_2\rangle,$$

$$a_1^+ a_1 |n_1, n_2\rangle = \sqrt{n_1} a_1^+ |n_1 - 1, n_2\rangle = n_1 |n_1, n_2\rangle, \text{ etc,}$$

we find

$$\hat{J}^2 |n_1, n_2\rangle = \frac{1}{4} [(n_1 + 1) n_2 + (n_2 + 1) n_1 + n_1^2 + n_2^2 + n_1 + n_2] |n_1, n_2\rangle$$

$$= \frac{1}{2} (n_1 + n_2) \left[\frac{1}{2} (n_1 + n_2) + 1 \right] |n_1, n_2\rangle$$

and

$$\hat{J}_z |n_1, n_2\rangle = \frac{1}{2}(a_1^+ a_1 - a_2^+ a_2)|n_1, n_2\rangle$$

$$= \frac{1}{2}(n_1 - n_2)|n_1, n_2\rangle.$$

Thus the eigenvalues of $\hat{J}^2$, J_z are respectively

$$J^2 = \frac{1}{2}(n_1 + n_2)\left[\frac{1}{2}(n_1 + n_2) + 1\right]$$

$$J_z = \frac{1}{2}(n_1 - n_2).$$

Furthermore with

$$n_1 = j + m, \ n_2 = j - m,$$

the above give

$$\hat{J}^2 |j, m\rangle = j(j+1)|j, m\rangle,$$
$$J_z |j, m\rangle = m|j, m\rangle.$$

c. Energy levels with the same value of J are degenerate. The situation is exactly analogous to that of the general angular momemtum. Adding the perturbation $\mathbf{V} \cdot \mathbf{J}$ will remove the degeneracy because the different energy levels have different value of $\mathbf{J}\mathbf{V}$ in the direction of the vector $\mathbf{V}$.

5048

Consider a system in the unperturbed state described by a Hamiltonian $H^{(0)} = \begin{pmatrix} 1 & 0 \\ 0 & 1 \end{pmatrix}$. The system is subjected to a perturbation of the form $H' = \begin{pmatrix} \delta & \delta \\ \delta & \delta \end{pmatrix}$. Find the energy eigenvalues of the system using first order perturbation theory.

Sol:

$H^{(0)} = \begin{pmatrix} 1 & 0 \\ 0 & 1 \end{pmatrix}$, the eigenvalues of $H^{(0)}$ give the unperturbed energies.

As this is a diagonal matrix the eigenvalue is the diagonal elements of this matrix. The unperturbed energies are given by (1,1)

$H^1 = \begin{pmatrix} \delta & \delta \\ \delta & \delta \end{pmatrix}$, the eigenvalue H^1 gives the first order correction to the energies,

and to find the eigenvalues, $|H' - \lambda I| = 0$, i.e., $\begin{vmatrix} \delta - \lambda & \delta \\ \delta & \delta - \lambda \end{vmatrix} = 0$,

which gives the values 0, 2δ

Therefore, the final energies after the correction is just the sum of unperturbed and perturbed eigenvalues, which is given by 1, $1+2\delta$.

5049

A particle of mass m moves (nonrelativistically) in the three-dimensional potential

$$V = \frac{1}{2}k(x^2 + y^2 + z^2 + \lambda xy).$$

a. Consider λ as a small parameter and calculate the ground state energy through second order perturbation theory.

b. Consider λ as a small parameter and calculate the first excited energy levels to first order in perturbation theory.

Formulas from the standard solution of the one-dimensional harmonic oscillator:

$$\omega = (k/m)^{\frac{1}{2}}, \quad E_n = (n + \frac{1}{2})\hbar\omega, \quad n = 0, 1, 2, \ldots,$$

$$x = \left(\frac{\hbar}{2m\omega}\right)^{\frac{1}{2}}(a + a^+), \quad a\psi_n = \sqrt{n}\,\psi_{n-1},$$

$$[a, a^+] = 1, \quad a^+\psi_n = \sqrt{n+1}\,\psi_{n+1}.$$

(Berkeley)

Sol:

a. The ground state has wave function

$$\psi_{000}(x, y, z) = \psi_0(x)\psi_0(y)\psi_0(z)$$

and energy $\frac{3}{2}\hbar w$. Consider $\frac{1}{2}\lambda xy$ as perturbation, the first order energy correction is

$$E^{(1)} = \langle 000| \frac{k\lambda}{2}xy|000\rangle = 0,$$

as the integral is an odd function with respect to x or y. The second order energy correction is

$$E^{(2)} = \sum_{n_1, n_2} \left| \langle 000 | \frac{k\lambda}{2} xy | n_1 n_2 n_3 \rangle \right|^2 / (-n_1 - n_2) \hbar \omega$$

$$= -\frac{\lambda^2}{32} \hbar \omega,$$

as

$$\langle 0,0,0 | \frac{k\lambda}{2} xy | n_1 n_2 n_3 \rangle = \frac{\lambda}{4} \hbar \omega \delta_{1n_1} \delta_{1n_2} \delta_{0n_3}.$$

Therefore, the ground state energy corrected to second order is

$$E_0^{(1)} = \hbar \omega \left(\frac{3}{2} - \frac{\lambda^2}{32} \right).$$

b. The first excited energy level $E_1 = \frac{5}{2} \hbar \omega$ is three-fold degenerate, the three states being

$$|1,0,0\rangle, \ |0,1,0\rangle, \ |0,0,1\rangle.$$

The matrix of the perturbation $\frac{\lambda k}{2} xy$ is

$$\frac{\lambda \hbar \omega}{4} \begin{pmatrix} 0 & 1 & 0 \\ 1 & 0 & 0 \\ 0 & 0 & 0 \end{pmatrix},$$

and the secular equation

$$\begin{vmatrix} E_1^{(1)} & \dfrac{\lambda \hbar \omega}{4} & 0 \\[2mm] \dfrac{\lambda \hbar \omega}{4} & E_1^{(1)} & 0 \\[2mm] 0 & 0 & E_1^{(1)} \end{vmatrix} = 0$$

has roots

$$E_1^{(1)} = 0, \ \frac{\lambda \hbar \omega}{4}, \ -\frac{\lambda \hbar \omega}{4}.$$

Thus the first excited energy level splits into three levels

$$\left(\frac{5}{2} + \frac{\lambda}{4} \right) \hbar \omega, \ \frac{5}{2} \hbar \omega, \ \left(\frac{5}{2} - \frac{\lambda}{4} \right) \hbar \omega.$$

5050

Consider the following model for the Van der Waals force between two atoms. Each atom consists of one electron bound to a very massive nucleus by a potential $V(r_i) = \frac{1}{2}m\omega^2 r_i^2$. Assume that the two nuclei are $d \gg \sqrt{\frac{\hbar}{m\omega}}$ apart along the x-axis, as shown in Fig. 5.14, and that there is an interaction $V_{12} = \beta\frac{x_1 x_2 e^2}{d^3}$. Ignore the fact that the particles are indistinguishable.

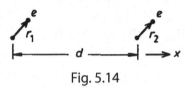

Fig. 5.14

a. Consider the ground state of the entire system when $\beta = 0$. Give its energy and wave function in terms of $\mathbf{r}_1$ and $\mathbf{r}_2$.

b. Calculate the lowest nonzero correction to the energy, ΔE, and to the wave function due to V_{12}.

c. Calculate the r.m.s. separation along the x direction of the two electrons to lowest order in β.

$$\psi_0(x) = \langle x | 0 \rangle = \left(\frac{1}{\sqrt{\pi}}\right)^{1/2} e^{-\frac{x^2 m\omega}{2\hbar}} ;$$

$$\psi_1(x) = \langle x | 1 \rangle = \left(\frac{1}{\sqrt{\pi}}\frac{2m\omega}{\hbar}\right)^{1/2} x e^{-\frac{x^2 m\omega}{2\hbar}} ,$$

$$\langle n | x | m \rangle = 0, \text{ for } |n - m| \neq 1,$$

$$\langle n-1 | x | n \rangle = (n\hbar/2m\omega)^{1/2},$$

$$\langle n+1 | x | n \rangle = ((n+1)\hbar/2m\omega)^{1/2},$$

(Wisconsin)

Sol:

a. The Schrödinger equation of the system is

$$\left\{-\frac{\hbar^2}{2m}(\nabla_1^2 + \nabla_2^2) + \frac{1}{2}m\omega^2(r_1^2 + r_2^2) + \beta\frac{x_1 x_2 e^2}{d^3}\right\}\psi = E\psi.$$

When $\beta = 0$ the system is equal to two independent three-dimensional harmonic oscillators and the energy and wave function of the ground state are

$$E_0^{(0)} = \frac{3}{2}\hbar\omega + \frac{3}{2}\hbar\omega = 3\hbar\omega,$$

$$\Psi_0^{(0)}(\mathbf{r}_1, \mathbf{r}_2) = \psi_0(x_1)\psi_0(y_1)\psi_0(z_1)\psi_0(x_2)\psi_0(y_2)\psi_0(z_2)$$

$$= \left(\frac{1}{\sqrt{\pi}}\right)^3 e^{-\frac{m\omega}{2\hbar}(r_1^2 + r_2^2)},$$

where $r_1^2 = x_1^2 + y_1^2 + z_1^2$, etc.

b. Treating

$$H' = \frac{\beta e^2}{d^3} x_1 x_2$$

as perturbation we have the first order energy correction

$$\Delta E_0^{(1)} = \frac{\beta e^2}{d^3}\langle 00 | x_1 x_2 | 00 \rangle$$

$$= \frac{\beta e^2}{d^3}\langle \psi_0(x_1)|x_1|\psi_0(x_1)\rangle \langle \psi_0(x_2)|x_2|\psi_0(x_2)\rangle = 0$$

as $\langle n|x|k\rangle = 0$ for $k \neq n \pm 1$.

For the second order energy correction, we have

$$\langle 00 | H' | n_1 n_2 \rangle = \frac{\beta e^2}{d^3}\langle 0|x_1|n_1\rangle \langle 0|x_2|n_2\rangle$$

$$= \frac{\beta e^2}{d^3}\frac{\hbar}{2m\omega}\delta_{n_1,1}\delta_{n_2,1},$$

and hence

$$\Delta E_0^{(2)} = \sum_{n_1,n_2 \neq 0}' \frac{|\langle 00 | H' | n_1 n_2 \rangle|^2}{E_0 - E_{n_1 n_2}}$$

$$= \frac{|\langle 00 | H' | 11 \rangle|^2}{E_0 - E_{11}} = -\frac{1}{2\hbar\omega}\left(\frac{\beta e^2}{d^3}\frac{\hbar}{2m\omega}\right)^2,$$

as $E_{n_1 n_2} = \left(n_1 + \frac{3}{2}\right)\hbar\omega + \left(n_2 + \frac{3}{2}\right)\hbar\omega$. Thus the energy corrected to lowest order is

$$E_0 = 3\hbar\omega - \frac{1}{8}\left(\frac{e^2}{d^3}\right)^2 \frac{\hbar}{m^2\omega^3}\beta^2,$$

and the corrected wave function is

$$\Psi_0 = \Psi_0^{(0)} + \frac{\langle 00 | H' | 11 \rangle}{E_0 - E_{11}} \Psi_1^{(0)}$$

$$= \Psi_0^{(0)} - \frac{\beta e^2}{4d^3} \frac{1}{m\omega^2} \Psi_1^{(0)},$$

where $\Psi_1^{(0)} = \psi_1(x_1)\psi_0(y_1)\psi_0(z_1)\psi_1(x_2)\psi_0(y_2)\psi_0(z_2)$.

c. Let $S_{12} = x_2 - x_1$. Then $\langle S_{12} \rangle = \langle x_2 \rangle - \langle x_1 \rangle = 0$ as Ψ_0 remains the same when
1 and 2 are interchanged showing that $\langle x_1 \rangle = \langle x_2 \rangle$. Consider

$$\langle S_{12}^2 \rangle = \langle x_1^2 + x_2^2 - 2x_1 x_2 \rangle = 2\langle x_1^2 \rangle - 2\langle x_1 x_2 \rangle.$$

We have

$$\langle x_1 x_2 \rangle = \langle \Psi_0^{(0)} - \lambda \Psi_1^{(0)} | x_1 x_2 | \Psi_0^{(0)} - \lambda \Psi_1^{(0)} \rangle$$

$$= -\lambda \left\{ \langle \Psi_0^{(0)} | x_1 x_2 | \Psi_1^{(0)} \rangle + \text{complex conjugate} \right\}$$

$$= -2\lambda (\langle 0 | x | 1 \rangle)^2 = -\lambda \frac{\hbar}{m\omega},$$

where

$$\lambda = \frac{\beta e^2}{4d^3} \frac{1}{m\omega^2},$$

and

$$\langle x_1^2 \rangle = \langle \Psi_0^{(0)} - \lambda \Psi_1^{(0)} | x_1^2 | \Psi_0^{(0)} - \lambda \Psi_1^{(0)} \rangle$$

$$= \langle 0 | x^2 | 0 \rangle + \lambda^2 \langle 1 | x^2 | 1 \rangle$$

$$= \langle 0 | x^2 | 0 \rangle + O(\lambda^2).$$

Also according to the virial theorem

$$\frac{1}{2} m\omega^2 \langle 0 | x^2 | 0 \rangle = \frac{1}{4} \hbar \omega,$$

or

$$2\langle x_1^2 \rangle = \frac{\hbar}{m\omega} + O(\lambda^2).$$

Hence

$$\langle S_{12}^2 \rangle = 2\langle x_1^2 \rangle - 2\langle x_1 x_2 \rangle$$

$$= \frac{\hbar}{m\omega} + \frac{2\hbar}{m\omega} \lambda + O(\lambda^2) \approx \frac{\hbar}{m\omega}(1 + 2\lambda).$$

Thus the root-mean-square distance between the two electrons in x direction is

$$\sqrt{\langle(d+S_{12})^2\rangle}=\sqrt{d^2+2d\langle S_{12}\rangle+\langle S_{12}^2\rangle}$$

$$\approx\sqrt{d^2+\frac{\hbar}{m\omega}(1+2\lambda)}=d\sqrt{1+\frac{\hbar}{m\omega d^2}\left(1+\frac{\beta e^2}{2m\omega^2 d^3}\right)}.$$

5051

The first excited state of three-dimensional isotropic harmonic oscillator (of natural angular frequency ω_0 and mass m) is three-dold degenerate, Use the perturbation method to calculate the splitting (to the first order) of this three-fold degenerate state due to a small perturbation of the form $H'=bxy$, where b is a constant. Give the first-order wave functions of the three split levels in terms of the wave functions of the unperturbed three-dimensional harmonic oscillator, given that, for a one-dimensional harmonic oscillator,

$$\langle n|x|n+1\rangle=\sqrt{\frac{(n+1)\hbar}{2m\omega_0}}.$$

(Wisconsin)

Sol: Write the unperturbed energy eigenstate as

$$|n_x n_y n_z\rangle=|n_x\rangle|n_y\rangle|n_z\rangle,$$

where $|n\rangle$ is the nth eigenstate of a one-dimensional harmonic oscillator. The first excited state of the 3-dimensional isotropic harmonic oscillator is degenerate in the states

$$|\psi_1\rangle=|100\rangle,\ |\psi_2\rangle=|010\rangle,\ |\psi_3\rangle=|001\rangle.$$

Calculating the matrix elements

$$H'_{ij}=b\langle\psi_i|xy|\psi_j\rangle,$$

we find

$$H'_{11}=b\langle100|xy|100\rangle=b\langle1|x|1\rangle\langle0|y|0\rangle=0=H'_{22}=H'_{33},$$

$$H'_{12}=b\langle100|xy|010\rangle=b\langle1|x|0\rangle\langle0|y|1\rangle=b\langle0|x|1\rangle^*\langle0|y|1\rangle=\frac{\hbar b}{2m\omega_0}=H'_{21},$$

$$H'_{13}=b\langle100|xy|001\rangle=b\langle1|x|0\rangle\langle0|y|0\rangle=0=H'_{31},$$

$$H'_{23}=b\langle010|xy|001\rangle=b\langle0|x|0\rangle\langle1|y|0\rangle=0=H'_{32}.$$

Thus

$$H' = \frac{\hbar b}{2m\omega_0} \begin{pmatrix} 0 & 1 & 0 \\ 1 & 0 & 0 \\ 0 & 0 & 0 \end{pmatrix}.$$

The secular equation

$$\det|H' - E^{(1)}| = 0$$

has roots $E^{(1)} = 0$, $E^{(1)} = \pm\dfrac{\hbar b}{2m\omega_0}$. The unperturbed oscillator has energy $E_n^{(0)} = \left(n + \dfrac{3}{2}\right)\hbar\omega$. The first excited state, $n = 1$, now splits into three levels with the wave functions indicated:

$$E_0^{(1)} = \frac{5}{2}\hbar\omega + 0 = \frac{5}{2}\hbar\omega, \quad |\psi_0'\rangle = |\psi_3\rangle = |001\rangle.$$

$$E_\pm^{(1)} = \frac{5}{2}\hbar\omega \pm \frac{\hbar b}{2m\omega_0}, \quad |\psi_\pm^{(1)}\rangle = \frac{1}{\sqrt{2}}(|100\rangle \pm |010\rangle).$$

5052

A quantum mechanical system is described by the Hamiltonian $H = H_0 + H'$, where $H' = i\lambda[A, H_0]$ is a perturbation on the unperturbed Hamiltonian H_0, A is a Hermitian operator and λ is a real number. Let B be a second Hermitian operator and let $C = i[B, A]$.

a. You are given the expectation values of the operators A, B and C in the unperturbed (and nondegenerate) ground states; call these $\langle A \rangle_0$, $\langle B \rangle_0$ and $\langle C_0 \rangle$. With perturbation switched on, evaluate the expectation value of B in the perturbed ground state to first order in λ.

b. Test this result on the following three-dimensional problem.

$$H_0 = \sum_{i=1}^{3}\left(\frac{p_i^2}{2m} + \frac{1}{2}m\omega^2 x_i^2\right), \quad H' = \lambda x_3,$$

by computing the ground state expectation value $\langle x_i \rangle$, $(i = 1, 2, 3)$ to lowest order in λ. Compare this result with an exact evaluation of $\langle x_i \rangle$.

(Princeton)

Sol:

a. Assume that the eigenstates and the corresponding energies of an unperturbed system are respectively

$$|k\rangle^{(0)}, E_k^{(0)},$$

then

$$\hat{H}_0|k\rangle^{(0)} = E_k^{(0)}|k\rangle^{(0)}.$$

The matrix elements of the perturbation Hamiltonian are

$$H'_{n0} = {}^{(0)}\left\langle n\left|\widehat{H}'\right|0\right\rangle^{(0)} = {}^{(0)}\left\langle n\left|i\lambda AH_0 - i\lambda H_0 A\right|0\right\rangle^{(0)}$$

$$= i\lambda\left(E_0^{(0)} - E_n^{(0)}\right){}^{(0)}\langle n|A|0\rangle^{(0)}.$$

Then, the ground state wave function with first order perturbation correction is

$$|0\rangle = |0\rangle^{(0)} + \sum_{n\neq 0}^{\infty} \frac{H'_{n0}}{E_0^{(0)} - E_n^{(0)}}|n\rangle^{(0)}$$

$$= |0\rangle^{(0)} + \sum_{n\neq 0}^{\infty} i\lambda{}^{(0)}\langle n|A|0\rangle^{(0)}|n\rangle^{(0)}$$

$$= (1 - i\lambda\langle A\rangle_0)|0\rangle^{(0)} + \sum_{n=0}^{\infty} i\lambda{}^{(0)}\langle n|A|0\rangle^{(0)}|n\rangle^{(0)}.$$

Hence

$$\langle 0|B|0\rangle = \Big[{}^{(0)}\langle 0|(1 + i\lambda\langle A\rangle_0) + (-i\lambda)$$

$$\times \sum_{n=0}^{\infty}\left({}^{(0)}\langle n|A|0\rangle\right)^{*(0)}\langle n|\Big]B\Big[(1 - i\lambda\langle A\rangle_0)|0\rangle^{(0)}$$

$$+ i\lambda\sum_{m=0}^{(0)}{}^{\infty}\langle m|A|0\rangle^{(0)}|m\rangle^{(0)}\Big]$$

$$\approx \langle B\rangle_0 - \lambda{}^{(0)}\langle 0|iAB - iBA|0\rangle^{(0)}$$

$$= \langle B\rangle_0 + \lambda{}^{(0)}\langle 0|C|0\rangle^{(0)} = \langle B\rangle_0 + \lambda\langle C\rangle_0,$$

(to first order in λ).

Note that the completness of $|k\rangle^{(0)}$,

$$\sum_k |k\rangle^{(0)}\langle k|^{(0)} = 1,$$

has been assumed in the above calculation,

b. The given Hamiltonians

$$H_0 = \sum_{i=1}^{3}\left(\frac{p_i^2}{2m} + \frac{1}{2}m\omega^2 x_i^2\right), \quad H' = \lambda x_3$$

satisfy $H' = i\lambda[A, H_0] = \lambda x_3$ if we set $A = \dfrac{p_3}{m\omega^2\hbar}$. Using the results of (a) we have the following: For $B = x_1$, as

$$C_1 = i[B,A] = \frac{i}{m\omega^2\hbar}[x_1, p_3] = 0,$$

we have

$$\langle x_1 \rangle = \langle B \rangle \approx \langle B \rangle_0 + \lambda \langle C_1 \rangle_0 = \langle x_1 \rangle_0 + \lambda \langle C_1 \rangle_0 = 0.$$

Note that $\langle x_1 \rangle = 0$ as the integral is an odd function of x_i. For $B = x_2$, a similar calculation gives $\langle x_2 \rangle = 0$. For $B = x_3$, as

$$C_3 = i[B,A] = i\left[x_3, \frac{p_3}{m\omega^2\hbar}\right] = -\frac{1}{m\omega^2},$$

and so

$$\langle C_3 \rangle_0 = -\frac{1}{m\omega^2},$$

we have

$$\langle x_3 \rangle = \langle B \rangle \approx \langle x_3 \rangle_0 + \lambda \langle C_3 \rangle_0$$
$$= -\frac{\lambda}{m\omega^2}.$$

For an accurate solution for $\hat{H} = \hat{H}_0 + H'$, write

$$\hat{H} = \sum_{i=1}^{3}\left(\frac{p_i^2}{2m} + \frac{1}{2}m\omega^2 x_i^2\right) + \lambda x_3$$
$$= \hat{H}_{01}(x_1) + \hat{H}_{02}(x_2) + \hat{H}_{03}\left(x_3 + \frac{\lambda}{m\omega^2}\right) - \frac{\lambda^2}{2m\omega^2},$$

where $\hat{H}_{0i}(x_i) = -\dfrac{\hbar^2}{2m}\dfrac{d^2}{dx_i^2} + \dfrac{1}{2}m\omega^2 x_i^2$ is the Hamiltonian of a onedimensional harmonic oscillator. As the constant term $-\lambda^2/2m\omega^2$ does not affect the dynamics of the system, the accurate wave function of the ground state is just that for an isotropic harmonic oscillator:

$$|0\rangle = \left(\frac{m\omega}{\pi\hbar}\right)^{3/4}\exp\left(-\frac{m\omega}{2\hbar}x_1^2\right)\exp\left(-\frac{m\omega}{2\hbar}x_2^2\right)$$
$$\times \exp\left[-\frac{m\omega}{2\hbar}\left(x_3 + \frac{\lambda}{m\omega^2}\right)^2\right].$$

It follows that

$$\langle x_1 \rangle = 0, \ \langle x_2 \rangle = 0, \ \langle x_3 \rangle = -\frac{\lambda}{m\omega^2}.$$

These results are exactly the same as those obtained in a)

5053

A particle of mass m is moving in the three-dimensional harmonic oscillator potential $V(x,y,z) = \frac{1}{2}m\omega^2(x^2 + y^2 + z^2)$. A weak perturbation is applied in the form of the function $\Delta V = kxyz + \frac{k^2}{\hbar\omega}x^2y^2z^2$, where k is a small constant. Note the same constant k appears in both terms.

a. Calculate the shift in the ground state energy to second order in k.

b. Using an argument that does not depend on perturbation theory, what is the expectation value of **x** in the ground state of this system?

Note: You may wish to know the first few wave functions of the onedimensional harmonic oscillator:

ground state

$$\psi_0(x) = \left(\frac{m\omega}{\pi\hbar}\right)^{1/4} \exp\left(-\frac{m\omega}{2\hbar}x^2\right),$$

first excited state

$$\psi_1(x) = \left(\frac{m\omega}{\pi\hbar}\right)^{1/4} \sqrt{\frac{2m\omega}{\hbar}} x \exp\left(-\frac{m\omega}{2\hbar}x^2\right),$$

second excited state

$$\psi_2(x) = \left(\frac{m\omega}{\pi\hbar}\right)^{1/4} \frac{1}{\sqrt{2}}\left(\frac{2m\omega}{\hbar}x^2 - 1\right)\exp\left(-\frac{m\omega}{2\hbar}x^2\right).$$

(Princeton)

Sol: The ground state of a particle in the potential well of three-dimensional harmonic oscillator is

$$\Phi_0(x,y,z) = \psi_0(x)\psi_0(y)\psi_0(z)$$

$$= \left(\frac{m\omega}{\pi\hbar}\right)^{3/4} \exp\left[-\frac{m\omega^2}{2\hbar}(x^2 + y^2 + z^2)\right].$$

The first order energy correction is

$$\langle \Delta E \rangle_1 = \int \Phi_0(x,y,z)\left(kxyz + \frac{k^2}{\hbar\omega}x^2y^2z^2 \right)\Phi_0(x,y,z)d^3x$$

$$= \left(\frac{m\omega}{\pi\hbar}\right)^{3/2}\frac{k^2}{\hbar\omega}\left[\int_{-\infty}^{+\infty} x^2\exp\left(-\frac{m\omega}{\hbar}x^2\right)dx\right]^3$$

$$= \left(\frac{\hbar}{2m\omega}\right)^3\frac{k^2}{\hbar\omega}.$$

While the perturbation $\Delta V' = kxyz$ does not give rise to first order correction, it is to be considered for second order perturbation in order to calculate the energy shift accurate to k^2. Its perturbation Hamiltonian has matrix elements

$$\langle n|\Delta V'|0\rangle = \int \Phi_n(x,y,z)\, kxyz\, \Phi_0(x,y,z)d^3x$$

$$= k\int_{-\infty}^{+\infty}\psi_{n_1}(x)x\psi_0(x)dx \int_{-\infty}^{+\infty}\psi_{n_2}(y)y\psi_0(y)dy$$

$$\times \int_{-\infty}^{+\infty}\psi_{n3}(z)z\psi_0(z)dz$$

$$= k\left(\frac{\hbar}{2m\omega}\right)^{3/2}\delta(n_1-1)\delta(n_2-1)\delta(n_3-1),$$

where $n = n_1 + n_2 + n_3$, and so the second order energy correction is

$$\langle \Delta E \rangle_2 = \sum_{n\neq 0}\frac{|\langle n|\Delta V'|0\rangle|^2}{E_0 - E_n}$$

$$= \frac{k^2\left(\dfrac{\hbar}{2m\omega}\right)^3}{E_0 - E_3} = -\left(\frac{\hbar}{2m\omega}\right)^3\frac{k^2}{3\hbar\omega}.$$

Therefore the energy shift of the ground state accurate to k^2 is

$$\Delta E = \langle \Delta E \rangle_1 + \langle \Delta E \rangle_2 = \frac{2}{3}\frac{k^2}{\hbar\omega}\left(\frac{\hbar}{2m\omega}\right)^3.$$

b. $V + \Delta V$ is not changed by the inversion $x \to -x$, $y \to -y$, i.e.,

$$H(x,y,z) = H(-x, -y, z).$$

Furthermore the wave function of the ground state is not degenerate, so $\psi(-x, -y, z) = \psi(x,y,z)$ and, consequently,

$$\langle x \rangle = (\psi, x\psi)$$

$$= \int_{-\infty}^{+\infty} dz' \int_{-\infty}^{+\infty} \int_{-\infty}^{+\infty} \psi^*(x', y', z')x'\psi(x', y', z')dx'\, dy'$$

$$= -\int_{-\infty}^{+\infty} dz \int_{-\infty}^{+\infty} \int_{-\infty}^{+\infty} \psi^*(x, y, z) \cdot x \cdot \psi(x, y, z)dx\, dy = -\langle x \rangle,$$

where we have applied the transformation $x' = -x$, $y' = -y$, $z' = z$. Hence $\langle x \rangle = 0$. In the same way we find $\langle y \rangle = 0$, $\langle z \rangle = 0$. Thus (x) = 0.

5054

A spin $-\frac{1}{2}$ particle of mass m moves in spherical harmonic oscillator potential $V = \frac{1}{2}m\omega^2 r^2$ and is subject to an interaction $\lambda\boldsymbol{\sigma}\cdot\mathbf{r}$ (spin orbit forces are to be ignored). The net Hamiltonian is therefore

$$H = H_0 + H',$$

where

$$H_0 = \frac{p^2}{2m} + \frac{1}{2}m\omega^2 r^2, \quad H' = \lambda\boldsymbol{\sigma}\cdot\mathbf{r}.$$

Compute the shift of the ground state energy through second order in the perturbation H'.

(Princeton)

Sol: The unperturbed ground state is two-fold degenerate with spin up and spin down along the z-axis.

Using the perturbation method for degenerate states, if the degeneracy does not disappear after diagonalizing the perturbation Hamiltonian, one has to diagonalize the following matrix to find the energy positions:

$$\langle n|V|n'\rangle + \sum_m \frac{\langle n|V|m\rangle\langle m|V|n'\rangle}{E_n^{(0)} - E_m^{(0)}} \equiv \langle n|W|n'\rangle.$$

Let $|n_x n_y n_z \uparrow\rangle$ and $|n_x n_y n_z \downarrow\rangle$ be the unperturbation quantum states, where n_x, n_y and n_z are the oscillation quantum numbers in the x, y and z direction, $\uparrow(\downarrow)$ represents the spin up (down) state. As

$$E_{n_x n_y n_z}^{(0)} = \left(n_x + n_y + n_z + \frac{3}{2} \right)\hbar\omega,$$

the matrix has elements

$$\langle 000 \uparrow | W | 000 \uparrow \rangle = \frac{\lambda^2 \langle 000 \uparrow | \sigma \cdot \mathbf{r} | 001 \uparrow \rangle \langle 001 \uparrow | \sigma \cdot \mathbf{r} | 000 \uparrow \rangle}{\frac{3}{2}\hbar\omega - \frac{5}{2}\hbar\omega}$$

$$+ \frac{\lambda^2 \langle 000 \uparrow | \sigma \cdot \mathbf{r} | 100 \downarrow \rangle \langle 100 \downarrow | \sigma \cdot \mathbf{r} | 000 \uparrow \rangle}{\frac{3}{2}\hbar\omega - \frac{5}{2}\hbar\omega}$$

$$+ \frac{\lambda^2 \langle 000 \uparrow | \sigma \cdot \mathbf{r} | 010 \downarrow \rangle \langle 010 \downarrow | \sigma \cdot \mathbf{r} | 000 \uparrow \rangle}{\frac{3}{2}\hbar\omega - \frac{5}{2}\hbar\omega}$$

$$= \lambda^2 \left[\frac{|\langle 000 \uparrow | \sigma_z z | 001 \uparrow \rangle|^2}{-\hbar\omega} + \frac{|\langle 000 \uparrow | \sigma_x x | 100 \downarrow \rangle|^2}{-\hbar\omega} \right.$$

$$\left. + \frac{|\langle 000 \uparrow | \sigma_y y | 010 \downarrow \rangle|^2}{-\hbar\omega} \right]$$

$$= -\frac{3\lambda^2}{2m\omega^2},$$

$$\langle 000 \downarrow | W | 000 \downarrow \rangle = -\frac{3\lambda^2}{2m\omega^2},$$

$$\langle 000 \uparrow | W | 000 \downarrow \rangle = 0.$$

In the above calculation we have used the fact that

$$\langle n_i | x_i | n_i + 1 \rangle = \sqrt{\frac{(n_i+1)\hbar}{2m\omega}}, \quad x_i = x, y, z,$$

all other elements being zero. It is seen that a two fold degeneracy still exists for the eigenvalue $\dfrac{-3\lambda^2}{2m\omega^2}$. This means that the degeneracy will not disappear until at least the second order approximation is used. The ground state, which is still degenerate, has energy $\dfrac{3}{2}\hbar\omega - \dfrac{3\lambda^2}{2m\omega^2}$.

5055

Consider a spinless particle of mass m and charge e confined in a spherical cavity of radius R: that is, the potential energy is zero for $|x| \leq R$ and infinite for $|x| > R$.

a. What is the ground state energy of this system?

b. Suppose that a weak uniform magnetic field of strength $|\mathbf{B}|$ is switched on. Calculate the shift in the ground state energy.

c. Suppose that, instead, a weak uniform electric field of strength $|\mathbf{E}|$ is switched on. Will the ground state energy increase or decrease? Write down, but do not attempt to evaluate, a formula for the shift in the ground state energy due to the electric field.

d. If, instead, a very strong magnetic field of strength $|\mathbf{B}|$ is turned on, approximately what would be the ground state energy?

(Princeton)

Sol: The radial part of the Schrödinger equation for the particle in the potential well is

$$R'' + \frac{2}{r}R' + \left[k^2 - \frac{l(l+1)}{r^2} \right] R = 0, \ (r < R),$$

where $k = \sqrt{2mE/\hbar^2}$, the boundary condition being $R(r)|_{r=R_0} = 0$. Introducing a dimensionless variable $\rho = kr$, we can rewrite the equation as

$$\frac{d^2 R}{d\rho^2} + \frac{2}{\rho}\frac{dR}{d\rho} + \left[1 - \frac{l(l+1)}{\rho^2} \right] R = 0.$$

This equation has solutions $j_l(\rho)$ that are finite for $\rho \to 0$, $j_l(\rho)$ being related to Bessel's function by

$$j_l(\rho) = \left(\frac{\pi}{2\rho} \right)^{\frac{1}{2}} J_{l+\frac{1}{2}}(\rho).$$

Thus the radial wave function is

$$R_{kl}(r) = C_{kl} j_l(kr),$$

where C_{kl} is the normalization constant.

The boundary condition requires

$$j_l(kR_0) = 0$$

which has solutions

$$kR_0 = \alpha_{nl}, \quad n = 1, 2, 3 \ldots$$

Hence the bound state of the particle has energy levels

$$E_{nl} = \frac{\hbar^2}{2mR_0^2}\alpha_{nl}^2, \quad n = 1, 2, 3 \ldots$$

For the ground state, $\rho = 0$ and $j_0(\rho) = \dfrac{\sin\rho}{\rho}$, so that $\alpha_{10} = \pi$ and the energy of the ground state is

$$E_{10} = \hbar^2 \pi^2 / 2mR_0^2.$$

b. Take the direction of the magnetic field as the z direction. Then the vector potential **A** has components

$$A_x = -\frac{B}{2}y, \ A_y = \frac{B}{2}x, \ A_z = 0,$$

and the Hamiltonian of the system is

$$\hat{H} = \frac{1}{2m}\left[\left(p_x + \frac{eB}{2c}y\right)^2 + \left(p_y - \frac{eB}{2c}x\right)^2 + p_z^2\right] + V(r)$$

$$= \frac{1}{2m}\left[\mathbf{p}^2 - \frac{eB}{c}(xp_y - yp_x) + \frac{e^2B^2}{4c^2}(x^2 + y^2)\right] + V(r)$$

$$= \frac{1}{2m}\left[\mathbf{p}^2 - \frac{eB}{c}l_z + \frac{e^2B^2}{4c^2}(x^2 + y^2)\right] + V(r).$$

As the magnetic field is weak we can treat $-\frac{eB}{c}l_z + \frac{e^2B^2}{4c^2}(x^2 + y^2)$ as a perturbation. When the system is in the ground state $l = 0$, $l_z = 0$ we only need to consider the effect of the term $\frac{e^2B^2}{8mc^2}(x^2 + y^2)$. The wave function of the ground state is

$$\psi(r,\theta,\varphi) = \sqrt{\frac{2k^2}{R_0}} j_0(kr) Y_{(00)}(\theta,\varphi)$$

$$= \sqrt{\frac{k^2}{2\pi R_0}} j_0(kr) = \frac{\sin(kr)}{\sqrt{2\pi R_0} \, r},$$

and the first order energy correction is

$$E' = \left\langle \psi(r,\theta,\varphi) \left| \frac{e^2B^2}{8mc^4}(x^2 + y^2) \right| \psi(r,\theta,\varphi) \right\rangle$$

$$= \frac{e^2B^2}{8mc^2} \cdot \frac{1}{2\pi R_0} \int_0^{R_0} r^2 \sin^2(kr)\,dr \int_0^\theta 2\pi \sin^3\theta\,d\theta$$

$$= \left(\frac{1}{3} - \frac{1}{2\pi^2}\right) \frac{e^2B^2R_0^2}{12mc^2}.$$

Note that in the above calculation we have used

$$x^2 + y^2 = r^2\sin^2\theta, \ \sin(kR_0) = 0 \ \text{ or } \ kR_0 = \pi.$$

c. Suppose a weak uniform electric field **E** is applied in the z direction, instead of the magnetic field. The corresponding potential energy of the particle is $V' = -eEz$, which is to be taken as the perturbation. The shift of the ground state energy is then

$$E_e' = \langle \psi(r,\theta,\varphi) | -eEr \cos\theta | \psi(r,\theta,\varphi) \rangle.$$

As E_e' is negative, the energy of the ground state decreases as a result.

d. If a strong magnetic field, instead of the weak one, is applied then

$$\hat{H} = \frac{P^2}{2m} + \frac{e^2 B^2}{8mc^2}(x^2 + y^2)$$

and the $\mathbf{B}^2$ term can no longer be considered as a perturbation. The particle is now to be treated as a two-dimensional harmonic oscillator with

$$\frac{1}{2}m\omega^2 = \frac{e^2 B^2}{8mc^2}, \text{ or } \omega = \frac{eB}{2mc}.$$

Hence, the ground state energy is approximately

$$E_0 = \hbar\omega = \frac{eB}{2mc}\hbar.$$

5056

A particle of mass m and electric charge Q moves in a three-dimensional isotropic harmonic-oscillator potential $V = \frac{1}{2}kr^2$.

a. What are the energy levels and their degeneracies?

b. If a uniform electric field is applied, what are the new energy levels and their degeneracies?

c. If, instead, a uniform magnetic field is applied, what are the energies of the four lowest states?

(Columbia)

Sol:

a. The Hamiltonian of the system is

$$H = -\frac{\hbar^2}{2m}\nabla^2 + \frac{1}{2}kr^2 = H_x + H_y + H_z,$$

where

$$H_i = -\frac{\hbar^2}{2m}\frac{\partial^2}{\partial x_i^2} + \frac{1}{2}kx_i^2 \quad (i = x, y, z).$$

The energy levels are given by

$$E_N = (N + 3/2)\hbar\omega_0,$$

where $\omega_0 = \sqrt{k/m}, N = n_x + n_y + n_z.$

The degeneracy of state N is

$$f = \sum_{n_x=0}^{N}(N-n_x+1)=\frac{1}{2}(N+1)(N+2).$$

b. Take the direction of the uniform electric field as the z direction. Then the Hamiltonian is

$$\hat{H} = \hat{\mathbf{P}}^2/2m+\frac{1}{2}kr^2-QEz$$

$$=\left(\frac{\hat{p}_x^2}{2m}+\frac{1}{2}kx^2\right)+\left[\hat{p}_y^2/2m+\frac{1}{2}ky^2\right]$$

$$+\left[\hat{p}_z^2/2m+\frac{1}{2}k(z-QE/k)^2\right]-Q^2E^2/2k.$$

Comparing this with the Hamiltonian in (a), we get

$$E_N=(N+3/2)\hbar\omega_0-Q^2E^2/2k,$$

$$f=\frac{1}{2}(N+1)(N+2).$$

c. Consider the case where a magnetic field, instead of the electric field, is applied. In cylindrical coordinates, the vector potential has components

$$A_\varphi=\frac{1}{2}B\rho,\ A_\rho=A_z=0.$$

Thus the Hamiltonian is

$$\hat{H}=\frac{1}{2m}\left(\hat{\mathbf{p}}-\frac{Q}{c}\mathbf{A}\right)^2+V$$

$$=\frac{1}{2m}\hat{\mathbf{p}}^2-\frac{Q}{mc}\hat{\mathbf{p}}\cdot\mathbf{A}+\frac{Q^2}{2mc^2}\mathbf{A}^2+\frac{1}{2}k\rho^2+\frac{1}{2}kz^2,$$

where we have used $r^2=\rho^2+z^2$, and $\nabla\cdot\mathbf{A}=0$ which means $\hat{\mathbf{p}}\cdot\mathbf{A}=\mathbf{A}\cdot\hat{\mathbf{p}}$. Write

$$\hat{H}=\left[-\frac{\hbar^2}{2m}\nabla_t^2+\frac{1}{2}m\omega'^2\rho^2\right]+\left[\frac{\hat{p}_z^2}{2m}+\frac{1}{2}kz^2\right]-\frac{Q}{|Q|}\omega\hat{L}_z$$

$$=\hat{H}_t+\hat{H}_z\mp\omega\hat{L}_z,$$

where

$$\nabla_t^2=\nabla_x^2+\nabla_y^2,\omega=|Q|B/2mc,\omega'^2=\omega^2+\omega_0^2,\ \hat{L}_z=\hat{p}_z\rho,$$

and the symbols $\mp$ correspond to positive and negative values of Q. Of the partial Hamiltonians, $\hat{H}_t$ corresponds to a two-dimensional harmonic oscillator normal to the z-axis, $\hat{H}_z$ corresponds to a one-dimensional harmonic oscillator along the z-axis. Therefore, the energy levels of the system are given by

$$E_{n_p n_z m} = (2n_p + 1 + |m|)\hbar w' + \left(n_z + \frac{1}{2}\right)\hbar\omega_0 \mp m\hbar\omega$$

$$= \left(\hbar\omega' + \frac{1}{2}\hbar\omega_0\right) + 2n_p\hbar\omega' + |m|\hbar\omega' \mp m\hbar\omega + n_z\hbar\omega_0,$$

where

$$n_p = 0, 1, 2, \ldots,$$
$$n_z = 0, 1, 2, \ldots,$$
$$m = 0, \pm 1, \pm 2, \ldots.$$

The four lowest energy levels are thus

$$E_{000} = \hbar\omega' + \frac{1}{2}\hbar\omega_0,$$

$$E_{001} = \hbar\omega' + \frac{1}{2}\hbar\omega_0 + \hbar(\omega' - \omega) = 2\hbar\omega' - \hbar\omega + \frac{1}{2}\hbar\omega_0,$$

$$E_{010} = \hbar\omega' + \frac{1}{2}\hbar\omega_0 + \hbar\omega_0 = \hbar\omega' + \frac{3}{2}\hbar\omega_0,$$

$$E_{002} = \hbar\omega' + \frac{1}{2}\hbar\omega_0 + 2\hbar(\omega' - \omega) = 3\hbar\omega' - 2\hbar\omega + \frac{1}{2}\hbar\omega_0.$$

5057

a. Describe the splitting of atomic energy levels by a weak magnetic field. Include in your discussion a calculation of the Landé g-factor (assume LS couplings).

b. Describe the splitting in a strong magnetic field (Paschen-Back effect).

(Columbia)

Sol: In LS coupling the magnetic moment of an atomic system is the sum of contributions of the orbital and spin angular momenta:

$$\hat{\mu} = \mu_0(g_l\hat{\mathbf{l}} + g_s\hat{\mathbf{s}})$$
$$= \mu_0[g_l\hat{\mathbf{j}} + (g_s - g_l)\hat{\mathbf{s}}],$$

where μ_0 is the Bohr magneton. Taking the direction of the magnetic field as the z direction, the change of the Hamiltonian caused by the magnetic field is

$$\hat{H}_1 = -\mu \cdot B = -g_l\mu_0 B j_z - (g_s - g_l)\mu_0 B s_z.$$

a. The Hamiltonian of system is

$$\hat{H}\hat{H} = \hat{H}_0 + \hat{H}_1 = \hat{p}^2/2m + V(r) + \xi(r)\hat{s}\cdot\hat{\mathbf{l}} + \hat{H}_1.$$

Considering $-(g_s - g_l)\mu_0 B s_z$ as perturbation and operating $(H_0 - g_l\mu_0 B j_z)$ on the common state of $\hat{\mathbf{l}}^2, \hat{\mathbf{j}}^2$ and $\hat{j}_z$ we have

$$(\hat{H}_0 - g_l\mu_0 B\hat{j}_z)\psi_{nljm_j} = (E_{nlj} - g l\mu_0 B m_j)\psi_{nljm_j}.$$

If B very weak, then

$$\bar{s}_z = \frac{\hbar}{2}\langle jm_j | \hat{\sigma}_z | jm_j \rangle = \begin{cases} \dfrac{m_j}{2j} & \text{for } j = l + \dfrac{1}{2}, \\[2mm] \dfrac{-m_j}{2(j+1)} & \text{for } j = l - \dfrac{1}{2}, \end{cases}$$

where we have used the relations

$$|jm_j\rangle = \sqrt{\frac{l \pm m_j + \frac{1}{2}}{2l+1}}\left|m_j - \frac{1}{2}, \frac{1}{2}\right\rangle \mp \sqrt{\frac{l \mp m_j + \frac{1}{2}}{2l+1}}\left|m_j + \frac{1}{2}, -\frac{1}{2}\right\rangle$$

for $j = l \pm \frac{1}{2}$, and

$$\sigma_z\left|m_j \mp \frac{1}{2}, \pm\frac{1}{2}\right\rangle = \pm\left|m_j \mp \frac{1}{2}, \pm\frac{1}{2}\right\rangle.$$

Hence, the energy level of the system becomes

$$E_{nljm_j} \approx E_{nlj} - g l\mu_0 B m_j$$

$$-(g_s - g_l)\mu_0 B \begin{cases} \dfrac{m_j}{2j}, & j = l + \dfrac{1}{2}, \\[2mm] \dfrac{-m_j}{(2j+2)}, & j = l - \dfrac{1}{2}. \end{cases}$$

As $g_l = -1$, $g_s = -2$ we can write

$$E_{nljm_j} \approx E_{nlj} - g\mu_0 B m_j,$$

where

$$g = -\left[1 + \frac{j(j+1) + s(s+1) - l(l+1)}{2j(j+1)}\right]$$

is the Landé g-factor. Thus an energy level of the atom splits into $(2j+1)$ levels.

b. If the magnetic field is very strong, we can neglect the term $\xi(r)\mathbf{s}\cdot\mathbf{l}$ and the Hamiltonian of the system is

$$\hat{H} = \hat{\mathbf{p}}^2 / 2m + V(r) + \hat{H}_1 = \hat{H}_0 + \hat{H}_1.$$

Operating on the common eigenstate of $(\hat{H}_0, \hat{l}^2, \hat{l}_z, \hat{s}^2, s_z)$, we get

$$\hat{H}\psi_{nlm,ms} = E_{nlmpms}\psi_{nlm,ms},$$

where

$$E_{nlm,ms} = E_{nl} - g_l\mu_0 Bm_l - g_s\mu_0 Bm_S$$
$$= E_{nl} + \mu_0 B(m_l + 2m_S),$$

as $g_l = -1$, $g_s = -2$. For an electron, $m_S = \pm\dfrac{1}{2}$. Then, due to the selection rule $\Delta m_S = 0$, transitions can only occur within energy levels of $m_S = +\dfrac{1}{2}$ and within energy levels of $m_S = -\dfrac{1}{2}$. The split levels for a given l are shown in Fig. 5.15. For the two sets of energy levels with $m_l + 2m_S = -l+1$ to $l-1$ (one set with $m_S = \dfrac{1}{2}$, the other with $m_S = -\dfrac{1}{2}$, i.e., $2l-1$ levels for each set), there is still a two-fold degeneracy. So the total number of energy levels is $2(2l+1)-(2l-1) = 2l+3$ as shown.

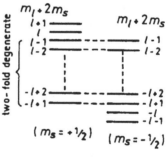

Fig. 5.15

5058

Consider an atom with a single valence electron. Its fine-structure Hamiltonian is given by $\xi\,\mathbf{L}.\mathbf{S}$.

a. Determine the difference in the energies of the levels characterized by $J = L+1/2$ and $J = L-1/2$ (fine structure interval) in terms of ξ.

b. This atom is placed in a weak external magnetic field of magnitude H. Use perturbation theory to determine the energy splitting between adjacent magnetic (Zeeman) sublevels of the atom.

c. Describe qualitatively how things change in a very strong magnetic field.

(Columbia)

ol:

a. In the representation of $(\hat{J}^2, \hat{L}^2, j_z)$,

$$\hat{S} \cdot \hat{L} = \frac{1}{2}(\hat{J}^2 - \hat{L}^2 - \hat{S}^2)$$

$$= \frac{\hbar^2}{2}\left[j(j+1) - l(l+1) - s(s+1)\right].$$

$$= \frac{\hbar^2}{2}\left[j(j+1) - l(l+1) - \frac{3}{4}\right].$$

As $\left(l + \frac{1}{2}\right)\left(l + \frac{3}{2}\right) - \left(l - \frac{1}{2}\right)\left(l + \frac{1}{2}\right) = 2l + 1$, the energy difference is

$$\Delta E = E_{j=l+\frac{1}{2}} - E_{j=l-\frac{1}{2}} = \overline{\xi(r)} \cdot \frac{\hbar^2}{2}(2l + 1).$$

b. If the atom is placed in a weak magnetic field of magnitude H whose direction is taken to be the z direction, the Hamiltonian of the system is

$$\hat{H} = \frac{\hat{P}^2}{2m} + V(r) + \frac{eH}{2mc}(\hat{L}_z + 2\hat{S}_z) + \xi(r)\hat{S} \cdot \hat{L}$$

$$= \frac{\hat{P}^2}{2m} + V(r) + \frac{eH}{2mc}\hat{J}_z + \xi(r)\hat{S} \cdot \hat{L} + \frac{eH}{2mc}\hat{S}_z$$

$$\equiv \hat{H}_0 + \frac{eH}{2mc}\hat{S}_z,$$

where $\hat{J}_z = \hat{L}_z + \hat{S}_z$.

Let ψ_{nljm_j} be a common eigenstate of $(\hat{J}^2, \hat{L}^2, j_z)$ for the unperturbed Hamiltonian $\hat{H}_0$. Then

$$\hat{H}_0 \psi_{nljm_j} = \left(E_{nlj} + m_j \frac{eH\hbar}{2mc}\right)\psi_{nljm_j}.$$

To consider the effect of the perturbation term $\frac{eH}{2mc}\hat{S}_z$ use spherical coordinates and write $\psi_{nljm_j} = R_n(r)\varphi_{ljm_j}(\theta, \varphi)$, where the angular wave functions are

$$\varphi_{ljm_j} = \sqrt{\frac{j+m_j}{2j}}\, \alpha Y_{j-\frac{1}{2},m_j-\frac{1}{2}}$$

$$+ \sqrt{\frac{j-m_j}{2j}}\, \beta Y_{j-\frac{1}{2},m_j+\frac{1}{2}} \qquad \text{for } j = l+\frac{1}{2},$$

$$\varphi_{ljm_j} = -\sqrt{\frac{j-m_j+1}{2j+2}}\, \alpha Y_{j+\frac{1}{2},m_j-\frac{1}{2}}$$

$$+ \sqrt{\frac{j+m_j+1}{2j+2}}\, \beta Y_{j+\frac{1}{2},m_j+\frac{1}{2}} \qquad \text{for } j = l-\frac{1}{2},$$

α, β being eigenstates of $\hat{S}_z$ of eigenvalues $\dfrac{\hbar}{2}$ and $-\dfrac{h}{2}$ respectively. Thus for $j = l + \dfrac{1}{2}$,

$$\hat{S}_z \varphi_{ljm_j} \equiv \hat{S}_z \left| jm_j \right\rangle = \sqrt{\frac{j+m_j}{2j}}\, \frac{\hbar}{2}\, \alpha Y_{j-\frac{1}{2},m_j-\frac{1}{2}}$$

$$- \sqrt{\frac{j-m_j}{2j}}\, \frac{\hbar}{2}\, \beta Y_{j-\frac{1}{2},m_j+\frac{1}{2}},$$

and

$$\langle jm_j | \hat{S}_z | jm_j \rangle = \frac{j+m_j}{2j}\frac{\hbar}{2} - \frac{j-m_j}{2j}\frac{\hbar}{2} = \frac{\hbar}{2}\frac{m_j}{j},$$

and for $j = l - \dfrac{1}{2}$

$$\langle jm_j | \hat{S}_z | jm_j \rangle = -\frac{\hbar}{2}\frac{m_j}{(j+1)}.$$

Hence,

$$\frac{eH}{2mc}\langle S_z \rangle = \frac{eH\hbar}{4mc}
\begin{cases}
m_j/j, & j = l+\dfrac{1}{2}, \\[2mm]
-m_j/(j+1), & j = l-\dfrac{1}{2},
\end{cases}$$

and so

$$E_{nljm_j} = E_{nlj} + \frac{eH\hbar}{2mc}
\begin{cases}
\left(1+\dfrac{1}{2j}\right)m_j, & j = l+\dfrac{1}{2}, \\[3mm]
\left(1-\dfrac{1}{2j+2}\right)m_j, & j = l-\dfrac{1}{2}.
\end{cases}$$

Therefore,

$$\Delta E = \begin{cases} \mu_B H\left(1+\dfrac{1}{2j}\right), & j=l+\dfrac{1}{2}, \\[2ex] \mu_B H\left(1-\dfrac{1}{2j+2}\right), & j=l-\dfrac{1}{2}, \end{cases}$$

where $\mu_B = \dfrac{e\hbar}{2mc}$ is the Bohr magneton.

c. If the magnetic field is very strong, $\xi(r)\mathbf{S}\cdot\mathbf{L} \ll \mu_B B$ and the Hamiltonian of the system is

$$\hat{H} = \frac{\mathbf{p}^2}{2m} + V(r) + \frac{eH}{2mc}(\hat{l}_z + 2\hat{S}_z).$$

Since $(\hat{H}_0, \hat{\mathbf{L}}^2, \hat{L}_z, \hat{\mathbf{S}}^2, S_z)$ form a complete set of dynamical variables,

$$\begin{aligned} E_{nlmm_s} &= E_{nl} + \frac{eH}{2mc}\hbar(m+2m_s) \\ &= E_{nl} + \mu_B H(m\pm 1), \end{aligned}$$

as $m_s = \pm\tfrac{1}{2}$. Hence

$$\Delta E = \mu_B H.$$

5059

Positronium is a hydrogen-like system consisting of a positron and an electron. Consider positronium in its ground state ($l=0$). The Hamiltonian H can be written: $H = H_0 + H_S + H_B$, where H_0 is the usual spin-independent part due to the Coulomb forces, $H_S = A s_p \cdot s_e$ is the part due to the interaction of the spins of the positron and the electron, and $H_B = -(\mu_p + \mu_e)\cdot \mathbf{B}$ is the part due to the interaction with an externally applied magnetic field $\mathbf{B}$.

a. In the absence of an externally applied field what choice of spin and angular momentum eigenfunctions is most convenient? Calculate the energy shifts for each of these states due to H_S.

b. A very weak magnetic field is applied ($H_B \ll H_S$). What are the allowed energies for this system in this case?

c. Now suppose the applied magnetic field is increased such that $H_B \gg H_S$. What kind of eigenfunctions are now most appropriate? What are the energy shifts for each of these states due to H_B?

d. Indicate how you would solve this problem for the energies and the corresponding eigenfunctions in the general case; however, do not try to carry out the algebra unless you have nothing better to do. No long essays, please.

(Berkeley)

Sol:

a. In the absense of the external field **B** take H_s as perturbation. The total spin of the system is $\mathbf{S} = \mathbf{s}_p + \mathbf{s}_e$ and so

$$s_p \cdot s_e = \frac{1}{2}(S^2 - s_p^2 - s_e^2) = \frac{1}{2}[S(S+1) - s_p(s_p+1) - s_e(s_e+1)].$$

It is most convenient to choose the eigenfunctions in the form $|lmSS_z\rangle$. The ground state $l=0$ consists of four states $S=1, S_z=0,\pm1; S=0, S_z=0$. For $l=0, S=1$, as

$$As_p \cdot s_e |lm1S_z\rangle = \frac{A}{2}\left[1(1+1) - \frac{3}{2}\right]\hbar^2|lm1S_z\rangle,$$

we have

$$E_s = \langle lm1S_z | H_s | lm1S_z\rangle = \frac{A\hbar^2}{4}, \quad (S_z = 0, \pm1),$$

which is the energy shift for the triplet state. For $l=0, S=0$, as

$$H_S|lm00\rangle = \frac{A}{2}\left(0 - \frac{3}{2}\right)\hbar^2|lm00\rangle,$$

the energy shifts is $-\frac{3}{4}A\hbar^2$.

b. The external field **B** is switched on, but as $H_B \ll H_S$ we can consider the effect of the external field as a perturbation $(H = (H_0 + H_s) + H_B)$. Taking $|lmSS_z\rangle$ as the state vector and the direction of **B** as the z direction, i.e., we have $\mathbf{B} = Be_z$,

$$H_B = -(\mu_p + \mu_e)\cdot\mathbf{B} = \frac{eB}{mc}(\hat{s}_{ez} - \hat{s}_{pz}),$$

and (cf. **Problem 5066** (b))

$$H_B|0000\rangle = \frac{eB\hbar}{mc}|0010\rangle, \quad H_B|0011\rangle = 0,$$

$$H_B|0010\rangle = \frac{eB\hbar}{mc}|0000\rangle, \quad H_B|001,-1\rangle = 0.$$

Hence, $\langle lmSS_z | H_B | lmSS_z \rangle = 0$ and the energy levels do not change further for first order perturbation, in addition to the splitting into singlet and triplet states described in (a).

We have

$$E_n = E_n^{(0)} + \sideset{}{'}\sum_{i \neq n} \frac{|\langle n | H_B | i \rangle|^2}{E_n^{(0)} - E_i^{(0)}}.$$

As only the following matrix elements are nonzero:

$$\langle 0010 | H_B | 0000 \rangle = \frac{eB\hbar}{mc}$$

$$\langle 0000 | H_B | 0010 \rangle = \frac{eB\hbar}{mc},$$

the energy levels from second order perturbation are, for $|0000\rangle$:

$$\begin{aligned} E_1 &= -\frac{3}{4}A\hbar^2 + \left(\frac{eB\hbar}{mc}\right)^2 \bigg/ \left(-\frac{3}{4}A\hbar^2 - \frac{1}{4}A\hbar^2\right) \\ &= -\frac{3}{4}A\hbar^2 - \frac{1}{A}\left(\frac{eB}{mc}\right)^2, \end{aligned}$$

for $|0011\rangle$ and $|001,-1\rangle$:

$$E_2 = E_4 = \frac{1}{4}A\hbar^2,$$

for $|0010\rangle$:

$$\begin{aligned} E_3 &= \frac{1}{4}A\hbar^2 + \left(\frac{eB\hbar}{mc}\right)^2 \bigg/ \left(\frac{1}{4}A\hbar^2 + \frac{3}{4}A\hbar^2\right) \\ &= \frac{1}{4}A\hbar^2 + \frac{1}{A}\left(\frac{eB}{mc}\right)^2. \end{aligned}$$

c. As now $H_B \gg H_S$, we can neglect the H_s term and consider only a perturbation consisting of $H_B = \dfrac{eB}{mc}(\hat{s}_{ez} - \hat{s}_{pz})$. It is then convenient to choose $|lms_{ez}s_{pz}\rangle$ as the eigenfunctions. Then for states $\left|lm, \pm\dfrac{1}{2}, \pm\dfrac{1}{2}\right\rangle$, we have

$$\begin{aligned} (\hat{s}_{ez} - \hat{s}_{pz})\left|\pm\frac{1}{2}, \pm\frac{1}{2}\right\rangle &= \hat{s}_{ez}\left|\pm\frac{1}{2}, \pm\frac{1}{2}\right\rangle - \hat{s}_{pz}\left|\pm\frac{1}{2}, \pm\frac{1}{2}\right\rangle \\ &= \left[\pm\frac{\hbar}{2} - \left(\pm\frac{\hbar}{2}\right)\right]\left|\pm\frac{1}{2}, \pm\frac{1}{2}\right\rangle = 0 \end{aligned}$$

and the corresponding energy shifts are zero. For states $lm, \pm\frac{1}{2}, \mp\frac{1}{2}\rangle$, we have

$$H_B\left|\pm\frac{1}{2},\mp\frac{1}{2}\right\rangle=\pm\frac{eB\hbar}{mc}\left|\pm\frac{1}{2},\mp\frac{1}{2}\right\rangle,$$

and hence the energy shifts $\pm\dfrac{eB\hbar}{mc}$.

d. In the general case, take $\left|lmSS_z\right\rangle$ as the state vector and $H'=H_S+H_B$ as the perturbation Hamiltonian. Then treat the problem using the perturbation method for degenerate states, i.e., solve the secular equation $\det\left|H'_{mn}-E\delta_{mn}\right|=0$ to find the energy corrections and the corresponding wave functions.

5060

An atom is in a state of total electron spin S, total orbital angular momentum L, and total angular momentum J. (The nuclear spin can be ignored for this problem). The z component of the atomic total angular momentum is Jz. By how much does the energy of this atomic state change if a weak magnetic field of strength B is applied in the z direction? Assume that the interaction with the field is small compared with the fine structure interaction.

The answer should be given as an explicit expression in terms of the quantum numbers J, L, S, Jz and natural constants.

(Princeton)

Sol: The Hamiltonian and eigenfunctions before the introduction of magnetic field are as follows:

$$\hat{H}=\hat{H}_0,$$
$$\psi_{nLJM_J}=R_{nLJ}(r_1,\ldots,r_n)\phi_{SLJM_J},$$

where the subscripts of r, 1, 2, …, n, represent the different electrons in the atom, and ϕ_{SLJM_J} is the common eigenstate of $(\mathbf{L}^2,\mathbf{S}^2,\mathbf{J}^2,J_z)$, i.e,

$$\phi_{SLJM_J}=\sum_{M_L}\left\langle LM_LS,M_J-M_L|JM_J\right\rangle Y_{LM_L}\Theta_{S,M_L-M_L},$$

$\left\langle LM_LS,M_J-M_L|JM_J\right\rangle$ being Clebsch-Gordan coefficients. The corresponding unperturbed energy is E_{nSLJ}.

After switching on the weak magnetic field, the Hamiltonian becomes

$$\hat{H} = \hat{H}_0 + \frac{eB}{2m_e c}\hat{j}_z + \frac{eB}{2m_e c}\hat{S}_z.$$

As B is very small, we can still consider $(\mathbf{L}^2, \mathbf{S}^2, \mathbf{J}^2, J_z)$ as conserved quantities and take the wave function of the system as approximately ψ_{nLJM_J}. The energy change caused by the term $\frac{eB}{2mc}J_z$ is $\Delta E_1 = M_J \hbar\frac{eB}{2mc}$ as J_z has eigen value $M_J\hbar$. The matrix of $\frac{eB}{2mc}S_z$ is diagonal in the subspace of the $2J+1$ state vectors for the energy E_{nLJ}, and hence the energy change caused by it is

$$\Delta E_2 = \frac{eB}{2mc}\langle \hat{S}_z \rangle = \frac{eB}{2mc}\langle JM_J | \hat{S}_z | JM_J \rangle,$$

where

$$\langle JM_J | \hat{S}_z | JM_J \rangle = \sum_{M_L} \hbar(M_J - M_L)\left[\langle LM_L S, M_J - M_L | JM_J \rangle\right]^2.$$

The total energy change is then

$$\Delta E = M_J \hbar\omega_l - \hbar w_l + \sum_{M_L}(M_J - M_L)\cdot\left[\langle LM_L S, M_J - M_L | JM_J \rangle\right]^2,$$

where

$$w_l = \frac{eB}{2mc}.$$

5061

The deuteron is a bound state of a proton and a neutron. The Hamiltonian in the center of mass system has the form

$$H = \frac{\mathbf{p}^2}{2\mu} + V_1(r) + \sigma_p\cdot\sigma_n V_2(r) + \left[\left(\sigma_p\cdot\frac{\mathbf{x}}{r}\right)\left(\sigma_n\cdot\frac{\mathbf{x}}{r}\right) - \frac{1}{3}(\sigma_p\cdot\sigma_n)\right]V_3(r).$$

Here $\mathbf{x} = \mathbf{x}_n - \mathbf{x}_p$, $r = |\mathbf{x}|$, σ_p and σ_n are the Pauli matrices for the spins of the proton and neutron, μ is the reduced mass, and $\mathbf{p}$ is conjugate to $\mathbf{x}$

a. The total angular momentum $\mathbf{J}^2 = J(J+1)$ and parity are good quantum numbers. Show that if $V_3 = 0$, the total orbital angular momentum $\mathbf{L}^2 = L(L+1)$, total spin $\mathbf{S}^2 = S(S+1)$ and $\mathbf{S} = \frac{1}{2}\sigma_p + \frac{1}{2}\sigma_n$ are good quantum numbers. Show that If $V_3 \neq 0$, S is still a good quantum number. (It may help to consider interchange of proton and neutron spins.)

b. The deuteron has $J=1$ and positive parity. What are the possible values of L? What is the value of S?

c. Assume that V_3 can be treated as a small perturbation. Show that in zeroth order (i.e. $V_3 = 0$) the wave function of the state with $J_z = +1$ is of the form $\psi_0(r)|\alpha,\alpha\rangle$ where $|\alpha,\alpha\rangle$ is the spin state with $s_{pz} = s_{nz} = +1/2$. What is the differential equation satisfied by $\psi_0(r)$?

d. What is the first order shift in energy due to the term in V_3? Suppose that to first order the wave function is

$$\psi_0(r)|\alpha,\alpha\rangle + \psi_1(x)|\alpha,\alpha\rangle + \psi_2(x)\big(|\alpha,\beta\rangle + |\beta,\alpha\rangle\big) + \psi_3(x)|\beta,\beta\rangle,$$

where $|\beta\rangle$ is a state with $S_z = -1/2$ and ψ_0 is as defined in part (c). By selecting out the part of the Schrödinger equation that is first order in V_3 and proportional to $|\alpha,\alpha\rangle$ find the differential equation statisfied by $\psi_1(x)$. Separate out the angular dependence of $\psi_1(x)$ and write down a differential equation for its radial dependence.

(MIT)

Sol:

a. Use units such that $\hbar = 1$. First consider the case where $V_3 = 0$. As

$$\mathbf{P}^2 = -\frac{1}{r}\frac{\partial^2}{\partial r^2}r - \frac{1}{r^2}\nabla_{\theta,\varphi}^2, \quad \mathbf{L}^2 = -\nabla_{\theta,\varphi}^2,$$

we have

$$[\mathbf{L}^2, \mathbf{P}^2] = 0,$$

$$[\mathbf{L}^2, V_1 + (\sigma_p \cdot \sigma_n)V_2] = [\mathbf{L}^2, V_1] + (\sigma_p \cdot \sigma_n)[\mathbf{L}^2, V_2]$$
$$= 0 + (\sigma_p \cdot \sigma_n) \cdot 0 = 0,$$

and so

$$[\mathbf{L}^2, H_{V_3=0}] = 0.$$

Thus $\mathbf{L}^2$ is a good quantum number. Now consider the total spin $\mathbf{S}^2$. As

$$\mathbf{s}_p \cdot \mathbf{s}_n = \frac{1}{2}(\mathbf{S}^2 - \mathbf{s}_p^2 - \mathbf{s}_n^2) = \frac{1}{2}\left(\mathbf{S}^2 - \frac{1}{2}\cdot\frac{3}{2} - \frac{1}{2}\cdot\frac{3}{2}\right) = \frac{1}{2}\left(\mathbf{S}^2 - \frac{3}{2}\right),$$

$$\mathbf{s}_p = \frac{1}{2}\sigma_p, \qquad \mathbf{s}_n = \frac{1}{2}\sigma_n,$$

we have

$$[\mathbf{S}^2, \sigma_p \cdot \sigma_n] = \frac{1}{2}[\sigma_p \cdot \sigma_n, \sigma_p \cdot \sigma_n] = 0,$$

Furthermore,

$$[\mathbf{S}^2, \mathbf{P}^2] = 0, \qquad [\mathbf{S}^2, V_1(r)] = 0.$$

Hence, $[\mathbf{S}^2, H_{V_3=0}] = 0$ and $\mathbf{S}^2$ is also a good quantum number.

If $V_3 \neq 0$, we have to consider also

$$\left(\sigma_p \cdot \frac{\mathbf{x}}{r}\right)\left(\sigma_n \cdot \frac{\mathbf{x}}{r}\right) = \frac{1}{2}\left\{\left[(\sigma_p + \sigma_n) \cdot \frac{\mathbf{x}}{r}\right]^2 - \left(\sigma_p \cdot \frac{\mathbf{x}}{r}\right)^2 - \left(\sigma_n \cdot \frac{\mathbf{x}}{r}\right)^2\right\}.$$

As $\sigma_p + \sigma_n = 2(\mathbf{s}_p + \mathbf{s}_n) = 2\mathbf{S}$ and

$$\left(\sigma_p \cdot \frac{\mathbf{x}}{r}\right)^2 = \left(\sigma_p \cdot \frac{\mathbf{x}}{r}\right)\left(\sigma_p \cdot \frac{\mathbf{x}}{r}\right) = \frac{\mathbf{x}}{r} \cdot \frac{\mathbf{x}}{r} = 1 = \left(\sigma_n \cdot \frac{\mathbf{x}}{r}\right)^2,$$

using the formula

$$(\sigma \cdot \mathbf{A})(\sigma \cdot \mathbf{B}) = \mathbf{A} \cdot \mathbf{B} + i\sigma \cdot (\mathbf{A} \times \mathbf{B}),$$

the above becomes

$$\left(\sigma_p \cdot \frac{\mathbf{x}}{r}\right)\left(\sigma_n \cdot \frac{\mathbf{x}}{r}\right) = 2\left(\mathbf{S} \cdot \frac{\mathbf{x}}{r}\right)^2 - 1.$$

Then as

$$\left[\mathbf{S}^2, \left(\mathbf{S} \cdot \frac{\mathbf{x}}{r}\right)\right] = \sum_i [\mathbf{S}^2, S_i]\left(\frac{x_i}{r}\right) = 0,$$

we have

$$\left[\mathbf{S}^2, \left(\sigma_p \cdot \frac{\mathbf{x}}{r}\right)\left(\sigma_n \cdot \frac{\mathbf{S}}{r}\right)\right] = 0,$$

and so

$$[\mathbf{S}^2, H] = 0, \qquad [S, H] = 0.$$

Hence, S is still a good quantum number if $V_3 \neq 0$.

b. The parity of the deuteron nucleus is

$$P = P(p) \cdot P(n)P_L = (+1) \cdot (+1) \cdot (-1)^L = (-1)^L.$$

Since the parity is positive, $L = $ even. Then as S can only be 0 or 1 and $J = 1$, we must have S equal to 1 and L equal to 0 or 2.

c. In zeroth order perturbation $V_3 = 0, L, S$ are good quantum numbers.

For the lowest state, $L=0$ and so $L_z=0$. Then for the state $J_z = L_z + S_z = 1$, $S_z = 1$ and hence $S=1$, $s_{pz} = +\frac{1}{2}$, $s_{nz} = +\frac{1}{2}$. Because $L=0$, the spatial wave function is spherically symmetric i.e., $\psi_0 = \psi_0(r)$. Thus the wave function of state $J_z = 1$ is $\psi_0(r)$, and

$$H\psi_0(r)|\alpha,\alpha\rangle = E\psi_0(r)|\alpha,\alpha\rangle.$$

As

$$\mathbf{r}_p \cdot \mathbf{r}_n = 4\mathbf{s}_p \cdot \mathbf{s}_n = 2S^2 - 2s_p^2 - 2s_n^2 = 2S^2 - 3,$$

and so

$$\mathbf{r}_p \cdot \mathbf{r}_n |\alpha,\alpha\rangle = (2S^2-3)|\alpha,\alpha\rangle = (2\times 2-3)|\alpha,\alpha\rangle = |\alpha,\alpha\rangle,$$

we have

$$H\psi_0(r)|\alpha,\alpha\rangle = \left[-\frac{\hbar^2}{2\mu}\frac{1}{r^2}\frac{\partial}{\partial r}\left(r^2\frac{\partial}{\partial r}\right) + V_1(r)\right]$$
$$\times \psi_0(r)|\alpha,\alpha\rangle + V_2(r)\psi_0(r)|\alpha,\alpha\rangle.$$

Hence the differential equation satisfied by ψ_0 is

$$\frac{1}{r^2}\frac{d}{dr}\left(r^2\frac{d\psi_0}{dr}\right) + \left\{\frac{2\mu}{\hbar^2}[E - V_1(r) - V_2(r)]\right\}\psi_0 = 0.$$

For $L\neq 0$, the wave functions of states with $J_z = 1$ do not have the above form.

d. In first order approximation, write the Hamiltonian of the system as

$$H = H_0 + H',$$

where

$$H' = \left[\left(\sigma_p \cdot \frac{\mathbf{x}}{r}\right)\left(\sigma_n \cdot \frac{\mathbf{x}}{r}\right) - \frac{1}{3}\sigma_p \cdot \sigma_n\right]V_3(r), \qquad H_0 = H_{V_3=0},$$

and the wave function as

$$\psi = \psi_0|\alpha,\alpha\rangle.$$

The energy correction is given by

$$\Delta E = \langle\psi|H'|\psi\rangle.$$

As

$$\sigma\cdot\frac{\mathbf{x}}{r} = \sigma_x\sin\theta\cos\varphi + \sigma_y\sin\theta\sin\varphi + \sigma_z\cos\theta = \begin{pmatrix} \cos\theta & \sin\theta e^{-i\varphi} \\ \sin\theta e^{i\varphi} & -\cos\theta \end{pmatrix},$$

$$\left\langle\alpha\left|\sigma\cdot\frac{\mathbf{x}}{r}\right|\alpha\right\rangle = (1 \quad 0)\begin{pmatrix} \cos\theta & \sin\theta e^{-i\varphi} \\ \sin\theta e^{i\varphi} & \cos\theta \end{pmatrix}\begin{pmatrix} 1 \\ 0 \end{pmatrix} = \cos\theta,$$

we have

$$\langle \alpha, \alpha | \left(\sigma_p \cdot \frac{\mathbf{x}}{r} \right) \left(\sigma_n \cdot \frac{\mathbf{x}}{r} \right) - \frac{1}{3} \sigma_p \cdot \sigma_n | \alpha, \alpha \rangle = \cos^2 \theta - \frac{1}{3},$$

and hence

$$\Delta E = \langle \psi | H' | \psi \rangle$$

$$= \int |\psi_0|^2 \, V_3(r) \left(\cos^2 \theta - \frac{1}{3} \right) dx$$

$$= \int_0^\infty V_3(r) |\psi_0|^2 \, r^2 \, dr \int_0^\pi \int_0^{2\pi} \left(\cos^2 \theta - \frac{1}{3} \right) \sin \theta \, d\theta \, d\varphi$$

$$= 0.$$

Note that **S** as is conserved and **L** is not, the wave function is a superposition of the spin-triplet states $(S = 1)$:

$$\psi(\mathbf{x}) = \psi_0(r) | \alpha, \alpha \rangle + \psi_1(\mathbf{x}) | \alpha, \alpha \rangle + \psi_2(\mathbf{x}) (| \alpha, \beta \rangle$$
$$+ | \beta, \alpha \rangle) + \psi_3(\mathbf{x}) | \beta, \beta \rangle,$$

and

$$H\psi = (H_0 + H')\psi = (E + E^{(1)} + \cdots)\psi.$$

Therefore, in first order approximation, making use of

$$E^{(1)} = \Delta E = 0$$

and

$$H_0 \psi_0 | \alpha, \alpha \rangle = E \psi_0 | \alpha, \alpha \rangle,$$

we obtain

$$H_0 \big[\psi_1 | \alpha, \alpha \rangle + \psi_2 (| \alpha, \beta \rangle + | \beta, \alpha \rangle) + \psi_3 | \beta, \beta \rangle \big] + H' \psi_0 | \alpha, \alpha \rangle$$

$$= E[\psi_1 | \alpha, \alpha \rangle + \psi_2 (| \alpha, \alpha \rangle + | \alpha, \alpha \rangle) + \psi_3 | \beta, \beta \rangle].$$

To calculate the perturbation term $H' \psi_0 | \alpha, \alpha \rangle$:

$$H' \psi_0 | \alpha, \alpha \rangle = V_3(r) \psi_0 \left[\left(\sigma_p \cdot \frac{\mathbf{x}}{r} \right) \left(\sigma_n \cdot \frac{\mathbf{x}}{r} \right) - \frac{1}{3} (\sigma_p \cdot \sigma_n) \right] | \alpha, \alpha \rangle$$

$$= V_3(r) \psi_0 \left[\begin{pmatrix} \cos \theta \\ \sin \theta e^{i\varphi} \end{pmatrix}_p \begin{pmatrix} \cos \theta \\ \sin \theta e^{i\varphi} \end{pmatrix}_n - \frac{1}{3} | \alpha \alpha \rangle \right].$$

As

$$\begin{pmatrix} \cos \theta \\ \sin \theta e^{i\varphi} \end{pmatrix} = \cos \theta \begin{pmatrix} 1 \\ 0 \end{pmatrix} + \sin \theta e^{i\varphi} \begin{pmatrix} 0 \\ 1 \end{pmatrix}$$

$$= \alpha \cos \theta + \beta \sin \theta e^{i\varphi},$$

$$H'\psi_0|\alpha,\alpha\rangle = V_3(r)\psi_0\left[\cos^2\theta - \frac{1}{3}\right]|\alpha,\alpha\rangle + V_3(r)\psi_0\sin^2\theta e^{i2\varphi}|\beta,\beta\rangle.$$

By considering the terms in the above equation that are proportional to $|\alpha,\alpha\rangle$, we can obtain the equation for the wave function $\psi_1(x)$:

$$-\frac{\hbar^2}{2\mu}\left(\frac{1}{r^2}\frac{\partial}{\partial r}\left(r^2\frac{\partial}{\partial r}\right) - \frac{\hat{L}^2}{\hbar^2 r^2}\right)\psi_1(x) + V_1(r)\psi_1$$

$$+V_2(r)\psi_1 + V_3(r)\psi_0(r)\left(\cos^2\theta - \frac{1}{3}\right) = E\psi_1.$$

Writing $\psi_1(x) = R_1(r)\Phi_1(\theta,\varphi)$, we can obtain from the above

$$\Phi_1(\theta,\varphi) = \cos^2\theta - \frac{1}{3} = \frac{1}{3}\sqrt{\frac{16\pi}{5}}Y_{20}(\theta,\varphi),$$

and thus $\hat{L}^2\Phi_1 = 2(2+1)\hbar^2\Phi_1$. The equation for R_1 is

$$\frac{1}{r^2}\frac{d}{dr}\left(r^2\frac{dR_1}{dr}\right) + \frac{2\mu}{\hbar^2}\left[E - V_1(r) - V_2(r) - \frac{6}{r^2}\right]R_1$$

$$= \frac{2\mu}{\hbar^2}V_3(r)\psi_0(r).$$

Here, it should be noted, even though the normalization factor of Φ_1 will affect the normalization factor of $R1$, their product will remain the same. It is noted also that $\psi_1(x)$ corresponds to $L=2$, $L_z=0$. By considering the term in $H'\psi_0|\alpha\alpha\rangle$ that is proportional to $|\beta,\beta\rangle$, we see that $\psi_3(x)$ corresponds to $L=2$, $L_z=2$. Then from the given $J_z=1$, we can see that $\psi_2(x)$ corresponds to $L=2$, $L_z=1$ (note that J_z is also conserved if $V_3\neq0$). In other words, the existence of V_3 requires the ground state of deuteron to be a combination of the $L=0$ and $L=2$ states, so that $J=1$, $S=1$, $J_z=1$ and parity = positive.

5062

Consider the bound states of a system of two non-identical, nonrelativistic, spin one-half particles interacting via a spin-independent central potential. Focus in particular on the $^3P_2$ and $^1P_1$ levels ($^3P_2$: spin-triplet, $L=1$, $J=2$; $^1P_1$: spin-singlet, $L=1$, $J=1$). A tensor force term $H' = \lambda[3\sigma(1)\cdot\hat{r}\sigma(2)\cdot\hat{r} - \sigma(1)\cdot\sigma(2)]$ is added to the Hamiltonian as a perturbation, where λ is a constant, $\hat{r}$ is a unit vector along the line joining the two particles, $\sigma(1)$ and $\sigma(2)$ are the Pauli spin operators for particles 1 and 2.

a. Using the fact that H' commutes with all components of the total angular momentum, show that the perturbed energies are independent of m, the eigenvalues of J_z.

b. The energy is most easily evaluated for the triplet state when the eigenvalue of J_z takes on its maximum value $m = j = 2$. Find the perturbation energy $\Delta E(^3P_2)$.

c. Find $\Delta E(^1P_1)$.

(Princeton)

ol:

a. Use units for which $\hbar = 1$. As $\mathbf{S} = \frac{1}{2}[\sigma(1) + \sigma(2)]$, the perturbation Hamiltonian H' can be written as

$$H' = \lambda \left\{ \left[\frac{3}{2}(2\mathbf{S}\cdot\hat{\mathbf{r}})^2 - (\sigma(1)\cdot\hat{\mathbf{r}})^2 - (\sigma(2)\cdot\hat{\mathbf{r}})^2 \right] \right.$$
$$\left. - \frac{1}{2}\left[4\mathbf{S}^2 - \sigma(1)^2 - \sigma(2)^2 \right] \right\}$$
$$= \lambda \left[6(\mathbf{S}\cdot\hat{\mathbf{r}})^2 - 2\mathbf{S}^2 \right],$$

where we have also used the relation

$$(\sigma\cdot\hat{\mathbf{r}})^2 = (\sigma_x + \sigma_y + \sigma_z)(\sigma_x + \sigma_y + \sigma_z)$$
$$= \sigma_x^2 + \sigma_y^2 + \sigma_z^2 = \sigma^2,$$

on account of

$$\sigma_i\sigma_j + \sigma_j\sigma_i = 2\delta_{ij}.$$

To prove that H' commutes with all components of the total angular momentum $\mathbf{J}$, we show for example $[J_z, H'] = 0$. As $[S_z, \mathbf{S}^2] = 0$, $[L_z, \mathbf{S}^2] = 0$, we have $[J_z, \mathbf{S}^2] = 0$. Also, as $[S_x, S_y] = i\hbar S_z$, etc, $L_z = -i\hbar\frac{\partial}{\partial\varphi}$, we have

$$[J_z, \mathbf{S}\cdot\hat{\mathbf{r}}] = [S_z, \mathbf{S}\cdot\hat{\mathbf{r}}] + [L_z, \mathbf{S}\cdot\hat{\mathbf{r}}]$$
$$= [S_z, \sin\theta\cos\varphi S_x + \sin\theta\sin\varphi S_y + \cos\theta S_z]$$
$$+ [L_z, \sin\theta\cos\varphi S_x + \sin\theta\sin\varphi S_y + \cos\theta S_z]$$
$$= i\hbar\sin\theta\cos\varphi S_y - i\hbar\sin\theta\sin\varphi S_x$$
$$- i\hbar\frac{\partial}{\partial\varphi}\left(\sin\theta\cos\varphi S_x + \sin\theta\sin\varphi S_y + \cos\theta S_z \right)$$
$$= 0.$$

and hence

$$[J_z,(\mathbf{S}\cdot\hat{\mathbf{r}})^2]=(\mathbf{S}\cdot\hat{\mathbf{r}})[J_z,\mathbf{S}\cdot\hat{\mathbf{r}}]+[J_z,\mathbf{S}\cdot\hat{\mathbf{r}}](\mathbf{S}\cdot\hat{\mathbf{r}})=0$$

Combining the above results, we have $[J_z,H']=0$.

Similarly we can show

$$[J_x,H']=[J_y,H']=0.$$

It follows that $J_+ = J_x + iJ_y$ also commutes with H'. $J+$ has the property

$$J_+|j,m\rangle=a|j,m+1\rangle.$$

Where a is a constant. Suppose there are two unperturbed states

$|j,m_1\rangle$ and $|j,m_2\rangle$, where $m_2 = m_1 + 1$,

which are degenerate and whose energies to first order perturbation are E_1 and E_2 respectively. Then

$$\langle j,m_2|[J_+,H_0+H']|j,m_1\rangle$$
$$=\langle j,m_2|J_+(H_0+H')|j,m_1\rangle-\langle j,m_2|(H_0+H')J_+|j,m_1\rangle$$
$$=(E_1-E_2)\langle j,m_2|J_+|J,m_1\rangle$$
$$=a(E_1-E_2)=0$$

as

$$(H_0+H')|j,m_1\rangle=E_1|j,m_1\rangle,$$
$$(H_0+H')J+|j,m_1\rangle=(H_0+H')a|j,m_2\rangle$$
$$=E_2a|j,m_2\rangle=E_2J_+|j,m_1\rangle.$$

Since the matrix element $a\neq0, E_1=E_2$, i.e., the perturbation energies are independent of m.

b. The perturbation energy is

$$\Delta E(^3P_2)=\langle j=2,m=2|H'|j=2,m=2\rangle.$$

Since

$$|j=2,m=2\rangle=|l=1,m_l=1\rangle|S=1,m_S=1\rangle,$$
$$\Delta E(^3P_2)=\int d\Omega Y_{11}^*(\theta,\varphi)\langle S=1,m_s=1|H'|S=1,m_S=1\rangle Y_{11}(\theta,\varphi)$$
$$=\frac{3}{8\pi}\lambda\int\sin^2\theta(3\cos^2\theta-1)2\pi\sin\theta d\theta$$
$$=\frac{3}{4}\lambda\int_0^\pi\sin^3(3\cos^2\theta-1)d\theta$$
$$=-\frac{2}{5}\lambda.$$

c. For the state $^1P_1$, as $S=0$, $m_s=0$ and so $H'=0$, we have $\Delta E(^1P_1)=0$.

5063

hydrogen atom is initially in its absolute ground state, the $F=0$ state of the hyperfine structure. (F is the sum of the proton spin I, the electron spin s and the orbital angular momentum L.) The $F=0$ state is split from the $F=1$ state by a small energy difference ΔE.

A weak, time-dependent magnetic field is applied. It is oriented in the z direction and has the form $B_z = 0, t < 0, B_z = B_0 \exp(-\gamma t), t > 0$. Here B_0 and γ are constants.

a. Calculate the probability that in the far future when the field dies away, the atom will be left in the $F=1$ state of the hyperfine structure.

b. Explain why, in solving this problem, you may neglect the interaction of the proton with the magnetic field.

(Princeton)

Sol:

a. When considering the hyperfine structure of the hydrogen atom, we write the Hamiltonian of the system as

$$H = H_0 + f(r)\sigma_p \cdot \sigma_e,$$

where H_0 is the Hamiltonian used for considering the fine structure of the hydrogen atom, $f(r)\sigma_p \cdot \sigma_e$ is the energy correction due to the hyperfine structure, σ_p and σ_e being the Pauli spin matrices for the proton and electron respectively.

When the atom is in its absolute ground state with $L=0, j = s = \dfrac{1}{2}$, the hyperfine structure states with $F=0$ and $F=1$ respectively correspond to spin parallel and spin antiparallel states of the proton and electron.

The initial wave function of the system is

$$\psi(r, F) = R_{10} Y_{00}(\theta, \varphi)\Theta_{00}.$$

Letting α, β represent the spinors $\begin{pmatrix} 1 \\ 0 \end{pmatrix}, \begin{pmatrix} 0 \\ 1 \end{pmatrix}$, we have the spin wave function

$$\Theta_{00} = \frac{1}{\sqrt{2}}(-\alpha_p \beta_e + \alpha_e \beta_p),$$

which makes ψ antisymmetric. When $t>0$, a weak magnetic field $B_z = B_0 e^{-\gamma t}$ acts on the system and the Hamiltonian becomes

$$\hat{H} = H_0 + f(r)\sigma_p \cdot \sigma_e + \frac{\hbar e B_z}{2\mu c}\hat{\sigma}_{ez},$$

neglecting any interaction between the magnetic field and the proton.

Suppose the wave function of the system at $t>0$ is

$$\psi(r, F, t) = R_{10}(r)Y_{00}[C_1(t)\Theta_{00} + C_2(t)\Theta_{11} + C_3(t)\Theta_{10} + C_4(t)\Theta_{1,-1}],$$

where

$$\Theta_{11} = \alpha_p \alpha_e,$$

$$\Theta_{1,-1} = \beta_p \beta_e,$$

$$\Theta_{1,0} = \frac{1}{\sqrt{2}}(\alpha_p \beta_e + \alpha_e \beta_p).$$

Then the probability that the system is at hyperfine structure state $F=1$ at time t is

$$A(t) = 1 - |C_1(t)|^2,$$

and the initial conditions are

$$C_1(0) = 1, \; C_2(0) = C_3(0) = C_4(0) = 0.$$

As

$$\sigma_x \alpha = \beta, \qquad \sigma_x \beta = \alpha,$$
$$\sigma_y \alpha = i\beta, \qquad \sigma_y \beta = -i\alpha,$$
$$\sigma_z \alpha = \alpha, \qquad \sigma_z \beta = -\beta,$$

we have

$$\mathrm{r}_p \cdot \mathrm{r}_e \Theta_{00} = (\sigma_{px}\sigma_{ex} + \sigma_{py}\sigma_{ey} + \sigma_{pz}\sigma_{ez})\frac{1}{\sqrt{2}}(-\alpha_p \beta_e + \alpha_e \beta_p)$$

$$= -3\Theta_{00},$$

and similarly

$$\mathrm{r}_p \cdot \sigma_e \Theta_{1m} = \Theta_{1m}.$$

Finally, from the Schrödinger equation $i\hbar\frac{\partial}{\partial t}\psi = \hat{H}\psi$ we obtain

$$i\hbar R_{10}(r)Y_{00}\left(\frac{dC_1}{dt}\Theta_{00} + \frac{dC_2}{dt}\Theta_{11} + \frac{dC_3}{dt}\Theta_{10} + \frac{dC_4}{dt}\Theta_{1,-1}\right)$$

$$= R_{10}(r)Y_{00}\{[E_{10} - 3f(a_0)]C_1(t)\Theta_{00}$$

$$+ [E_{10} + f(a_0)][C_2(t)\Theta_{11} + C_3(t)\Theta_{10} + C_4(t)\Theta_{1,-1}]\}$$

$$+ R_{10}(r)Y_{00}\mu_0 B_z[C_3(t)\Theta_{00} + C_2(t)\Theta_{11}$$

$$+ C_1(t)\Theta_{10} - C_4(t)\Theta_{1,-1}].$$

Comparing the coefficient of Θ_{00} and of Θ_{10} on the two sides of the equation, we get

$$\begin{cases} i\hbar\dfrac{d}{dt}C_1(t)=[E_{10}-3f(a_0)]C_1(t)+C_3(t)\mu_0 B_0 e^{-\gamma t}, \\[2ex] i\hbar\dfrac{d}{dt}C_3(t)=[E_{10}+f(a_0)]C_3(t)+C_1(t)\mu_0 B_0 e^{-\gamma t}. \end{cases}$$

As the energy of the hyperfine structure, $f(a_0)$, is much smaller than E_{10}, energy of the fine structure, it can be neglected when calculating $C_1(t)$. Then the above equations give

$$C_1(t)=\frac{1}{2}e^{-i\frac{E_{10}}{\hbar}t}\left\{ e^{\frac{i\mu_0 B_0}{\gamma\hbar}\left(1-e^{-\gamma t}\right)} + e^{-\frac{i\mu_0 B_0}{\gamma\hbar}\left(1-e^{-\gamma t}\right)} \right\}.$$

As $t\to\infty$,

$$|C_1(t)|^2 \to \cos^2\left(\frac{\mu_0 B_0}{\gamma\hbar}\right).$$

Hence

$$A(t)|_{t\to\infty}=1-\cos^2\left(\frac{\mu_0 B_0}{\gamma\hbar}\right)=\sin^2\left(\frac{\mu_0 B_0}{\gamma\hbar}\right),$$

which is the probability that the hydrogen atom stays in state $F=1$.

b.   The interaction between the magnetic moment of the proton and the magnetic field can be neglected because the magnetic moment of the proton is only $\dfrac{1}{1840}$ of the magnetic moment of the electron.

5064

A spinless nonrelativistic particle in a central field is prepared in an s-state, which is degenerate in energy with a p-level ($m_\ell=0,\pm1$). At time $t=0$ an electric field $\mathbf{E}=E_0\sin\omega t\,\hat{\mathbf{z}}$ is turned on. Ignoring the possibility of transitions to other than the above-mentioned states but making no further approximations, calculate the occupation probability for each of the four states at time t, in terms of the nonzero matrix elements of $\hat{\mathbf{z}}$.

(Wisconsin)

Sol: Choose the four given states as the state vectors and the level of energy such that the degenerate energy $E=0$. The Hamiltonian of the system is

$$\hat{H}=\hat{H}_0+\hat{H}',\ \hat{H}'=-g\mathbf{E}\cdot\mathbf{r}=-gE_0 z\sin\omega t.$$

To find the perturbation energy matrix one notes that only elements $\langle l+1, m_l | z | l, m_l \rangle$ are non-vanishing. Thus we have

$$
\begin{array}{cccc}
|00\rangle & |10\rangle & |1,-1\rangle & |11\rangle
\end{array}
$$

$$
\hat{H}' = \begin{array}{c} \langle 00| \\ \langle 10| \\ \langle 1,-1| \\ \langle 11| \end{array}
\begin{pmatrix}
0 & 1 & 0 & 0 \\
1 & 0 & 0 & 0 \\
0 & 0 & 0 & 0 \\
0 & 0 & 0 & 0
\end{pmatrix}
(-gE_0 \langle 00|z|10\rangle \sin\omega t).
$$

Suppose the wave function is $\psi = (x_1, x_2, x_3, x_4)$, and initially $\psi(t=0) = (1, 0, 0, 0)$, where x_1, x_2, x_3, x_4 are the four state vectors. The Schrödinger equation $i\hbar \frac{\partial}{\partial t}\psi = \hat{H}\psi$ can then be written as

$$
i\hbar \frac{d}{dt}\begin{pmatrix} x_1 \\ x_2 \end{pmatrix} = -\lambda \sin\omega t \begin{pmatrix} 0 & 1 \\ 1 & 0 \end{pmatrix}\begin{pmatrix} x_1 \\ x_2 \end{pmatrix}, \tag{1}
$$

$$
i\hbar \frac{d}{dt}\begin{pmatrix} x_3 \\ x_4 \end{pmatrix} = 0, \tag{2}
$$

where $\lambda = gE_0 \langle 00|z|10\rangle$ is a real number.

Equation (2) and the initial condition give

$$
\begin{pmatrix} x_3 \\ x_4 \end{pmatrix} = \begin{pmatrix} x_3 \\ x_4 \end{pmatrix}_{t=0} = \begin{pmatrix} 0 \\ 0 \end{pmatrix},
$$

which means that the probability that the system occupies the states $m_l = \pm 1$ of the p-level is zero. To solve Eq. (1), we first diagonalize the matrix by solving the secular equation

$$
\begin{vmatrix} \lambda & 1 \\ 1 & \lambda \end{vmatrix} = 0,
$$

which gives $\lambda = \pm 1$. Hence, the first two components of ψ are to be transformed to

$$
\psi = \frac{1}{\sqrt{2}}\begin{pmatrix} x_1 + x_2 \\ x_1 - x_2 \end{pmatrix} \equiv \begin{pmatrix} a \\ b \end{pmatrix}.
$$

Then Eq. (1) becomes

$$
i\hbar \frac{d}{dt}\begin{pmatrix} a \\ b \end{pmatrix} = -\lambda \sin\omega t \begin{pmatrix} 1 & 0 \\ 0 & -1 \end{pmatrix}\begin{pmatrix} a \\ b \end{pmatrix},
$$

subject to the initial condition

$$\binom{a}{b}_{t=0}=\frac{1}{\sqrt{2}}\binom{1}{1}.$$

Solving the equation we find

$$\binom{a}{b}=\frac{1}{\sqrt{2}}\begin{pmatrix}\exp\left\{\dfrac{i\lambda}{\hbar\omega}(1-\cos\omega t)\right\}\\[2mm]\exp\left\{-\dfrac{i\lambda}{\hbar\omega}(1-\cos\omega t)\right\}\end{pmatrix}.$$

To get back to the original state vectors,

$$\psi=\binom{x_1}{x_2}=\frac{1}{\sqrt{2}}\binom{a+b}{a-b}$$

$$=\begin{pmatrix}\cos\left[\dfrac{\lambda}{\hbar\omega}(1-\cos\omega t)\right]\\[3mm]i\sin\left[\dfrac{\lambda}{\hbar\omega}(1-\cos\omega t)\right]\end{pmatrix}.$$

Therefore, the occupation probability for each of the four states at time t is

$$P_s(t)=|x_1|^2=\cos^2\left[\frac{gE_0\langle 00|z|10\rangle}{\hbar\omega}(1-\cos\omega t)\right]$$

$$P_{p(m_l=0)}(t)=|x_2|^2=\sin^2\left[\frac{gE_0\langle 00|z|10\rangle}{\hbar\omega}(1-\cos\omega t)\right]$$

$$P_{p(m_l=\pm 1)}(t)=0,$$

where $\langle 00|z|10\rangle$ is to be calculated using wave functions $R_{nl}Y_{lm}(\theta,\varphi)$ for a particle in a central field and is a real number.

5065

An ion of a certain atom has $L=1$ and $S=0$ when it is in free space. The ion is implanted in a crystalline solid (at $x=y=z=0$) and sees a local environment of 4 point charges as shown in the Fig. 5.16. One can show by applying the Wigner-Eckart theorem (DON'T TRY) that the effective perturbation to the Hamiltonian caused by this environment can be written as

$$H_1=\frac{\alpha}{\hbar^2}(L_x^2-L_y^2),$$

Where Lx and Ly are the x and y components of the orbital angular momentum operator, and α is a constant. In addition, a magnetic field is applied in the z direction and causes a further perturbation $H_z = \dfrac{\beta B}{\hbar} L_z$, where Lz is the z component of the angular momentum operator and β is a constant.

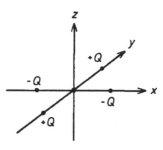

Fig. 5.16

a. Express the perturbation Hamiltionian $H' = H_1 + H_2$ in terms of L_+ and L_-, the "raising" and "lowering" operators for orbital angular momentum.

b. Find the matrix of the perturbation Hamiltonian in the basis set using the three states $|1,0\rangle, |1,1\rangle$ and $|1,-1\rangle$.

c. Find the energy levels of the ion as a function of B. Make a careful sketch of your result.

d. When $B = 0$, what are the eigenfunctions describing the ion?

(MIT)

Sol:

a. From the definitions $L_\pm = L_x \pm iL_y$, we get

$$[L_+, L_-] = -2i[L_x, L_y] = 2\hbar L_z,$$

$$L_x^2 - L_y^2 = \frac{1}{4}(L_+ + L_-)^2 + \frac{1}{4}(L_+ - L_-)^2 = \frac{1}{2}(L_+^2 + L_-^2).$$

Thus the perturbation Hamiltonian is

$$H' = \frac{\alpha}{\hbar^2}(L_x^2 - L_y^2) + \frac{\beta B}{\hbar} L_z$$

$$= \frac{\alpha}{2\hbar^2}(L_+^2 + L_-^2) + \frac{\beta B}{2\hbar^2}(L_+L_- - L_-L_+).$$

b. Using the formula

$$L_{\pm}|L, M_L\rangle = \hbar\sqrt{L(L+1)- M_L(M_L \pm 1)}\,|L, M_L \pm 1\rangle,$$

we find the following non-vanishing elements

$$L_-^2|1,1\rangle = \sqrt{2}\hbar L_-|1,0\rangle$$
$$= 2\hbar^2|1,-1\rangle,$$
$$L_+^2|1,-1\rangle = 2\hbar^2|1,1\rangle,$$
$$L_+L_-|1,1\rangle = 2\hbar^2|1,1\rangle,$$
$$L_+L_-|1,0\rangle = 2\hbar^2|1,0\rangle,$$
$$L_-L_+|1,0\rangle = 2\hbar^2|1,0\rangle,$$
$$L_-L_+|1,-1\rangle = 2\hbar^2|1,-1\rangle$$

and hence the matrix $\left(\langle L', M_L'|H'|L, M_L\rangle\right)$ as follows:

$$
\begin{array}{cc}
 & \begin{array}{ccc} |1,1\rangle & |1,0\rangle & |1,-1\rangle \end{array} \\
H' = \begin{array}{c} \langle 1,1| \\ \langle 1,0| \\ \langle 1,-1| \end{array} & \begin{pmatrix} \beta B & 0 & \alpha \\ 0 & 0 & 0 \\ \alpha & 0 & -\beta B \end{pmatrix}.
\end{array}
$$

c. The perturbation energy E is determined by the matrix equation

$$
\begin{pmatrix} \beta B - E & 0 & \alpha \\ 0 & -E & 0 \\ \alpha & 0 & -\beta B - E \end{pmatrix} \begin{pmatrix} a \\ b \\ c \end{pmatrix} = 0,
$$

whose secular equation

$$
\begin{vmatrix} \beta B - E & 0 & \alpha \\ 0 & -E & 0 \\ \alpha & 0 & -\beta B - E \end{vmatrix}
$$

gives the corrections

$$E_1 = -\sqrt{(\beta B)^2 + \alpha^2},$$
$$E_2 = 0,$$
$$E_3 = \sqrt{(\beta B)^2 + \alpha^2}.$$

The perturbation energy levels are shown in Fig. 5.17 as functions of B, where the dashed lines are the asymptotes.

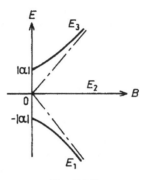

Fig. 5.17

d. If $B = 0$, the energy levels are

$$E_1 = -\alpha, \quad E_2 = 0, \quad E_3 = \alpha,$$

and the eigenstates are given, respectively, by

$$a = -c \neq 0, \quad b = 0;$$
$$a = c = 0, \quad b \neq 0;$$
$$a = c \neq 0, \quad b = 0.$$

Thus the corresponding energy eigenstates are

$$\begin{pmatrix} a \\ b \\ c \end{pmatrix}_{(1)} = \frac{1}{\sqrt{2}} \begin{pmatrix} -1 \\ 0 \\ 1 \end{pmatrix}, \quad \begin{pmatrix} a \\ b \\ c \end{pmatrix}_{(2)} = \begin{pmatrix} 0 \\ 1 \\ 0 \end{pmatrix},$$

$$\begin{pmatrix} a \\ b \\ c \end{pmatrix}_{(3)} = \frac{1}{\sqrt{2}} \begin{pmatrix} 1 \\ 0 \\ 1 \end{pmatrix}.$$

In terms of the state vectors $|1,1\rangle$, $|1,0\rangle$, $|1,-1\rangle$ the wave functions are

$$|E_1 = -\alpha\rangle = -\frac{1}{\sqrt{2}}|1,1\rangle + \frac{1}{\sqrt{2}}|1,-1\rangle,$$
$$|E_2 = 0\rangle = |1,0\rangle,$$
$$|E_3 = \alpha\rangle = \frac{1}{\sqrt{2}}|1,1\rangle + \frac{1}{\sqrt{2}}|1,-1\rangle.$$

5066

he spin-dependent part of the effective Hamiltonian for a positronium
bound state of electron and positron) in a magnetic field B may be written as

$$H_{\text{spin}} = A\sigma_e \cdot \sigma_p + \mu_B B(\sigma_{ez} - \sigma_{pz}),$$

where σ_e and σ_p are the Pauli spin matrices for the electron and the positron,
nd μ_B is the Bohr magneton.

a. At zero magnetic field the singlet state lies 8×10^{-4} eV below the triplet state. What is the value of A?

b. Illustrate by a sketch the dependence of the energy of each of the four
spin states on the magnetic field B, including both the weak and strong
field cases.

c. If the positronium atom is in its lowest energy state in a strong magnetic field and the field is instantaneously switched off, what is the
probability of finding the atom in the singlet state?

d. How would the result of (c) be changed if the field is switched off very
slowly?

(Wisconsin)

ol:

a. When $B=0$, the effective Hamiltonian has the expectation value

$$\left\langle H_{\text{spin}} \right\rangle = A\left\langle F'm'_F \middle| \sigma_e \cdot \sigma_p \middle| Fm_F \right\rangle,$$

where $\mathbf{F} = \mathbf{s}_e + \mathbf{s}_p$.

As

$$\sigma_e \cdot \sigma_p = \frac{2}{\hbar^2}(\mathbf{F}^2 - \mathbf{s}_e^2 - \mathbf{s}_p^2),$$

$$E = \left\langle H_{\text{spin}} \right\rangle = 2A[F(F+1) - s_e(s_e+1) - s_p(s_p+1)] \cdot \delta_{F'F} \cdot \delta_{m'_F m_F}.$$

For the triplet state, $F=1$, so $E_{F=1} = 2A(1\cdot2 - \frac{1}{2}\cdot\frac{3}{2} - \frac{1}{2}\cdot\frac{3}{2}) = A$.

For the singlet state, $F=0$, so $E_{F=0} = -3A$.

Hence, $E_{F=1} - E_{F=0} = 4A = 8\times10^{-4}$ eV, giving $A = 2\times10^{-4}$ eV.

b. We first transform from representation in coupling state vectors to that in non-coupling state vectors:

$$\left|F=1, m_F=1\right\rangle = \left|s_e=\frac{1}{2}, m_{se}=\frac{1}{2}, s_p=\frac{1}{2}, m_{sp}=-\frac{1}{2}\right\rangle,$$

$$\left|F=1, m_F=1\right\rangle = \frac{1}{\sqrt{2}}\left\{\left|s_e=\frac{1}{2}, m_{se}=\frac{1}{2}, s_p=\frac{1}{2},\right.\right.$$
$$\left.\left. m_{sp}=-\frac{1}{2}\right\rangle + \left|s_e=\frac{1}{2}, m_{se}=-\frac{1}{2}, s_p=\frac{1}{2}, m_{sp}=-\frac{1}{2}\right\rangle\right\}.$$

$$\left|F=1, m_F=-1\right\rangle = \left|s_e=\frac{1}{2}, m_{se}=\frac{1}{2}, s_p=\frac{1}{2}, m_{sp}=-\frac{1}{2}\right\rangle,$$

$$\left|F=0, m_F=0\right\rangle = \frac{1}{\sqrt{2}}\left\{\left|s_e=\frac{1}{2}, m_{se}=\frac{1}{2}, s_p=\frac{1}{2},\right.\right.$$
$$\left.\left. m_{sp}=-\frac{1}{2}\right\rangle - \left|s_e=\frac{1}{2}, m_{se}=-\frac{1}{2}, s_p=\frac{1}{2}, m_{sp}=\frac{1}{2}\right\rangle\right\}.$$

Then as $\mu_B B(\sigma_{ez}-\sigma_{pz})=\frac{2\mu_B B}{\hbar}(s_{ez}-s_{pz})$ and

$$(s_{ez}-s_{pz})\left|F=0, m_F=0\right\rangle = \frac{1}{\sqrt{2}}(s_{ez}-s_{pz})\left(\left|\frac{1}{2},\frac{1}{2},\frac{1}{2},-\frac{1}{2}\right\rangle\right.$$
$$\left. -\left|\frac{1}{2},-\frac{1}{2},\frac{1}{2},\frac{1}{2}\right\rangle\right)$$
$$= \frac{\hbar}{\sqrt{2}}\left[\left(\frac{1}{2}+\frac{1}{2}\right)\left|\frac{1}{2},\frac{1}{2},\frac{1}{2},-\frac{1}{2}\right\rangle\right.$$
$$\left. -\left(-\frac{1}{2}-\frac{1}{2}\right)\left|\frac{1}{2},-\frac{1}{2},\frac{1}{2},\frac{1}{2}\right\rangle\right]$$
$$= \hbar\left|F=1,\ m_F=0\right\rangle,$$

$$(s_{ez}-s_{pz})\left|F=1, m_F=0\right\rangle = \hbar\left|F=0, m_F=0\right\rangle,$$

$$(s_{ez}-s_{pz})\left|F=1, m_F=1\right\rangle = (s_{ez}-s_{pz})\left|F=1, m_F=-1\right\rangle = 0,$$

and using the results of (a), we have in the order of $|1,1\rangle$, $|1,-1\rangle$, $|1,0\rangle$, $|0,0\rangle$,

$$H_{spin} = \begin{pmatrix} A & 0 & 0 & 0 \\ 0 & A & 0 & 0 \\ 0 & 0 & A & 2\mu_B B \\ 0 & 0 & 2\mu_B B & -3A \end{pmatrix}.$$

The secular equation

$$\begin{vmatrix} A-E & 0 & 0 & 0 \\ 0 & A-E & 0 & 0 \\ 0 & 0 & A-E & 2\mu_B B \\ 0 & 0 & 2\mu_B B & -3A-E \end{vmatrix} = 0$$

then gives the spin-energy eigenvalues

$$E_1 = E_2 = A,$$

$$E_3 = -A + 2\sqrt{A^2 + \mu_B^2 B^2},$$

$$E_4 = -A - 2\sqrt{A^2 + \mu_B^2 B^2}.$$

The variation of E with B is shown in Fig. 5.18. If the magnetic field B is weak we can consider the term $\mu_B B(\sigma_{ez} - \sigma_{pz})$ as perturbation. The energy correction is zero in first order perturbation and is proportional to B^2 in second order perturbation. When the magnetic field B is very strong, the energy correction is linear in B.

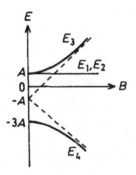

Fig. 5.18

c. When the positronium is placed in a strong magnetic field $(\mu_B B \gg A)$, the lowest energy state, i.e., the eigenstate whose energy is approximately $-A - 2\mu_B B$, is

$$\left| m_{se} = -\frac{1}{2},\ m_{sp} = \frac{1}{2} \right\rangle = \frac{1}{\sqrt{2}}\{|1,0\rangle - |0,0\rangle\},$$

where we have considered $A\boldsymbol{\sigma}_e \cdot \boldsymbol{\sigma}_p$ as perturbation. If the magnetic field is switched off suddenly, the probability that the atom is in state $|F = 0,\ m_F = 0\rangle$ is

$$P = \left| \left\langle 0,0 \middle| \frac{1}{2}, -\frac{1}{2}, \frac{1}{2}, \frac{1}{2} \right\rangle \right|^2 = \left| \frac{1}{\sqrt{2}}\langle 0,0|1,0\rangle - \frac{1}{\sqrt{2}}\langle 0,0|0,0\rangle \right|^2 = \frac{1}{2}.$$

d. If the magnetic field is switched off very slowly, no transition occurs and the atom will remain in the state $|\frac{1}{2},-\frac{1}{2},\frac{1}{2}\frac{1}{2}\rangle$, and the energy of the system is $E=-A$.

5067

Positronium consists of an electron and a positron bound by their Coulomb attraction.

a. What is the radius of the ground state? The binding energy of the ground state?

b. The singlet and triplet ground states are split by their spin-spin interaction such that the singlet state lies about 10^{-3} volts below the triplet state. Explain the behavior of positronium in a magnetic field. Draw an energy level diagram to illustrate any dependence on the magnetic field.

(Berkeley)

Sol:

a. The hydrogen atom has ground state radius and binding energy
$$a_0 = \hbar^2/\mu e^2 \approx 0.53 \text{ Å},$$
$$E_1 = \mu e^4/2\hbar^2 \approx 13.6 \text{ eV},$$

where $\mu = m_e m_p/(m_e+m_p)$, the Coulomb potential $V(r)=-e^2/|\mathbf{r}_1-\mathbf{r}_2|$ having been used.

The results may be applied to any hydrogen-like atom of nuclear charge Ze with the replacements
$$V(r)\rightarrow V'(r)=-Ze^2/|\mathbf{r}_1-\mathbf{r}_2|,$$
$$\mu\rightarrow\mu'=m_1m_2/(m_1+m_2),$$

m_1,m_2 being the mass of the nucleus and that of the orbiting electron. For positronium, $\mu'=\frac{m_e}{2}$, $Z=1$, and so
$$a_0'=\hbar^2/\mu'e^2=a_0\mu/\mu'=2a_0\approx1 \text{ Å},$$
$$E_1'=\mu'e^4/2\hbar^2=\mu'E_1/\mu=\frac{1}{2}E_1\approx6.8 \text{ eV}.$$

b. Choose $|0, 0, S, S_z\rangle$ as the eigenstate and take as the perturbation Hamiltonian

$$H' = A\mathbf{s}_e \cdot \mathbf{s}_p + \frac{eB}{mc}(s_{ez} - s_{pz}),$$

where $A = \frac{10^{-3}\text{eV}}{\hbar^2}$. Using the results of **Problem 5066** we have the perturbation energy matrix

$$H'_{mn} = \begin{pmatrix} A\hbar^2/4 & 0 & 0 & 0 \\ 0 & A\hbar^2/4 & 0 & 0 \\ 0 & 0 & A\hbar^2/4 & -\hbar\dfrac{eB}{mc} \\ 0 & 0 & -\hbar\dfrac{eB}{mc} & -\dfrac{3}{4}A\hbar^2 \end{pmatrix} \begin{matrix} (1,1) \\ (1,-1) \\ (1,0) \\ (0,0) \end{matrix},$$

$$\qquad\qquad (1,1) \quad (1,-1) \quad (1,0) \quad (0,0)$$

from which we find the perturbation energies

$$E_1 = E_2 = A\hbar^2/4,$$

$$E_3 = -A\hbar^2/4 + \sqrt{\left(\frac{A\hbar^2}{2}\right)^2 + \hbar^2\left(\frac{eB}{mc}\right)^2},$$

$$E_4 = -A\hbar^2/4 - \sqrt{\left(\frac{A\hbar^2}{2}\right)^2 + \hbar^2\left(\frac{eB}{mc}\right)^2}.$$

The dependence of the energy levels on B is shown in Fig. 5.19.

The result shows that the energy levels of the hyperfine structure is further split in the presence of a weak magnetic field, whereas the hyperfine structure is destroyed in the presence of a strong magnetic field. These limiting situations have been discussed in **Problem 5059**.

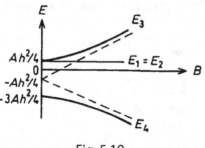

Fig. 5.19

5068

Estimate the magnetic susceptibility of a He atom in its ground state. Is it para-magnetic or diamagnetic?

(Chicago)

Sol: Suppose the He atom is in an external uniform magnetic field $\mathbf{H} = H\hat{e}_z$ (H being a constant). The vector potential $\mathbf{A}$, defined by $\mathbf{H} = \nabla \times \mathbf{A}$, can be taken to be $\mathbf{A} = \frac{1}{2}H(-y\hat{e}_x + x\hat{e}_y)$. The Hamiltonian of the system is then

$$\hat{H} = \sum_{i=1}^{2} \frac{1}{2m}\left(\mathbf{P}_i - \frac{e}{c}\mathbf{A}\right)^2 - \mathbf{1}_J \cdot \mathbf{H}.$$

As the helium atom is in the ground state, $\mathbf{1}_J = \mathbf{C}$. Thus

$$\hat{H} = \sum_{i=1}^{2} \frac{1}{2m}\left(\mathbf{P}_i - \frac{e}{c}\mathbf{A}\right)^2.$$

Where m and e are respectively the mass and charge of an electron.

The magnetic susceptibility is given by $\chi = -4\pi \frac{\partial^2 E}{\partial H^2}\big|_{H=0}$. For helium atom in its ground state,

$$\hat{H} = \sum_{i=1}^{2} \frac{1}{2m}\left(\mathbf{P}_i^2 - \frac{e}{c}\mathbf{A}\cdot\mathbf{P}_i - \frac{e}{c}\mathbf{P}_i\cdot\mathbf{A} + \frac{e^2}{c^2}\mathbf{A}^2\right),$$

$$\frac{\partial^2 \hat{H}}{\partial H^2} = \sum_{i=1}^{2} \frac{e^2}{4mc^2}(x^2 + y^2) = \frac{e^2}{2mc^2}(x^2 + y^2),$$

and so

$$\chi = -4\pi \frac{\partial^2 E}{\partial H^2}\bigg|_{H=0}$$

$$= -4\pi \frac{\partial^2}{\partial H^2}\left\langle He\ \text{ground state}\,\middle|\,\hat{H}\,\middle|\,He\ \text{ground state}\right\rangle$$

$$= -4\pi \frac{e^2}{2mc^2}\left\langle He\ \text{ground state}\,\middle|\,x^2 + y^2\,\middle|\,He\ \text{ground state}\right\rangle$$

$$= -4\pi \frac{e^2}{2mc^2}\left(\overline{x^2} + \overline{y^2}\right).$$

As $x^2 + y^2 + z^2 = r^2$, we can take

$$\overline{x^2} = \overline{y^2} \approx \frac{1}{3}r^2_{Heg.s.},$$

where $r^2_{Heg.s.}$ is the mean-square distance of each electron from the He nucleus in the ground state, which gives

$$\chi \approx -\frac{4\pi}{c^2} \frac{e^2 r^2 H_{eg.s}}{3m}$$

in Gaussian units. Note that as $\chi < 0$, helium atom in its ground state is diamagnetic.

5069

An atom with no permanent magnetic moment is said to be diamagnetic. In this problem the object is to calculate the induced diamagnetic moment for a hydrogen atom in a weak magnetic field **B** ignoring the spins of the electron and proton.

a. Write down the nonrelativistic Hamiltonian for a particle of mass m and charge q in a combined vector and scalar electromagnetic field.

b. Using the fact that a satisfactory vector potential **A** for a uniform magnetic field is $\mathbf{A} = -\frac{1}{2}\mathbf{r} \times \mathbf{B}$, write down the steady-state Schrödinger equation for the hydrogen atom. Assume the proton to be infinitely massive so center of mass motion can be ignored.

c. Treat the terms arising from the magnetic field as a perturbation and calculate the shift in the ground state energy due to the existence of the magnetic field.

d. Calculate the induced diamagnetic moment per atom.

(MIT)

Sol:

a. A particle of mass m and charge q in an electromagnetic field of potentials $(\phi, \mathbf{A})$ has Hamiltonian

$$H = \frac{1}{2m}\left(\mathbf{p} - \frac{q}{c}\mathbf{A}\right)^2 + q\phi.$$

b. If the motion of the proton is neglected, the Schrödinger equation for a hydrogen atom in a uniform magnetic field **B** is

$$\left[\frac{1}{2m_e}(\mathbf{p} + \frac{e}{2c}\mathbf{B}\times\mathbf{r})^2 - \frac{e^2}{r}\right]\psi(\mathbf{r}) = E\psi(\mathbf{r}),$$

where $q = -e$ for an electron, or

$$\left\{\frac{1}{2m_e}\left[\mathbf{p}^2 + \frac{e}{2c}\mathbf{p}\cdot\mathbf{B}\times\mathbf{r} + \frac{e}{2c}\mathbf{B}\times\mathbf{r}\cdot\mathbf{p} + \frac{e^2}{4c^2}(\mathbf{B}\times\mathbf{r})^2\right] - \frac{e^2}{r}\right\}\psi(\mathbf{r})$$
$$= E\psi(\mathbf{r}).$$

As

$$\mathbf{p}\cdot\mathbf{B}\times\mathbf{r} - \mathbf{B}\times\mathbf{r}\cdot\mathbf{p} = -i\hbar\nabla\cdot(\mathbf{B}\times\mathbf{r})$$
$$= -i\hbar(\nabla\times\mathbf{B})\cdot\mathbf{r} + i\hbar\mathbf{B}\cdot\nabla\times\mathbf{r} = 0$$

because $\mathbf{B}$ is uniform and $\nabla\times\mathbf{r} = 0$, we have

$$\mathbf{p}\cdot\mathbf{B}\times\mathbf{r} = \mathbf{B}\times\mathbf{r}\cdot\mathbf{p} = \mathbf{B}\cdot\mathbf{r}\times\mathbf{p} = \mathbf{B}\cdot\mathbf{L}.$$

Taking the direction of the uniform magnetic field $\mathbf{B}$ as the z direction, we have $\mathbf{B} = B\hat{e}_z$, $\mathbf{B}\cdot\mathbf{L} = BL_z = i\hbar B\dfrac{\partial}{\partial\varphi}$ and

$$(\mathbf{B}\times\mathbf{r})^2 = (-By\hat{e}_x + Bx\hat{e}_y)^2 = B^2(x^2 + y^2) = B^2 r^2\sin^2\theta$$

in spherical coordinates.

The Schrödinger equation can then be written as

$$\left(-\frac{\hbar^2}{2m_e}\nabla^2 - \frac{e^2}{r} - \frac{ieB\hbar}{2m_e c}\frac{\partial}{\partial\varphi} + \frac{e^2 B^2}{8m_e c^2}r^2\sin^2\theta\right)\psi(r,\theta,\varphi)$$
$$= E\psi(r,\theta,\varphi)$$

in spherical coordinates.

c.   Treat

$$H' = \frac{eB}{2m_e c}L_z + \frac{e^2 B^2}{8m_e c^2}r^2\sin^2\theta$$

as a perturbation. The energy shift of the ground state is

$$\Delta E = \langle 100|H'|100\rangle$$

with

$$|100\rangle = R_{10}(r)Y_{00}(\theta,\varphi) = \sqrt{\frac{1}{\pi a^3}}e^{-\frac{r}{a}},$$

where $a = \dfrac{\hbar^2}{m_e e^2}$. Thus

$$\Delta E = \left[2\pi\int_0^\pi \sin\theta\, d\theta\int_0^\infty r^2\, dr\frac{1}{\pi}\left(\frac{1}{a}\right)^3 e^{-2r/a}r^2\sin^2\theta\right]\cdot\frac{e^2 B^2}{8m_e c^2}$$
$$= \frac{1}{a^3}\int_0^\infty r^4 e^{-2r/a}\, dr\cdot\frac{e^2 B^2}{3m_e c^2} = \frac{a^2 e^2 B^2}{4m_e c^2}.$$

Note that for the ground state, $l=0$, $m_l=0$, and the first term of H' makes no contribution to ΔE.

d. The above energy shift is equal to the energy of a magnetic dipole in a magnetic field **B**,

$$\Delta E = -\mu \cdot \mathbf{B},$$

if the dipole moment is

$$\mu = -\frac{e^2 a^2}{4m_e c^2}\mathbf{B}.$$

This can be considered as the dipole moment induced by the field. The atom is diamgnetic as μ is antiparallel to **B**.

5070

The magnetic polarizability of an atom is defined by $\alpha_H = -\left.\dfrac{\partial^2 E(H)}{\partial H^2}\right|_{H=0}$, where $E(H)$ is the energy of the atom in a constant external magnetic field.

a. Estimate the magnetic polarizability of the $F=0$, 1s hyperfine ground state of a hydrogen atom.

b. Estimate the magnetic polarizability of the ground state $1s^2$ of a helium.

(CUSPEA)

Sol:

a. If the magnetic field **H** is very weak, the perturbation Hamiltonian is $H' = -\mu \cdot \mathbf{H}$.

Taking the direction of **H** as that of the z-axis and letting the spins of the electron and proton be S and I respectively we have

$$\mu \cdot \mathbf{H} = \hbar^{-1}(g_e \mu_B S_z + g_p \mu_p I_z)H$$

$$= \hbar^{-1}\left[\frac{1}{2}(g_e \mu_B + g_p \mu_p)(S_z + I_z) + \frac{1}{2}(g_e \mu_B - g_p \mu_p)(S_z - I_z)\right]H.$$

The first order perturbation makes no contribution to α_H as (cf. **Problem 5066**). $\langle F=0, m_F=0|S_z \pm I_z|F=0, m_F=0\rangle = 0$. We then consider the energy correction of second order perturbation for the ground state $F=0$, 1s:

$$E^{(2)}(H) = \sum_{m=-1}^{1}\frac{|\langle F=1|-\mu\cdot\mathbf{H}|F=0\rangle|^2}{E_{F=0} - E_{F=1}}$$

where m is the quantum number of the projection on z axis of **F**.

$$E^{(2)}(H) = \sum_{m=-1}^{1} \left| \langle F=1, m | -\boldsymbol{\mu} \cdot \mathbf{H} | F=0,0 \rangle \right|^2 / \left(E_{F=0} - E_{F=1} \right)$$

$$= H^2 \hbar^{-2} \sum_{m=-1}^{1} \left| \langle F=1, m | \frac{1}{2} \left(g_e \mu_B - g_p \mu_p \right) \left(S_z - I_z \right) | F=0,0 \rangle \right|^2$$

$$/ \left(E_{F=0} - E_{F=1} \right)$$

as $\left(S_z + I_z \right) | 00 \rangle = 0$. Then as $\left(S_z - I_z \right) | 0,0 \rangle = \hbar | 1,0 \rangle$, the matrix elements are all zero except for $m = 0$. Thus

$$E^{(2)}(H) = \left| \langle F=1, m=0 | \frac{1}{2} (g_e \mu_B - g_p \mu_p)(S_z - I_z) \right.$$

$$\left| F=0,0 \rangle \right|^2 H^2 \hbar^{-2} / \left(E_{F=0} - E_{F=1} \right).$$

As $\mu_p \ll \mu_B, g_e = 1$, and the spectral line $F=1 \to F=0$ has frequency 140 MHz, corresponding to $E_{F=1} - E_{F=0} = 0.58 \times 10^{-6}$ eV,

$$\alpha(H) = \frac{\mu_B^2}{2(E_{F=1} - E_{F=0})} = \left(5.8 \times 10^{-9} \, \text{eV/Gs} \right)^2 / \left(2 \times 5.8 \times 10^{-7} \, \text{eV} \right)$$

$$= 2.9 \times 10^{-11} \, \text{eV/Gs}^2 .$$

b. Consider a helium atom in a uniform magnetic field $\mathbf{H}$. The vector potential is $\mathbf{A} = \frac{1}{2} \mathbf{H} \times \mathbf{r}$ and it contributes $e^2 A^2 / 2mc^2$ per electron to the perturbation Hamiltonian that gives rise to the magnetic polarizabihty $\alpha(H)$ (**Problem 5068**). If the helium atom is in the ground state $1 s^2$, then $\mathbf{L} = \mathbf{S} = \mathbf{J} = 0$. Taking the direction of $\mathbf{H}$ as the z direction, we have

$$H' = 2 \cdot \frac{e^2 A^2}{2mc^2} = \frac{e^2 H^2}{4mc^2} (x^2 + y^2),$$

the factor of 2 being added to account for the two electrons of helium atom. The energy correction is thus

$$E(H') = \left\langle \psi_0 \left| \frac{e^2 H^2}{4mc^2} (x^2 + y^2) \right| \psi_0 \right\rangle$$

$$= \frac{e^2 H^2}{4mc^2} \overline{x^2 + y^2} = \frac{e^2 H^2}{4mc^2} \cdot \frac{2}{3} r_0^2 ,$$

where r_0 is the root-mean-square radius of helium atom in the ground state, as $r_0^2 = \overline{x^2} + \overline{y^2} + \overline{z^2}$ and $\overline{x^2} = \overline{y^2} = \overline{z^2}$. Since $r_0 = \frac{\hbar^2}{zme^2} = \frac{a_0}{2}$, a_0 being the Bohr radius of hydrogen atom,

$$\alpha(H) = -\frac{\partial^2 E}{\partial H^2}\bigg|_{H=0} = -\frac{e^2}{2mc^2} \cdot \frac{2}{3}\left(\frac{a_0}{2}\right)^2 = -\frac{1}{6}\left(\frac{e\hbar}{2mc}\right)^2 \frac{2a_0}{e^2}$$

$$= -\frac{\mu_B^2}{6E_I},$$

where $\mu_B = \dfrac{e\hbar}{2mc}$ is the Bohr magneton, $E_I = \dfrac{e^2}{2a_0}$ is the ionization potential of hydrogen. Thus

$$\alpha(H) = -\frac{(0.6\times10^{-8})^2}{6\times13.6} = -4.4\times10^{-19} \text{ eV/Gs}^2.$$

5071

A particle of mass m moves in a three-dimensional harmonic oscillator well. The Hamiltonian is

$$H = \frac{\mathbf{p}^2}{2m} + \frac{1}{2}kr^2.$$

a. Find the energy and orbital angular momentum of the ground state and the first three excited states.

b. If eight identical non-interacting (spin$-\frac{1}{2}$) particles are placed in such a harmonic potential, find the ground state energy for the eight-particle system.

c. Assume that these particles have magnetic moment of magnitude μ. If a magnetic field B is applied, what is the approximate ground state energy of the eight-particle system as a function of B. Plot the magnetization $\left(-\frac{\partial E}{\partial B}\right)$ for the ground state as a function of B.

(Columbia)

Sol:

a. A three-dimensional harmonic oscillator has energy levels

$$E_N = \left(N + \frac{3}{2}\right)\hbar\omega,$$

where $\omega = \sqrt{\frac{k}{m}}$,

$$N = 2n_r + l, \ N = 0,1,2,\ldots, n_r = 0,1,2,\ldots,$$
$$l = N - 2n_r.$$

For the ground state, $N=0$ and the energy is $E_0 = \frac{3}{2}\hbar\omega$, the orbit angular momentum is $L=0$.

For the first exited state, $N=1$, $E_1 = \frac{5}{2}\hbar\omega$, $L=\hbar$. As $l=1$ the level contains three degenerate states.

b. For spin $-\frac{1}{2}$ particles, two can fill up a state. Thus when fully filled. The ground state contains two particles and the first three excited states contain six particles. Thus the ground state energy of the eight-particle system is
$$E_0 = 2\times\frac{3}{2}\hbar\omega + 6\times\frac{5}{2}\hbar\omega = 18\hbar\omega.$$

c. The Hamiltonian of the system is
$$\hat{H} = \sum_{i=1}^{8}\left(\frac{\mathbf{p}_i^2}{2m}+\frac{1}{2}kr_i^2\right) - \sum_{i=1}^{8}\boldsymbol{\mu}_i\cdot\mathbf{B} - \sum_{i=1}^{8}\frac{e}{2mc}\mathbf{L}_i\cdot\mathbf{B}$$
$$+\sum_{i=1}^{8}\frac{e^2}{2mc^2}A^2(\mathbf{r}_i)+\sum_{i=1}^{8}\frac{1}{2m^2c^2r_i^2}\frac{dV(r_i)}{dr_i}\mathbf{L}_i\cdot\mathbf{s}_i,$$

where $V(r_i)=\frac{1}{2}kr_i^2$, $\mathbf{A}$ is the vector potential $\frac{1}{2}\mathbf{B}\times\mathbf{r}$ giving rise to $\mathbf{B}$.

As the eight particles occupy two shells, all the shells are full and we have $S=0, L=0, j=0$.

The wave functions of the system are the products of the following functions (excluding the radial parts):

$$\begin{cases}
Y_{00}(\mathbf{e}_1)Y_{00}(\mathbf{e}_2)\frac{1}{\sqrt{2}}\{\alpha(1)\beta(2)-\alpha(2)\beta(1)\}, \\[2mm]
Y_{11}(\mathbf{e}_3)Y_{11}(\mathbf{e}_4)\frac{1}{\sqrt{2}}\{\alpha(3)\beta(4)-\alpha(4)\beta(3)\}, \\[2mm]
Y_{10}(\mathbf{e}_5)Y_{10}(\mathbf{e}_6)\frac{1}{\sqrt{2}}\{\alpha(5)\beta(6)-\alpha(6)\beta(5)\}, \\[2mm]
Y_{1-1}(\mathbf{e}_7)Y_{1-1}(\mathbf{e}_8)\frac{1}{\sqrt{2}}\{\alpha(7)\beta(8)-\alpha(8)\beta(7)\},
\end{cases}$$

where $\mathbf{e}_i = \mathbf{r}_i/r_i$. Note that the two space sub-wave functions in each are same. Then as the total space wave function is symmetric, the total spin wave function must be antisymmetric. As
$$\sigma_x\alpha=\beta, \sigma_y\alpha=i\beta, \qquad \sigma_z\alpha=\alpha,$$
$$\sigma_x\beta=\alpha, \sigma_y\beta=-i\alpha, \qquad \sigma_z\beta=-\beta,$$

we have

$$\sigma_{1x}\frac{1}{\sqrt{2}}\{\alpha(1)\beta(2)-\alpha(2)\beta(1)\}=\frac{1}{\sqrt{2}}\{\beta(1)\beta(2)-\alpha(2)\alpha(1)\},$$

$$\sigma_{2x}\frac{1}{\sqrt{2}}\{\alpha(1)\beta(2)-\alpha(2)\beta(1)\}=-\frac{1}{\sqrt{2}}\{\beta(1)\beta(2)-\alpha(2)\alpha(1)\},$$

$$\sigma_{1y}\frac{1}{\sqrt{2}}\{\alpha(1)\beta(2)-\alpha(2)\beta(1)\}=\frac{i}{\sqrt{2}}\{\beta(1)\beta(2)+\alpha(2)\alpha(1)\},$$

$$\sigma_{2y}\frac{1}{\sqrt{2}}\{\alpha(1)\beta(2)-\alpha(2)\beta(1)\}=-\frac{i}{\sqrt{2}}\{\beta(1)\beta(2)+\alpha(2)\alpha(1)\},$$

$$\sigma_{1z}\frac{1}{\sqrt{2}}\{\alpha(1)\beta(2)-\alpha(2)\beta(1)\}=\frac{1}{\sqrt{2}}\{\alpha(1)\beta(2)+\alpha(2)\beta(1)\},$$

$$\sigma_{2z}\frac{1}{\sqrt{2}}\{\alpha(1)\beta(2)-\alpha(2)\beta(1)\}=-\frac{1}{\sqrt{2}}\{\alpha(1)\beta(2)+\alpha(2)\beta(1)\}.$$

Their inner products with the bra $\frac{1}{\sqrt{2}}\{\alpha(1)\beta(2)-\alpha(2)\beta(1)\}^{+}$ will result in $\langle\sigma_{1x}\rangle,\langle\sigma_{2x}\rangle,\langle\sigma_{1y}\rangle,\langle\sigma_{2y}\rangle$, as well as $\langle\sigma_{1z}+\sigma_{2z}\rangle$ being zero. Hence

$$\langle\sigma_{1z}L_{1z}+\sigma_{2z}L_{2z}\rangle=-i\hbar\left\langle\frac{\partial}{\partial\varphi_{1}}\cdot\sigma_{1z}+\frac{\partial}{\partial\varphi_{2}}\cdot\sigma_{2z}\right\rangle$$

$$=-i\hbar\left[\left\langle\frac{\partial}{\partial\varphi_{1}}\right\rangle\langle\sigma_{1z}\rangle+\left\langle\frac{\partial}{\partial\varphi_{2}}\right\rangle\langle\sigma_{2z}\rangle\right]$$

$$=-i\hbar\left\langle\frac{\partial}{\partial\varphi}\right\rangle\langle\sigma_{1z}+\sigma_{2z}\rangle=0.$$

Thus the ground state energy is

$$E=18\hbar\omega+\frac{e^{2}}{8mc^{2}}\sum_{i=1}^{8}\langle(\mathbf{B}\times\mathbf{r}_{i})^{2}\rangle$$

$$=18\hbar\omega+e^{2}B^{2}\Big/8mc^{2}\sum_{i=1}^{8}\langle r_{i}^{2}\sin^{2}\theta_{i}\rangle,$$

and the magnetization is

$$-\frac{\partial E}{\partial B}=\frac{-e^{2}B}{4mc^{2}}\sum_{i=1}^{8}\langle r_{i}^{2}\sin^{2}\theta_{i}\rangle=\chi B,$$

giving $\chi=-\frac{e^{2}}{4mc^{2}}\sum_{i=1}^{8}\langle r_{i}^{2}\sin^{2}\theta_{i}\rangle$ as the diamagnetic susceptibility. $\frac{\partial E}{\partial B}$ as a function of B is shown in Fig. 5.20.

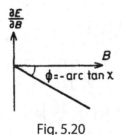

Fig. 5.20

5072

Suppose one has an electron in an S state of a hydrogen atom which is in a magnetic field along the z direction of strength Hz. At time $t=0$ a magnetic field along the x direction is turned on; its strength increases uniformly from zero to H_x at time $t=T(H_x \ll H_z)$ and then remains constant after $t=T$. Assume the nucleus has no spin and that

$$\left(\frac{H_x}{H_z}\right)^2 \ll \frac{H_x}{H_z}.$$

Consider only the interaction between the electron spin and the magnetic field. Neglect all terms of order $\left(\frac{H_x}{H_z}\right)^2$ or higher. If the electron has its spin in the z direction at $t=0$, find the state of the electron when $t=T$. Show that the state is an eigenstate of the Hamiltonian due to the combined magnetic fields $H=(H_x,0,H_z)$ provided that T is sufficiently long. Explain what sufficiently long means in this case.

(Berkeley)

Sol: Treat the potential energy $H'=-\dfrac{t}{T}\mu \cdot H_x\hat{e}_x = -\dfrac{t(-e)}{Tmc}\mathbf{s}\cdot\hat{e}_x H_x = \dfrac{teH_x}{Tmc}s_x$ as a perturbation. Before H_x is turned on, the electron in the S state of a hydrogen has two spin states, $\left|\dfrac{1}{2}\right\rangle$ with energy $E_+ = -\dfrac{(-e)}{mc}\cdot\dfrac{1}{2}\hbar H_z = \dfrac{e\hbar}{2mc}H_z$, and $\left|-\dfrac{1}{2}\right\rangle$ with energy $E_- = -\dfrac{e\hbar}{2mc}H_z$.

Use time-dependent perturbation theory, taking the wave function as

$$\psi(t)=e^{-iE_+t/\hbar}\left|\frac{1}{2}\right\rangle+a_-e^{-iE_-t/\hbar}\left|-\frac{1}{2}T\right\rangle,$$

where a_- is given by

$$
\begin{aligned}
a_- &= \frac{1}{i\hbar}\int_0^T\left\langle-\frac{1}{2}\left|\hat{H}'\right|\frac{1}{2}\right\rangle e^{-i(E_+-E_-)t/\hbar}\,dt\\
&= \frac{1}{i\hbar}\int_0^T\frac{eH_x}{mc}\frac{t}{T}\left\langle-\frac{1}{2}\left|s_z\right|\frac{1}{2}\right\rangle\exp\left(-i\frac{eH_z}{mc}t\right)dt\\
&= \frac{eH_x}{2imcT}\int_0^T t\exp\left(-i\frac{eH_z}{mc}t\right)dt\\
&= \frac{1}{2}\left(\frac{H_x}{H_z}\right)\exp\left(-i\frac{eH_z}{mc}T\right)-\frac{imcH_x}{2eTH_z^2}\left[\exp\left(-i\frac{eH_z}{mc}T\right)-1\right],
\end{aligned}
$$

where we have used $s_x |\frac{1}{2}\rangle = \frac{\hbar}{2}|-\frac{1}{2}\rangle$.

Thus the spin state of the electron at time T is

$$\psi(T) = \exp\left(-i\frac{eH_z}{2mc}T\right)\left|\frac{1}{2}\right\rangle + \left\{\frac{1}{2}\left(\frac{H_x}{H_z}\right)\exp\left(-i\frac{eH_z}{mc}T\right)\right.$$

$$\left.-\frac{imcH_x}{2eTH_z^2}\left[\exp\left(-i\frac{eH_z}{mc}T\right)-1\right]\right\}$$

$$\times \exp\left(i\frac{eH_z}{2mc}T\right)\left|-\frac{1}{2}\right\rangle.$$

If the time T is sufficiently long so that $\frac{mc}{eT} \leq H_x$, we can neglected the second term of a_- and obtain

$$\psi(T) = \exp\left(-i\frac{eH_z}{2mc}T\right)\left(\left|\frac{1}{2}\right\rangle + \frac{1}{2}\frac{H_x}{H_z}\left|-\frac{1}{2}\right\rangle\right).$$

The Hamiltonian due to the combined magnetic field $\mathbf{H} = (H_x, 0, H_z)$ is

$$\hat{H} = -\boldsymbol{\mu} \cdot \mathbf{B} = \frac{eH_x}{mc}s_x + \frac{eH_z}{mc}s_z.$$

Let $\alpha = \frac{e}{mc}$ and consider $H\psi(T)$. As $s_x\left|\pm\frac{1}{2}\right\rangle = \frac{\hbar}{2}\left|\mp\frac{1}{2}\right\rangle$, $s_z\left|\pm\frac{1}{2}\right\rangle = \pm\frac{\hbar}{2}\left|\pm\frac{1}{2}\right\rangle$ we have for $T \to \infty$,

$$\hat{H}\psi(T) = \alpha\exp\left(-\frac{i\alpha H_z T}{2}\right)(H_x s_x + H_z s_z)\left(\left|\frac{1}{2}\right\rangle + \frac{1}{2}\frac{H_x}{H_z}\left|-\frac{1}{2}\right\rangle\right)$$

$$= \frac{\alpha\hbar}{2}\exp\left(-\frac{i\alpha H_z T}{2}\right)\left(H_x\left|-\frac{1}{2}\right\rangle + \frac{1}{2}\frac{H_x^2}{H_z}\left|\frac{1}{2}\right\rangle\right.$$

$$\left.+H_z\left|\frac{1}{2}\right\rangle - \frac{1}{2}H_x\left|-\frac{1}{2}\right\rangle\right)$$

$$= \frac{\alpha\hbar H_z}{2}\exp\left(\frac{-i\alpha H_z T}{2}\right)\left\{\left[1 + \frac{1}{2}\left(\frac{H_x}{H_z}\right)^2\right]\left|\frac{1}{2}\right\rangle\right.$$

$$\left.+\frac{1}{2}\frac{H_x}{H_z}\left|-\frac{1}{2}\right\rangle\right\}$$

$$\approx \frac{\alpha\hbar H_z}{2}\exp\left(-\frac{i\alpha H_z T}{2}\right)\left(\left|\frac{1}{2}\right\rangle + \frac{1}{2}\frac{H_x}{H_z}\left|-\frac{1}{2}\right\rangle\right) = \frac{e\hbar H_z}{2mc}\psi(T).$$

The result shows that when T is sufficiently large $\psi(T)$ is an eigenstate of the Hamiltonian due to the combined magnetic field with eigenvalue $\frac{e\hbar H_z}{2mc}$.

5073

An electron is in the $n=1$ eigenstate of a one-dimensional infinite square-well potential which extends from $x=-a/2$ to $x=a/2$. At $t=0$ a uniform electric field E is applied in the x direction. It is left on for a time τ and then removed. Use time-dependent perturbation theory to calculate the probabilities P_2 and P_3 that the electron will be, respectively, in the $n=2$ and $n=3$ eigenstates at $t>\tau$. Assume that τ is short in the sense that $\tau \ll \dfrac{\hbar}{E_1-E_2}$, where E_n is the energy of the eigenstate n. Specify any requirements on the parameters of the problem necessary for the validity of the approximations made in the application of time-dependent perturbation theory.

(Columbia)

Sol: The electron in the $n=1$ eigenstate of the potential well

$$V = \begin{cases} 0, & |x| \le a/2, \\ \infty, & \text{otherwise} \end{cases}$$

has wave functions and corresponding energies

$$\psi_n(x) = \sqrt{\frac{2}{a}} \sin\left[\frac{\pi n}{a}\left(\frac{a}{2}+x\right)\right],$$

$$E_n = \hbar^2 \pi^2 n^2 / 2ma^2, \quad n=1, 2,\ldots$$

The uniform electric field $E\hat{e}_x$ has potential $\phi = -\int E dx = -E_x$. The potential energy of the electron (charge $-e$) due to E, $H' = eEx$, is considered as a perturbation. We have

$$H'_{n_2 n_1} = \langle n_2 | H' | n_1 \rangle$$

$$= \frac{2}{a}\int_{-\frac{a}{2}}^{\frac{a}{2}} \sin\left[\frac{n_1\pi}{a}\left(x+\frac{a}{2}\right)\right]\sin\left[\frac{n_2\pi}{a}\left(x+\frac{a}{2}\right)\right] eEx\, dx$$

$$= \frac{eE}{a}\int_{-\frac{a}{2}}^{\frac{a}{2}}\left\{\cos\left[\frac{(n_1-n_2)}{a}\pi\left(x+\frac{a}{2}\right)\right]\right.$$

$$\left.- \cos\left[\frac{(n_1+n_2)}{a}\pi\left(x+\frac{a}{2}\right)\right]\right\}x\, dx$$

$$= \frac{eE}{a}\left\{\frac{a^2}{(n_1-n_2)^2\pi^2}\left[(-1)^{n_1-n_2}-1\right]\right.$$

$$-\frac{a^2}{(n_1+n_2)^2\,\pi^2}[(-1)^{n_1+n_2}-1]\bigg\}$$

$$=\frac{4eEa}{\pi^2}\frac{n_1 n_2}{(n_1^2-n_2^2)^2}[(-1)^{n_1+n_2}-1],$$

$$\omega_{n_2 n_1}=\frac{1}{\hbar}(E_{n_2}-E_{n_1})=\frac{\hbar\pi^2}{2ma^2}(n_2^2-n_1^2),$$

$$C_{k'k}(t)=\frac{1}{i\hbar}\int_0^\tau H'_{k'k}e^{i\omega_{k'k}t}\,dt=\frac{1}{\hbar}H'_{k'k}(1-e^{i\omega_{k'k}\tau})\frac{1}{\omega_{k'k}}.$$

For the transition $1\to 2$,

$$H'_{21}=\langle 2|H'|1\rangle=-\frac{16eEa}{9\pi^2},\qquad \omega_{21}=3\hbar\pi^2/2ma^2,$$

and so the probability of finding the electron in the $n=2$ state at $t>\tau$ is

$$P_2=|C_{21}(t)|^2=\frac{1}{\hbar^2\omega_{21}^2}H'^2_{21}(1-e^{i\omega 21\tau})(1-e^{-i\omega 21\tau})$$

$$=\left(\frac{16a^2}{9\pi^2}\right)^3\left[\frac{eEm}{\hbar^2\pi}\sin\left(\frac{3\hbar\pi^2}{4ma^2}\tau\right)\right]^2\approx\left(\frac{16}{9\pi^2}\frac{eEa}{\hbar}\tau\right)^2$$

for $\tau\ll\dfrac{\hbar}{E_1-E_2}$.

For the transition $1\to 3$,

$$H'_{31}=\langle 3|H'|1\rangle=0,$$

and so

$$P_3=|\langle C_{31}(t)|^2=0.$$

The validity of the time-dependent perturbation theory requires the time τ during which the perturbation acts should be small. The perturbation potential itself should also be small.

5074

For a particle of mass m in a one-dimensional box of length l, the eigenfunctions and energies are

$$\psi_n(x)=\sqrt{\frac{2}{l}}\sin\frac{n\pi x}{l},\quad 0\le x\le l,$$

$$E_n=\frac{1}{2m}\left(\frac{n\pi\hbar}{l}\right)^2\quad n=\pm 1,\ \pm 2,\dots$$

Suppose the particle is originally in a state $|n\rangle$ and the box length is increased to a length of $2l(0 \le x \le 2l)$ in a time $t \ll h/E_n$. Afterwards what is the probability that the particle will be found in an energy eigenstate with energy E_n?

(MIT)

Sol: First consider the process in which the box length is increased from l to $2l$. As $t \ll \dfrac{h}{E_n}$, it is reasonable to assume that the state of the particle in the box is unable to respond to the change during such a short time. Therefore the wave function of the particle after the change is completed is

$$\psi(x) = \begin{cases} \sqrt{\dfrac{2}{l}} \sin \dfrac{n\pi x}{l}, & 0 \le x \le l, \\[2mm] 0, & l \le x \le 2l. \end{cases}$$

On the other hand, the eigenstates and eigenvalues of the same particle in the one-dimensional box of length $2l$ would be, respectively,

$$\phi_{n'}(x) = \sqrt{\dfrac{1}{l}} \sin \dfrac{n'\pi x}{2l}, \quad (0 \le x \le 2l),$$

$$E_{n'} = \dfrac{1}{2m} \left(\dfrac{n'\pi\hbar}{2l} \right)^2, \quad (n' = \pm1, \pm2, \ldots).$$

The energy E_n of the particle corresponds to the energy level $E_{n'}$ in the $2l$ box, where $\dfrac{n}{l} = \dfrac{n'}{2l}$, i.e., $n' = 2n$. The corresponding eigenstate is then ϕ_{2n}. Thus the probability amplitude is

$$A = \int_{-\infty}^{\infty} \phi_{2n}(x)\psi(x)dx = \dfrac{\sqrt{2}}{t} \int_0^l \sin^2 \dfrac{n\pi x}{l} dx = \dfrac{1}{\sqrt{2}},$$

and the probability of finding the particle in an eigenstate with energy E_n is

$$P = |A|^2 = \dfrac{1}{2}.$$

5075

A particle is initially in its ground state in a box with infinite walls at 0 and L. The wall of the box at $x = L$ is suddenly moved to $x = 2L$.

a. Calculate the probability that the particle will be found in the ground state of the expanded box.

b. Find the state of the expanded box most likely to be occupied by the particle.

c. Suppose the walls of the original box $[0, L]$ are suddenly dissolved and that the particle was in the ground state. Construct the probability distribution for the momentum of the freed particle.

(Berkeley)

ol:

a. The wave function of the particle before the box expands is

$$\psi(x) = \begin{cases} \sqrt{\dfrac{2}{L}}\sin\dfrac{\pi x}{L}, & x \in [0, L], \\ 0 & \text{otherwise.} \end{cases}$$

The wave function for the ground state of the system after the box has expanded is

$$\phi_1(x) = \begin{cases} \sqrt{\dfrac{1}{L}}\sin\dfrac{\pi x}{2L}, & x \in [0, 2L], \\ 0 & \text{otherwise.} \end{cases}$$

The probability required is then

$$P_1 = \left| \int_{-\infty}^{\infty} \phi_1^*(x)\psi(x)dx \right|^2 = \left| \frac{\sqrt{2}}{L}\int_0^L \sin\frac{\pi x}{2L}\sin\frac{\pi x}{L}dx \right|^2 = \frac{32}{9\pi^2}.$$

b. The probability that the particle is found in the first exited state of the expanded box is

$$P_2 = \left| \int_{-\infty}^{\infty} \phi_2^*(x)\psi(x)dx \right|^2 = \left| \frac{\sqrt{2}}{L}\int_0^L \sin^2\frac{\pi x}{L}dx \right|^2 = \frac{1}{2},$$

where

$$\phi_2(x) = \begin{cases} \sqrt{\dfrac{1}{L}}\sin\dfrac{\pi x}{L}, & x \in [0, 2L], \\ 0 & \text{otherwise.} \end{cases}$$

For the particle to be found in a state $n \geq 3$, the probability is

$$P_n = \left| \frac{\sqrt{2}}{L}\int_0^L \sin\frac{n\pi x}{2L}\sin\frac{\pi x}{L}dx \right|^2$$

$$= \frac{2}{\pi^2}\left| \frac{\sin\left(\dfrac{n}{2}-1\right)\pi}{(n-2)} - \frac{\sin\left(\dfrac{n}{2}+1\right)\pi}{(n+2)} \right|^2.$$

$$= \frac{32}{\pi^2} \frac{\sin^2\left(\frac{n}{2}+1\right)\pi}{(n^2-4)^2}$$

$$\leq \frac{32}{25\pi^2} < \frac{1}{2}.$$

Hence, the particle is most likely to occupy the first excited state of the expanded box.

c.   The wave function of the freed particle with a momentum p is $\frac{1}{\sqrt{2\pi\hbar}}e^{ipx/\hbar}$.

The probability amplitude is then

$$\Phi(p) = \int_0^L \frac{1}{\sqrt{2\pi\hbar}}e^{-ipx/\hbar}\sqrt{\frac{2}{L}}\sin\frac{\pi x}{L}dx$$

$$= \frac{1}{\sqrt{\pi\hbar L}}[1+e^{-ipL/\hbar}]\frac{L/\pi}{1-(pL/\hbar\pi)^2}.$$

The probability distribution for the momentum is therefore

$$|\Phi(p)|^2 = \frac{2\pi\hbar^3 L}{(\hbar^2\pi^2 - p^2 L^2)^2}\left(1+\cos\frac{pL}{\hbar}\right).$$

5076

A particle of mass M is in a one-dimensional harmonic oscillator potential $V_1 = \frac{1}{2}kx^2$.

a.   It is initially in its ground state. The spring constant is suddenly doubled $(k \to 2k)$ so that the new potential is $V_2 = kx^2$. The particle's energy is then measured. What is the probability for finding that particle in the ground state of the new potential V_2?

b.   The spring constant is suddenly doubled as in part (a), so that V_1 suddenly becomes V_2, but the energy of the particle in the new potential V_2 is not measured. Instead, after a time T has elapsed since the doubling of the spring constant, the spring constant is suddenly restored back to the original value. For what values of T would the initial ground state in V_1 be restored with 100% certainty?

(CUSPEA)

ol:

a. The wave function of the system before k change is

$$\psi(x)=\frac{1}{\sqrt{\pi}}\left(\frac{M\omega_0}{\hbar}\right)^{1/4}e^{-\frac{1}{2}M\omega_0 x^2/\hbar}.$$

Suppose that the particle is also in the ground state of the new potential well after k change. Then the new wave function is

$$\psi'(x)=\frac{1}{\sqrt{\pi}}\left(\frac{M\omega_1}{\hbar}\right)^{1/4}e^{-\frac{1}{2}M\omega_1 x^2/\hbar}.$$

The transition matrix element is

$$\langle\psi'|\psi\rangle=\int\frac{1}{\pi}\left(\frac{M}{\hbar}\right)^{1/2}(\omega_0\omega_1)^{1/4}\exp\left[-\frac{1}{2}M\frac{(\omega_0+\omega_1)x^2}{\hbar}\right]dx$$

$$=\frac{1}{\pi}\left(\frac{M}{\hbar}\right)^{1/2}(\omega_0\omega_1)^{1/4}\cdot\frac{\pi}{\sqrt{\frac{1}{2}\frac{M(\omega_0+\omega_1)}{\hbar}}}$$

$$=\frac{(\omega_1\omega_0)^{1/4}}{\sqrt{\frac{1}{2}(\omega_0+\omega_1)}}.$$

When k changes into $2k$, ω_0 changes into $\omega_1=\sqrt{2}\omega_0$, thus

$$|\langle\psi'|\psi\rangle|^2=\frac{(\omega_1\omega_0)^{1/2}}{\frac{1}{2}(\omega_0+\omega_1)}=\frac{(\sqrt{2}\omega_0^2)^{1/2}}{\frac{1}{2}(\sqrt{2}+1)\omega_0}$$

$$=\frac{2^{1/4}}{\frac{1}{2}(\sqrt{2}+1)}=2\cdot2^{1/4}(\sqrt{2}-1).$$

Hence, the probability that the particle is in the state $\psi'(x)$ is

$$2^{\frac{5}{4}}\left(\sqrt{2}-1\right).$$

b. The quantum state is not destroyed as the energy is not measured. At $t=0$, $\psi(x,0)=\psi_0(x)$, $\psi_n(x)$ being the eigenstates of V_1. We expand $\psi(x,0)$ in the set of eigenstates of V_2:

$$\psi(x,0)=\langle\psi'_m|\psi_0\rangle|\psi'_m(x)\rangle.$$

Here and below we shall use the convention that a repeated index implies summation over that index. Then

$$\psi(x,t)=e^{-iH_2 t/\hbar}\psi(x,0)=\langle\psi'_m|\psi_0\rangle|\psi'_m(x)\rangle e^{-iE'_m t/\hbar},$$

where H_2 is the Hamiltonian corresponding to V_2. Since $\psi_0(x)$ has even parity, parity conservation gives

$$\langle \psi'_m(x) | \psi_0(x) \rangle = \begin{cases} 0, & m = 2n+1, \\ \neq 0, & m = 2n, \end{cases}$$

and so

$$|\psi(x,\tau)\rangle = \psi_{2m}(x)\langle \psi_{2m} | \psi_0 \rangle e^{-iE_{2m}\tau/\hbar}.$$

Hence $|\psi(x,\tau)\rangle = |\psi_0(x)\rangle$ can be expected only if $E_{2m}\tau/\hbar = 2N\pi + c$, where N is a natural number and c is a constant, for any m. As

$$E_{2m} = \left(2m + \frac{1}{2}\right)\hbar\omega'_1,$$

we require

$$\left(2m + \frac{1}{2}\right)\omega'_1\tau = 2N\pi + c.$$

Setting

$$c = \frac{1}{2}\omega'_1\tau,$$

we require

$$2m\omega'_1\tau = 2N\pi,$$

or

$$2\omega'_1\tau = 2N'\pi$$

i.e.,

$$\tau = \frac{\pi N'}{\omega'_1},$$

where $N' = 0, 1, 2, \ldots$

Thus only if $\tau = N'\pi\sqrt{\dfrac{M}{2k}}$, will the state change into $\psi_0(x)$ with 100% certainty.

5077

A particle which moves only in the x direction is confined between vertical walls at $x = 0$ and $x = a$. If the particle is in the ground state, what is the energy? Suppose the walls are suddenly separated to infinity; what is the

robability that the particle has momentum of magnitude between p and
$p+dp$? What is the energy of such a particle? If this does not agree with the
round state energy, how do you account for energy conservation?

(Chicago)

ol: When the particle is confined between $x=0$ and $x=a$ in the ground state, its
wave function is

$$\psi_0 = \begin{cases} \sqrt{\dfrac{2}{a}}\sin\dfrac{\pi x}{a}, & 0 \le x \le a, \\ 0 & \text{otherwise,} \end{cases}$$

and its energy is

$$E = \frac{\pi^2 \hbar^2}{2ma^2}.$$

When the walls are suddenly removed to infinity, the wave function of the particle
cannot follow the change in such a short time but will remain in the original form.
However, the Hamiltonian of the system is now changed and the original wave
function is not an eigenstate of the new Hamiltonian. The original wave function
is to be taken as the initial condition in solving the Schrödinger equation for the
freed particle. The wave packet of the ground state in the original potential well
will expand and become uniformly distributed in the whole space when $t \to \infty$.
Transforming the original wave packet to one in momentum $(p = \hbar k)$, representation, we have

$$\psi(p) = \frac{1}{\sqrt{2\pi\hbar}} \int_0^a \sqrt{\frac{2}{a}} \sin\left(\frac{\pi x}{a}\right) \cdot e^{ikx} \, dx$$

$$= -\sqrt{\frac{a\pi}{\hbar}} \frac{1 + e^{ika}}{(ka)^2 - \pi^2}.$$

During the short time period of separating the walls the probability that the
momentum is in the range $p \to p + dp$ is given by

$$f(p)dp = \{|\psi(p)|^2 + |\psi(-p)|^2\}dp$$

$$= 8\frac{a\pi}{\hbar} \frac{\cos^2\left(\dfrac{ka}{2}\right)dp}{[(ka)^2 - \pi^2]^2} \qquad \text{if } p \ne 0;$$

$$f(0)dp = |\psi(0)|^2 \, dp = 4\frac{a}{\pi^3\hbar} \, dp \qquad \text{if } p = 0.$$

Because the new Hamiltonian is not time-dependent, we can calculate the average value of the energy using the original wave function:

$$\bar{E} = \int_0^\infty \frac{p^2}{2m} f(p)\,dp = \int_0^\infty \frac{\hbar^2 k^2}{2m} \cdot 8 \frac{a\pi}{\hbar} \frac{\cos^2\left(\dfrac{ka}{2}\right)}{[(ka)^2 - \pi^2]^2}\,d(\hbar k)$$

$$= \frac{4\hbar^2}{ma^2} \int_0^\infty \frac{y^2 \cos^2\left(\dfrac{\pi y}{2}\right)}{(y^2 - 1)^2}\,dy = \frac{\pi^2 \hbar^2}{2ma^2},$$

where $y = \dfrac{ka}{\pi}$. This means that the energy of the system is not changed during the short time period of separating walls, which is to be expected as

$$\langle \psi_0 | H_{\text{before}} | \psi_0 \rangle = \int_0^a \psi_0^* \frac{p}{2m} \psi_0\,dx,$$

$$\langle \psi(t) | H_{\text{after}} | \psi(t) \rangle = \langle \psi_0 | \exp(iH_{\text{after}}t/\hbar) \times H_{\text{after}} \exp\left(\frac{-iH_{\text{after}}t}{\hbar}\right) | \psi_0 \rangle$$

$$= \langle \psi_0 | H_{\text{after}} | \psi_0 \rangle = \int_0^a \psi_0^* \frac{p^2}{2m} \psi_0\,dx$$

$$= \langle \psi_0 | H_{\text{before}} | \psi_0 \rangle.$$

If the walls are separating to an infinite distance slowly or if the walls are not infinite high, there would be energy exchange between the particle and the walls. Consequently, the energy of the particle would change during the time of wall separation.

5078

A nucleus of charge Z has its atomic number suddenly changed to $Z+1$ by β-decay as shown in Fig. 5.21. What is the probability that a K-electron before the decay remains a K-electron around the new nucleus after the $\beta-$ decay? Ignore all electron-electron interactions.

(CUSPEA)

Sol:

The wave function of a K-electron in an atom of nuclear charge Z is

$$\psi(r) = NZ^{3/2} e^{-rZ/a}.$$

As $\int_0^\infty r^2\, dr N^2 e^{-2r/a} = 1$, the probability that the K-electron remains in the original orbit is

$$P = \left|\left\langle \psi_{Z+1}(r) \middle| \psi_Z(r) \right\rangle\right|^2 = \frac{\left(1+\frac{1}{Z}\right)^3}{\left(1+\frac{1}{2Z}\right)^6}.$$

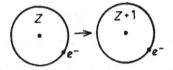

Fig. 5.21

5079

A tritium atom (^3H) can undergo spontaneous radioactive decay into a helium-3 ion ($^3He^+$) by emission of a beta particle. The departure of the electron is so fast that to the orbital electron the process appears as simply an instantaneous change in the nuclear charge from $Z=1$ to $Z=2$. Calculate the probability that the He ion will be left in its ground state.

(Berkeley)

Sol: The wave function of the ground state of He^+ is

$$\psi_{1s}^{He^+} = \frac{1}{\sqrt{\pi}}\left(\frac{2}{a}\right)^{3/2} \exp\{-2r/a\},$$

where a is the Bohr radius. Let the wave function of ^3H be $\varphi(\mathbf{r})$.

As the process of β decay takes place very fast, during the time period in which the ^3H becomes $^3He^+$ the wave function does not have time to change. Hence the probability that the $^3He^+$ is in the ground state is

$$P = \frac{\left|\left\langle \psi_{1s}^{He^+} \middle| \varphi \right\rangle\right|^2}{\left|\left\langle \varphi \middle| \varphi \right\rangle\right|^2}.$$

Initially, the ^3H is in the ground state so that

$$\varphi(\mathbf{r}) = \frac{1}{\sqrt{\pi}}\left(\frac{1}{a}\right)^{3/2} e^{-r/a}.$$

Therefore,

$$P = \left| 4\left(\frac{2}{a^2}\right)^{3/2} \int_0^\infty r^2 \exp\left\{-\frac{3r}{a}\right\} dr \right|^2$$

$$= \frac{2^7}{3^6} \left| \int_0^\infty x^2 e^{-x} dx \right|^2 = \frac{2^9}{3^6} = 0.702.$$

5080

Tritium (hydrogen of mass 3 i.e. ^3H) is beta-radioactive and decays into a helium nucleus of mass 3 (^3He) with the emission of an electron and a neutrino. Assume that the electron, originally bound to the tritium atom, was in its ground state and remains associated with the ^3He nucleus resulting from the decay, forming a $^3He^+$ ion.

a. Calculate the probability that the $^3He^+$ ion is found in its 1s state.

b. What is the probability that it is found in a 2p state?

(MIT)

Sol: Neglect the small difference in reduced mass between the hydrogen atom and the helium atom systems. The radius of the ion $^3He^+$ is $a_0/2$, where a_0 is the Bohr radius, so the wave functions are

$$\psi_{1s}^H = Y_{00} \frac{2}{a_0^{3/2}} e^{-r/a_0},$$

$$\psi_{1s}^{He^+} = Y_{00} \frac{2}{\left(\frac{a_0}{2}\right)^{3/2}} e^{-2r/a_0},$$

$$\psi_{2p}^{He^+} = Y_{1m} \frac{1}{2\sqrt{6}(a_0/2)^{3/2}} \frac{2r}{a_0} e^{-r/a_0}$$

$$(m = 1, 0, -1).$$

a. The amplitude of the probability that the ion He^+ is in the state 1s is

$$A = \int \left(\psi_{1s}^{He^+}\right)^* \psi_{1s}^H d^3x = \frac{2^{7/2}}{a_0^3} \int_0^\infty r^2 e^{-3r/a_0} dr = \frac{16\sqrt{2}}{27}.$$

Hence the probability is $|A|^2 = 2\left(\frac{16}{27}\right)^2 = 0.702$.

b. On account of the orthonormality of spherical harmonics, the probability that the ion $^3He^+$ is in a state 2p is zero $(\langle Y_{1m}|Y_{00}\rangle = 0)$.

5081

beam of excited hydrogen atoms in the 2s state passes between the plates f a capacitor in which a uniform electric field **E** exists over a distance *l*. The ydrogen atoms have velocity v along the *x*-axis and the **E** field is directed long the *z*-axis, as shown in Fig. 5.22.

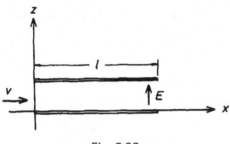

Fig. 5.22

ll the $n=2$ states of hydrogen are degenerate in the absence of the **E** field, ut certain of them mix when the field is present.

a. Which of the $n=2$ states are connected in first order via the perturbation?

b. Find the linear combination of $n=2$ states which removes the degeneracy as much as possible.

c. For a system which starts out in the 2s state at $t=0$, express the wave function at time $t \le \dfrac{l}{v}$.

d. Find the probability that the emergent beam contains hydrogen in the various $n=2$ states.

(MIT)

ol: Consider the potential energy eEz of the electron (charge $-e$) of a hydrogen atom in the external electric field $E\hat{e}_z$ as a perturbation. As the $n=2$ states are degenerate, we calculate $\langle 2\ell'm'|H'|2\ell m \rangle$, where $H' = eEz$, and $\ell = 0$, $m = 0$. It is known that only the following matrix elements are non-vanishing:

$$\langle 2,\ \ell+1,\ m|z|2,\ell,m \rangle = -\sqrt{\frac{(\ell-m+1)(\ell+m+1)}{(2\ell+1)(2l+3)}}\langle 2,\ \ell+1|r|2,\ \ell \rangle,$$

with

$$\langle 2, \ell+1 | r | 2, \ell \rangle = \frac{3a}{2} n \sqrt{n^2 - \ell^2}.$$

Thus all the matrix elements are zero except

$$\langle 210 | H' | 200 \rangle = -3eEa.$$

a. The $2s$ and $2p$ states are connected via the perturbation in first order since for the H' matrix only elements with $\Delta\ell = \pm 1$ are nonzero.

b. The perturbation Hamiltonian is

$$H' = \begin{pmatrix} 0 & -3eEa \\ -3eEa & 0 \end{pmatrix},$$

whose secular equation

$$\begin{vmatrix} -\lambda & -3eEa \\ -3eEa & -\lambda \end{vmatrix} = 0$$

gives eigenvalues $\pm 3eEa$, the corresponding eigenstate vectors being $\frac{1}{\sqrt{2}}\begin{pmatrix} 1 \\ \mp 1 \end{pmatrix}$.

The degeneracy of the state $n = 2$ is now removed.

c. As $t = 0$, just before the atoms enter the electric field,

$$\psi(0) = \frac{1}{\sqrt{2}}(|+\rangle + |-\rangle),$$

where

$$|+\rangle = \frac{1}{\sqrt{2}}\begin{pmatrix} 1 \\ 1 \end{pmatrix}, \qquad |-\rangle = \frac{1}{\sqrt{2}}\begin{pmatrix} 1 \\ -1 \end{pmatrix}$$

are the state vectors obtained in (b).

At time $0 < t \leq \frac{l}{v}$ when the atoms are subject to the electric field,

$$|\psi(t)\rangle = \frac{1}{\sqrt{2}}\left(e^{i3eEat/\hbar}|+\rangle + e^{-i3eEat/\hbar}|-\rangle\right)$$

$$= \begin{pmatrix} \cos(3eEat/\hbar) \\ i\sin(3eEat/\hbar) \end{pmatrix} = \cos\left(\frac{3eEat}{\hbar}\right)|2s\rangle + i\sin\left(\frac{3eEat}{\hbar}\right)|2p\rangle,$$

where

$$|2s\rangle = \frac{1}{\sqrt{2}}(|+\rangle + |-\rangle) = \begin{pmatrix} 1 \\ 0 \end{pmatrix},$$

$$|2p\rangle = \frac{1}{\sqrt{2}}(|+\rangle - |-\rangle) = \begin{pmatrix} 0 \\ 1 \end{pmatrix}.$$

d. For $t \geq \dfrac{l}{v}$, we find from (c) the probabilities

$$\left|\langle 200|\psi(t)\rangle\right|^2 = \cos^2\frac{3eEat}{\hbar},$$

$$\left|\langle 210|\psi(t)\rangle\right|^2 = \sin^2\frac{3eEat}{\hbar}.$$

5082

a. Consider a particle of mass m moving in a time-dependent potential $V(x, t)$ in one dimension. Write down the Schrödinger equations appropriate for two reference systems (x, t) and (x', t) moving with respect to each other with velocity v (i.e. $x = x' + vt$).

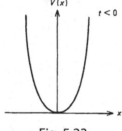

Fig. 5.23

b. Imaging that a particle sits in a one-dimensional well (Fig. 5.23) such that the well generates a potential of the form $m\omega^2 x^2/2$. At $t = 0$ the well is instantly given a kick and moves to the right with velocity v (see Fig. 5.24). In other words, assume that $V(x, t)$ has the form

$$V(x, t) = \begin{cases} \dfrac{1}{2}m\omega^2 x^2, & \text{for } t < 0, \\[2mm] \dfrac{1}{2}m\omega^2 x'^2, & \text{for } t > 0. \end{cases}$$

If for $t < 0$ the particle is in the ground state as viewed from the (x, t) coordinate system, what is the probability that for $t > 0$ it will be in the ground state as viewed from the (x', t) system?

(Columbia)

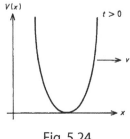

Fig. 5.24

Sol:

a. Both (x, t) and (x', t) are inertial systems, and so the Schrödinger equations are: for the (x, t) system,

$$\left[-\frac{\hbar^2}{2m}\frac{d^2}{dx^2} + V(x, t)\right]\psi(x, t) = i\hbar\frac{\partial}{\partial t}\psi(x, t),$$

for the (x', t) system,

$$\left[-\frac{\hbar^2}{2m}\frac{d^2}{dx'^2} + V'(x', t)\right]\psi(x', t) = i\hbar\frac{\partial}{\partial t}\psi(x', t),$$

where $V'(x', t) = V'(x - \upsilon t, t) = V(x, t)$.

b. This problem is the same as **Problem 6052** for the following reason. Consider an observer at rest in the (x', t) system. At $t < 0$ he sees the particle as sitting in the ground state of the potential well V. At $t = 0$, the potential well V instantly requires a velocity υ to the right along the x direction. The situation is the same as if V remains stationary while the particle acquires a velocity $-\upsilon$ along the $-x$ direction. It is required to find the probability that the particle remains in the ground state. **Problem 6052** deals with an Al nucleus which by emitting a y to the right acquires a uniform velocity to the left. The physics involved is exactly the same as the present problem and we can just make use of the results there.

5083

If the baryon number is conserved, the transition $n \leftrightarrow \bar{n}$ known as "neutron oscillation" is forbidden. The experimental limit on the time scale of such oscillations in free space and zero magnetic field is $\tau_{n-\bar{n}} \geq 3 \times 10^6$ sec. Since neutrons occur abundantly in stable nuclei, one would naively think it possible to obtain a much better limit on $\tau_{n-\bar{n}}$. The object of this problem is to undertand why the limit is so poor. Let H_0 be the Hamiltonian of the world in the bsence of any interaction which mixes n and $\bar{n}$. Then

$$H_0|n\rangle = m_n c^2|n\rangle \quad \text{and} \quad H_0|\bar{n}\rangle = m_n c^2|\bar{n}\rangle$$

or states at rest. Let H' be the interaction which turns n into $\bar{n}$ and vice ersa:

$$H'|n\rangle = \varepsilon|\bar{n}\rangle \quad \text{and} \quad H'|\bar{n}\rangle = \varepsilon|n\rangle,$$

here ε is real and H' does not flip spin.

a. Start with a neutron at $t=0$ and calculate the probability that it will be observed to be an antineutron at time t. When the probability is first equal to 50%, call that time $\tau_{n-\bar{n}}$. In this way convert the experimental limit on $\tau_{n-\bar{n}}$ into a limit on ε. Note $m_n c^2 = 940$ MeV.

b. Now reconsider the problem in the presence of the earth's magnetic field $\left(B_0 \geq \dfrac{1}{2}\text{gauss}\right)$. The magnetic moment of the neutron is $\mu_n \approx -6\times10^{-18}$ MeV/gauss. The magnetic moment of the antineutron is opposite. Begin with a neutron at $t=0$ and calculate the probability it will be observed to be an antineutron at time t. (Hint: work to lowest order in small quantities.) Ignore possible radiative transitions.

c. Nuclei with spin have non-vanishing magnetic fields. Explain briefly and qualitatively, in sight of part (b), how neutrons in such nuclei can be so stable while $\tau_{n-\bar{n}}$ is only bounded by $\tau_{n-\bar{n}} \geq 3\times10^6$ sec.

d. Nuclei with zero spin have vanishing average magnetic field. Explain briefly why neutron oscillation in such nuclei is also suppressed.

(MIT)

ol:

a. To find the eigenstates of the Hamiltonian $H = H_0 + H'$ we introduce in the neutron-antineutron representation the state vectors

$$|n\rangle \sim \begin{pmatrix} 1 \\ 0 \end{pmatrix}, |\bar{n}\rangle \sim \begin{pmatrix} 0 \\ 1 \end{pmatrix}.$$

As

$$\langle n | H_0 + H' | n \rangle = m_n c^2,$$
$$\langle \bar{n} | H_0 + H' | n \rangle = \varepsilon$$

we have energy eigenvalue equation

$$\begin{pmatrix} m_n c^2 - E & \varepsilon \\ \varepsilon & m_n c^2 - E \end{pmatrix} \begin{pmatrix} a \\ b \end{pmatrix} = 0.$$

Solving the equation, we get

$$E_+ = m_n c^2 + \varepsilon, \quad \begin{pmatrix} a \\ b \end{pmatrix}_+ = \frac{1}{\sqrt{2}} \begin{pmatrix} 1 \\ 1 \end{pmatrix},$$

$$E_- = m_n c^2 - \varepsilon, \quad \begin{pmatrix} a \\ b \end{pmatrix}_- = \frac{1}{\sqrt{2}} \begin{pmatrix} 1 \\ -1 \end{pmatrix}.$$

At $t = 0$, the system is in the neutron state and so

$$|n\rangle = \frac{1}{\sqrt{2}} |E_+\rangle + \frac{1}{\sqrt{2}} |E_-\rangle,$$

where

$$|E_+\rangle \sim \frac{1}{\sqrt{2}} \begin{pmatrix} 1 \\ 1 \end{pmatrix}, |E_-\rangle \sim \frac{1}{\sqrt{2}} \begin{pmatrix} 1 \\ -1 \end{pmatrix}.$$

At time t, the state of the system becomes

$$|\psi, t\rangle = \frac{1}{\sqrt{2}} e^{-iE_+ t/\hbar} |E_+\rangle + \frac{1}{\sqrt{2}} e^{-iE_- t/\hbar} |E_-\rangle.$$

In the neutron-antineutron representation, we thus have

$$|\psi, t\rangle = e^{-im_n c^2 t/\hbar} \begin{pmatrix} \cos \dfrac{\varepsilon t}{\hbar} \\ -i \sin \dfrac{\varepsilon t}{\hbar} \end{pmatrix},$$

and hence

$$|\psi, t\rangle = e^{-im_n c^2 t/\hbar} \cos \frac{\varepsilon t}{\hbar} |n\rangle - i e^{-im_n c^2 t/\hbar} \sin \frac{\varepsilon t}{\hbar} |\bar{n}\rangle.$$

The probability that at time t the particle is observed as an antineutron is therefore

$$P(t) = \left| i e^{-im_n c^2 t/\hbar} \sin \frac{\varepsilon t}{\hbar} \right|^2 = \sin^2 (\varepsilon t / \hbar).$$

$\tau_{n-\bar{n}}$ is defined as the time at which $P = \dfrac{1}{2}$, i.e.,

$$\tau_{n-\bar{n}} = \frac{\hbar}{\varepsilon}\arcsin\frac{1}{\sqrt{2}} = \frac{\pi\hbar}{4\varepsilon}.$$

Then as $T_{n-\bar{n}} \geq 3\times10^6$ s, the experimental limit on ε is $\varepsilon \leq \dfrac{h}{8\times3\times10^6} = 1.7\times10^{-28}$ MeV.

b. Noting that H' does not change the spin, after introducing the magnetic field one can take the neutron-antineutron representation

$$n\uparrow\sim\begin{pmatrix}1\\0\\0\\0\end{pmatrix}, \bar{n}\uparrow\sim\begin{pmatrix}0\\1\\0\\0\end{pmatrix}, n\downarrow\sim\begin{pmatrix}0\\0\\1\\0\end{pmatrix}, \bar{n}\downarrow\sim\begin{pmatrix}0\\0\\0\\1\end{pmatrix}$$

and calculate the matrix elements of $H' + \mu B_0$. Thus one obtains the perturbation Hamiltonian

$$\begin{pmatrix} -\mu_n B_0 & \varepsilon & 0 & 0 \\ \varepsilon & -\mu_{\bar{n}} B_0 & 0 & 0 \\ 0 & 0 & \mu_n B_0 & \varepsilon \\ 0 & 0 & \varepsilon & \mu_{\bar{n}} B_0 \end{pmatrix}$$

with $\mu_n \approx -6\times10^{-18}$ MeV/Gs, $\mu_{\bar{n}} \approx 6\times10^{-18}$ MeV/Gs.

This gives rise to two eigenequations:

$$\begin{pmatrix} -\mu_n B_0 - E^{(1)} & \varepsilon \\ \varepsilon & \mu_n B_0 - E^{(1)} \end{pmatrix}\begin{pmatrix} a\uparrow \\ b\uparrow \end{pmatrix} = 0,$$

$$\begin{pmatrix} \mu_n B_0 - E^{(1)} & \varepsilon \\ \varepsilon & -\mu_n B_0 - E^{(1)} \end{pmatrix}\begin{pmatrix} a\downarrow \\ b\downarrow \end{pmatrix} = 0,$$

where we have used the relation $\mu_{\bar{n}} = -\mu_n$.

Solving the two equations, we obtain

$$E_{\pm}^{(1)} = \pm\lambda = \pm\sqrt{\epsilon^2 + (\mu_n B_0)^2},$$

and hence

$$\begin{pmatrix} a\uparrow \\ b\uparrow \end{pmatrix}_+ = \frac{1}{\sqrt{2\lambda}}\begin{pmatrix} \sqrt{\lambda-\mu_n B_0} \\ \sqrt{\lambda+\mu_n B_0} \end{pmatrix}, \begin{pmatrix} a\uparrow \\ b\uparrow \end{pmatrix}_- = \frac{1}{\sqrt{2\lambda}}\begin{pmatrix} \sqrt{\lambda+\mu_n B_0} \\ -\sqrt{\lambda-\mu_0 B_0} \end{pmatrix},$$

$$\begin{pmatrix} a\uparrow \\ b\uparrow \end{pmatrix}_+ = \frac{1}{\sqrt{2\lambda}}\begin{pmatrix} \sqrt{\lambda+\mu_n B_0} \\ \sqrt{\lambda-\mu_n B_0} \end{pmatrix}, \begin{pmatrix} a\uparrow \\ b\uparrow \end{pmatrix}_- = \frac{1}{\sqrt{2\lambda}}\begin{pmatrix} \sqrt{\lambda-\mu_n B_0} \\ -\sqrt{\lambda+\mu_0 B_0} \end{pmatrix}.$$

As $t=0$, the system is in the neutron state

$$n\uparrow \sim \sqrt{\frac{\lambda-\mu_n B_0}{2\lambda}}\begin{pmatrix} a\uparrow \\ b\uparrow \end{pmatrix}_+ + \sqrt{\frac{\lambda+\mu_n B_0}{2\lambda}}\begin{pmatrix} a\uparrow \\ b\uparrow \end{pmatrix}_-,$$

$$n\downarrow \sim \sqrt{\frac{\lambda-\mu_n B_0}{2\lambda}}\begin{pmatrix} a\uparrow \\ b\uparrow \end{pmatrix}_+ + \sqrt{\frac{\lambda-\mu_n B_0}{2\lambda}}\begin{pmatrix} a\uparrow \\ b\uparrow \end{pmatrix}_-.$$

At time t, the states of the system are

$$(\uparrow) \sim e^{-im_nc^2t/\hbar}\frac{1}{2\lambda}$$
$$\times \begin{pmatrix} (\lambda-\mu_n B_0)e^{-i\lambda t/\hbar}+(\lambda+\mu_n B_0)e^{i\lambda t/\hbar} \\ \sqrt{(\lambda^2-(\mu_n B_0)^2}(e^{-i\lambda t/\hbar}-e^{i\lambda t/\hbar}) \end{pmatrix},$$

$$(\downarrow) \sim e^{-im_nc^2t/\hbar}\frac{1}{2\lambda}$$
$$\times \begin{pmatrix} (\lambda+\mu_n B)e^{-i\lambda t/\hbar}+(\lambda-\mu_n B_0)e^{i\lambda t/\hbar} \\ \sqrt{(\lambda^2-(\mu_n B_0)^2}(e^{-i\lambda t/\hbar}-e^{i\lambda t/\hbar}) \end{pmatrix}.$$

Therefore the probability of $n\uparrow \to \bar{n}\uparrow$ is

$$P_{n\uparrow\to\bar{n}\uparrow}(t) = \frac{\varepsilon^2}{\lambda^2}\sin^2\frac{\lambda t}{\hbar}$$
$$= \frac{\varepsilon^2}{\varepsilon^2+(\mu_n B_0)^2}\sin^2\frac{\sqrt{\varepsilon^2+(\mu_n B_0)^2}\,t}{\hbar},$$

and that of $n\downarrow \to \bar{n}\downarrow$ is

$$P_{n\downarrow\to\bar{n}\downarrow}(t) = \frac{\varepsilon^2}{\lambda^2}\sin^2\frac{\lambda t}{\hbar}$$
$$= \frac{\varepsilon^2}{\varepsilon^2+(\mu_n B_0)^2}\sin^2\frac{\sqrt{\varepsilon^2+(\mu_n B_0)^2}\,t}{\hbar}.$$

Finally, if the neutron is not polarized the probability of $n \to \bar{n}$ is

$$P(t) = \frac{1}{2} P_{n\uparrow \to \bar{n}\uparrow}(t) + \frac{1}{2} P_{n\downarrow \to \bar{n}\downarrow}(t)$$

$$= \frac{\varepsilon^2}{\varepsilon^2 + (\mu_n B_0)^2} \sin^2 \frac{\sqrt{\varepsilon^2 + (\mu_n B_0)^2}\, t}{\hbar},$$

which means that the polarization of the neutron has no effect on the transition probability.

As $\mu_n B_0 \gg \varepsilon$,

$$P(t) \le \left(\frac{1.65 \times 10^{-28}}{6 \times 10^{-18} \times 1/2} \right)^2 \approx 0.3 \times 10^{-20},$$

which shows that the transition probability is extremely small.

c. If nuclear spin is not zero, the magnetic field inside a nucleus is strong, much larger than 0.5 Gs. Then the result of (b) shows that

$$P_{n \to \bar{n}} \ll 10^{-20},$$

which explains why the neutron is stable inside a nucleus.

d. If nuclear spin is zero, then the average magnetic field in the nucleus is zero. Generally this means that the magnetic field outside the nucleus is zero while that inside the nucleus may not be zero, but may even be very large, with the result that $P_{n \to \bar{n}}$ is very small. Besides, even if the magnetic field inside the nucleus averaged over a long period of time is zero, it may not be zero at every instant. So long as magnetic field exists inside the nucleus, $P_{n \to \bar{n}}$ becomes very small. Neutron oscillation is again suppressed.

Part VI

Scattering
Theory & Quantum
Transitions

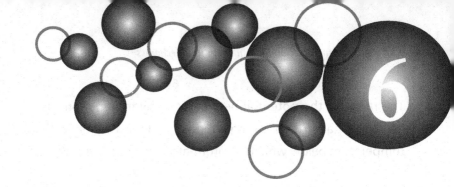

6001

Derive the quantum mechanical expression for the s-wave cross section for scattering from a hard sphere of radius R.

(MIT)

ol: The effect of the rigid sphere is equal to that of the potential

$$V(r)=\begin{cases} \infty & (r<R), \\ 0 & (r>R). \end{cases}$$

Let the radial wave function of the s-wave be $R_0(r)=\chi_0(r)/r$. Then the Schröödinger equation can be written as

$$\chi_0''(r)+k^2\chi_0(r)=0$$

with

$$\chi_0(r)=0 \quad (r<R), \quad k=\frac{1}{\hbar}\sqrt{2mE}.$$

The solution for $r>R$ is $\chi_0(r)=\sin(kr+\delta_0)$. The continuity of the wave function at $r=R$ gives

$$\sin(kR+\delta_0)=0,$$

which requires $\delta_0=n\pi-kR$, or $\sin\delta_0=(-1)^{n+1}\sin kR$ $(n=0,1,2,\ldots)$. Therefore the total cross-section of the s-wave is

$$\sigma_t=\frac{4\pi}{k^2}\sin^2\delta_0=\frac{4\pi}{k^2}\sin^2 kR$$

For low energies, $k\to 0$, $\sin kR\approx kR$, and so $\sigma_t=4\pi R^2$. For high energies, $k\to\infty$ and $\sigma_t\approx 0$.

6002

The range of the potential between two hydrogen atoms is approximately 4Å. For a gas in thermal equilibrium, obtain a numerical estimate of the temperature below which the atom-atom scattering is essentially *s*-wave.

(MIT)

Sol: The problem concerns atom-atom scatterings inside a gas. If mainly s partial waves are involved, the uncertainty principle requires $\mu v_r a \leq \hbar$ where $\mu = \frac{1}{2} m_p$ is the reduced mass of the two atoms, $\mathbf{v}_r = \mathbf{v}_1 - \mathbf{v}_2$ is the relative velocity between the two atoms, of velocities $\mathbf{v}_1$, $\mathbf{v}_2$, $a = 4$Å. When thermal equilibrium is reached,

$$\langle v \rangle = 0, \quad \frac{1}{2} m_p \langle v^2 \rangle = \frac{3}{2} kT,$$

k being Boltzmann's constant and T the absolute temperature. The meansquare value of the relative speed v_r is

$$\langle v_r^2 \rangle = \langle (\mathbf{v}_1 - \mathbf{v}_2)^2 \rangle = \langle v_1^2 + v_2^2 - 2\mathbf{v}_1 \cdot \mathbf{v}_2 \rangle = 2 \langle v^2 \rangle = \frac{6kT}{m_p},$$

since on average $\mathbf{v}_1 \cdot \mathbf{v}_2 = 0$, $\langle v_1^2 \rangle = \langle v_2^2 \rangle = \langle v^2 \rangle$. Thus

$$\mu a v_r \approx \frac{m_p a}{2} \sqrt{\frac{6kT}{m_p}} \leq \hbar,$$

i.e.,

$$T \leq \frac{2\hbar^2}{3m_p c^2} \left(\frac{c}{a} \right)^2 \frac{1}{k} = \frac{2 \times (6.58 \times 10^{-16})^2}{3 \times 938 \times 10^6} \times \left(\frac{3 \times 10^{10}}{4 \times 10^{-8}} \right)^2 \frac{1}{8.62 \times 10^{-5}}$$

$$= 2^\circ \text{K}.$$

Hence, under normal temperatures the scattering of other partial waves must also be taken into account.

6003

A nonrelativistic particle of mass m and energy E scatters quantummechanically in a central potential $V(r)$ given by

$$V(r) = \frac{\hbar^2}{2m} U(r), \quad U(r) = -2 \left(\frac{\lambda}{\cosh \lambda r} \right)^2,$$

where λ is a parameter. This particle potential has the property that the cross section $\sigma(E)$ grows larger and larger as $E \to 0$, diverging for $E = 0$.

clearly, for *E* very small the cross section is dominated by the *s*-wave ($l = 0$) contribution. Thus for small *E* one needs to compute only the $l = 0$ partial wave amplitude. In connection with this, to save you mathematical efforts, you are told mathematically that the equation

$$\frac{d^2\phi}{dr^2} + A\phi = U(r)\phi,$$

where *A* is a positive constant, has a general solution

$$\phi = \alpha(\lambda \tanh \lambda r - ik)e^{ikr} + \beta (\lambda \tanh \lambda r + ik)e^{-ikr}$$

where $k = \sqrt{A}$ and α and β are integration constants. Recall $\tanh x = \frac{e^x - e^{-x}}{e^x + e^{-x}}$

and compute $\sigma(E)$ for $E \rightarrow 0$.

(CUS)

Sol: The *s* partial wave function is spherically symmetric, its equation being

$$-\frac{\hbar^2}{2m} \cdot \frac{1}{r^2} \cdot \frac{d}{dr}\left(r^2 \frac{d}{dr}\phi(r)\right) + \frac{\hbar^2}{2m}U(r)\phi(r) = E\phi(r).$$

With $R(r) = \phi(r)r$, the above becomes

$$R''(r) + \frac{2m}{\hbar^2}\left[E - \frac{\hbar^2}{2m}U(r)\right]R(r) = 0,$$

i.e.,

$$R''(r) + \frac{2mE}{\hbar^2}R(r) = U(r)R(r).$$

The solution is

$$R(r) = \alpha(\lambda \tanh \lambda r - ik)e^{ikr} + \beta(\lambda \tanh \lambda r + ik)e^{-ikr},$$

where

$$k = \sqrt{\frac{2mE}{\hbar^2}}.$$

Consider $r \rightarrow 0$. As $\phi(0)$ is finite, $R \rightarrow 0$. Then as $\tanh \lambda r \rightarrow \lambda r$, $e^{ikr} \rightarrow 1$, we have for $r \rightarrow 0$

$$R(r) \approx \alpha(\lambda^2 r - ik) + \beta(\lambda^2 r + ik) \rightarrow \alpha(-ik) + \beta(ik) = 0,$$

giving $\alpha = \beta$. Consider $r \to \infty$. As $\tanh \lambda r \to 1$, we have for $r \to \infty$

$$R(r) \to \alpha(\lambda - ik)e^{ikr} + \beta(\lambda + ik)e^{-ikr}$$
$$= \alpha\left[(\lambda - ik)e^{ikr} + (\lambda + ik)e^{-ikr}\right]$$
$$= \alpha\sqrt{\lambda^2 + k^2}\left(e^{ikr - i\alpha_1} + e^{-ikr + i\alpha_1}\right)$$
$$= \alpha\sqrt{\lambda^2 + k^2} \cdot 2\cos(kr - \alpha_1)$$
$$= 2\alpha\sqrt{\lambda^2 + k^2}\sin\left(kr + \frac{\pi}{2} - \alpha_1\right) \sim \sin(kr + \delta_0),$$

where

$$\delta_0 = \frac{\pi}{2} - \alpha_1,$$

and α_1 is defined by

$$\tan\alpha_1 = k/\lambda, \quad \text{or} \quad \alpha_1 = \tan^{-1}k/\lambda.$$

Thus the total cross section for scattering is

$$\sigma_t = \frac{4\pi}{k^2}\sin^2\delta_0 = \frac{4\pi}{k^2}\cos^2\alpha_1.$$

For low energies, $E \to 0$, $k \to 0$, $\alpha_1 \to 0$, and so

$$\sigma_t = \frac{4\pi}{k^2} = \frac{2\pi\hbar^2}{mE}.$$

6004

A particle of mass m is interacting in three dimensions with a spherically symmetric potential of the form $V(\mathbf{r}) = -C\delta(|\mathbf{r}| - a)$.

In other words, the potential is a delta function that vanishes unless the particle is precisely a distance "a" from the center of the potential. Here C is a positive constant.

a. Find the minimum value of C for which there is bound state.

b. Consider a scattering experiment in which the particle is incident on the potential with a low velocity. In the limit of small incident velocities what is the scattering cross section? What is the angular distribution?

(Princeton)

Sol:

a. Suppose the eigenfunction of a bound state of the single-particle system has the form

$$\psi(\mathbf{r}) = R(r)Y_{lm}(\theta, \varphi).$$

Then the radial function $R(r)$ satisfies

$$R'' + \frac{2}{r}R' + \left[(ik)^2 + \frac{2mC}{\hbar^2}\delta(|\mathbf{r}| - a) - \frac{l(l+1)}{r^2}\right]R = 0, \tag{1}$$

where $k = \sqrt{-2mE/\hbar^2}$. Note $E < 0$ for a bound state. If $r \neq a$, the equation is an imaginary-variable spherical Bessel equation. For $r < a$ it has the solution that is finite at $r = 0$

$$R(r) = A_k j_l(ikr),$$

where j_l is spherical Bessel function of the first kind of order l. For $r > a$ it has the solution that is finite for $r \to \infty$

$$R(r) = B_k h_l^{(1)}(ikr),$$

where $h_l^{(1)}$ is spherical Bessel function of the third kind, (spherical Hankel function) of order l. The wave function is continuous at $r = a$. Thus

$$A_k j_l(ika) = B_k h_l^{(1)}(ika).$$

Integrating Eq. (1) from $a - \varepsilon$ to $a + \varepsilon$, where ε is a small positive number, and then letting $\varepsilon \to 0$, we have

$$R'(a+0) - R'(a-0) = -C'R(a), \tag{2}$$

where

$$C' = \frac{2mC}{\hbar^2}.$$

Suppose there is at least a bound state. Consider the ground state $l = 0$, for which

$$R(r) = \begin{cases} A j_0(ikr) = A\dfrac{\sin(ikr)}{ikr}, & r < a, \\[4mm] B h_0^{(1)}(ikr) = B\dfrac{(-1)e^{-kr}}{kr}, & r > a. \end{cases}$$

Differentiating $R(r)$ and letting $r \to a$, we have

$$R'(a+0) = \frac{B}{k} \cdot \frac{e^{-ka}}{a}\left(k + \frac{1}{a}\right),$$

$$R'(a-0) = \frac{A}{k}\left[\frac{k\cosh(ka)}{a} - \frac{\sinh(ka)}{a^2}\right].$$

Substituting these in Eq. (2) gives

$$aC' = \frac{2ka}{1 - e^{-2ka}}.$$

As for $x \geq 0$, $x \geq 1 - e^{-x}$, we have $aC' \geq 1$ and

$$C'_{min} = 1/a, \quad \text{or} \quad C_{min} = \frac{\hbar^2}{2ma}.$$

This is the minimum value of C for which there is a bound state.

b. We use the method of partial waves. When the particle is incident on the potential with a low velocity, only the $\ell = 0$ partial wave is important, for which the radial wave equation

$$R'' + \frac{2}{r}R' + \left[k^2 + \frac{2mC}{\hbar^2}\delta(r-a) \right] R = 0.$$

On setting $R(r) = \chi_0(r)/r$ it becomes

$$\chi_0'' + \left[k^2 + \frac{2mC}{\hbar^2}\delta(r-a) \right]\chi_0 = 0, \tag{3}$$

which has solutions finite at $r \to 0$ and $r \to \infty$

$$\chi_0(r) = \begin{cases} A\sin kr, & r < a, \\ \sin(kr + \delta_0), & r > a. \end{cases}$$

As $\chi_0(r)$ is continuous at $r = a$, we require

$$A\sin ka = \sin(ka + \delta_0).$$

Integrating Eq. (3) from $a - \varepsilon$ to $a + \varepsilon$ gives

$$\chi_0'(a+\varepsilon) - \chi_0'(a-\varepsilon) = -\frac{2mC}{\hbar^2}\chi_0(a).$$

Substituting in the expressions for $\chi_0(r)$, it becomes

$$\frac{ka}{\tan(ka+\delta_0)} - \frac{ka}{\tan ka} = -\frac{2mCa}{\hbar^2}.$$

For $k \to 0$, the above becomes

$$\frac{ka}{\tan \delta_0} - 1 = -\frac{2mCa}{\hbar^2},$$

or

$$\tan \delta_0 = \frac{ka}{1 - \frac{2maC}{\hbar^2}},$$

i.e.,

$$\sin\delta_0 = \frac{ka}{\sqrt{k^2a^2 + \left(1 - \frac{2maC}{\hbar^2}\right)^2}} \approx \frac{ka}{\left(1 - \frac{2maC}{\hbar^2}\right)}.$$

Hence, the total scattering cross section is

$$\sigma_t = \frac{4\pi}{k^2}\sin^2\delta_0 \approx \frac{4\pi a^2}{\left(1 - \frac{2maC}{\hbar^2}\right)^2}.$$

Note that for low velocities only s-waves $(l=0)$ need be considered and the differential cross section is simply

$$\sigma(\theta) = \frac{1}{k^2}\sin^2\delta_0 = a^2\left(1 - \frac{2maC}{\hbar^2}\right)^{-2},$$

which is independent of the angles. Thus the angular distribution is isotropic.

6005

a. Find the s-wave phase shift, as a function of wave number k, for a spherically symmetric potential which is infinitely repulsive inside of a radius r_0, and vanishes outside of r_0.

b. For $k \to 0$ discuss the behavior of the phase shifts in the higher partial waves.

(Wisconsin)

Sol:

a. This is a typical scattering problem that can be readily solved by the method of partial waves. The potential can be expressed as

$$V(r) = \begin{cases} \infty, & r < r_0, \\ 0, & r > r_0. \end{cases}$$

The radial wave function for the ℓ partial wave is

$$R_\ell(kr) = \begin{cases} 0, & r < r_0, \\ j_\ell(kr)\cos\delta_\ell - n_\ell(kr)\sin\delta_\ell, & r > r_0. \end{cases} \tag{2}$$

Here j_ℓ and n_ℓ are spherical Bessel function and spherical Neumann function of order ℓ. These functions have the asymptotic forms

$$j_\ell(x) \xrightarrow{x\to\infty} \frac{1}{x}\sin(x - \ell\pi/2),$$

$$n_\ell(x) \xrightarrow{x\to\infty} -\frac{1}{x}\cos(x - \ell\pi/2).$$

Hence, for $r > r_0$ we have

$$R_\ell(kr) \xrightarrow{kr\to\infty} \sin\left(kr - \frac{\ell\pi}{2} + \delta_\ell\right).$$

The phase shift δ_ℓ can be determined by the continuity of the wave function at $r = r_0$. Writing $kr_0 = x$, the continuity condition

$$R_\ell(x) = j_\ell(x)\cos\delta_\ell - n_\ell(x)\sin\delta_\ell = 0$$

gives

$$\tan\delta_\ell = \frac{j_\ell(x)}{n_\ell(x)}.$$

In the low-energy limit $x \to 0$, the functions have the asymptotic forms

$$j_\ell(x) \xrightarrow{x\to 0} \frac{x^\ell}{(2\ell+1)!!},$$

$$n_\ell(x) \xrightarrow{x\to 0} -\frac{(2\ell-1)!!}{x^{\ell+1}},$$

so that

$$\tan\delta_\ell = \frac{j_\ell(x)}{n_\ell(x)} \xrightarrow{x\to 0} -\frac{x^{2\ell+1}}{\left[(2l-1)!!\right]^2 (2\ell+1)}.$$

Thus the s-wave ($\ell = 0$) phase shift is

$$\tan\delta_0 = -x = -kr_0.$$

It gives a finite contribution to the scattering and the corresponding total cross section is

$$\sigma_t = \frac{4\pi}{k^2}\sin^2\delta_0 \approx \frac{4\pi}{k^2}\delta_0^2 \approx 4\pi r_0^2.$$

The scattering is spherically symmetric, and the total cross section is four times the classical value πr_0^2.

b.   Consider the low-energy limit $k \to 0$.

As

$$\tan\delta_l \approx -\frac{x^{2\ell+1}}{\left[(2\ell-1)!!\right]^2 (2\ell+1)},$$

δ_ℓ falls off very rapidly as ℓ increases. All the phase shifts vanish as $k \to 0$, except for the $l = 0$ partial wave. Hence, s-waves predominate in low-energy scattering. Physically, particles with higher partial waves are farther away from the force center so the effect of the force on such particles is smaller, causing $|\delta_\ell|$ to be smaller.

6006

A particle of mass m is scattered by the central potential

$$V(r) = -\frac{\hbar^2}{ma^2}\frac{1}{\cosh^2(r/a)},$$

where a is a constant. Given that the equation

$$\frac{d^2y}{dx^2} + k^2 y + \frac{2}{\cosh^2 x} y = 0$$

has the solutions $y = e^{\pm ikx}(\tanh x \mp ik)$, calculate the s-wave contribution to the total scattering cross section at energy E.

(MIT)

Sol:

Letting $\chi_0(r) = rR(r)$ we have for the radial part of the Schrödinger equation for s-waves ($l = 0$)

$$\frac{d^2\chi_0(r)}{dr^2} + \frac{2m}{\hbar^2}\left[E + \frac{\hbar^2}{ma^2}\frac{1}{\cosh^2(r/a)}\right]\chi_0(r) = 0.$$

With $x = r/a$, $y(x) = \chi_0(r)$ and $k = \sqrt{\frac{2mE}{\hbar^2}}$, the above becomes

$$\frac{d^2y(x)}{dx^2} + k^2 a^2 y(x) + \frac{2}{\cosh^2(x)} y(x) = 0.$$

This equation has solutions $y = e^{\pm iakx}(\tanh x \mp iak)$. For R finite at $r = 0$ we require $y(0) = 0$. The solution that satisfies this condition has the form

$$y(x) = e^{iakx}(\tanh x - iak) + e^{-iakx}(\tanh x + iak)$$
$$= 2\cos(akx)\tanh x + 2ak\sin(akx),$$

or

$$\chi_0(r) = 2\cos(kr)\tanh\left(\frac{r}{a}\right) + 2ak\,\sin(kr).$$

Thus

$$\frac{1}{\chi_0}\frac{d\chi_0}{dr} = \frac{ak^2\cos(kr) - k\sin(kr)\tanh\left(\dfrac{r}{a}\right) + \dfrac{1}{a}\cos(kr)\mathrm{sech}^2\left(\dfrac{r}{a}\right)}{ak\sin(kr) + \cos(kr)\tanh\left(\dfrac{r}{a}\right)}$$

$$\xrightarrow{r\to\infty} \frac{ak^2\cos(kr) - k\sin(kr)}{ak\sin(kr) + \cos(kr)} = k\frac{ak\cot(kr) - 1}{\cot(kr) + ak}.$$

On the other hand if we write χ_0 in the form

$$\chi_0(r) = \sin(kr + \delta_0),$$

then as

$$\frac{1}{\chi_0}\frac{d\chi_0}{dr} = k\cot(kr + \delta_0) = k\frac{\cot(kr)\cot\delta_0 - 1}{\cot(kr) + \cot\delta_0},$$

we have to put

$$\cot\delta_0 = ak,$$

or

$$\sin^2\delta_0 = \frac{1}{1 + a^2k^2}.$$

Hence, the s-wave contribution to the total scattering cross section is

$$\sigma_t = \frac{4\pi}{k^2}\sin^2\delta_0 = \frac{4\pi}{k^2}\frac{1}{1 + a^2k^2} = \frac{2\pi\hbar^2}{mE}\frac{1}{1 + \dfrac{2a^2mE}{\hbar^2}}.$$

6007

A spinless particle of mass m, energy E scatters through angle θ in an attractive square-well potential $V(r)$:

$$V(r) = \begin{cases} -V_0, & 0 < r < a, \qquad V_0 > 0, \\ 0, & r > a. \end{cases}$$

a. Establish a relation among the parameters V_0, a, m and universal constants which guarantees that the cross section vanishes at zero energy $E = 0$. This will involve a definite but transcendental equation, which you must derive but need not solve numerically. For parameters meeting the above condition, the differential cross section, as $E \to 0$, will behave like

$$\frac{\partial \sigma}{\partial \Omega} \xrightarrow[E \to 0]{} E^{\lambda} F(\cos \theta).$$

b. What is the numerical value of the exponent λ ?

c. The angular distribution function $F(\cos \theta)$ is a polynomial in $\cos \theta$. What is the highest power of $\cos \theta$ in this polynomial?

(Princeton)

Sol:

a. When the energy is near zero, only the partial wave with $l = 0$ is important. Writing the radial wave function as $R(r) = \chi(r)/r$, then $\chi(r)$ must satisfy the equations

$$\begin{cases} \chi'' + \dfrac{2mE}{\hbar^2} \chi = 0, & r > a, \\[2mm] \chi'' + \dfrac{2m}{\hbar^2}(E + V_0)\chi = 0, & 0 < r < a \end{cases}$$

with $k = \sqrt{\dfrac{2mE}{\hbar^2}}$, $K = \sqrt{\dfrac{2m(E + V_0)}{\hbar^2}}$. The above has solutions

$$\chi(r) = \sin(kr + \delta_0), \quad r > a,$$
$$\chi(r) = A \sin(Kr), \quad 0 < r < a.$$

As both $\chi(r)$ and $\chi'(r)$ are continuous at $r = a$, we require

$$\sin(ka + \delta_0) = A \, \sin(Ka),$$
$$k \, \cos(ka + \delta_0) = K A \, \cos(Ka),$$

or

$$K \tan(ka + \delta_0) = k \tan(Ka),$$

and hence

$$\delta_0 = \tan^{-1}\left[\frac{k}{K} \tan(Ka)\right] - ka.$$

For $E \to 0$

$$k \to 0, \quad K \to k_0 \equiv \sqrt{\frac{2mV_0}{\hbar^2}},$$

and so

$$\delta_0 \to k\left[\frac{\tan(k_0 a)}{k_0} - a\right].$$

For the total cross section to be zero at $E = 0$, we require

$$\frac{4\pi}{k^2}\sin^2\delta_0 \to 4\pi a^2\left[\frac{\tan(k_0 a)}{k_0 a} - 1\right]^2 = 0,$$

or

$$\tan(k_0 a) = k_0 a,$$

i.e.

$$\tan\left(\frac{\sqrt{2mV_0}}{\hbar}a\right) = \frac{\sqrt{2mV_0}}{\hbar}a,$$

which is the transcendental equation that the parameters V_0, a, m and universal constant $\hbar$ must satisfy.

b & c When $k \to 0$, the partial wave with $l = 0$ is still very important for the differential cross-section, although its contribution also goes to zero. Expanding $\tan(Ka)$ as a Taylor series in k, we have

$$\tan(Ka) = \tan\left(a\sqrt{k^2 + k_0^2}\right) = \tan(k_0 a) + \frac{ak^2}{2k_0\cos^2(k_0 a)} + \cdots.$$

Neglecting terms of orders higher than k^2, we have

$$\begin{aligned}\delta_0 &= \tan^{-1}\left[\frac{k}{K}\tan(Ka)\right] - ka \\ &\approx \tan^{-1}\left\{\frac{k}{k_0}\left(1 - \frac{k^2}{2k_0^2}\right)\left[\tan(k_0 a) + \frac{k^2 a}{2k_0\cos^2(k_0 a)}\right]\right\} - ka \\ &\approx \tan^{-1}\left\{\frac{k}{k_0}\tan(k_0 a) - \frac{k^3}{2k_0^3}\tan(k_0 a) + \frac{k^3 a}{2k_0^2\cos^2(k_0 a)}\right\} - ka \\ &\approx \tan^{-1}\left\{ka - \frac{k^3 a}{2k_0^2} + \frac{k^3 a}{2k_0^2\cos^2(k_0 a)}\right\} - ka \\ &\approx \frac{k^3 a}{2k_0^2\cos^2(k_0 a)} - \frac{k^3 a}{2k_0^2} - \frac{k^3 a^3}{3}.\end{aligned}$$

Hence,

$$\frac{d\sigma}{d\Omega} \approx \frac{1}{k^2} \sin^2 \delta_0$$

$$\approx k^4 F(k_0, a)$$

$$= E^2 \left(\frac{2m}{\hbar^2}\right)^2 F(k_0, a).$$

Thus the differential cross section per unit solid angle is approximately iso-tropic and proportional to E^2 for $E \to 0$. To find the contribution of partial wave with $\ell = 1$, consider its radial wave equations

$$\frac{1}{r^2}\frac{d}{dr}\left(r^2\frac{d}{dr}R\right) + \left(K^2 - \frac{2}{r^2}\right)R = 0 \quad (r < a),$$

$$\frac{1}{r^2}\frac{d}{dr}\left(r^2\frac{d}{dr}R\right) + \left(k^2 - \frac{2}{r^2}\right)R = 0 \quad (r > a).$$

The solutions have the form of the first-order spherical Bessel function $j_1(\rho) = \frac{\sin\rho}{\rho^2} - \frac{\cos\rho}{\rho}$, or

$$R_1 = \begin{cases} \dfrac{\sin(Kr)}{(Kr)^2} - \dfrac{\cos(Kr)}{Kr}, & 0 < r < a, \\[3mm] A\left[\dfrac{\sin(kr+\delta_1)}{(kr)^2} - \dfrac{\cos(kr+\delta_1)}{kr}\right], & r > a. \end{cases}$$

The continuity of R_1 and the first derivative of r^2R at $r = a$ gives

$$\frac{\sin Ka}{(Ka)^2} - \frac{\cos Ka}{Ka} = A\left[\frac{\sin(ka+\delta_1)}{(ka)^2} - \frac{\cos(ka+\delta_1)}{ka}\right],$$

$$\sin Ka = A\,\sin(ka+\delta_1).$$

Taking the ratios we have $k^2\left[1 - Ka\cot(Ka)\right] = K^2\left[1 - ka\cot(ka+\delta_1)\right]$,

or

$$\tan(ka+\delta_1) = ka\left[1 + \frac{k^2}{k_0^2} - \frac{k^2a\cot(Ka)}{k_0} + O(k^4)\right]$$

$$= ka + O(k^3),$$

as

$$K = \sqrt{k^2 + k_0^2} \simeq k_0\left[1 + \frac{1}{2}\frac{k^2}{k_0^2} + O(k^4)\right].$$

Hence,

$$\delta_1 = \tan^{-1}\left[ka + O(k^3)\right] - ka = -\frac{1}{3}(ka)^3 + O(k^3) = O(k^3).$$

Thus its contribution to $\frac{\partial \sigma}{\partial \Omega}$,

$$\frac{9}{k^2} \sin^2 \delta_1 \cos^2 \theta,$$

is also proportional to k^4. Similarly, for $l = 2$,

$$R_2 = \begin{cases} \left[\dfrac{3}{(Kr)^3} - \dfrac{1}{Kr}\right] \sin(Kr) - \dfrac{3 \cos(Kr)}{(Kr)^2}, & 0 < r < a, \\[3mm] \left[\dfrac{3}{(kr)^3} - \dfrac{1}{kr}\right] \sin(kr + \delta_2) - \dfrac{3 \cos(kr + \delta_2)}{(kr)^2}, & r > a. \end{cases}$$

The continuity of R and $(r^3 R_2)'$ at $r = a$ then gives

$$\frac{k^2}{K^2} \frac{[\tan(ka + \delta_2) - ka]}{[\tan(Ka) - Ka]} = \frac{[3 - (ka)^2]\tan(ka + \delta_2) - 3ka}{[3 - (Ka)^2]\tan(Ka) - 3Ka}.$$

Let $y = \tan(ka + \delta_2) - ka$. The above becomes

$$\left[\frac{3}{(ka)^3} - \frac{1}{ka}\right] y - 1 = \frac{y(-1 + O(k))}{bK^2(1 + O \cdot (k^2))},$$

where

$$b = \frac{a}{2\cos^2(k_0 a) - k_0}.$$

Therefore

$$\begin{aligned} y &= \frac{1}{\dfrac{3}{(ka)^3} - \dfrac{1}{ka} + \dfrac{1}{bK^2} + O(k)} \\[3mm] &= \frac{1}{\dfrac{3}{(ka)^3}\left[1 - \dfrac{(ka)^2}{3} + \dfrac{(ka)^3}{3bK^2} + O(k^4)\right]} \\[3mm] &= \frac{(ka)^3}{3}\left[1 + \dfrac{(ka)^2}{3} - \dfrac{(ka)^3}{3bK^2} + O(k^4)\right] \\[3mm] &= \frac{(ka)^3}{3} + \frac{(ka)^5}{9} + O(k^6), \end{aligned}$$

and

$$\begin{aligned} \delta_2 &= \tan^{-1}(y + ka) - ka \\ &\approx y = O(k^3). \end{aligned}$$

Thus the contribution of partial waves with $l = 2$ to $\frac{d\sigma}{d\Omega}$ is also proportional to k^4.
This is true for all ℓ for $E \to 0$.

Hence

$$\frac{\partial\sigma}{\partial\Omega}=|f(\theta)|^2=\frac{1}{k^2}\left|\sum_{l=0}^{\infty}(2l+1)e^{i\delta_l}\sin\delta_l P_l(\cos\theta)\right|^2$$

$$\xrightarrow[E\to0]{} k^4 F(\cos\theta)\sim E^2 F(\cos\theta),$$

and the exponent of E is $\lambda=2$. The highest power of $\cos\theta$ in the angular distribution function is also 2 since the waves consist mainly of $\ell=0$ and $l=1$ partial waves.

6008

1. The shell potential for the three-dimensional Schrödinger equation is
    $$V(r)=\alpha\delta(r-r_0).$$

a. Find the *s*-state ($l=0$) wave function for $E>0$. Include an expression that determines the phase shift δ. With $\hbar k=\sqrt{2mE}$ show that in the limit $k\to0$, $\delta\to Ak$, where A is a constant (called the scattering length). Solve for A in terms of α and r_0.

b. How many bound states can exist for $l=0$ and how does their existence depend on α? (Graphical proof is acceptable)

c. What is the scattering length A when a bound state appears at $E=0$? Describe the behavior of A as α changes from repulsive ($\alpha>0$) to attractive, and then when α becomes sufficiently negative to bind. Is the range of A distinctive for each range of α? Sketch A as a function of α.

                                                                    (*MIT*)

Sol:

a. The radial part of the Schrödinger equation for $\ell=0$ is
    $$-\frac{\hbar^2}{2m}\frac{1}{r^2}\frac{\partial}{\partial r}\left(r^2\frac{\partial}{\partial r}\psi\right)+V(r)\psi=E\psi.$$
    With $\psi=\mu/r$, $V(r)=\alpha\delta(r-r_0)$ it becomes
    $$-\frac{\hbar^2}{2m}\mu''+\alpha\delta(r-r_0)\mu=E\mu,$$

i. e.,

$$\mu'' - \beta\delta(r - r_0)\mu = -k^2\mu, \tag{1}$$

where

$$\beta = \frac{2m\alpha}{\hbar^2}, \qquad k = \sqrt{\frac{2mE}{\hbar^2}}.$$

The equation has solution for which $\mu = 0$ at $r = 0$ and $\mu =$ finite for $r \to \infty$

$$\mu = \begin{cases} \sin kr, & r < r_0, \\ a \sin(kr + \delta), & r > r_0. \end{cases}$$

Integrating Eq. (1) from $r_0 - \epsilon$ to $r_0 + \epsilon$ and letting $\epsilon \to 0$ give

$$\mu'(r_0+) - \mu'(r_0-) = \beta\mu(r_0).$$

The continuity of μ at $r = r_0$ and this condition give

$$\begin{cases} \sin kr_0 = a \sin(kr_0 + \delta), \\ \dfrac{\beta}{k}\sin kr_0 = a \cos(kr_0 + \delta) - \cos kr_0 \end{cases}$$

Hence,

$$a^2[\sin^2(kr_0 + \delta) + \cos^2(kr_0 + \delta)] = a^2 = 1 + \frac{\beta}{k}\sin 2kr_0 + \frac{\beta^2}{k^2}\sin^2 kr_0,$$

$$\tan(kr_0 + \delta) = \frac{\tan kr_0}{1 + \dfrac{\beta}{k}\tan kr_0}, \tag{2}$$

which determine a and the phase shift δ. In the limiting case of $k \to 0$, the above equation becomes

$$\frac{kr_0 + \tan\delta}{1 - kr_0 \tan\delta} \approx \frac{kr_0}{1 + \beta r_0},$$

or

$$\tan\delta \approx -\frac{\beta r_0^2}{1 + \beta r_0}k,$$

neglecting $O(k^2)$. Then, as $k \to 0$, we have $\tan\delta \to 0$ and so

$$\delta \approx -\frac{r_0 k}{1 + \dfrac{1}{\beta r_0}} = Ak,$$

where

$$A = \frac{-r_0}{1 + \dfrac{\hbar^2}{2m\alpha r_0}}$$

is the scattering length.

b. For bound states, $E < 0$ and Eq. (1) can be written as

$$\mu'' - \beta\delta(r - r_0)\mu = k^2\mu$$

with

$$\beta = \frac{2m\alpha}{\hbar^2}, \qquad k^2 = -\frac{2mE}{\hbar^2}.$$

The solution in which $\mu = 0$ at $r = 0$ and $\mu =$ finite for $r \to \infty$ is

$$\mu = \begin{cases} \sinh kr, & r < r_0, \\ ae^{-kr}, & r > r_0. \end{cases}$$

The continuity conditions, as in (a), give

$$\mu = \begin{cases} \sinh kr_0 = ae^{-kr_0}, \\ \beta a e^{-kr_0} = -ake^{-kr_0} - k\cosh kr_0. \end{cases}$$

Eliminating a we have

$$(\beta + k)\sinh kr_0 = -k\cosh kr_0,$$

or

$$c^{-2kr_0} = 1 + \frac{2kr_0}{\beta r_0}.$$

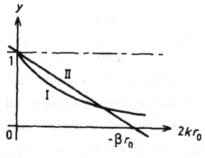

Fig. 6.1

For bound states $E < 0$. Between $E = -\infty$ and $E = 0$, or between $2kr0 = \infty$ and $2kr_0 = 0$, there is one intersection between the curves (I) $y = e^{-2kr_0}$ and

(II) $y = 1 + \dfrac{2kr_0}{\beta r_0}\, if -1 < \dfrac{1}{\beta ro} < 0$, as shown in Fig. 6.1. Thus if this condition is

satisfied there will be one bound state with $\ell = 0$. This condition requires

$$-1 < \frac{-\hbar^2}{2m|\alpha| r_0}, \quad or \quad \alpha < \frac{-\hbar^2}{2mr_0} = \alpha_0.$$

c.   In (a) it is found that

$$A = -\frac{r_0}{1 + \dfrac{1}{\beta r_0}} = -\frac{r_0}{1 + \dfrac{\hbar^2}{2mr_0\alpha}}.$$

The behavior of A is shown in Fig. 6.2, where it is seen that for $\alpha = 0$, $A = 0$;

for $\alpha = \alpha_0 = \dfrac{-\hbar^2}{2mro}$, $A = \pm\infty$; $\alpha = \pm\infty$, $A = -r_0$. With $E \to +0$, a bound state

appears at $E = 0$. At this energy $\alpha = \alpha_0$, $\delta = \pm\pi / 2$ and $A = \infty$.

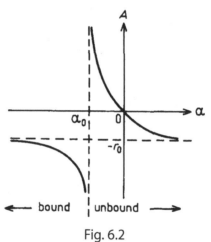

Fig. 6.2

6009

In a scattering experiment, the potential is spherically symmetric and the particles are scattered at such energy that only s and p waves need to be considered. Show that the differential cross-section can be written in the form. $\sigma(\theta) = a + b \cos(\theta) + c \cos^2(\theta)$ and find the values of a, b and c in terms of phase shifts.

Sol:

Since only the s and p waves need to be considered, we only need the $l = 0$ and $l = 1$ terms.

The scattering amplitude

$$f(\theta)=\frac{1}{k}\sum_l (2l+1)e^{i\delta_l} p_l(\cos\theta)\sin\delta_l$$

Since $p_0(\cos\theta)=1$, $p_1(\cos\theta)=(\cos\theta)$

$$f(\theta)=\frac{1}{k}[e^{i\delta_0}\sin\delta_0+3e^{i\delta_1}\cos\theta\sin\delta_1]$$

Because

$$\sigma(\theta)=|f(\theta)|^2$$

we get the form

$$\sigma(\theta)=\frac{1}{k^2}[\sin^2\delta_0+6\sin\delta_0\sin\delta_1\cos(\delta_0-\delta_1)\cos\theta+9\sin^2\delta_1\cos^2\theta]$$

From this, we got the form $\sigma(\theta)=a+b\cos(\theta)+c\cos^2(\theta)$ where

$$a=\frac{1}{k^2}\sin^2\delta_0,\ b=\frac{1}{k^2}[6\sin\delta_0\sin\delta_1\cos(\delta_0-\delta_1)]\ ,\ \text{and}\ c=\frac{9}{k^2}\sin^2\delta_1.$$

6010

Consider the quantum-mechanical scattering problem in the presence of inelastic scattering. Suppose one can write the partial wave expansion of the scattering amplitude for the elastic channel in the form

$$f(k,\theta)=\sum_{l=0}^{\infty}(2l+1)\frac{\eta le^{2i\delta_l}-1}{2ik}P_l(\cos\theta),$$

where $\delta_l(k)$ and $\eta_l(k)$ are real quantities with $0\le\eta_l\le1$, the wave number is denoted by k, and θ is the scattering angle. For a given partial wave, obtain the lower and upper bounds for the elastic cross section $\sigma_{elastic}^{(l)}$ in terms of $\sigma_{inelastic}^{(l)}$.

(Chicago)

Sol: As

$$\sigma_{elastic}^{(l)}=\pi\lambdabar^2(2l+1)|1-\eta_l e^{2i\delta_l}|^2,$$

$$\sigma_{inelastic}^{(l)}=\pi\lambdabar^2(2l+1)(1-|\eta_l e^{2i\delta_l}|^2),$$

where

$$\lambdabar^2=\frac{\lambda}{2\pi}=\frac{1}{k},$$

we have

$$\sigma_{elastic}^{(l)} = \frac{|1-\eta_l e^{2i\delta_l}|^2}{1-|\eta_l e^{2i\delta_l}|^2} \cdot \sigma_{inelastic}^{(l)}.$$

As η_l, δ_l are real numbers and $0 \le \eta_l \le 1$, we have

$$\frac{(1-\eta_l)^2}{1-\eta_l^2} \le \frac{|1-\eta_l e^{2i\delta_l}|^2}{1-|\eta_l e^{2i\delta_l}|^2} \le \frac{(1+\eta_l)^2}{1-\eta_l^2},$$

or

$$\frac{(1-\eta_l)^2}{1-\eta_l^2} \sigma_{inelastic}^{(l)} \le \sigma_{elastic}^{(l)} \le \frac{(1+\eta_l)^2}{1-\eta_l^2} \sigma_{inelastic}^{(l)}.$$

Therefore the upper and lower bounds of $\sigma_{elastic}^{(l)}$ are respectively

$$\frac{(1+\eta_l)^2}{1-\eta_l^2} \sigma_{inelastic}^{(l)} \quad \text{and} \quad \frac{(1-\eta_l)^2}{1-\eta_l^2} \sigma_{inelastic}^{(l)}.$$

6011

A slow electron of wave number k is scattered by a neutral atom of effective (maximum) radius R, such that $kR \ll 1$.

a. Assuming that the electron-atom potential is known, explain how the relevant phase shift δ is related to the solution of a Schrödinger equation.

b. Give a formula for the differential scattering cross section in terms of δ and k. (If you do not remember the formula, try to guess it using dimensional reasoning.)

c. Explain, with a diagram of the Schrödinger-equation solution, how a non-vanishing purely attractive potential might, at a particular k, give no scattering.

d. Explain, again with a diagram, how a potential that is attractive at short distances but repulsive at large distances might give resonance scattering near a particular k.

e. What is the maximum value of the total cross section at the center of the resonance?

(Berkeley)

ol:

a. We need only consider the s partial wave as $kR \ll 1$. The solution of the Schrödinger equation has for $r \to \infty$ the asymptotic form

$$\psi(r) \to \frac{\sin(kr + \delta)}{kr}.$$

The phase shift δ is thus related to the solution of the Schrödinger equation.

b. The differential scattering cross section is given by

$$\sigma(\theta) = \frac{\sin^2 \delta}{k^2}.$$

c. The phase shift δ in general is a function of the wave number k. When $\delta = n\pi$, $\sigma(\theta) = 0$, $\delta_t = 0$ and no scattering takes place. The asymptotic solution of the Schrödinger equation with $\ell = 0$ is shown in Fig. 6.3(a)

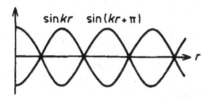

Fig. 6.3(a)

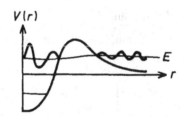

Fig. 6.3(b)

d. Consider a potential well as that given in Fig. 6.3(b). If the energy of the incident particle is near an eigenvalue of the well (a bound state), its wave function inside the well will be strongly coupled with its wave function outside and the wave function in the well will have a large amplitude, resulting in resonance scattering.

e. The maximum value of the total cross section at the center of the resonance peak is $4\pi R^2$, where R is the range of the force of interaction.

6012

For an attractive square-well potential $(V = -V_0, r < a; V = 0, r > a)$ find the "matching equation" at a positive energy, which determines the energy dependence of the $\ell = 0$ phase shift δ_0. From this show that at high energies $\delta(k) \to \dfrac{maV_0}{\hbar^2 k}$, and obtain this result from the Born approximation.

(Wisconsin)

Sol: Let $\chi = rR$. For the $\ell = 0$ partial wave, the Schrödinger equation becomes

$$\chi'' + k'^2 \chi = 0, \quad k'^2 = k^2\left(1 + \frac{V_0}{E}\right), \qquad r < a,$$

$$\chi'' + k^2 \chi = 0, \quad k^2 = \frac{2mE}{\hbar^2}, \qquad r > a.$$

The solutions are

$$\chi = \begin{cases} \sin(k'r), & r < a, \\ A\,\sin(kr + \delta_0), & r > a. \end{cases}$$

The continuity condition

$$(\ln \chi)'\big|_{r=a^-} = (\ln \chi)'\big|_{r=a^+}$$

gives an equation for determining δ_0:

$$k'\tan(ka + \delta_0) = k\tan(k'a).$$

As

$$k'^2 = k^2\left(1 + \frac{V_0}{E}\right) \text{ and } k^2 = \frac{2mE}{\hbar^2},$$

when $k \to \infty$, $k' \to k$. Hence

$$\delta_0 = \arctan\left[\frac{k}{k'}\tan(k'a)\right] - ka \to (k' - k)a \text{ as } k \to \infty.$$

Thus, letting $k \to \infty$ we obtain

$$\delta_0 \to \left(\frac{k'^2 - k^2}{k' + k}\right)a \approx \frac{kV_0}{2E}a = \frac{maV_0}{\hbar^2 k}.$$

The Born approximation expression for the phase shift for $l = 0$ is

$$\delta_0 \approx -\frac{2mk}{\hbar^2}\int_0^\infty V(r)j_0^2(kr)r^2 dr = \frac{2mkV_0}{\hbar^2}\int_0^a \frac{\sin^2 kr}{k^2}dr$$

$$= \frac{mV_0}{\hbar^2 k^2}\left[ka - \frac{1}{2}\sin(2ka)\right],$$

whence

$$\delta_0 \to \frac{mV_0 a}{\hbar^2 k}.$$

as $k \to \infty$, in agreement with the partial-wave calculation.

6013

Calculate the scattering cross section for a low energy particle from a potential given by $V = -V_0$ for $r < a$, $V = 0$ for $r > a$. Compare this with the Born approximation result.

(Columbia)

Sol: The radial Schrödinger equation can be written in the form

$$\chi_l''(r) + \left[k^2 - \frac{l(l+1)}{r^2}\right]\chi_l(r) = 0, \qquad r > a,$$

$$\chi_l''(r) + \left[k'^2 - \frac{l(l+1)}{r^2}\right]\chi_l(r) = 0, \qquad r < a,$$

where $\chi = rR(r)$,

$$k^2 = \frac{2mE}{\hbar^2}, \qquad k'^2 = \frac{2m(E+V_0)}{\hbar^2}.$$

Scattering at low energies is dominated by the s partial wave, for which $\ell = 0$, and the above become

$$\chi_l''(r) + k^2 \chi_l(r) = 0, \qquad r > a,$$
$$\chi_l''(r) + k'^2 \chi_l(r) = 0, \qquad r < a,$$

whose solutions are

$$\chi_l(r) = \begin{cases} A\sin(k'r), & r < a, \\ \sin(kr + \delta_0), & r > a. \end{cases}$$

The continuity condition $(\ln \chi_l)'\big|_{r=a^-} = (\ln \chi_l)'\big|_{r=a^+}$ gives

$$k\tan(k'a) = k'\tan(ka + \delta_0),$$

or

$$\delta_0 = \arctan\left[\frac{k}{k'}\tan(k'a)\right] - ka.$$

For low energies

$$k \to 0, k' \to k_0 = \sqrt{\frac{2mV_0}{\hbar^2}},$$

and the above becomes

$$\delta_0 \approx ka\left[\frac{\tan(k_0 a)}{k_0 a} - 1\right].$$

The total scattering cross section is then

$$\sigma \approx \frac{4\pi}{k^2}\sin^2\delta_0 \approx \frac{4\pi}{k^2}\delta_0^2 = 4\pi a^2\left[\frac{\tan(k_0 a)}{k_0 a} - 1\right]^2.$$

If $k_0 a \ll 1$,

$$\sigma \approx 4\pi a^2\left[\frac{k_0 a}{k_0 a} + \frac{(k_0 a)^3}{3k_0 a} - 1\right]^2 = \frac{16\pi a^6 m^2 V_0^2}{9\hbar^4}.$$

In the Born approximation,

$$f(\theta) = -\frac{m}{2\pi\hbar^2}\int e^{-i\mathbf{k}\cdot\mathbf{r}}V(r)e^{i\mathbf{k}'\cdot\mathbf{r}}d^3r,$$

where $\mathbf{k}'$, $\mathbf{k}$ are respectively the wave vectors of the incident and scattered waves. Let $\mathbf{q} = \mathbf{k} - \mathbf{k}'$, with $|\mathbf{k}'| = |\mathbf{k}| = k$ for elastic scattering. Then $q = 2k\sin\frac{\theta}{2}$, where θ is the scattering angle. Thus

$$\begin{aligned}
f(\theta) &= -\frac{m}{2\pi\hbar^2}\int_0^\infty V(r)r^2 dr \int_0^\pi e^{-iqr\cos\theta'}2\pi\sin\theta' d\theta'\\
&= -\frac{2m}{\hbar^2}\int_0^\infty V(r)\frac{\sin(qr)}{qr}r^2 dr = \frac{2mV_0}{\hbar^2 q}\int_0^a r\sin(qr)dr\\
&= \frac{2mV_0}{\hbar^2 q^3}[\sin(qa) - qa\cos(qa)].
\end{aligned}$$

Hence

$$\sigma(\theta) = |f(\theta)|^2 = \frac{4m^2 V_0^2}{\hbar^4 q^6}[\sin(qa) - qa\cos(qa)]^2.$$

For low energies $k \to 0$, $q \to 0$,

$$\sin(qa) \approx qa - \frac{1}{3!}(qa)^3, \quad \cos(qa) \approx 1 - \frac{1}{2!}(qa)^2,$$

and hence

$$\sigma(\theta) \approx \frac{4m^2 V_0^2 a^6}{9\hbar^4}.$$

The total cross section for scattering at low energies is then

$$\sigma = \int \sigma(\theta) \, d\Omega = \frac{16\pi m^2 V_0^2 a^6}{9\hbar^4}.$$

Therefore at low energies for which $k \to 0, ka \ll 1$, the two methods give the same result.

6014

In scattering from a potential $V(r)$, the wave function may be written as an incident plane wave plus an outgoing scattered wave: $\psi = e^{ikz} + v(r)$. Derive a differential equation for $v(r)$ in the first Born approximation.

(Wisconsin)

Sol: Two methods may be used for this problem.

Method 1:

For a particle of mass m in a central field $V(r)$, the Schrödinger equation can be written as

$$(\nabla^2 + k^2)\psi = U\psi,$$

where

$$U = \frac{2m}{\hbar^2} V, \qquad k = \sqrt{\frac{2mE}{\hbar^2}}.$$

Define Green's function $G(\mathbf{r} - \mathbf{r}')$ by

$$(\nabla^2 + k^2)G(\mathbf{r} - \mathbf{r}') = -4\pi\delta(\mathbf{r} - \mathbf{r}').$$

This is satisfied by the function

$$G(\mathbf{r} - \mathbf{r}') = \frac{\exp(ik|\mathbf{r} - \mathbf{r}'|)}{|\mathbf{r} - \mathbf{r}'|},$$

and the Schrödinger equation is satisfied by

$$\psi(\mathbf{r}) = \psi_0(\mathbf{r}) - \frac{1}{4\pi} \int G(\mathbf{r} - \mathbf{r}') U(\mathbf{r}') \psi(\mathbf{r}') d^3 r'.$$

As the incident wave is a plane wave $e^{ikz'}$, we replace $U(\mathbf{r}')\psi(\mathbf{r}')$ by $U(\mathbf{r}')e^{ikz'}$ in the first Born approximation:

$$\psi(r) = e^{ikz} - \frac{1}{4\pi} \int \frac{\exp(ik|\mathbf{r} - \mathbf{r}'|)}{|\mathbf{r} - \mathbf{r}'|} U(\mathbf{r}')e^{ikz'} d^3 r'.$$

Hence, the scattered wave is

$$v(r)=-\frac{1}{4\pi}\int\frac{\exp(ik|\mathbf{r}-\mathbf{r}'|)}{|\mathbf{r}-\mathbf{r}'|}U(\mathbf{r}')e^{ikz'}d^3\mathbf{r}'.$$

Applying the operator (∇^2+k^2) to the two sides of the equation, we get

$$(\nabla^2+k^2)v(r)=-\frac{1}{4\pi}\int(\nabla^2+k^2)\frac{\exp(ik(r-r'))}{|r-r'|}U(r')e^{ikz'}d^3r'$$

$$=\int\delta(r-r')U(r')e^{ikz'}d^3r'=U(r)e^{ikz}.$$

Hence, the differential equation for $v(r)$ is

$$(\nabla^2+k^2)v(r)=U(r)e^{ikz}.$$

Method 2:

Writing the radial Schrödinger equation as

$$(\nabla^2+k^2)\psi=U\psi,$$

where

$$k^2=\frac{2mE}{\hbar^2},\ U=\frac{2m}{\hbar^2}V,$$

and substituting in $\psi=e^{ikz}+v(r)$, we get

$$(\nabla^2+k^2)e^{ikz}+(\nabla^2+k^2)v(r)=U[e^{ikz}+v(r)],$$

or

$$(\nabla^2+k^2)v(r)=\frac{2m}{\hbar^2}V[e^{ikz}+v(r)],$$

as $(\nabla^2+k^2)e^{ikz}=0$. In the first Born approximation, $e^{ikz}+v(r)\approx e^{ikz}$, and so the differential equation for $v(r)$ is

$$(\nabla^2+k^2)\,v(r)\approx\frac{2m}{\hbar^2}Ve^{ikz}.$$

6015

In the quantum theory of scattering from a fixed potential, we get the following expression for the asymptotic form of the wave function

$$\psi(\mathbf{r})\xrightarrow[r\to\infty]{}e^{ikz}+f(\theta,\varphi)\frac{e^{ikr}}{r}.$$

a. If the entire Hamiltonian is rotationally invariant, give the argument that the scattering amplitude f should be independent of the angle φ.

b. Why cannot this argument be extended (considering rotation about any axis) to conclude that f should be independent of θ as well?

c. Reconsider part(b) in the case where the incident energy approaches zero.

d. What is the formula for the scattering cross section in terms of f?

e. What is the formula for the first Born approximation for f? (Be sure to define all quantities introduced. You need not worry about simple dimensionless factors like 2 or π).

f. Under what conditions is the Born approximation valid?

(*Berkeley*)

Sol:

a. The incident wave $e^{ikz} = e^{ikr\cos\theta}$ is the eigenstate of $\hat{l}_z$, third component of the angular momentum **L**, with eigenvalue $m = 0$. If the Hamiltonian is rotationally invariant, the angular momentum is conserved and the outgoing wave is still the eigenstate of $\hat{l}_z$ with eigenvalue $m = 0$, that is,

$$\hat{l}_z f(\theta, \varphi) = mf(\theta, \varphi) = 0.$$

Since $\hat{l}_z = \dfrac{\hbar}{i}\dfrac{\partial}{\partial\varphi}$, =this means that $\partial f(\theta, \varphi)/\partial\varphi = 0$.

b. As the asymptotic form of the wave function $\psi(\mathbf{r})$ is not an eigenfunction of $\hat{L}^2$, we cannot extend the above argument to conclude that f is independent of θ.

c. When the energy $E \to 0$, i.e., $k \to 0$, the incident wave consists mainly only of the $l = 0$ partial wave; other partial waves have very small amplitudes and can be neglected. Under such conditions, the rotational invariance of H results in the conservation of $\hat{L}^2$. Then the outgoing wave must also be the eigenstate of $\hat{L}^2$ with eigenvalue $l = 0$ (approximately). As

$$\hat{L}^2 = -\hbar^2\left[\frac{1}{\sin\theta}\frac{\partial}{\partial\theta}\left(\sin\theta\frac{\partial}{\partial\theta}\right) + \frac{1}{\sin^2\theta}\frac{\partial^2}{\partial\varphi^2}\right],$$

we have

$$\frac{1}{\sin\theta}\frac{d}{d\theta}\left[\sin\theta\frac{df(\theta)}{d\theta}\right] = 0.$$

As $f(\theta)$ must be a wave function with all the appropriate properties, this means

$$df(\theta)/d\theta = 0.$$

d. The differential scattering cross section is given by

$$\frac{d\sigma}{d\Omega} = |f(\theta, \varphi)|^2.$$

e. In the first Born approximation, for scattering from a central field $V(r')$, f is given by

$$f(\theta, \varphi) = -\frac{m}{2\pi\hbar^2} \int V(r') \exp(-i\mathbf{q} \cdot \mathbf{r}') d^3 r'$$

$$= -\frac{2m}{\hbar^2 q} \int_0^\infty r' V(r') \sin(qr') dr',$$

where $\mathbf{q} = \mathbf{k} - \mathbf{k}_0$, $\mathbf{k}$ and $\mathbf{k}_0$ being respectively the momenta of the particle before and after scattering.

f. The validity of Born approximation requires that the interaction potential is small compared with the energy of the incident particle.

6016

Consider a particle of mass m which scatters off a potential $V(x)$ in one dimension.

a. Show that

$$G_E(x) = \frac{1}{2\pi} \int_{-\infty}^{\infty} dk \frac{e^{ikx}}{E - \frac{\hbar^2 k^2}{2m} + i\varepsilon},$$

with ε positive infinitesimal, is the free-particle Green's function for the time-independent Schrödinger equation with energy E and outgoing-wave boundary conditions.

b. Write down an integral equation for the energy eigenfunction corresponding to an incident wave traveling in the positive x direction. Using this equation find the reflection probability in the first Born approximation for the potential

$$V(x) = \begin{cases} V_0, & |x| < a/2, \\ 0, & |x| > a/2. \end{cases}$$

For what values of E do you expect this to be a good approximation?

(Buffalo)

ol:

a. To solve the one-dimensional time-independent Schrödinger equation

$$\left(\frac{\hbar^2}{2m} \frac{d^2}{dx^2} + E \right) \psi = V\psi,$$

we define a Green's function $G_E(x)$ by

$$\left(\frac{\hbar^2}{2m} \frac{d^2}{dx^2} + E \right) G_E(x) = \delta(x).$$

Expressing $G_E(x)$ and $\delta(x)$ as Fourier integrals

$$G_E(x) = \frac{1}{2\pi} \int_{-\infty}^{\infty} f(k) e^{ikx} dk,$$

$$\delta(x) = \frac{1}{2\pi} \int_{-\infty}^{\infty} e^{ikx} dk$$

and substituting these into the equation for $G_E(x)$, we obtain

$$\left(-\frac{\hbar^2 k^2}{2m} + E \right) f(k) = 1,$$

or

$$f(k) = \frac{1}{E - \dfrac{\hbar^2 k^2}{2m}}.$$

As the singularity of $f(k)$ is on the path of integration and the Fourier integral can be understood as an integral in the complex k-plane, we can add $i\varepsilon$, where ε is a small positive number, to the denominator of $f(k)$. We can then let $\varepsilon \to 0$ after the integration. Consider

$$G_E(k) = \frac{1}{2\pi} \int_{-\infty}^{\infty} dk \frac{e^{ikx}}{E - \dfrac{\hbar^2 k^2}{2m} + i\varepsilon}.$$

The intergral is singular where

$$(E + i\varepsilon) - \frac{\hbar^2 k^2}{2m} = 0,$$

i.e., at

$$k = \pm k_1,$$

where

$$k_1 = \frac{\sqrt{2m(E+i\varepsilon)}}{\hbar}.$$

When $x > 0$, the integral becomes a contour integral on the upper half-plane with a singularity at k_1 with residue

$$a_1 = -\frac{me^{ik_1 z}}{\pi\hbar^2 k_1}.$$

Cauchy's integral formula then gives

$$G_E(x) = 2\pi i a_1 = -i\frac{m}{\hbar^2 k_1}e^{ik_1 x} \quad (x > 0).$$

As

$$\varepsilon \to 0, k_1 \to \frac{\sqrt{2mE}}{\hbar}.$$

This is the value of k_1 to be used in the expression for $G_E(x)$. Similarly, when $x < 0$, we can carry out the integration along a contour on the lower half-plane and get

$$G_E(x) = -i\frac{m}{\hbar^2 k_1}e^{-ik_1 x} \qquad (x < 0).$$

Here $k_1 = \frac{\sqrt{2mE}}{\hbar}$ also. Thus the free-particle Green's function $G_E(x)$ represents the outgoing wave whether $x > 0$ or $x < 0$.

b. The solution of the stationary Schrödinger equation satisfies the integral equation

$$\psi_E(x) = \psi^0(x) + G_E(x) * [V(x)\psi_E(x)]$$
$$= \psi^0(x) + \int_{-\infty}^{\infty} G_E(x-\xi)V(\xi)\psi_E(\xi)d\xi,$$

where $\psi^0(x)$ is a solution of the equation

$$\left(\frac{\hbar^2}{2m}\frac{d^2}{dx^2} + E\right)\psi(x) = 0.$$

In the first-order Born approximation we replace ψ_0 and ψ_E on the right side of the integral equation by the incident wave function and get

$$\psi_E(x) = e^{ikx} + \int_{-\infty}^{\infty} G_E(x-\xi)V(\xi)e^{ik\xi}d\xi = e^{ikx}$$
$$+ \int_{-\infty}^{x}(-i)\frac{m}{\hbar^2 k}e^{ik(x-\xi)}V(\xi)e^{ik\xi}d\xi$$
$$+ \int_{x}^{\infty}(-i)\frac{m}{\hbar^2 k}e^{-ik(x-\xi)}V(\xi)e^{ik\xi}d\xi.$$

For reflection we require $\psi_E(x)$ for $x \to -\infty$. For $x \to -\infty$,

$$\int_{-\infty}^{x}(-i)\frac{m}{\hbar^2 k}e^{ik(x-\xi)}V(\xi)e^{ik\xi}d\xi=0,$$

$$\int_{x}^{\infty}(-i)\frac{m}{\hbar^2 k}e^{-ikx}e^{2ik\xi}V(\xi)d\xi=\int_{-a/2}^{a/2}(-i)\frac{m}{\hbar^2 k}e^{-ikx}V_0 e^{2ik\xi}d\xi$$

$$=-i\frac{mV_0}{\hbar^2 k^2}\sin(ka)e^{-ikx}.$$

Hence, the reflection probability is

$$R=|\psi_E(-\infty)|^2/|\psi_0|^2=\frac{m^2 V_0^2}{\hbar^4 k^4}\sin^2 ka.$$

When the energy is high,

$$\left|\int_{-\infty}^{\infty}G_E(x-\xi)V(\xi)e^{ik\xi}d\xi\right| \ll |e^{ikx}|,$$

and replacing $\psi(x)$ by e^{ikx} is a good approximation.

6017

Calculate, using the Born approximation, the differential and total cross sections for the scattering from the potential that is given by $V_0\exp(-\alpha^2 r^2)$.

Sol:

The differential cross section is given by $\dfrac{d\sigma}{d\Omega}=|f(\theta)|^2$

where $f(\theta)$ is given by

$$f(\theta)=\frac{-2m}{\hbar^2 q}\int_0^{\infty}r\sin(qr)V(r)dr$$

Plugging in the above given potential, we get

$$f(\theta)=\frac{-2m}{\hbar^2 q}\int_0^{\infty}r\sin(qr)V_0 e^{-\alpha^2 r^2}\,dr$$

to evaluate the integral $\int_0^{\infty}\sin(qr)re^{-\alpha^2 r^2}\,dr=(1/2)\int_{-\infty}^{\infty}\dfrac{r(e^{iqr}-e^{-iqr})e^{-\alpha^2 r^2}\,dr}{2i}$

$$=-(1/4)\frac{\partial}{\partial q}\left\{e^{-q^2/4\alpha^2}\int_{-\infty}^{\infty}(e^{-(\alpha r+iq/2\alpha)^2}+e^{-(\alpha r-iq/2\alpha)^2})dr\right\}$$

$$=-(1/4)\frac{\partial}{\partial q}e^{-q^2/4\alpha^2}\left(2\frac{\sqrt{\pi}}{\alpha}\right)=\frac{\sqrt{\pi}q}{4\alpha^3}e^{\frac{-q^2}{4\alpha^2}}$$

So plugging this integral in $f(\theta)$, we get $f(\theta) = \dfrac{-\sqrt{\pi} m V_0}{2\hbar^2 \alpha^3} e^{\frac{-q^2}{4\alpha^2}}$

The relation between momentum transfer and the scattering angle in the Born approximation is $q = 2k \sin(\theta/2)$. In this approximation, we get

$$\frac{d\sigma}{d\Omega} = |f(\theta)|^2 = \frac{\pi m^2 V_0^2}{4\hbar^4 \alpha^6} e^{\frac{-2k^2}{\alpha^2}\sin^2(\theta/2)}$$

And the total cross section is given by $\sigma(\theta) = \int d\Omega \dfrac{d\sigma}{d\Omega}$

$$= \frac{\pi m^2 V_0^2}{4\hbar^4 \alpha^6} \int_0^{2\pi} d\phi \int_{-1}^{+1} d\cos\theta\, e^{\frac{-2k^2}{\alpha^2}(1-\cos\theta)}$$

So the total cross section is given by $\sigma(\theta) = \dfrac{\pi^2 m^2 V_0^2}{2\hbar^4 \alpha^6}(1 - e^{-2k^2/a^2})$

6018

A particle of energy E scatters off a repulsive spherical potential

$$\begin{cases} V(r) = V_0 & \text{for } r < a \\ 0 & \text{for } r > a. \end{cases}$$

In the low energy limit the total cross section is given by $4\pi a^2 \left(\dfrac{\tanh(ka)}{ka} - 1\right)^2$

where $k^2 = \dfrac{2m}{\hbar^2}(V_0 - E) > 0$. In the limit $V_0 \to \infty$, find the ratio of σ to the classical scattering cross section off a sphere of radius a.

Sol:

$$\sigma = 4\pi a^2 \left(\frac{\tanh(ka)}{ka} - 1\right)^2$$

In the low energy limit, as the potential $V_0 \to \infty$ and E is negligibly small $k^2 = \dfrac{2m}{\hbar^2}(V_0 - E)$ tends to infinity and when $ka \to \infty$, $\tanh(ka) \to 1$.

So $\sigma = 4\pi a^2 \left(\dfrac{1}{ka} - 1\right)^2$ and approximating on the limit, $\lim\limits_{ka\to\infty} \sigma = 4\pi a^2$.

The classical scattering cross section is given by $\sigma_c = \pi a^2$.

The ratio of these cross section is $\dfrac{\sigma}{\sigma_c} = 4$.

6019

A nucleon is scattered elastically from a heavy nucleus. The effect of the heavy nucleus can be represented by a fixed potential

$$V(r) = \begin{cases} -V_0, & r < R, \\ 0, & r > R, \end{cases}$$

where V_0 is a positive constant. Calculate the deferential cross section to the lowest order in V_0.

(Berkeley)

Sol: Let μ be the reduced moves of the nucleon and the nucleus, $\mathbf{q} = \mathbf{k} - \mathbf{k}'$ where $\mathbf{k}'$, $\mathbf{k}$ are respectively the wave vectors of the nucleon before and after the scattering. In the Born approximation, as in **Problem 6013**, we have

$$\sigma(\theta) = |f(\theta)|^2 = \frac{4\mu^2}{\hbar^4 q^2} \left| \int_0^\infty r' V(r') \sin qr' dr' \right|^2$$

$$= \frac{4\mu^2 V_0^2}{\hbar^4 q^6} (\sin qR - qR\cos qR)^2,$$

where $q = 2k\sin(\theta/2)$, $k = |\mathbf{k}| = |\mathbf{k}'|$.

6020

A particle of mass m, charge e, momentum p scatters in the electrostatic potential produced by a spherically symmetric distribution of charge. You are given the quantity $\int r^2 \rho d^3 x = A$, $\rho(r) d^3 x$ being the charge in a volume element $d^3 x$. Supposing that ρ vanishes rapidly as $r \to \infty$ and that $\int \rho d^3 x = 0$; working in the first Born approximation, compute the differential cross section for forward scattering. (That is $\frac{d\sigma}{d\Omega}\big|_{\theta=0}$, where θ is the scattering angle.)

(Princeton)

Sol: In the first Born approximation, we have

$$\frac{d\sigma}{d\Omega} = \frac{m^2 e^2}{4\pi^2 \hbar^4} \left| \int U(r) \exp(i\mathbf{q} \cdot \mathbf{r}) d^3 x \right|^2,$$

464   *Problems and Solutions in Quantum Mechanics*

where

$$\mathbf{q}=\mathbf{k}-\mathbf{k}',\ q=2k\sin\frac{\theta}{2}=\frac{2p}{\hbar}\sin\frac{\theta}{2},\ \mathbf{k}'\text{ and }\mathbf{k}$$

being the wave vectors of the particle before and after the scattering, $U(r)$ is the electrostatic Coulomb potential and satisfies the Poission equation

$$\nabla^2 U=-4\pi\rho(r).$$

Let

$$F(q)=\int\rho(r)\ \exp(i\mathbf{q}\cdot\mathbf{r})d^3x,$$

where $F(q)$ is the Fourier transform of $\rho(r)$. Using the Poisson equation we have

$$\int U(r)\ \exp(i\mathbf{q}\cdot\mathbf{r})d^3x=\frac{2\pi}{q^2}F(q).$$

Hence

$$\frac{d\sigma}{d\Omega}=\frac{m^2e^2}{4\pi^2\hbar^4}\cdot\frac{(4\pi)^2}{q^4}\left|F(q)\right|^2=\frac{4m^2e^2}{\hbar^4q^4}\left|F(q)\right|^2.$$

For forward scattering, θ is small and so q is small also. Then

$$F(q)=\int\rho(r)\ \exp(i\mathbf{q}\cdot\mathbf{r})\ d^3x$$

$$=\int\rho(r)\left[1+i\mathbf{q}\cdot r+\frac{1}{2!}(i\mathbf{q}\cdot\mathbf{r})^2+\cdots\right]d^3x$$

$$=\int\rho(r)d^3x-\frac{1}{2!}\int\rho(r)(\mathbf{q}\cdot\mathbf{r})^2d^3x+\cdots$$

$$\approx-\frac{1}{6}q^2\int\rho(r)r^2d^3x=-\frac{Aq^2}{6},$$

since as $\int\rho d^3x=0$, $\int_0^\pi\cos^{2n+1}\theta\cdot\sin\theta\ d\theta=0$, the lowest order term for $\theta\to0$ is

$$\frac{1}{2!}\int\rho(r)\ (i\mathbf{q}\cdot\mathbf{r})^2\ d^3x\approx-\frac{q^2}{6}\int\rho(r)r^2d^3x.$$

Hence

$$\left.\frac{d\sigma}{d\Omega}\right|_{\theta=0}=\frac{A^2m^2e^2}{9\hbar^4}.$$

6021

Use Born approximation to find, up to a multiplicative constant, the differential scattering cross section for a particle of mass m moving in a repulsive potential

$$V = Ae^{-r^2/a^2}.$$

(Berkeley)

Sol: In Born approximation we have (**Problem 6013**)

$$f(\theta) = -\frac{2m}{\hbar^2 q} \int_0^\infty rV(r)\,\sin(qr)\,dr,$$

where $q = 2k\sin(\theta/2)$, $\hbar k$ being the momentum of incident particle. As

$$f(\theta) = -\frac{2mA}{\hbar^2 q} \int_0^\infty r e^{-r^2/a^2} \sin(qr)dr = -\frac{mA}{\hbar^2 q} \int_{-\infty}^\infty r e^{-r^2/a^2} \sin(qr)dr$$

$$= \frac{mAa^2}{2\hbar^2 q} \int_{-\infty}^\infty (e^{-r^2/a^2})' \sin(qr)dr = -\frac{mAa^2}{2\hbar^2} \int_{-\infty}^\infty e^{-r^2/a^2} \cos(qr)dr$$

$$= -\frac{mAa^3}{2\hbar^2} \int_{-\infty}^\infty e^{-(r/a)^2} \cos\left(qa\frac{r}{a}\right)d\left(\frac{r}{a}\right)$$

$$= -\frac{mAa^3}{2\hbar^2} \int_{-\infty}^\infty e^{-r^2} \cos(qar)dr$$

$$= -\frac{mAa^3}{4\hbar^2} \int_{-\infty}^\infty \left\{ \exp\left[-\left(r - \frac{iqa}{2}\right)^2\right] + \exp\left[-\left(r + \frac{iqa}{2}\right)^2\right] \right\} e^{-q^2a^2/4} dr$$

$$= -\frac{mAa^3}{2\hbar^2} \sqrt{\pi} e^{-q^2a^2/4},$$

$$\sigma(\theta) = |f(\theta)|^2 = \frac{m^2 A^2 a^6}{4\hbar^4} \pi e^{-q^2a^2/2}.$$

6022

A nonrelativistic particle is scattered by a square-well potential

$$V(r) = \begin{cases} -V_0, & r < R, \ (V_0 > 0) \\ 0, & r > R. \end{cases}$$

a. Assuming the bombarding energy is sufficiently high, calculate the scattering cross section in the first Born approximation (normalization is not essential), and sketch the shape of the angular distribution, indicating angular units.

b. How can this result be used to measure R?

c. Assuming the validity of the Born approximation, if the particle is a proton and $R = 5 \times 10^{-13}$ cm, roughly how high must the energy be in order for the scattering to be sensitive to R?

(Wisconsin)

Sol:

a. Using Born approximation we have (**Problem 6013**)

$$f(\theta) \propto \frac{-1}{q} \int_0^\infty rV(r) \sin(qr)\, dr$$

$$= \frac{V_0}{q} \int_0^R r \sin(qr)\, dr = \frac{V_0}{q^3}(\sin qR - qR \cos qR).$$

Hence

$$\frac{d\sigma}{d\Omega} \propto \left(\frac{\sin x - x\cos x}{x^3} \right)^2,$$

where $x = qR = 2kR\sin\frac{\theta}{2}$.

The angular distribution is shown in Fig. 6.4

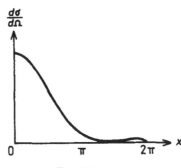

Fig. 6.4

b. The first zero of $\frac{d\sigma}{d\Omega}$ occurs at x for which $x = \tan x$, whose solution is $x \approx 1.43\pi$. This gives

$$R = \frac{1.43\pi}{2k\sin\dfrac{\theta_1}{2}}.$$

By measuring the minimum angle θ_1 for which $\frac{d\sigma}{d\Omega} = 0$, R can be determined.

c. In order that R may be determined from the zero points of $\frac{d\sigma}{d\Omega}$, we require that the maximum value of x, $2kR$, is larger than 1.43π, or

$$E \geq \frac{\hbar^2}{2m_p}\left(\frac{1.43\pi}{2R}\right)^2 = \frac{(1.43\pi)^2}{8}\frac{\hbar^2}{m_p c^2}\left(\frac{c}{R}\right)^2$$

$$= \frac{(1.43\pi)^2}{8} \times \frac{(6.58\times10^{-22})^2}{938} \times \left(\frac{3\times10^{10}}{5\times10^{-13}}\right)^2$$

$$= 4.2 \text{ MeV.}$$

6023

Consider the potential $V(r)=\sum_i V_0 a^3 \delta^{(3)}(r-r_i)$, where r_i are the position vectors of a cube of length a centred at the origin and V_0 is a constant. If $V_0 a^2 \ll \frac{\hbar^2}{m}$, find the total scattering cross section in the low energy limit.

Sol:

(Note: As this question is involving the delta functions and the definition of delta functions is in three dimensions and it indicates a unit cube not a sphere as if it were sphere we will be using spherical polar coordinates and there will be an extra volumetric term.)

$$V(r)=\sum_i V_0 a^3 \delta^{(3)}(r-r_i)=V(r)=\sum_i V_0 a^3 \delta(x-x_i)\delta(y-y_i)\delta(z-z_i),$$ where x_i, y_i, and z_i are the coordinates at the eight-corners of the cube whose centre is at the origin.

$$f(\theta)=\frac{-m}{2\pi\hbar^2}\int V(r)d^3r = \frac{-m}{2\pi\hbar^2}V_0 a^3 \iiint \sum_{i=1\text{ to }8} V_0 a^3\delta(x-x_i)\delta(y-y_i)\delta(z-z_i)$$

$$f(\theta)=\frac{-m}{2\pi\hbar^2}V_0 a^3[1+1+1+1+1+1+1+1]=\frac{-4m}{\pi\hbar^2}V_0 a^3 \text{ total scattering cross section}$$

is given by $\sigma = \int |f(\theta)|^2 \sin\theta\, d\theta\, d\phi$.

$$\sigma = \frac{64a^2}{\pi}\left[\frac{mV_0 a^2}{\hbar^2}\right]^2 \text{ is the total cross section.}$$

6024

A convenient model for the potential energy V of a particle of charge q scattering on an atom of nuclear charge Q is $V=\frac{qQ}{r}e^{-\alpha r}$, where α^- represents the screening length of the atomic electrons.

a. Use the Born approximation

$$f = -\frac{1}{4\pi} \int e^{-i\Delta \mathbf{k} \cdot \mathbf{r}} \frac{2m}{\hbar^2} V(\mathbf{r}) d^3\mathbf{r}$$

to calculate the scattering cross section σ.

b. How should α depend on the nuclear charge Z?

(Columbia)

Sol:

For the elastic scattering, $|\mathbf{k}| = |\mathbf{k}_0| = k$, and $|\Delta \mathbf{k}| = |\mathbf{k} - \mathbf{k}_0| = 2k \sin\frac{\theta}{2}$, θ being the scattering angle. Thus

$$f = -\frac{1}{4\pi} \int e^{-i\Delta \mathbf{k} \cdot \mathbf{r}} \frac{2m}{\hbar^2} V(r) d^3\mathbf{r}$$

$$= -\frac{m}{2\pi\hbar^2} \int_0^\infty r^2 dr \int_0^{2\pi} d\varphi \int_0^\pi e^{-i\Delta k r \cos\theta'} V(r) \sin\theta' d\theta'$$

$$= -\frac{2mqQ}{\Delta k \hbar^2} \int_0^\infty e^{-\alpha r} \sin(\Delta k r) dr$$

$$= -\frac{2mqQ}{\hbar^2} \frac{1}{(\alpha^2 + \Delta k^2)}.$$

Therefore

$$\sigma = |f(\theta)|^2 = \frac{4m^2 q^2 Q^2}{\hbar^4 [\alpha^2 + 4k^2 \sin^2(\theta/2)]^2}.$$

a. In the Thomas-Fermi approximation, when Z is large, the atomic electrons can be regarded as a Fermi gas. As such an electron is in a bound state in the atom; its energy is lower than $E(\infty) = 0$. Then its maximum possible momentum p_{max} at r must satisfy

$$\frac{1}{2m} p_{max}^2(r) - e\phi(r) = 0, \tag{1}$$

where $\phi(r)$ is the potential at distance r from the nucleus, since its energy is negative. Thus the Fermi momentum at r is

$$p_f(r) = p_{max}(r) = [2me\phi(r)]^{1/2}.$$

For a Fermi gas,

$$p_f = \hbar(3\pi^2 n)^{1/3},$$

where n is the number density. Comparing the above expressions we have

$$n(r) = \frac{1}{3\pi^2 \hbar^3} [2me\phi(r)]^{3/2}$$

$$= \frac{1}{3\pi^2 \hbar^3} \left(2me \frac{Ze}{r} e^{-\alpha r} \right)^{\frac{3}{2}},$$

where Ze is the nuclear charge. As the atom is neutral,

$$Z = \int n d^3 \mathbf{r} = 4\pi \int_0^\infty n(r) r^2 dr$$

$$= \frac{4}{3\pi \hbar^3} (2mZe^2)^{3/2} \int_0^\infty e^{-\frac{3}{2}\alpha r} r^{1/2} dr$$

$$= \frac{2}{3\sqrt{\pi} \hbar^3} \left(\frac{4mZe^2}{3\alpha} \right)^{3/2}.$$

Hence

$$\alpha = \frac{4me^2}{3\hbar^2} \left(\frac{4}{9\pi} \right)^{1/3} Z^{1/3} = \frac{4}{3} \left(\frac{4}{9\pi} \right)^{1/3} \frac{1}{a_0} Z^{1/3},$$

where $a_0 = \hbar^2/me^2$ is the Bohr radius.

6025

A particle of mass m is scattered by a potential $V(\mathbf{r}) = V_0 \exp(-r/a)$.

a. Find the differential scattering cross section in the first Born approximation. Sketch the angular dependence for small and large k, where k is the wave number of the particle being scattered. At what k value does the scattering begin to be significantly non-isotropic? Compare this value with the one given by elementary arguments based on angular momentum.

b. The criterion for the validity of the Born approximation is

$$\left| \Delta \psi^{(1)}(0) / \psi^{(0)}(0) \right| \ll 1,$$

where $\Delta \psi^{(1)}$ is the first order correction to the incident plane wave $\psi^{(0)}$. Evaluate this criterion explicitly for the present potential. What is the low-k limit of your result? Relate it to the strength of the attractive potential required for the existence of bound states (see the statement of problem). Is the high-k limit of the criterion less or more restrictive on the strength of the potential?

(Berkeley)

Sol:

a. The first Born approximation gives

$$f(\theta) = -\frac{2m}{\hbar^2 q} \int_0^\infty r' V(r') \sin(qr') dr'$$

$$= -\frac{2mV_0}{\hbar^2 q} \int_0^\infty r' \sin(qr') \exp(-r'/a) dr'$$

$$= -\frac{4mV_0 a^3}{\hbar^2 (1 + q^2 a^2)^2},$$

where $q = 2k \sin(\theta/2)$, $q\hbar$ being the magnitude of the momentum transfer in the scattering.

Hence

$$\sigma(\theta) = |f(\theta)|^2 = \frac{16 m^2 V_0^2 a^6}{\hbar^4 \left(1 + 4k^2 a^2 \sin^2 \dfrac{\theta}{2}\right)^4}.$$

The angular distribution $\sigma(\theta)/\sigma(0)$ is plotted in Fig. 6.5 for $ka = 0$ and $ka = 1$. It can be seen that for $ka \gtrsim 1$, the scattering is significantly nonisotropic. The angular momentum at which only s-wave scattering, which is isotropic, is important must satisfy

$$a \cdot k\hbar \leq \hbar, \quad \text{i.e.,} \quad ka \leq 1.$$

When $ka \sim 1$, the scattering begins to be significantly non-isotropic. This is in agreement with the result given by the first Born approximation,

b. The wave function to the first order is

$$\psi(r) = e^{ikz} - \frac{1}{4\pi} \int \frac{e^{-ik|\mathbf{r}-\mathbf{r}'|}}{|\mathbf{r}-\mathbf{r}'|} \frac{2m}{\hbar^2} V(\mathbf{r}') e^{ikz'} dV'.$$

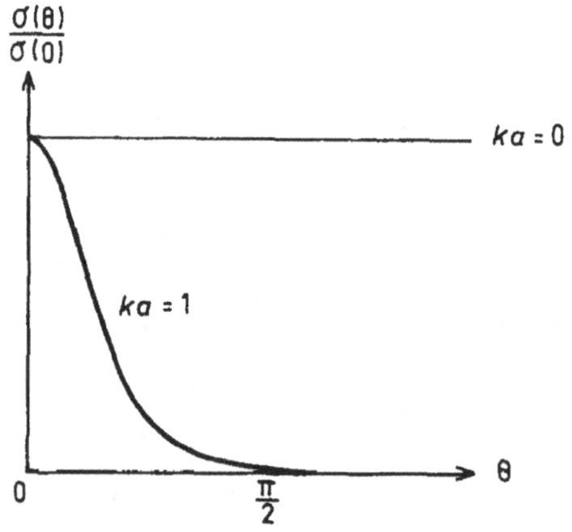

Fig. 6.5

Hence

$$\left|\frac{\Delta\psi^{(1)}(0)}{\psi^{(0)}(0)}\right| = \left|\frac{V_0}{4\pi}\int\frac{e^{ikr'}}{r'}\frac{2m}{\hbar^2}e^{-r'/a+ikz'}dV'\right|$$

$$= \left|\frac{mV_0}{\hbar^2}\int r'e^{ikr'-r'/a+ikr'\cos\theta'}\sin\theta' d\theta' dr'\right|$$

$$= \frac{2m|V_0|a^2\sqrt{1+4k^2a^2}}{\hbar^2(4k^2a^2+1)} = \frac{2m|V_0|a^2}{\hbar^2\sqrt{1+4k^2a^2}}.$$

The criterion for the validity of the first Born approximation is then

$$\frac{2m|V_0|a^2}{\hbar^2\sqrt{1+4k^2a^2}} \ll 1.$$

In the low-k, limit, $ka \ll 1$, the above becomes

$$\frac{2m|V_0|a^2}{\hbar^2} \ll 1, \quad \text{or} \quad |V_0| \ll \frac{\hbar^2}{2ma^2}.$$

In the high-k limit, $ka \gg 1$, the criterion becomes

$$\frac{m|V_0|a}{\hbar^2 k} \ll 1, \quad \text{or} \quad |V|_0 \ll \frac{\hbar^2 k}{ma}.$$

Since in this case $k \gg \frac{1}{a}$ the restriction on $|V_0|$ is less than for the low-k limit.

6026

For an interaction $V(r) = \beta r^{-1}\exp(-\alpha r)$ find the differential scattering cross section in Born approximation. What are the conditions for validity? Suggest one or more physical applications of this model.

(Berkeley)

Sol:

In Born approximation, we first calculate (**Problem 6013**)

$$f(\theta) = -\frac{2m}{\hbar^2 q}\int_0^\infty r'V(r')\sin qr' dr'$$

$$= -\frac{2m\beta}{\hbar^2 q}\int_0^\infty e^{-\alpha r}\sin qr\, dr = \frac{-2m\beta}{\hbar^2(q^2+\alpha^2)},$$

where $q = 2k\sin\frac{\theta}{2}$ and m is the mass of the particle, and then the differential cross section

$$\sigma(\theta) = |f(\theta)|^2 = \frac{4m^2\beta^2}{\hbar^4(q^2+\alpha^2)^2}.$$

The derivation is based on the assumption that the interaction potential can be treated as a perturbation, so that the wave function of the scattered particle can be written as

$$\psi(\mathbf{r}) = \psi_0(\mathbf{r}) + \psi_1(\mathbf{r}), \qquad \text{where} \quad |\psi_1| \ll |\psi_0|,$$

$$\psi_0(\mathbf{r}) = e^{ikz},$$

$$\psi_1(\mathbf{r}) \approx -\frac{m}{2\pi\hbar^2} \int \frac{e^{ik|\mathbf{r}-\mathbf{r}'|}}{|\mathbf{r}-\mathbf{r}'|} V(\mathbf{r}')\,\psi_0(\mathbf{r}')\,d^3r'.$$

Specifically, we shall consider two cases, taking a as the extent of space where the potential is appreciable.

i. The potential is sufficiently weak or the potential is sufficiently localized. As

$$|\psi_1| \le \frac{m}{2\pi\hbar^2} \int \frac{|V(\mathbf{r}')|}{|\mathbf{r}-\mathbf{r}'|} |\psi_0(\mathbf{r}')|\,d^3r'$$

$$\approx \frac{m}{2\pi\hbar^2} |V||\psi_0| \int_0^a \frac{4\pi\bar{r}^2\,d\bar{r}}{\bar{r}}$$

$$\approx m|V|a^2|\psi_0|/\hbar^2,$$

for $|\psi_1| \ll |\psi_0|$ we require

$$\frac{m|V|a^2}{\hbar^2} \ll 1,$$

i.e.,

$$|V| \ll \frac{\hbar^2}{ma^2} \text{ or } a \ll \frac{\hbar}{\sqrt{m|V|}}.$$

This means that where the potential is weak enough or where the field is sufficiently localized, the Born approximation is valid. Note that the condition does not involve the velocity of the incident particle, so that as long as the interaction potential satisfies this condition the Born approximation is valid for an incident particle of any energy.

ii. High energy scattering with $ka \gg 1$. The Born approximation assumes $\psi_0 = e^{ikz}$ and a ψ_1 that satisfies

$$\nabla^2\psi_1 + k^2\psi_1 = \frac{2m}{\hbar^2} V e^{ikz}.$$

Let $\psi_1 = e^{ikz} f(\theta, \varphi)$. The above becomes

$$\frac{\partial f}{\partial z} = -\frac{im}{\hbar^2 k} V,$$

and so

$$\psi_1 = e^{ikz} f = -\frac{im}{\hbar^2 k} e^{ikz} \int V dz .$$

Then as

$$|\psi_1| \sim \frac{m}{\hbar^2 k} |V| a |\psi_0|,$$

for $|\psi_1| \ll |\psi_0| = 1$ we require

$$|V| \ll \frac{\hbar^2 k}{ma} = \frac{\hbar v}{a},$$

where v is the speed of the incident particle, $v = \frac{p}{m} = \frac{\hbar k}{m}$. Thus as long as the energy of the incident particle is large enough, the Born approximation is valid.

From the above results we see that if a potential field can be treated as a small perturbation for incident particles of low energy, it can always be so treated for incident particles of high energy. The reverse, however, is not true. In the present problem, the range of interaction can be taken to be $a \approx \frac{1}{\alpha}$, so that $V(a) \sim \frac{\beta}{a}$. The conditions then become

i. $|\beta| \ll \dfrac{\alpha \hbar^2}{m}$,

ii. $|\beta| \ll \hbar v = \dfrac{\hbar^2 k}{m}$, where $k = \sqrt{\dfrac{2mE}{\hbar^2}}$.

The given potential was used by Yukawa to represent the interaction between two nuclei and explain the short range of the strong nuclear force.

6027

Consider the scattering of a 1 keV proton by a hydrogen atom.

a. What do you expect the angular distribution to look like? (Sketch a graph and comment on its shape).

b. Estimate the total cross section. Give a numerical answer in cm², m² or barns $= 10^{-24}$ cm², and a reason for you answer.

(Wisconsin)

Sol:

The problem is equivalent to the scattering of a particle of reduced mass $\mu \approx \frac{1}{2} m_p = 470$ MeV, energy $E_r = 0.5$ keV by a potential which, on account of

474   *Problems and Solutions in Quantum Mechanics*

electron shielding, can be roughly represented by $\frac{e^2}{r}e^{-r/a}$, where a is the range of interaction given by the Bohr radius 0.53 Å. As

$$ka = \frac{a}{\hbar c}\sqrt{2\mu c^2 E_r}$$

$$= \frac{0.53\times10^{-8}\times\sqrt{2\times470\times10^6\times0.5\times10^3}}{6.58\times10^{-16}\times3\times10^{10}} = 1.84\times10^2 \gg 1,$$

for Born approximation to be valid we require (**Problem 6026**)

$$|V| \sim \frac{e^2}{a} \ll \frac{\hbar v}{a}, \quad \text{i.e.,} \quad \frac{e^2}{\hbar c} \ll \frac{v}{c}.$$

Since

$$\text{LHS} = \frac{1}{137} = 7.3\times10^{-3},$$

$$\text{RHS} = \sqrt{\frac{2E_r}{\mu c^2}} = \sqrt{\frac{2\times0.5\times10^3}{470\times10^6}} = 1.5\times10^{-3},$$

the condition is not strictly satisfied. But in view of the roughness of the estimates, we still make use of the Born approximation.

a.   When the proton collides with the hydrogen atom, it experiences a repulsive Coulomb interaction with the nucleus, as well as an attractive one with the orbital electron having the appearance of a cloud of charge density $e\rho(r)$. The potential energy is then

$$V(r) = \frac{e^2}{r} - e^2 \int \frac{\rho(\mathbf{r}')}{|\mathbf{r}-\mathbf{r}'|} d\mathbf{r}'.$$

Using Born approximation and the formula

$$\int \frac{e^{i\mathbf{q}\cdot\mathbf{r}}}{r} d\mathbf{r} = \frac{4\pi}{q^2},$$

we obtain

$$f(\theta) = -\frac{\mu e^2}{2\pi\hbar^2} \int e^{i\mathbf{q}\cdot\mathbf{r}} \left[\frac{1}{r} - \int \frac{\rho(\mathbf{r}')}{|\mathbf{r}-\mathbf{r}'|} d\mathbf{r}' \right] d\mathbf{r}$$

$$= -\frac{2\mu e^2}{\hbar^2 q^2}[1 - F(\theta)],$$

where

$$\mu = \frac{m_p}{2}, \quad q = 2k\,\sin\frac{\theta}{2},$$

$$F(\theta) = \frac{q^2}{4\pi} \int\int \frac{e^{i\mathbf{q}\cdot\mathbf{r}}\rho(r')}{|\mathbf{r}-\mathbf{r}'|} d\mathbf{r}'d\mathbf{r}$$

$$= \frac{q^2}{4\pi} \int e^{i\mathbf{q}\cdot\mathbf{r}'}\rho(r')d\mathbf{r}' \int \frac{e^{i\mathbf{q}\cdot\bar{\mathbf{r}}}}{\bar{r}} d\bar{\mathbf{r}}$$

$$= \int e^{i\mathbf{q}\cdot\mathbf{r}}\rho(r)d\mathbf{r}.$$

For the ground state of the hydrogen atom, we have

$$\rho(r) = |\psi_{100}|^2 = \frac{1}{\pi a^3}e^{-2r/a},$$

and so

$$F(\theta) = \frac{1}{\pi a^3}\int e^{i\mathbf{q}\cdot\mathbf{r}-\frac{2r}{a}}d\mathbf{r}$$

$$= \left(1 + \frac{a^2 q^2}{4}\right)^{-2}.$$

Hence

$$f(\theta) = -\frac{\mu e^2}{2\hbar^2 k^2}\cdot\frac{1}{\sin^2(\theta/2)}\left[1 - \frac{1}{(1+a^2 k^2 \sin^2\theta/2)^2}\right].$$

Taking into account the identical nature of the two colliding particles (two protons), we have for the singlet state: $\sigma_s = |f(\theta) + f(\pi-\theta)|^2$, the triplet state: $\sigma_A = |f(\theta) - f(\pi-\theta)|^2$.

Hence, the scattering cross section (not considering polarization) is

$$\sigma = \frac{1}{4}\sigma_s + \frac{3}{4}\sigma_A.$$

Some special cases are considered below,

i. $\theta \approx 0$:

$$\sigma_s = \frac{\mu^2 e^4}{4\hbar^4 k^4}\left[\frac{1-F(\theta)}{\sin^2\theta/2} + \frac{1-F(\pi-\theta)}{\cos^2\theta/2}\right]^2$$

$$\approx \frac{\mu^2 e^4}{4\hbar^4 k^4}\left(2a^2 k^2 + \frac{1}{\cos^2\theta/2}\right)^2$$

$$\approx \frac{\mu^2 e^4}{\hbar^4}a^4 = \left(\frac{m_p}{2m_e}\right)^2 a^2,$$

use having been made of the approximation for $x \approx 0$

$$1/(1+x)^2 \approx 1 - 2x,$$

as well as the expression

$$a = \frac{\hbar^2}{m_e e^2} = \frac{\hbar^2}{\mu e^2}\left(\frac{m_p}{2me}\right).$$

Similarly we obtain

$$\sigma_A = \left(\frac{m_p}{2m_e}\right)^2 a^2.$$

ii. $\theta \approx \pi$: A similar calculation gives

$$\sigma_S = \sigma_A = \left(\frac{m_p}{2m_e}\right)^2 a^2.$$

iii. $a^2 k^2 \sin^2\frac{\theta}{2} = 10$ or $\theta \approx 0.07\pi$: For $0.07\pi \le \theta \le 0.93\pi$, we have

$$\sigma(\theta) = \frac{1}{4}\frac{\mu^2 e^4}{4\hbar^4 k^4}\left(\frac{1}{\sin^2\theta/2} + \frac{1}{\cos^2\theta/2}\right)^2 + \frac{3}{4}\frac{\mu^2 e^4}{4\hbar^4 k^4}\left(\frac{1}{\sin^2\theta/2} - \frac{1}{\cos^2\theta/2}\right)^2$$

$$= \frac{\mu^2 e^4}{\hbar^4 k^4}\left(\frac{3\cos^2\theta+1}{\sin^4\theta}\right)$$

$$= \sigma_0\left(\frac{3\cos^2\theta+1}{\sin^4\theta}\right),$$

where $\sigma_0 = \frac{\mu^2 e^4}{\hbar^4 k^4}$. The angular distribution is shown in Fig. 6.6.

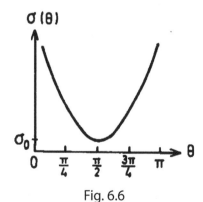

Fig. 6.6

d.   As $f(\theta) \to \infty$ for $\theta \to 0$, $\theta \to \pi$, to estimate the total scattering cross section, consider the total cross section for large scattering angles ($0.07\pi \le \theta \le 0.93\pi$) and for small scattering angles:

$$\sigma_t \text{large} = 2\pi\sigma_0 \int_{0.07\pi}^{0.93\pi} \left(\frac{3\cos^2\theta + 1}{\sin^4\theta} \right) \sin\theta d\theta$$

$$= 2\pi\sigma_0 \int_{0.07\pi}^{0.93\pi} \frac{4 - 3\sin^2\theta}{\sin^3\theta} d\theta$$

$$= 2\pi\sigma_0 \left[-\ln\left(\tan\frac{\theta}{2}\right) - 2\frac{\cot\theta}{\sin\theta} \right] \Bigg|_{0.07\pi}^{0.93\pi}$$

$$= 155\pi\sigma_0,$$

$$\sigma_{t\,\text{small}} > 2\pi \int_0^{0.07\pi} \sigma(0.07\pi)\sin\theta\, d\theta \times 2$$

$$= 2\pi \times 1703\sigma_0 [1 - \cos(0.07\pi)] \times 2$$

$$= 164\pi\sigma_0.$$

6028

The study of the scattering of high-energy electrons from nuclei has yielded much interesting information about the charge distributions in nuclei and nucleons. We shall here consider a simple version of the theory, in which the "electron" is assumed to have zero spin. We also assume that the nucleus, of charge Ze, remains fixed in space (i.e., its mass is assumed infinite). Let $\rho(\mathbf{x})$ denote the charge density in the nucleus. The charge distribution is assumed to be spherically symmetric, but otherwise arbitrary.

Let $f_e(\mathbf{p}_i, \mathbf{p}_f)$, where $\mathbf{p}_i$ is the initial, and $\mathbf{p}_f$ is the final momentum, be the scattering amplitude in the first Born approximation for the scattering of an electron from a point nucleus of charge Ze. Let $f(\mathbf{p}_i, \mathbf{p}_f)$ be the scattering amplitude, also in the first Born approximation, for the scattering of an electron from a real nucleus of the same charge. Let $\mathbf{q} = \mathbf{p}_i - \mathbf{p}_f$ denote the momentum transfer. The quantity F defined by $f(\mathbf{p}_i, \mathbf{p}_f) = F(\mathbf{q}^2) f_e(\mathbf{p}_i, \mathbf{p}_f)$ is called the form factor: it is easily seen that F in fact depends on $\mathbf{p}_i$ and $\mathbf{p}_f$ only through the quantity $\mathbf{q}^2$.

a. The form factor $F(\mathbf{q}^2)$ and the Fourier transform of the charge density $\rho(x)$ are related in a very simple manner: state and derive this relationship within the framework of the nonrelativistic Schrödinger theory. The assumption that the electrons are "nonrelativistic" is here made so that the problem will appear as simple as possible, but if you think about the matter it will probably be clear that the assumption is irrelevant: the same result applies in the "relativistic" case of the actual experiments. It is also the case that the neglect of the electron spin does not affect the essence of what we are here concerned with.

b. The graph in Fig. 6.7 shows some experimental results pertaining to the form factor for the proton, and we shall regard our theory as applicable to these data. On the basis of the data shown, compute the root-mean-square (charge) radius of the proton.

Hint: Note that there is a simple relationship between the root-mean-square radius and derivative of $F(q^2)$ with respect to q^2 at $q^2 = 0$. Find this relationship, and then compute.

(Berkeley)

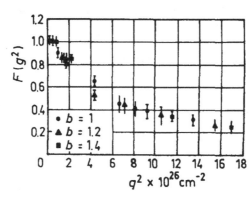

Fig. 6.7

Sol:

a. In the nonrelativistic Schrödinger theory the first Born approximation gives the scattering amplitude of an electron (charge $-e$) due to a central force field as

$$f(\mathbf{p}_i, \mathbf{p}_f) = -\frac{2m}{\hbar^2 q} \int_0^\infty r' V(r') \, \sin(qr') dr',$$

where q is the magnitude of the momentum transfer in the scattering. For a point nucleus, the scattering potential energy is

$$V_e(r) = -\frac{Ze^2}{r}.$$

For a real nucleus of charge density $\rho(\mathbf{r})$, the scattering potential energy is

$$V(r) = -e \int \frac{\rho(\mathbf{r}')d\mathbf{r}'}{|\mathbf{r}-\mathbf{r}'|} = -4\pi e \int_0^\infty \frac{\rho(r')r'^2 dr'}{|\mathbf{r}-\mathbf{r}'|},$$

which satisfies Poisson's equation

$$\nabla^2 V(r) = \frac{1}{r}\frac{d^2}{dr^2}(rV) = +4\pi e\rho(r). \tag{1}$$

Consider the integral in the expression for f:

$$\int_0^\infty rVe^{iqr}\,dr = \frac{1}{iq}\left[e^{iqr}\left(rV - \frac{1}{iq}(rV)'\right)\right]\Big|_0^\infty$$
$$-\frac{1}{q^2}\int_0^\infty (rV)'' e^{iqr}\,dr.$$

By a method due to Wentzel, the first term can be made to vanish and so

$$f(\mathbf{p}_i, \mathbf{p}_f) = -\frac{2m}{\hbar^2 q}\int_0^\infty rV(r)\sin(qr)\,dr$$

$$= -\frac{2m}{\hbar^2 q}\left(-\frac{1}{q^2}\right)\int_0^\infty (rV)'' \sin qr\,dr$$

$$= \frac{2m}{\hbar^2 q}\frac{4\pi e}{q^2}\int_0^\infty r\rho(r)\sin(qr)\,dr, \tag{2}$$

use having been made of Eq. (1).

In the case of a point-charge nucleus, only the region near $r = 0$ makes appreciable contribution to the integral and so

$$4\pi\int_0^\infty r\rho(r)\sin(qr)\,dr \approx 4\pi q\int_0^\infty r^2\rho(r)\,dr = qZe.$$

Hence for a point nucleus,

$$f_e(\mathbf{p}_i, \mathbf{p}_f) = \frac{2mZe^2}{\hbar^2 q^2}.$$

For an extended nucleus we can then write

$$f(\mathbf{p}_i, \mathbf{p}_f) = f_e(\mathbf{p}_i, \mathbf{p}_f)\cdot\frac{4\pi}{Ze}\int_0^\infty r^2\rho(r)\frac{\sin(qr)}{qr}\,dr.$$

By definition the form factor is

$$F(\mathbf{q}^2) = \frac{4\pi}{Ze} \int_0^\infty r^2 \rho(r) \frac{\sin(qr)}{qr} dr, \tag{3}$$

or

$$F(\mathbf{q}^2) = \frac{1}{Ze} \int \rho(r) e^{-\mathbf{q}\cdot\mathbf{r}} dr.$$

This is the required relation with the Fourier transform of the charge density

b.   Differentiating Eq. (3) with respect to q we have

$$\frac{dF}{dq} = \frac{4\pi}{Ze} \int_0^\infty r^2 \rho(r) \left[\frac{r\cos(qr)}{qr} - \frac{\sin(qr)}{q^2 r} \right] dr,$$

and hence

$$\frac{dF}{d(q^2)} = \frac{dF}{dq}\frac{dq}{d(q^2)} = \frac{1}{2q} \cdot \frac{4\pi}{Ze} \int_0^\infty r^2 \rho(r) \left[\frac{r\cos(qr)}{qr} - \frac{\sin(qr)}{q^2 r} \right] dr.$$

To find $\dfrac{dF}{d(q^2)}\bigg|_{q^2=0}$ we first calculate

$$\lim_{q\to 0} \left[\frac{r\cos(qr)}{q^2 r} - \frac{\sin(qr)}{q^3 r} \right]$$

$$= \lim_{q\to 0} \left[\frac{r\left[1 - \frac{1}{2}(qr)^2\right]}{q^2 r} - \frac{qr - \frac{1}{6}(qr)^3}{q^3 r} \right]$$

$$= \lim_{q\to 0} \left(-\frac{1}{3}r^2 \right) = -\frac{r^2}{3}.$$

Then

$$\frac{dF}{d(q^2)}\bigg|_{q^2=0} = -\frac{1}{6} \cdot \frac{1}{Ze} \int_0^\infty r^2 \rho(r) \cdot 4\pi r^2 dr = -\frac{1}{6}\langle r^2 \rangle.$$

From Fig. 6.7 can be found the slope of the curve $F(q^2)$ at $q^2 = 0$, whence the root-mean-square charge radius of the nucleus:

$$\sqrt{\langle r^2 \rangle} = \left(-6 \frac{dF}{d(q^2)}\bigg|_{q^2=0} \right)^{1/2}.$$

6029

In the early 1920's, Ramsauer (and independently Townsend) discovered that the scattering cross-section for electrons with an energy of ~ 0.4 eV was very much smaller than geometrical (πa^2, with a the radius of the atom) for scattering by argon atoms in gaseous form. It was also found that the cross section for 6-volt electrons was 3.5 times as great as the geometrical cross section and that the scattering was approximately isotropic. What is the origin of the "anomalous" cross sections? (What is the maximum possible cross section for scattering of low-energy electrons (with wavelength $\lambda \gg a$)?

(*Princeton*)

Sol:

If the attractive potential is strong enough, at a certain energy the partial wave with $\ell = 0$ has exactly a half-cycle more of oscillation inside the atomic potential. Then it has a phase shift of $\delta = \pi$ and so contributes nothing to $f(\theta)$ and hence the cross section. At low energies, the wavelength of the electron is large compared with a so the higher-ℓ partial waves' contribution is also negligible. This accounts for the Ramsauer-Townsend effect that the scattering cross section is very small at a certain low energy. For low-energy electrons, the maximum possible cross section for scattering is four times the geometrical cross section. It should be noted that a raregas atom, which consists entirely of closed shells, is relatively small, and the combined force of nucleus and orbital electrons exerted on an incident electron is strong and sharply defined as to range.

6030

Let $f(\omega)$ be the scattering amplitude for forward scattering of light at an individual scattering center in an optical medium. If the amplitude for incoming and outgoing light waves are denoted by $A_{in}(\omega)$ and $A_{out}(\omega)$ respectively, one has $A_{out}(\omega) = f(\omega)A_{in}(\omega)$. Suppose the Fourier transform

$$\bar{A}_{in}(x - t) = \frac{1}{\sqrt{2\pi}} \int_{-\infty}^{+\infty} e^{i\omega(x-t)} A_{in}(\omega) d\omega$$

vanishes for $x - t > 0$.

a. Use the causality condition (no propagation faster than the speed of light $c = 1$) to show that $f(\omega)$ is an analytic function in the half-plane $\text{Im}\,\omega > 0$.

b. Use the analyticity of $f(\omega)$ and the reality of $\bar{A}_{in}(\omega)$ and $\tilde{A}_{out}(\omega)$, and assume that $f(\omega)$ is bounded at infinity to derive the dispersion relation

$$Re\left[f(\omega+i\varepsilon)-f(0)\right]=\frac{2\omega^2}{\pi}P\int_0^\infty d\omega'\frac{Im\,f(\omega'+i\varepsilon)}{\omega'(\omega'^2-\omega^2)},$$

with ε arbitrarily small and positive.

(Chicago)

Sol:

a. $\bar{A}_{in}(x-t)=0$ for $t<x$ means $\tilde{A}_{out}(x-t)=0$ for $t<x$. Then

$$A_{in\atop out}(\omega)=\frac{1}{\sqrt{2\pi}}\int_{-\infty}^0 e^{-i\omega\tau}\bar{A}_{in\atop out}(\tau)d\tau$$

is a regular function when $Im\,\omega>0$, since when $\tau<0$ the factor $\exp(Im\,\omega\tau)$ of the integrand converges. As $A_{out}(\omega)=f(\omega)A_{in}(\omega)$, $f(\omega)$ is also analytic when $Im\,\omega>0$.

b. For $\omega\to\infty$, $0\le\arg\omega\le\pi$, we have $|f(\omega)|<M$, some positive number. Assume that $f(0)$ is finite (if not we can choose another point at which f is finite). Then $\chi(\omega)=\dfrac{f(\omega)-f(0)}{\omega}$ is sufficiently small at infinity, and so

$$\chi(\omega)=\frac{1}{2\pi i}\int_{-\infty}^{+\infty}\frac{\chi(\omega'+i0)}{\omega'-\omega}d\omega',\qquad Im\,\omega>0.$$

When ω is a real number, using

$$\frac{1}{\omega'-\omega-i0}=\frac{P}{\omega'-\omega}+i\pi\delta(\omega'-\omega),$$

we get

$$Re\chi(\omega)=\frac{P}{\pi}\int_{-\infty}^{+\infty}\frac{Im\chi(\omega'+i0)}{\omega'-\omega}d\omega'.$$

$\bar{A}_{in\atop out}(\omega)$ being a real number means that $A_{in\atop out}^*(-\omega^*)=A_{in\atop out}(\omega)$. Hence $f^*(\omega^*)=f(-\omega)$, and so $Im\,f(\omega+i0)=-Im\,f(-\omega+i0)$, and

$$Re\left[f(\omega+i0)-f(0)\right]=\frac{2\omega^2}{\pi}P\int_0^\infty\frac{Im f(\omega'+i0)}{\omega'(\omega'^2-\omega^2)}d\omega',$$

where P denotes the principal value of the integral.

6031

A spin-one-half projectile of mass m and energy $E = \frac{\hbar^2 k^2}{2m}$ scatters off an infinitely heavy spin-one-half target. The interaction Hamiltonian is

$$H_{int} = A\sigma_1 \cdot \sigma_2 \frac{e^{-\mu r}}{r} \qquad (\mu > 0),$$

where σ_1 and σ_2 are the Pauli spin operators of the projectile and target respectively. Compute the differential scattering cross section $\frac{d\sigma}{d\Omega}$ in lowest order Born approximation, averaging over initial and summing over final states of spin polarization. Express $\frac{d\sigma}{d\Omega}$ as a function of k and the scattering angle θ.

(Princeton)

Sol:

Suppose that the projectile is incident on the target along the z-axis, i.e., $\mathbf{k}_0 = k\mathbf{e}_z$. In lowest order Born approximation, the scattering amplitude is

$$
\begin{aligned}
f(\theta) &= -\frac{m}{2\pi\hbar^2} \int e^{i(\mathbf{k}_0 - \mathbf{k}) \cdot \mathbf{r}} A\sigma_1 \cdot \sigma_2 \frac{e^{-\mu r'}}{r'} d^3 r' \\
&= \frac{-m}{2\pi\hbar^2} \int e^{i\mathbf{q} \cdot \mathbf{r}} A\sigma_1 \cdot \sigma_2 \frac{e^{-\mu r'}}{r'} d^3 r' \\
&= \frac{-m}{\hbar^2} A\sigma_1 \cdot \sigma_2 \int_0^\infty e^{-\mu r'} r' dr' \int_0^\pi e^{iqr'\cos\theta} \sin\theta \, d\theta \\
&= -\frac{2m}{\hbar^2} A\sigma_1 \cdot \sigma_2 \frac{1}{\mu^2 + q^2},
\end{aligned}
$$

where $\mathbf{q} = \mathbf{k}_0 - \mathbf{k}$, $q = 2k\sin\frac{\theta}{2}$. Denote the total spin of the system by $\mathbf{S}$. Then $\mathbf{S} = \frac{1}{2}(\sigma_1 + \sigma_2)$ and

$$
\begin{aligned}
\sigma_1 \cdot \sigma_2 &= \frac{1}{2}(\sigma^2 - \sigma_1^2 - \sigma_2^2) = \frac{\hbar^2}{2}[4S(S+1) - 3 - 3] \\
&= \frac{\hbar^2}{2}[4S(S+1) - 6].
\end{aligned}
$$

For $S = 0$, $\sigma_1 \cdot \sigma_2 = -3\hbar^2$ and

$$f_0(\theta) = \frac{6Am}{\mu^2 + q^2}, \quad \frac{d\sigma_0}{d\Omega} = \frac{(6Am)^2}{(\mu^2 + q^2)^2}.$$

For $S=1$, $\sigma_1 \cdot \sigma_2 = \hbar^2$ and

$$f_1(\theta) = -\frac{2Am}{\mu^2 + q^2}, \quad \frac{d\sigma_1}{d\Omega} = \frac{(2Am)^2}{(\mu^2 + q^2)^2}.$$

If the initial states of spin of the projectile and target are $\begin{pmatrix} 1 \\ 0 \end{pmatrix} P = \alpha_P$, $\begin{pmatrix} 1 \\ 0 \end{pmatrix} T = \alpha_T$

respectively, then the initial state of spin of the system is $\Theta_{11} = \alpha_P \alpha_T$, the scattered wave function is $f_1(\theta)\frac{e^{ikr}}{r}\Theta_{11}$, and the corresponding differential scattering cross section is given by

$$\frac{d\sigma}{d\Omega}\left(\frac{1}{2}\frac{1}{2}; \frac{1}{2}\frac{1}{2}\right) = |f_1(\theta)|^2,$$

$$\frac{d\sigma}{d\Omega}\left(\frac{1}{2}\frac{1}{2}; \frac{1}{2}, -\frac{1}{2}\right) = \frac{d\sigma}{d\Omega}\left(\frac{1}{2}\frac{1}{2}; -\frac{1}{2}\frac{1}{2}\right)$$

$$= \frac{d\sigma}{d\Omega}\left(\frac{1}{2}\frac{1}{2}; -\frac{1}{2}, -\frac{1}{2}\right) = 0.$$

Noting that the triplet state vectors are

$$\Theta_{11} = \alpha_P \alpha_T, \ \Theta_{1,-1} = \beta_P \beta_T, \ \Theta_{10} = \frac{1}{\sqrt{2}}(\alpha_P \beta_T + \beta_P \alpha_T),$$

the singlet state vector is

$$\Theta_{00} = \frac{1}{\sqrt{2}}(\alpha_P \beta_T - \beta_P \alpha_T)$$

and that

$$\alpha_P \beta_T = \frac{1}{\sqrt{2}}(\Theta_{10} + \Theta_{00}), \text{ etc.,}$$

we can obtain the remaining differential cross sections:

$$\frac{d\sigma}{d\Omega}\left(\frac{1}{2}, -\frac{1}{2}; \frac{1}{2}, -\frac{1}{2}\right) = \frac{1}{4}|f_1(\theta) - f_0(\theta)|^2,$$

$$\frac{d\sigma}{d\Omega}\left(\frac{1}{2}, -\frac{1}{2}; -\frac{1}{2}\frac{1}{2}\right) = \frac{1}{4}|f_1(\theta) - f_0(\theta)|^2,$$

$$\frac{d\sigma}{d\Omega}\left(\frac{1}{2}, -\frac{1}{2}; \frac{1}{2}\frac{1}{2}\right) = \frac{d\sigma}{d\Omega}\left(\frac{1}{2}, -\frac{1}{2}; -\frac{1}{2}, -\frac{1}{2}\right) = 0;$$

$$\frac{d\sigma}{d\Omega}\left(-\frac{1}{2}\frac{1}{2}; \frac{1}{2}, -\frac{1}{2}\right) = \frac{1}{4}|f_1(\theta) - f_0(\theta)|^2,$$

$$\frac{d\sigma}{d\Omega}\left(-\frac{1}{2}\frac{1}{2}; -\frac{1}{2}\frac{1}{2}\right) = \frac{1}{4}|f_1(\theta) + f_0(\theta)|^2,$$

$$\frac{d\sigma}{d\Omega}\left(-\frac{1}{2}\frac{1}{2}; \frac{1}{2}\frac{1}{2}\right)=\frac{d\sigma}{d\Omega}\left(-\frac{1}{2}\frac{1}{2}; -\frac{1}{2}, -\frac{1}{2}\right)=0;$$

$$\frac{d\sigma}{d\Omega}\left(-\frac{1}{2}, -\frac{1}{2}; -\frac{1}{2}, -\frac{1}{2}\right)=\left|f_1(\theta)\right|^2,$$

$$\frac{d\sigma}{d\Omega}\left(-\frac{1}{2}, -\frac{1}{2}; \frac{1}{2}, \frac{1}{2}\right)=\frac{d\sigma}{d\Omega}\left(-\frac{1}{2}, -\frac{1}{2}; -\frac{1}{2}\frac{1}{2}\right)$$

$$=\frac{d\sigma}{d\Omega}\left(-\frac{1}{2}, -\frac{1}{2}; \frac{1}{2}, -\frac{1}{2}\right)=0.$$

Averaging over the initial states (i) and summing over the final states (f) of spin polarization, we obtain

$$\frac{d\sigma}{d\Omega}=\frac{1}{4}\sum_{s_{Pz}^{(i)},s_{Tz}^{(i)}}\sum_{s_{Pz}^{(f)},s_{Tz}^{(f)}}\frac{d\sigma}{d\Omega}\left(S_{Pz}^{(i)}S_{Tz}^{(i)}; S_{Pz}^{(f)}S_{Tz}^{(f)}\right)$$

$$=\frac{1}{4}\left[3f_1^2(\theta)+f_0^2(\theta)\right]=\frac{12A^2m^2}{\mu^2+4k^2\sin^2\frac{\theta}{2}}.$$

6032

Calculate in the Born approximation the differential scattering cross section for neutron-neutron scattering, assuming that the interaction potential responsible for scattering vanishes for the triplet spin state and is equal to $V(r)=V_0\dfrac{e^{-\mu r}}{r}$ for the singlet spin state. [Evaluate the cross section for an unpolarized (random spin orientation) initial state.]

(Berkeley)

Sol: The Born approximation gives (**Problem 6013**)

$$f_s(\theta)=-\frac{2m}{\hbar^2 q}\int_0^\infty rV(r)\sin qr\, dr$$

$$=-\frac{2m}{\hbar^2 q}\int_0^\infty V_0 e^{-\mu r}\sin qr\, dr$$

$$=-\frac{2mV_0}{\hbar^2 q}\frac{q}{q^2+\mu^2}, \quad q=2k\sin(\theta/2),$$

where **k** is the wave vector of the relative motion of the neutrons, $m=m_n/2$ is the reduced mass.

As the spin wave function of the spin singlet state is antisymmetric, its spatial wave function must be symmetric. Thus

$$\sigma_s = |f(\theta) + f(\pi - \theta)|^2$$

$$= \frac{4m^2 V_0^2}{\hbar^4} \left[\frac{1}{\mu^2 + 4k^2 \sin^2 \frac{\theta}{2}} + \frac{1}{\mu^2 + 4k^2 \cos^2 \frac{\theta}{2}} \right]^2$$

$$= \frac{16 m^2 V_0^2 (\mu^2 + 2k^2)^2}{\hbar^4 (\mu^2 + 4k^2 \sin^2 \frac{\theta}{2})^2 (\mu^2 + 4k^2 \cos^2 \frac{\theta}{2})^2}.$$

Because the neutrons are initially unpolarized, the scattering cross section is

$$\sigma(\theta) = \frac{1}{4}\sigma_s + \frac{3}{4}\sigma_t = \frac{1}{4}\sigma_s$$

$$= \frac{4m^2 V_0^2 (\mu^2 + 2k^2)^2}{\hbar^4 (\mu^2 + 4k^2 \sin^2 \frac{\theta}{2})^2 (\mu^2 + 4k^2 \cos^2 \frac{\theta}{2})^2}.$$

6033

The scattering of low-energy neutrons on protons is spin dependent. When the neutron-proton system is in the singlet spin state the cross section is $\sigma_1 = 78 \times 10^{-24}$ cm^2, whereas in the triplet spin state the cross section is $\sigma_3 = 2 \times 10^{-24}$ cm^2. Let f_1 and f_3 be the corresponding scattering amplitudes Express your answers below in terms of f_1 and f_3.

a. What is the total scattering cross section for unpolarized neutrons on unpolarized protons?

b. Suppose a neutron which initially has its spin up scatters from a proton which initially has its spin down. What is the probability that the neutron and proton flip their spins? (Assume s-wave scattering only.)

c. The H_2 molecule exists in two forms: ortho-hydrogen for which the total spin of the protons is 1 and para-hydrogen for which the total spin of the protons is 0. Suppose now a very low energy neutron ($\lambda_n \gg \langle d \rangle$, the average separation between the protons in the molecule) scatters from such molecules. What is the ratio of the cross section for scattering unpolarized neutrons from unpolarized ortho-hydrogen to that for scattering them from para-hydrogen?

(Berkeley)

ol:

a. The triplet and singlet spin states of a neutron-proton system can respectively be expressed as

$$\chi_1^3 = \alpha_n \alpha_p, \quad \chi_0^3 = \frac{1}{\sqrt{2}}(\alpha_n \beta_p + \alpha_p \beta_n), \quad \chi_{-1}^3 = \beta_n \beta_p,$$

$$\chi_{-1}^1 = \frac{1}{\sqrt{2}}(\alpha_n \beta_p - \alpha_p \beta_n),$$

with $\alpha = \begin{pmatrix} 1 \\ 0 \end{pmatrix}$, $\beta = \begin{pmatrix} 1 \\ 0 \end{pmatrix}$. If we define an operator $\hat{f}$ by

$$\hat{f} = \frac{3}{4}f_3 + \frac{1}{4}f_1 + \frac{1}{4}(f_3 - f_1)(\sigma_n \cdot \sigma_p),$$

then as

$$\sigma_n \cdot \sigma_p = \sigma_{nx}\sigma_{px} + \sigma_{ny}\sigma_{py} + \sigma_{nz}\sigma_{pz},$$
$$\sigma_x \alpha = \beta, \quad \sigma_y \alpha = i\beta, \quad \sigma_z \alpha = \alpha,$$
$$\sigma_x \beta = \alpha, \quad \sigma_y \beta = -i\alpha, \quad \sigma_z \beta = -\beta,$$

we have

$$\hat{f}\chi_1^3 = f_3\chi_1^3, \quad \hat{f}\chi_0^3 = f_3\chi_0^3, \quad \hat{f}\chi_{-1}^3 = f_3\chi_{-1}^3,$$
$$\hat{f}\chi_{-1}^1 = f_1\chi_{-1}^1,$$

i.e., the eigenvalues of $\hat{f}$ for the triplet and singlet spin states are f_3 and f_1 respectively.

Similarly, if we define

$$\hat{f}^2 = \frac{3}{4}f_3^2 + \frac{1}{4}f_1^2 + \frac{1}{4}(f_3^2 - f_1^2)(\sigma_n \cdot \sigma_p)$$

then $\hat{f}^2$ has eigenvalues f_3^2 and f_1^2 for the triplet and singlet states, and we can express the total cross section for the scattering as $\sigma_t = 4\pi \hat{f}^2$.

Assume the spin state of the incident neutron is

$$\begin{pmatrix} e^{-i\alpha} \cos\beta \\ e^{i\alpha} \sin\beta \end{pmatrix}.$$

Note here $(2\beta, 2\alpha)$ are the polar angles of the spin direction of the neutron. If the state of the polarized proton is $\begin{pmatrix} 1 \\ 0 \end{pmatrix}$, then the cross section is

$$\sigma_t = 4\pi \begin{pmatrix} e^{-i\alpha} \cos\beta \\ e^{i\alpha} \sin\beta \end{pmatrix}_n^+ \begin{pmatrix} 1 \\ 0 \end{pmatrix}_p^+ \hat{f}^2 \begin{pmatrix} 1 \\ 0 \end{pmatrix}_p \begin{pmatrix} e^{-i\alpha} \cos\beta \\ e^{i\alpha} \sin\beta \end{pmatrix}_n.$$

As

$$\begin{pmatrix}1\\0\end{pmatrix}_p^{\dagger}(\sigma_n\cdot\sigma_p)\begin{pmatrix}1\\0\end{pmatrix}_p$$

$$=(1\ \ 0)_p\left[\sigma_{nx}\begin{pmatrix}0\\1\end{pmatrix}_p+i\sigma_{ny}\begin{pmatrix}0\\1\end{pmatrix}_p+\sigma_{nz}\begin{pmatrix}1\\0\end{pmatrix}_p\right]$$

$$=\sigma_{nz},$$

$$\begin{pmatrix}e^{-i\alpha}\cos\beta\\e^{i\alpha}\sin\beta\end{pmatrix}_n^{\dagger}\sigma_{nz}\begin{pmatrix}e^{-i\alpha}\cos\beta\\e^{i\alpha}\sin\beta\end{pmatrix}_n$$

$$=(e^{i\alpha}\cos\beta\ \ e^{-i\alpha}\sin\beta)_n\begin{pmatrix}1&0\\0&-1\end{pmatrix}\begin{pmatrix}e^{-i\alpha}\cos\beta\\e^{i\alpha}\sin\beta\end{pmatrix}_n$$

$$=(e^{i\alpha}\cos\beta\ \ e^{-i\alpha}\sin\beta)_n\begin{pmatrix}e^{-i\alpha}\cos\beta\\-e^{-i\alpha}\sin\beta\end{pmatrix}_n$$

$$=\cos^2\beta-\sin^2\beta=\cos2\beta,$$

we have

$$\sigma_t=\pi\{3f_3^2+f_1^2-(f_3^2-f_1^2)\cos2\beta\}$$

$$=\left\{\frac{3}{4}\sigma_3+\frac{1}{4}\sigma_1-(\sigma_3-\sigma_1)\frac{\cos2\beta}{4}\right\}.$$

Since the incident neutrons are unpolarized, $\overline{\cos2\beta}=0$ and so

$$\sigma_t=\frac{3}{4}\sigma_3+\frac{1}{4}\sigma_1.$$

Because the direction of the z-axis is arbitrary, the total scattering cross section of unpolarized protons is as the same as that of polarized protons.

b. The state vector before interaction is

$$\begin{pmatrix}1\\0\end{pmatrix}_n\begin{pmatrix}0\\1\end{pmatrix}_p.$$

We can expand this in terms of the wave functions of the singlet and triplet states:

$$\begin{pmatrix}1\\0\end{pmatrix}_n\begin{pmatrix}0\\1\end{pmatrix}_p=\frac{1}{\sqrt2}\left\{\frac{1}{\sqrt2}\left[\begin{pmatrix}1\\0\end{pmatrix}_n\begin{pmatrix}0\\1\end{pmatrix}_p+\begin{pmatrix}0\\1\end{pmatrix}_n\begin{pmatrix}1\\0\end{pmatrix}_p\right]\right.$$

$$\left.+\frac{1}{\sqrt2}\left[\begin{pmatrix}1\\0\end{pmatrix}_n\begin{pmatrix}0\\1\end{pmatrix}_p-\begin{pmatrix}0\\1\end{pmatrix}_n\begin{pmatrix}1\\0\end{pmatrix}_p\right]\right\}.$$

The scattered wave is then

$$\frac{e^{ikr}}{r}\frac{1}{\sqrt{2}}\left\{f_s\frac{1}{\sqrt{2}}\left[\begin{pmatrix}1\\0\end{pmatrix}_n\begin{pmatrix}0\\1\end{pmatrix}_p+\begin{pmatrix}0\\1\end{pmatrix}_n\begin{pmatrix}1\\0\end{pmatrix}_p\right]\right.$$

$$\left.+f_1\frac{1}{\sqrt{2}}\left[\begin{pmatrix}1\\0\end{pmatrix}_n\begin{pmatrix}0\\1\end{pmatrix}_p-\begin{pmatrix}0\\1\end{pmatrix}_n\begin{pmatrix}1\\0\end{pmatrix}_p\right]\right\}$$

$$=\frac{e^{ikr}}{r}\left\{\frac{f_3+f_1}{2}\begin{pmatrix}1\\0\end{pmatrix}_n\begin{pmatrix}0\\1\end{pmatrix}_p+\frac{f_3-f_1}{2}\begin{pmatrix}0\\1\end{pmatrix}_n\begin{pmatrix}1\\0\end{pmatrix}_p\right\}.$$

Hence the probability that the neutron and proton both flip their spins is

$$\frac{(f_3-f_1)^2}{(f_3+f_1)^2+(f_3-f_1)^2}=\frac{1}{2}\frac{(f_3-f_1)^2}{f_3^2+f_1^2}.$$

c. Let

$$\hat{F}=\hat{f}_1+\hat{f}_2=\frac{3f_3+f_1}{2}+\frac{1}{2}(f_3-f_1)\sigma_n\cdot S$$

with

$$S=\frac{1}{2}\left(\sigma_{p_1}+\sigma_{p_2}\right)=\left(s_{p_1}+s_{p_2}\right).$$

As

$$\sigma_x\sigma_y=i\sigma_z,$$
$$\sigma_y\sigma_z=i\sigma_x,$$
$$\sigma_z\sigma_x=i\sigma_y,$$
$$\sigma_i\sigma_j+\sigma_j\sigma_i=2\delta_{ij},$$

we have

$$(\sigma_n\cdot S)^2=S^2-\sigma_n\cdot S,$$

and hence

$$\hat{F}^2=\frac{1}{4}\{(3f_3+f_1)^2+(5f_3^2-2f_1f_3-3f_1^2)\sigma_n\cdot S$$
$$+(f_3-f_1)^2S^2\}.$$

For para-hydrogen, $S=0$ and so

$$\sigma_p=\pi(3f_3+f_1)^2.$$

In this case, as there is no preferred direction the cross section is independent of the polarization of the incident neutrons.

For ortho-hydrogen, $S^2 = 1(1+1) = 2$. Taking the proton states as $\begin{pmatrix} 1 \\ 0 \end{pmatrix}_{p_1} \begin{pmatrix} 1 \\ 0 \end{pmatrix}_{p_2}$,

using

$$\sigma_n \cdot S = \frac{1}{2}(\sigma_n \cdot \sigma_{p_1} + \sigma_n \cdot \sigma_{p_2})$$

and following the calculation in (a) we have

$$\sigma \cdot S = \cos 2\beta.$$

Hence

$$\sigma_0 = \pi\{(3f_3 + f_1)^2 + (5f_3^2 - 2f_1 f_3 - 3f_1^2)\cos 2\beta$$
$$+ 2(f_3 - f_1)^2\},$$

where 2β is the angle between S and σ_n. If the neutrons are unpolarized, $\overline{\cos 2\beta} = 0$ and so

$$\sigma_0 = \pi\left[(3f_3 + f_1)^2 + 2(f_3 - f_1)^2\right].$$

This result is independent of the polarization of the hydrogen. The ratio we require is:

$$\frac{\sigma_0}{\sigma_P} = 1 + 2(f_3 - f_1)^2 / (f_1 + 3f_3)^2.$$

6034

Consider a hypothetical neutron-neutron scattering at zero energy. The interaction potential is

$$V(r) = \begin{cases} \sigma_1 \cdot \sigma_2 V_0, & r \le a, \\ 0, & r > a, \end{cases}$$

where σ_1 and σ_2 are the Pauli spin matrices of the two neutrons. Compute the total scattering cross section. Both the incident and target neutrons are unpolarized.

(CUS)

Sol: Consider the problem in the coupling representation. Let

$$S = s_1 + s_2 = \frac{1}{2}\sigma_1 + \frac{1}{2}\sigma_2.$$

Then

$$\sigma_1 \cdot \sigma_2 = \frac{1}{2}(4S^2 - \sigma_1^2 - \sigma_2^2)$$

$$= \frac{1}{2}\left[4S(S+1) - 3 - 3\right]$$

$$= 2S(S+1) - 3,$$

where $S = 1$ or 0. It is noted that an eigenstate of S is also an eigenstate of $V(r)$. For zero-energy scattering we need to consider only the s partial wave, which is symmetrical. The Pauli principle then requires the spin wave function to be antisymmetric. Thus we have $S = 0$ and

$$V = \begin{cases} -3V_0, & r < a, \\ 0, & r > a. \end{cases}$$

For s-waves, the wave equation for $r < a$ is

$$\frac{d^2 u}{dr^2} + k_0^2 u = 0,$$

where $u(r) = r\psi$, ψ being the radial wave function, $k_0^2 = 6mV_0/h^2$, and the solution is $u(r) = A\sin(k_0 r)$. For $r > a$, the wave equation is

$$\frac{d^2 u}{dr^2} + k_1^2 u = 0,$$

where $k_1^2 = \frac{2m}{\hbar^2}E$, and the solution is $u(r) = \sin(k_1 r + \delta_0)$.

The continuity of u and u' at $r = a$ gives

$$k_1 \tan(k_0 a) = k_0 \tan(k_1 a + \delta_0).$$

For $E \to 0$, $k_1 \to 0$ and

$$\frac{k_1}{k_0}\tan(k_0 a) = \frac{\tan(k_1 a) + \tan\delta_0}{1 - \tan(k_1 a)\tan\delta_0}$$

$$\to k_1 a + \tan\delta_0,$$

i.e.,

$$\delta_0 \approx k_1 a\left[\frac{\tan(k_0 a)}{k_0 a} - 1\right].$$

For collisions of identical particles,

$$\sigma(\theta) = \left| f(\theta) + f(\pi - \theta) \right|^2$$

$$= \frac{1}{k_1^2}\left| \sum_{l=0,2,4} 2(2l+1)e^{i\delta_l}\sin\delta_l P_l(\cos\theta) \right|^2.$$

Considering only the s partial wave, we have the differential cross section

$$\sigma(\theta) = \frac{4}{k_1^2} \sin^2 \delta_0 \approx \frac{4}{k_1^2} \delta_0^2$$

and the total cross section

$$\sigma_t = 4\pi\sigma = 16\pi a^2 \left[\frac{\tan(k_0 a)}{k_0 a} - 1 \right]^2.$$

As the incident and target neutrons are unpolarized the probability that $S = 0$ (i.e., opposite spins) is $\frac{1}{4}$. Hence

$$\sigma_t = 4\pi a^2 \left[\frac{\tan(k_0 a)}{k_0 a} - 1 \right]^2.$$

6035

A beam of spin-$\frac{1}{2}$ particles of mass m is scattered from a target consisting of heavy nuclei, also of spin 1/2. The interaction of a test particle with a nucleus is $c s_1 \cdot s_2 \delta^3(x_1 - x_2)$, where c is a small constant, s_1 and s_2 are the test particle and nuclear spins respectively, and x_1 and x_2 are their respective positions.

a. Calculate the differential scattering cross section, averaging over the initial spin states. What is the total cross section?

b. If the incident test particles all have spin up along the z-axis but the nuclear spins are oriented at random, what is the probability that after scattering the test particles still have spin up along the z-axis?

(Princeton)

Sol:

a. As the nuclear target, being heavy, acts as a fixed scattering center, the center-of-mass and laboratory frames coincide. Then the equation of relative motion is

$$\left[-\frac{\hbar^2}{2m} \nabla_r^2 + c s_1 \cdot s_2 \delta^{(3)}(\mathbf{r}) \right] \psi(\mathbf{r}) = E\psi(\mathbf{r}).$$

As c is a small constant, we can employ the Born approximation

$$f(\theta)=-\frac{m}{2\pi\hbar^2}\int e^{i(\mathbf{k}-\mathbf{k}')\cdot\mathbf{r}'}c\mathbf{s}_1\cdot\mathbf{s}_2\delta^{(3)}(\mathbf{r}')d^3\mathbf{r}'$$

$$=-\frac{cm}{2\pi\hbar^2}\mathbf{s}_1\cdot\mathbf{s}_2,$$

using the property of $\delta^{(3)}(\mathbf{r}')$. The differential scattering cross section is

$$\sigma(\theta)=|f(\theta)|^2=\frac{c^2m^2}{(2\pi\hbar^2)^2}|\mathbf{s}_1\cdot\mathbf{s}_2|^2,$$

where

$$\mathbf{s}_1\cdot\mathbf{s}_2=\frac{1}{2}\left[\mathbf{S}^2-s_1^2-s_2^2\right]$$

$$=\frac{1}{2}\left[S(S+1)-\frac{1}{2}\cdot\frac{3}{2}-\frac{1}{2}\cdot\frac{3}{2}\right]\hbar^2,$$

$$\mathbf{S}=\mathbf{s}_1+\mathbf{s}_2.$$

For the state of total spin $S=0$,

$$\sigma(\theta)=\frac{c^2m^2}{(2\pi\hbar^2)^2}\left|\frac{1}{2}\left(-\frac{3}{2}\right)\hbar^2\right|^2=\frac{(3mc)^2}{(8\pi)^2}.$$

For $S=1$,

$$\sigma_1(\theta)=\frac{c^2m^2}{(2\pi\hbar^2)^2}\left|\frac{1}{2}\cdot\frac{1}{2}\hbar^2\right|^2=\frac{(mc)^2}{(8\pi)^2}.$$

Averaging over the initial spin states, the differential scattering cross section is

$$\sigma_t(\theta)=\frac{1}{4}\sigma_0(\theta)+\frac{3}{4}\sigma_1(\theta)=\frac{3(mc)^2}{(8\pi)^2}.$$

Alternative solution:

Let $|\alpha_p\rangle$ denote the initial spin state of the incident particle. The spin of the target is unpolarized, so its state is a "mixture" of $|\alpha\rangle$ and $|\beta\rangle$ states (not "superposition"), $c_1(t)|\alpha N\rangle+c_2(t)|\beta_N\rangle$. Here $c_1(t)$ and $c_2(t)$ have no fixed phase difference, and so the mean-square values are each 1/2. In the coupling representation, the initial spin state is a mixture of the states

$$c_1(t)|\chi\rangle,\frac{c_2(t)}{\sqrt{2}}\left(|\chi_0^1\rangle+|\chi_0^0\rangle\right),$$

where $|\chi\rangle=|\alpha_p\rangle|\alpha_N\rangle$,

$$\left|\chi_0^1\right\rangle = \frac{1}{\sqrt{2}}\left(\left|\alpha_p\right\rangle\left|\beta_N\right\rangle + \left|\beta_p\right\rangle\left|\alpha_N\right\rangle\right),$$

$$\left|\chi_0^0\right\rangle = \frac{1}{\sqrt{2}}\left(\left|\alpha_p\right\rangle\left|\beta_N\right\rangle - \left|\beta_p\right\rangle\left|\alpha_N\right\rangle\right).$$

After scattering, the spin state of the system, with part of the asymptotic spatial states, becomes

$$\left|\psi(S)\right\rangle_f = c_1(t)f_1(\theta)\left|\chi_1^1\right\rangle + \frac{c_2(t)}{\sqrt{2}}\left[f_1(\theta)\left|\chi_0^1\right\rangle\right.$$

$$\left. + f_0(\theta)\left|\chi_0^0\right\rangle\right].$$

Taking the dot product of the different $\left|\chi\right\rangle$ states with $\left|\psi\right\rangle_f$, we obtain the corresponding probability amplitudes, which are then added together to give the total cross section

$$\sigma_t(\theta) = \left|c_1(t)\right|^2 \sigma_1(\theta) + \frac{1}{2}\left|c_2(t)\right|^2 \sigma_1(\theta)$$

$$+ \frac{1}{2}\left|c_2(t)\right|^2 \sigma_0(\theta).$$

As $c_1(t)$, $c_2(t)$ each has the mean-square value $\frac{1}{2}$, we have

$$\sigma_t(\theta) = \frac{1}{2}\sigma_1(\theta) + \frac{1}{4}\sigma_1(\theta) + \frac{1}{4}\sigma_0(\theta)$$

$$= \frac{3}{4}\sigma_1(\theta) + \frac{1}{4}\sigma_0(\theta),$$

same as that obtained before.

b. After scattering the two irrelevant spin states of the system are

$$\left|\chi_1^1\right\rangle \quad \text{and} \quad \frac{1}{\sqrt{2}}\left(\left|\chi_0^1\right\rangle + \left|\chi_0^0\right\rangle\right).$$

Taking the dot product of the two states with $\left|\psi_f\right\rangle$, we obtain the corresponding scattering amplitudes. Hence the scattering cross section is

$$\sigma^{(t)} = \left|c_1(t)\right|^2 \sigma_1(\theta) + \frac{1}{4}\left|c_2(t)\right|^2\left|\left[f_1(\theta) + f_0(\theta)\right]\right|^2.$$

Averaging over the ensemble, we have

$$\sigma^{(t)} = \frac{1}{2}\sigma_1(\theta) + \frac{1}{8}\left|f_1(\theta) + f_2(\theta)\right|^2 = \frac{m^2 c^2}{(8\pi)^2}.$$

The probability that the test particles still have spin up is

$$P = \frac{\sigma^{(t)}}{\sigma_t} = \frac{1}{3}.$$

6036

a. Two identical particles of spin $\frac{1}{2}$ and mass m interact through a screened Coulomb potential $V(r) = e^2 \exp(-\lambda r)/r$, where $1/\lambda$ is the screening length. Consider a scattering experiment in which each particle has kinetic energy E in the center-of-mass frame. Assume that E is large. The incoming spins are oriented at random. Calculate (in the center-of-mass frame) the scattering cross section $\frac{\partial \sigma}{\partial \Omega}$ for observation of a particle emerging at an angle θ relative to the axis of the incoming particles as shown in Fig. 6.8.

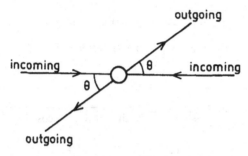

Fig. 6.8

b. Assuming that the outgoing particles are observed at an angle θ relative to the beam axis, what is the probability that after the scattering event the two particles are in a state of total spin one? What is the probability that, if one particle has spin up along the z-axis, the other particle also has spin up along the z-axis?

c. How large must the energy be for your approximations to be valid? Suppose that, instead, the energy is much less than this. In the limit of low energies, what is the probability that after being scattered the two particles are left in a state of $S = 1$?

(Princeton)

Sol: In the CM frame, the motions of the two particles are symmetric. The interaction is equivalent to a potential centered at the midpoint of the line joining the particles.

$$V(\rho) = e^2 \exp(-2\lambda\rho)/2\rho,$$

where $\rho = \frac{r}{2}, r$ being the separation of the two particles. As the energy E is large, we can use the first Born approximation

$$f(\theta) = -\frac{m}{2\pi\hbar^2}\int e^{-i\mathbf{q}\cdot\boldsymbol{\rho}}V(\rho)d^3\rho$$

$$= -\frac{2m}{\hbar^2 q}\int_0^\infty \frac{e^2}{2}\exp(-2\lambda\rho)\sin(q\rho)d\rho$$

$$= -\frac{me^2}{\hbar^2\left[q^2 + (2\lambda)^2\right]},$$

where $\mathbf{q}$ is the momentum transfer during the scattering, $q = 2k\sin(\theta/2)$. Considering the symmetry of the wave function of the two-identical-particle system, we have

$$\text{for } S = 0, \quad \sigma_s(\theta) = \left|f(\theta) + f(\pi - \theta)\right|^2,$$

$$\text{for } S = 1, \quad \sigma_a(\theta) = \left|f(\theta) - f(\pi - \theta)\right|^2.$$

Thus

$$\sigma_s(\theta) = \frac{1}{4}\left(\frac{me^2}{\hbar^2}\right)^2\left[\frac{1}{k^2\sin^2\frac{\theta}{2} + \lambda^2} + \frac{1}{k^2\cos^2\frac{\theta}{2} + \lambda^2}\right]^2$$

$$= \frac{1}{4}\left(\frac{me^2}{\hbar^2}\right)^2\left[\frac{k^2 + 2\lambda^2}{(k^2\sin^2\frac{\theta}{2} + \lambda^2)(k^2\cos^2\frac{\theta}{2} + \lambda^2)}\right]^2,$$

$$\sigma_a(\theta) = \frac{1}{4}\left(\frac{me^2}{\hbar^2}\right)^2\left[\frac{k^2\cos\theta}{(k^2\sin^2\frac{\theta}{2} + \lambda^2)(k^2\cos^2\frac{\theta}{2} + \lambda^2)}\right]^2.$$

Hence, the total cross section is

$$\sigma_t = \frac{1}{4}\sigma_s + \frac{3}{4}\sigma_a$$

$$= \frac{1}{16}\left(\frac{me^2}{\hbar^2}\right)^2 \frac{(k^2 + 2\lambda^2)^2 + 3(k^2\cos\theta)^2}{\left[\left(k^2\sin^2\frac{\theta}{2} + \lambda^2\right)\left(k^2\cos^2\frac{\theta}{2} + \lambda^2\right)\right]^2}.$$

a.   If the incident particle is unpolarized, then the probability that after the scattering the two particles are in a state of total spin one is

$$\frac{\frac{3}{4}\sigma_a}{\sigma_t} = \frac{3(k^2\cos\theta)^2}{(k^2 + 2\lambda^2)^2 + 3(k^2\cos\theta)^2}.$$

The probability that after the scattering both particles have spin up along the z-axis is

$$\frac{\frac{1}{4}\sigma_a}{\sigma_t} = \frac{(k^2\cos\theta)^2}{(k^2+2\lambda^2)^2+3(k^2\cos\theta)^2}.$$

b. The $\ell=0$ partial wave has the symmetry $f(\theta)=f(\pi-\theta)$. It makes no contribution to the $S=1$ state, while it is the main contributor to the $S=0$ state. Therefore the ratio of the scattering cross section of the $S=1$ state to that of the $S=0$ state is equal to the ratio of the scattering cross section of the $\ell=1$ partial wave to that of the $\ell=0$ partial wave. It tends to zero in the low energy limit.

6037

An electron (mass m) of momentum p scatters through angle θ in a spin-dependent (and parity-violating) potential $V=e^{-\mu r^2}(A+B\sigma\cdot\mathbf{r})$, where $\mu(>0)$, A, B are constants and $\sigma_x,\sigma_y,\sigma_z$ are the usual Pauli spin matrices. Let $\frac{d\sigma_i}{d\Omega}$ be the differential scattering cross section, summed over final spin states but for definite initial spin state, labeled by the index i, of the incident electron. In particular, quantizing spin along the line of flight of the incident electron, we may consider alternatively: incident spin "up" ($i=\uparrow$) or "down" ($i=\downarrow$). Compute $\frac{d\sigma\uparrow}{d\Omega}$ and $\frac{d\sigma\downarrow}{d\Omega}$ as functions of p and θ in lowest order Born approximation.

(Princeton)

Sol: Let the incident direction of the electron be along the x-axis. In a diagonal representation of σ_z, the spin wave function of the incident electron may be expressed as

$$\psi_+=\frac{1}{\sqrt{2}}\begin{pmatrix}1\\1\end{pmatrix},\quad\text{or}\quad\psi_-=\frac{1}{\sqrt{2}}\begin{pmatrix}1\\-1\end{pmatrix}.$$

Let $\mathbf{n}$ be the unit vector along the $\mathbf{r}$ direction, i.e., $\mathbf{n}=(\sin\theta\cos\varphi,\sin\theta\sin\varphi,\cos\theta)$. Then

$$\sigma \cdot \mathbf{n} = \begin{pmatrix} 0 & 1 \\ 1 & 0 \end{pmatrix} \sin\theta\cos\varphi + \begin{pmatrix} 0 & -i \\ i & 0 \end{pmatrix} \sin\theta\sin\varphi$$

$$+ \begin{pmatrix} 1 & 0 \\ 0 & -1 \end{pmatrix} \cos\theta = \begin{pmatrix} \cos\theta & \sin\theta e^{-i\varphi} \\ \sin\theta e^{i\varphi} & -\cos\theta \end{pmatrix}.$$

First consider ψ_+;

$$(\sigma \cdot \mathbf{n})\psi_+ = \frac{1}{\sqrt{2}} \begin{pmatrix} \cos\theta + \sin\theta \, e^{-i\varphi} \\ \sin\theta \, e^{i\varphi} - \cos\theta \end{pmatrix}$$

$$= \frac{1}{\sqrt{2}} (\cos\theta + \sin\theta \, e^{-i\varphi})\alpha + \frac{1}{\sqrt{2}} (\sin\theta \, e^{i\varphi} - \cos\theta)\beta,$$

where $\alpha = \begin{pmatrix} 1 \\ 0 \end{pmatrix}$, $\beta = \begin{pmatrix} 0 \\ 1 \end{pmatrix}$ are the eigenstates of σ_z in Pauli's representation. In first Born approximation the scattering amplitude (including spin) is given by

$$f(\theta) = -\frac{m}{2\pi\hbar^2} \int e^{-i\mathbf{q}\cdot\mathbf{r}} V(\mathbf{r}') \psi_+ d^3x',$$

where $\mathbf{q} = \frac{1}{\hbar}(\mathbf{p}_f - \mathbf{p})$, $|\mathbf{q}| = q = \frac{2p}{\hbar} \sin(\theta/2)$. This can be written as

$$f(\theta) = -\frac{m}{2\pi\hbar^2} \int e^{-iqr'\cos\theta'} e^{-\mu r'2}$$

$$\times \left[A\psi_+ + r'B(\sigma \cdot \mathbf{n})\psi_+ \right] d^3x' = I_1(\theta)\alpha + I_2(\theta)\beta,$$

where

$$I_1(\theta) = -\frac{m}{2\pi\hbar^2} \int e^{-iqr'\cos\theta'} e^{-\mu r'^2} \frac{1}{\sqrt{2}}$$

$$\times [A + r'B(\cos\theta' + \sin\theta' e^{-i\varphi'})] r'^2 \sin\theta' dr' d\theta' d\varphi',$$

$$I_2(\theta) = -\frac{m}{2\pi\hbar^2} \frac{B}{\sqrt{2}} \int e^{-iqr'\cos\theta'} e^{-\mu r'^2} (\sin\theta' e^{i\varphi'} - \cos\theta') r'^3$$

$$\times \sin\theta' dr' d\theta' d\varphi'.$$

As

$$\int_0^{2\pi} e^{\pm i\varphi'} d\varphi' = 0,$$

we have

$$I_1(\theta) = -\frac{m}{2\pi\hbar^2} \int_0^\pi \int_0^\infty e^{-iqr'\cos\theta'} e^{-\mu r'^2} \frac{1}{\sqrt{2}}$$

$$\times (A + r'B\cos\theta') r'^2 \sin\theta' d\theta' dr',$$

or, integrating over θ',

$$I_1(\theta) = -\frac{2m}{\hbar^2 q} \cdot \frac{A}{\sqrt{2}} \int_0^\infty r' e^{-\mu r'^2} \sin(qr')dr' - \frac{2mi}{\hbar^2 q}$$

$$\times \frac{3}{\sqrt{2}} \int_0^\infty r'^2 e^{-\mu r'^2} \cos(qr')dr'$$

$$+ \frac{2mi}{\hbar^2 q^2} \cdot \frac{B}{\sqrt{2}} \cdot \int_0^\infty r' e^{-\mu r'^2} \sin(qr')dr'.$$

As

$$f(\mu, q) = \int_0^\infty e^{-\mu r'^2} \cos(qr')dr' = \frac{1}{2}\sqrt{\frac{\pi}{\mu}} \exp\left(-\frac{q^2}{4\mu}\right),$$

we have

$$\frac{\partial f}{\partial q} = -\int_0^\infty r' e^{-\mu r'^2} \sin(qr')dr' = \frac{1}{2}\sqrt{\frac{\pi}{\mu}} \exp\left(-\frac{q^2}{4\mu}\right)\left(\frac{-2q}{4\mu}\right),$$

or

$$\int_0^\infty r' e^{-\mu r'^2} \sin(qr')dr' = \sqrt{\frac{\pi}{\mu}} \cdot \frac{q}{4\mu} \exp\left(-\frac{q^2}{4\mu}\right),$$

as well as

$$\frac{\partial f}{\partial \mu} = -\int_0^\infty r'^2 e^{-\mu r'^2} \cos(qr')dr' = \sqrt{\frac{\pi}{\mu}}$$

$$\times \left(-\frac{1}{4\mu}\right)\exp\left(-\frac{q^2}{4\mu}\right) + \frac{q^2}{8\mu^2}\sqrt{\frac{\pi}{\mu}} \exp\left(-\frac{q^2}{4\mu}\right),$$

or

$$\int_0^\infty r'^2 e^{-\mu r'^2} \cos qr'dr' = \frac{1}{4\mu}\sqrt{\frac{\pi}{\mu}} \exp\left(-\frac{q^2}{4\mu} - \frac{q^2}{8\mu^2}\right)$$

$$\times \sqrt{\frac{\pi}{\mu}} \exp\left(-\frac{q^2}{4\mu}\right).$$

Thus

$$I_1(\theta) = -\frac{2m}{\hbar^2 q} \cdot \frac{A}{\sqrt{2}} \sqrt{\frac{\pi}{\mu}} \cdot \frac{q}{4\mu} \exp\left(-\frac{q^2}{4\mu}\right)$$

$$-\frac{2mi}{\hbar^2 q} \frac{B}{\sqrt{2}} \sqrt{\frac{\pi}{\mu}} \exp\left(-\frac{q^2}{4\mu}\right)\left(\frac{1}{4\mu} - \frac{q^2}{8\mu^2}\right)$$

$$+\frac{2mi}{\hbar^2 q^2} \frac{B}{\sqrt{2}} \cdot \sqrt{\frac{\pi}{\mu}} \exp\left(-\frac{q^2}{4\mu}\right) \cdot \frac{q}{4\mu}$$

$$= \sqrt{\frac{\pi}{2\mu}} \cdot \frac{m}{2\hbar^2 \mu}\left(-A + i\frac{Bq}{2\mu}\right)\exp\left(-\frac{q^2}{4\mu}\right).$$

Hence,

$$\frac{\partial \sigma_\uparrow}{\partial \Omega} = |I_1(\theta)|^2 = \frac{\pi m^2}{8\mu^3 \hbar^4} \exp\left(-\frac{q^2}{2\mu}\right)\left(A^2 + \frac{q^2 B^2}{4\mu^2}\right).$$

Similarly we obtain $I_2(\theta) = -\frac{mA}{2\hbar^2 \mu}\sqrt{\frac{\pi}{2\mu}} \exp\left(-\frac{q^2}{4\mu}\right)$

$$- \frac{imBq}{4\hbar^2 \mu^2}\sqrt{\frac{\pi}{2\mu}} \exp\left(-\frac{q^2}{4\mu}\right),$$

and hence

$$\frac{\partial \sigma_\downarrow}{\partial \Omega} = |I_2(\theta)|^2 = \frac{\mu m^2}{8\mu^3 \hbar^4} \exp\left(-\frac{q^2}{2\mu}\right)\left(A^2 + \frac{q^2 B^2}{4\mu^2}\right).$$

The same results are found for ψ_- .

6038

A spinless charged particle P_1 is bound in a spherically symmetric state whose wave function is $\psi_1(\mathbf{r}) = (\pi a)^{-3/2} e^{-r^2/2a^2}$. If a spinless, nonrelativistic projectile P_2 interacts with P_1 via the contact potential $V(\mathbf{r}-\mathbf{r}') = V_0 b^3 \delta^3(\mathbf{r}-\mathbf{r}')$, calculate, in first Born approximation, the amplitude for the elastic scattering of P_2 from the above bound state of P_1 (without worrying about the overall normalization). Assuming P_1 is sufficiently massive that its recoil energy is negligible, sketch the shape of the angular distribution $\frac{d\sigma}{d\Omega}$ of the scattered projectiles. How does this shape change with bombarding energy, and how can it be used to determine the size of the P_1 bound state? What determines the minimum energy of P_2 necessary to measure this size?

(Wisconsin)

Sol: Because P_1 is very heavy and so can be considered as fixed, the Schrödinger equation of P_2 is

$$\left[-\frac{\hbar^2}{2m}\nabla^2 + \int d\mathbf{r}' \rho_1(\mathbf{r}') V_0 b^3 \delta^3(\mathbf{r}-\mathbf{r}')\right]\psi(\mathbf{r}) = E\psi(\mathbf{r}),$$

or

$$\left(-\frac{\hbar^2}{2m}\nabla^2 + V_0 b^3 \rho_1(\mathbf{r})\right)\psi(\mathbf{r}) = E\psi(\mathbf{r}),$$

where $p_1(\mathbf{r}) = |\psi_1(\mathbf{r})|^2$ is the probability density of the particle P_1 at $\mathbf{r}$ and m is the mass of P_2. Then Born approximation gives

$$f(\theta) = -\frac{2m}{\hbar^2 q} \int_0^\infty r' V_0 b^3 p_1(r') \sin(qr') dr' \,|.$$

Thus

$$f(\theta) \propto \frac{1}{q}\left(\frac{b}{a}\right)^3 \int_0^\infty r' \exp\left(-\frac{r'^2}{a^2}\right) \sin(qr') dr'$$

$$\propto b^3 \exp\left[-\frac{1}{4}(qa)^2\right],$$

and hence

$$\frac{d\sigma}{d\Omega} = |f(\theta)|^2 = \sigma_0 \exp\left[-\frac{1}{2}(qa)^2\right]$$

$$= \sigma_0 \exp\left[-2(ka)^2 \sin^2\frac{\theta}{2}\right],$$

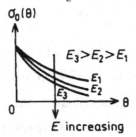

Fig. 6.9

where $k = \frac{p}{\hbar}$, $\sigma_0 = \sigma(\theta = 0)$. $\frac{d\sigma}{d\Omega}$ vs. θ for different energies are shown in the Fig. 6.9.

When the incident energy is increased, $\frac{d\sigma}{d\Omega}$ will more rapidly decrease with increasing θ. As

$$\ln\frac{d\sigma}{d\Omega} = -2k^2 a^2 \sin^2\frac{\theta}{2} + c,$$

where c is a constant, a plot of $\ln\frac{d\sigma}{d\Omega}$ against $\sin^2(\theta/2)$ will give a straight line with slope $-2k^2 a^2 = -2(\frac{p}{\hbar})^2 a^2$, which can be used to determine the size a of the P_1 bound state. The expression for $\frac{d\sigma}{d\Omega}$ does not appear to impose any restriction on the incident energy, except where the validity of Born approximation, on the basis of which the expression was derived, is concerned. In fact (**Problem 6026**), the validity of Born approximation requires

$$\frac{\hbar^2 k}{ma} \gg |V| \sim V_0 \left(\frac{b}{a}\right)^3,$$

i.e.,

$$k_{\text{min}} \sim \frac{mb^3 V_0}{\hbar^2 a^2}.$$

6039

a. State the electric-dipole selection rules for atomic states connected by emission or absorption of a photon.

b. Interpret the selection rule in terms of photon orbital angular momentum, spin, helicity and parity.

c. Make a semi-classical estimate of the lifetime of the $2p$ state of hydrogen, using the Bohr model and the classical formula

$$P = \frac{2}{3} \frac{q^2}{c^3} \dot{v}^2 \qquad \text{(c.g.s. units)}$$

for the power radiated by a particle of charge q and acceleration $\dot{v}$. Express your result in terms of $e, \hbar, c, a$ and ω, where a is the Bohr radius and ω is the angular velocity in the circular orbit.

d. Using the answer from (c), what is the width of the $2p$ state in electron volts?

(Berkeley)

Sol:

   a. The selection rules are

$$\Delta l = \pm 1, \quad \Delta m = \pm 1, 0.$$

   b. A photon has orbital angular momentum 0, spin 1, helicity ± 1, and negative parity. The conservation of angular momentum requires

$$\Delta l = 0, \pm 1, \qquad \Delta m = \pm 1, 0,$$

while the conservation of parity requires

$$(-1)^l = -(-1)^{l'}, \quad \text{i.e.,} \qquad l \neq l'.$$

Therefore,

$$\Delta l = \pm 1, \quad \Delta m = \pm 1, 0.$$

c. Classically, the power radiated by an electron of acceleration $\dot{v}$ is

$$P = \frac{2}{3}\frac{e^2}{c^3}|\dot{v}|^2 .$$

An electron in a circular orbit of radius a has acceleration

$$|\dot{v}| = \omega^2 a,$$

where $\omega^2 a$ is given by

$$\frac{e^2}{a^2} = m\omega^2 a .$$

For the $2p$ state, $n = 2$, $l = 1$, so the average radius of the electron orbit is

$$a = \frac{1}{2}[3n^2 - l(l+1)] = 5a_0,$$

where a_0 is the Bohr radius. This can also be obtained by a direct integration

$$a = \frac{\int R_{21}^2 r \cdot r^2 dr}{\int R_{21}^2 \cdot r^2 dr} = 5a_0,$$

where $R_{21} \propto r\exp\left(-\frac{r}{2a_0}\right)$. Thus the power radiated is

$$P = \frac{2}{3}\frac{e^6}{(5a_0)^4 m^2 c^3} .$$

In a transition to the ground state, the energy difference is

$$\Delta E = E_2 - E_1 = -\frac{e^2}{2a}\left(\frac{1}{2^2} - \frac{1}{1}\right) = \frac{3e^2}{8a_0} .$$

Hence the lifetime of the $2p$ state is

$$T = \Delta E / P = \left(\frac{75}{4}\right)^2 \frac{a_0^3 m^2 c^3}{e^4}$$

$$= \left(\frac{75}{4}\right)^2 \left(\frac{mc^2}{e^2}\right)\frac{a_0^3}{c}$$

$$= \left(\frac{75}{4}\right)^2 \left(\frac{1}{2.82\times10^{-13}}\right)^2 \times \frac{(0.53 \times 10^{-8})^3}{3\times10^{10}} = 2.2\times10^{-\varepsilon}$$

d. The width of the $2p$ state is

$$\Gamma = \hbar / T = 3.0\times10^{-8} \text{ eV} .$$

6040

he neutral K-meson states $|K^\circ\rangle$ and $|\bar{K}^\circ\rangle$ can be expressed in terms of states $K_L\rangle$, $|K_S\rangle$:

$$\left|K^{\circ}\right\rangle=\frac{1}{\sqrt{2}}\left(\left|K_{L}\right\rangle+\left|K_{S}\right\rangle\right),$$

$$\left|\overline{K}^{\circ}\right\rangle=\frac{1}{\sqrt{2}}\left(\left|K_{L}\right\rangle-\left|K_{S}\right\rangle\right),$$

where $\left|K_{L}\right\rangle$ and $\left|K_{S}\right\rangle$ are states with definite lifetimes $\tau_{L}=\frac{1}{\gamma_{L}}$ and $\tau_{S}=\frac{1}{\gamma_{S}}$ and distinct rest energies $m_{L}c^{2} \neq m_{S}c^{2}$. At time $t=0$, a meson is produced in the state $\left|\psi(t=0)\right\rangle=\left|K^{\circ}\right\rangle$. Let the probability of finding the system in state $\left|K^{\circ}\right\rangle$ at time t be $P_{0}(t)$ and that of finding the system in state $\left|\overline{K}^{\circ}\right\rangle$ at time t be $\overline{P}_{0}(t)$. Find an expression for $P_{0}(t)-\overline{P}_{0}(t)$ in terms of $\gamma_{L}, \gamma_{S}, m_{L}c^{2}$ and $m_{S}c^{2}$. Neglect CP violation.

(Columbia)

Sol: Suppose the K meson is a metastable state of width Γ at energy ε_{0}. In the region of energy

$$E = E_{0} - \frac{i}{2}\Gamma,$$

its wave function may be expressed as

$$\left|\psi(t)\right\rangle=\frac{1}{\sqrt{2}}\left(\left|K_{L}\right\rangle\exp\left[-i\left(m_{L}c^{2}/\hbar-\frac{i}{2}\gamma L\right)t\right]\right.$$

$$+\left|K_{S}\right\rangle\exp\left[-i(m_{S}c^{2}/\hbar-\frac{i}{2}\gamma_{S})t\right]$$

$$=\frac{1}{\sqrt{2}}\left[\left|K_{L}\right\rangle\exp(-im_{L}c^{2}t/\hbar)\exp\left(-\frac{1}{2}\gamma_{L}t\right)\right.$$

$$\left.+\left|K_{S}\right\rangle e^{-im_{S}c^{2}t/\hbar}e^{-\gamma_{S}t/2}\right].$$

The probability of its being in the $\left|K^{\circ}\right\rangle$ state at time t is

$$P_{0}(t)=\left|\left\langle K^{\circ}\middle|\psi(t)\right\rangle\right|^{2}$$

$$=\frac{1}{4}\left|e^{-im_{L}c^{2}t/\hbar}e^{-\gamma_{L}t/2}+e^{-im_{S}c^{2}t/\hbar}e^{-\gamma_{S}t/2}\right|^{2}$$

$$=\frac{1}{4}\{e^{-\gamma_{L}t}+e^{-\gamma_{S}t}+2e^{-(\gamma_{L}+\gamma_{S})t/2}\cos[(m_{L}-m_{S})c^{2}t/\hbar]\},$$

and the probability of its being in the $\bar{K}^\circ$ state is

$$\bar{P}_0(t) = \left| \langle \bar{K}^\circ | \psi(t) \rangle \right|^2$$

$$= \frac{1}{4} \{ e^{-\gamma_L t} + e^{-\gamma_S t} - 2e^{-(\gamma_L + \gamma_S)t/2} \cos[(m_L - m_S)c^2 t / \hbar] \}.$$

Thus

$$P_0(t) - \bar{P}_0(t) = e^{-(\gamma_L + \gamma_S)/2} \cos[(m_L - m_S)c^2 t / \hbar].$$

6041

The energy levels of the four lowest states of an atom are

$$E_0 = -14.0 \text{ eV},$$
$$E_1 = -9.0 \text{ eV},$$
$$E_2 = -7.0 \text{ eV},$$
$$E_3 = -5.5 \text{ eV}.$$

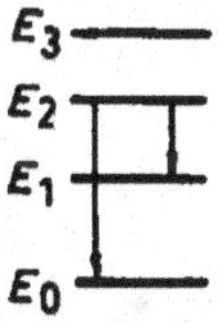

Fig. 6.10

and the transition rates A_{ij} (Einstein's A coefficients) for the $i \to j$ transitions are

$$A_{10} = 3.0 \times 10^8 \, s^{-1}, \quad A_{20} = 1.2 \times 10^8 \, s^{-1}, \quad A_{30} = 4.5 \times 10^7 \, s^{-1},$$
$$A_{21} = 8.0 \times 10^7 \, s^{-1}, \quad A_{31} = 0, \quad A_{32} = 1.0 \times 10^7 \, s^{-1}.$$

Imagine a vessel containing a substantial number of atoms in the level E_2.

a. Find the ratio of the energy emitted per unit time for the $E_2 \to E_0$ transition to that for the $E_2 \to E_1$ transition.

b. Calculate the radiative lifetime of the E_2 level.

(*Wisconsin*)

Sol:

a. The energy emitted per unit time is given by $(E_i - E_j)A_{ij}$, and so the ratio is

$$\frac{E_2 - E_0}{E_2 - E_1} \cdot \frac{A_{20}}{A_{21}} = \frac{7}{2} \cdot \frac{12}{8} = 5.25.$$

b. An atom at level E_2 can transit to E_0 or E_1 through spontaneous transition. So the decrease in the number of such atoms in the time period dt is

$$dN_2 = -(A_{20} + A_{21})N_2 dt.$$

i.e.,

$$\frac{dN_2}{N_2} = -(A_{20} + A_{21})dt,$$

or, by integration,

$$N_2 = N_{20} \exp\left[-(A_{20} + A_{21})t\right],$$

where N_{20} is the number of atoms at energy level E_2 at time $t = 0$. The mean lifetime is then

$$\tau = \frac{1}{N_{20}} \int_{t=0}^{\infty} t(-dN_2) = (A_{20} + A_{21}) \int_0^{\infty} t\exp\left[-(A_{20} + A_{21})t\right]dt$$

$$= \frac{1}{A_{20} + A_{21}} = 5.0 \times 10^{-9}\,s.$$

6042

The energy difference between the 3p and 3s levels in Na is 2.1eV. Spin-orbit coupling splits the 3p level, resulting in two emission lines differing by 6Å. Find the splitting of the 3p level.

Sol: The fine structure splitting of Na in ground and excited states is as follows:

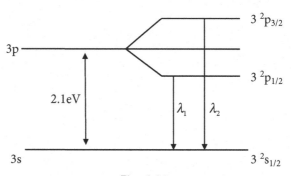

Fig. 6.11

The transition 3 $^2p_{3/2}$ to 3 $^2p_{1/2}$ produces a photon of wavelength λ_2 and the corresponding photon energy is as follows:

$$E_2 = \frac{12,400}{\lambda_2(\text{Å})}\,\text{eV}.$$

The transition 3 $^2p_{1/2}$ to 3 $^2s_{1/2}$ produces a photon of wavelength λ_1 and corresponding photon energy is $E_1 = \dfrac{12,400}{\lambda_1(\text{Å})}\,\text{eV}.$

The separation between $^2p_{3/2}$ and $^2p_{1/2}$ is given by

$$\Delta E = E_2 - E_1 = 12,400\left(\frac{1}{\lambda_2} - \frac{1}{\lambda_1}\right)\text{eV} = 12,400\left(\frac{\lambda_1 - \lambda_2}{\lambda_1\lambda_2}\right)\text{eV}$$

We are given $\Delta\lambda = \lambda_1 - \lambda_2 = 6\text{Å}$, and we can take $\lambda_1 = \lambda_2 = \lambda$ as a first order approximation and use the transition corresponding to 3p $\to$ 3s and to obtain $\lambda = (12,400/2.1)\text{Å}$. Therefore $\lambda_1\lambda_2 = \lambda^2 = \left(\dfrac{12,400}{2.1}\right)^2$

Thus, evaluating:

$$\Delta E = 12,400\left(\frac{6}{(12,400/2.1)^2}\right) = 2\text{ meV}.$$

The fine structure splitting is of the order of 10^{-3}eV.

6043

A hydrogen atom (with spinless electron and proton) in its ground state is placed between the plates of a condenser and subjected to a uniform weak electric field

$$E = E_0 e^{-\Gamma t}\theta(t),$$

where $\theta(t)$ is the step function: $\theta(t) = 0$ for $t < 0$ and $\theta(t) = 1$ for $t > 0$. Find the first order probability for the atom to be in any of the $n = 2$ states after a long time. Some hydrogenic wave functions in spherical coordinates are

$$\psi_{100} = \frac{1}{\sqrt{\pi a_0^3}}e^{-r/a_0}, \psi_{210} = \frac{1}{\sqrt{32\pi a_0^3}}e^{-r/2a_0}\frac{r}{a_0}\cos\theta,$$

$$\psi_{200} = \frac{1}{\sqrt{8\pi a_0^3}}\left(1 - \frac{r}{2a_0}\right)e^{-r/2a_0},$$

$$\psi_{21\pm1} = \mp\frac{1}{\sqrt{64\pi a_0^3}}\frac{r}{a_0}e^{-r/2a_0}\sin\theta e^{\pm i\phi}.$$

A useful integral is $\int_0^\infty x^n e^{-ax} dx = \dfrac{n!}{a^{n+1}}$.

(Wisconsin)

Sol: Take the direction of the electric field as the z direction. Then the perturbation Hamiltonian is

$$\hat{H}' = e\mathbf{r} \cdot \mathbf{E}(t) = ezE(t) .$$

The non-vanishing matrix elements of H' are those between states of opposite parities. Thus $P(1s \rightarrow 2s) = 0$. Consider $P(1s \rightarrow 2p)$.

The $2p$ state is three-fold degenerate, i.e.,

$$|2p, m\rangle, \quad \text{with} \quad m = 1, 0, -1 .$$

For $\langle m' | z | m'' \rangle$ not to vanish, the rule is $\Delta m = 0$. Thus

$$\langle 2p, 1 | H' | 1s, 0 \rangle = \langle 2p, -1 | H' | 1s, 0 \rangle = 0,$$

and finding the probability of the transition $1s \rightarrow 2p$ is reduced to finding that of $|1s, 0\rangle \rightarrow |2p, 0\rangle$. As

$$\langle 2p, 0 | H' | 1s, 0 \rangle = eE(t) \int \psi_{210}^* r \cos\theta \, \psi_{100} dr$$

$$= \frac{eE(t)}{4\sqrt{2}\pi a_0^4} \int_0^\infty \int_0^\pi \int_0^{2\pi} \exp\left(-\frac{3}{2} r / a_0\right)$$

$$\times r^4 \cos^2\theta \sin\theta \, dr \, d\theta \, d\varphi$$

$$= \frac{eE(t)}{4\sqrt{2}\pi a_0^4} 2\pi \cdot \frac{2}{3} \cdot \frac{4!}{\left(\dfrac{3}{2a_0}\right)^5} = \frac{2^7 \sqrt{2} a_0 e}{3^5} E(t),$$

the probability amplitude is

$$C_{2p0, 1s0} = \frac{1}{i\hbar} \int_{-\infty}^\infty \langle 2p, 0 | H' | 1s, 0 \rangle e^{i\omega_{21} t} dt$$

$$= \frac{1}{i\hbar} \frac{2^7 \sqrt{2} a_0 e}{3^5} E_0 \int_0^\infty e^{-\Gamma t} e^{i\omega_{21} t} dt$$

$$= \frac{2^7 \sqrt{2} a_0 e E_0}{3^5 i\hbar} \cdot \frac{1}{\Gamma - i\omega_{21}},$$

where $\omega_{21} = \frac{1}{\hbar}(E_2 - E_1)$. Hence the probability of the transition $|1s, 0\rangle \rightarrow |2p, 0\rangle$ is

$$P(1s \rightarrow 2p) = |C_{2p0,1s0}|^2 = \frac{2^{15} a_0^2 e^2 E_0^2}{3^{10} \hbar^2 (\Gamma^2 + \omega_{21}^2)} .$$

Note that

$$\omega_{21} = \frac{1}{\hbar}(E_2 - E_1) = \frac{e^2}{2a_0\hbar}\left(\frac{1}{2^2} - \frac{1}{1}\right) = \frac{3e^2}{8a_0\hbar}.$$

6044

A diatomic molecule with equally massive atoms, each with mass M, separated by D is electrically polarized, rotating about an axis perpendicular to D and running through the center of mass of the molecule.

a. Express the energy of the rotational state of the molecule with angular momentum quantum number J in terms of its mechanical properties.

b. What is the selection rule for electric dipole radiation emission from the molecule in one of its rotational states? (DERIVE ANSWER)

c. Determine the frequency of the electric dipole radiation emitted from the rotating molecule as a function of J. (Express answer as a function of J, M, D and any universal constants that may enter).

(Buffalo)

Sol:

a. As $\hat{H} = \frac{1}{2I}\mathbf{J}^2$, where $\mathbf{J}$ is the total angular momentum operator, $I = \frac{1}{2}MD^2$ is the moment of inertia of the molecule about the rotating axis, the energy of the rotating state of quantum number J is

$$E_J = \frac{1}{MD^2}J(J+1)\hbar^2.$$

b. The eigenfunctions of the rotational states are the spherical harmonic functions Y_{jm}. Take the z-axis along the electric field and consider the requirements for $\langle j''m'' | \cos\theta | j'm' \rangle \neq 0$. As

$$\langle J''m'' | \cos\theta | J'm' \rangle$$

$$= \sqrt{\frac{(J'+1-m')(J'+1+m')}{(2J'+1)(2J'+3)}}\delta_{J'',J'+1}\delta_{m''m'}$$

$$+ \sqrt{\frac{(J'+m')(J'-m')}{(2J'+1)(2J'-1)}}\delta_{J'',J'-1}\delta_{m''m'},$$

we require

$$\Delta J = j'' - j' = \pm 1,$$
$$\Delta m = m'' - m' = 0.$$

c.   For the transition from energy level J to $J-1$, we have

$$\hbar\omega = E_J - E_{J-1} = \frac{1}{MD^2} J(J+1)\hbar^2 - \frac{1}{MD^2}(J-1)J\hbar^2 = \frac{2J\hbar^2}{MD^2},$$

giving

$$\omega = \frac{2J\hbar}{MD^2}.$$

6045

a.   Find the energies above the bottom of the potential well of the ground state and first two excited states of a particle of mass m bound in a deep one-dimensional square-well potential of width l as shown in Fig. 6.12. Sketch the corresponding wave functions.

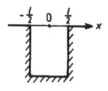

Fig. 6.12

b.   Calculate the matrix elements for electric dipole transitions from the first two excited states to the ground state, and explain any qualitative differences. [You need not carry out all the integration]

c.   Give the general selection rule for electric dipole transition between any two states in the system.

(*Wisconsin*)

Sol:

a.   The energy levels of the system are

$$E = \frac{\pi^2\hbar^2 n^2}{2ml^2}, \qquad \text{where } n = 1, 2, \ldots.$$

The wave functions of even parity are given by

$$\psi_n^+(x) = \sqrt{\frac{2}{l}} \cos\frac{n\pi x}{l}, \qquad \text{where } n = \text{an odd integer}.$$

The wave functions of odd parity are

$$\psi_n^-(x) = \sqrt{\frac{2}{l}}\sin\frac{n\pi x}{l}, \qquad\qquad \text{where } n = \text{an even integer}.$$

The ground state and the first two excited states are respectively

$$n = 1, \quad E_1 = \frac{\hbar^2 \pi^2}{2ml^2}, \quad \psi_1^+(x) = \sqrt{\frac{2}{l}}\cos\left(\frac{\pi x}{l}\right),$$

$$n = 2, \quad E_2 = 4E_1, \quad \psi_2^-(x) = \sqrt{\frac{2}{l}}\sin\left(\frac{2\pi x}{l}\right),$$

$$n = 3, \quad E_3 = 9E_1, \psi_3^+(x) = \sqrt{\frac{2}{l}}\cos\left(\frac{3\pi x}{l}\right).$$

These wave functions are sketched in Fig. 6.13.

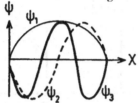

Fig. 6.13

b. The Einstein coefficient for electric dipole transition is

$$A_{kk'} = \frac{4e^2 \omega_{kk'}^3}{3\hbar c^3}\left|r_{kk'}\right|^2.$$

The matrix element for the transition of an electric dipole from the first excited state to the ground state is

$$\langle x \rangle_{21} = \int_{-l/2}^{l/2}\psi_1^*(x)\,x\psi_2(x)\,dx$$

$$= \frac{2}{l}\int_{-l/2}^{l/2} x\cos\frac{\pi x}{l}\sin\frac{2\pi x}{l}\,dx.$$

The matrix element for the transition of an electric dipole from the second excited state to the ground state is

$$\langle x \rangle_{31} = \int_{-l/2}^{l/2}\psi_1^*(x)\,x\psi_3(x)\,dx$$

$$= \frac{2}{l}\int_{-l/2}^{l/2} x\cos\frac{\pi x}{l}\cos\frac{3\pi x}{l}\,dx.$$

The second matrix element $\langle x \rangle_{31}$ is zero because the integrand is an odd function. Thus the second excited state cannot transit to the ground state by electric dipole transition. There is however no such restriction on electric dipole transition from the first excited state.

c.   The matrix element for electric dipole transition from a state k to a state k' of the system is

$$\langle x \rangle_{kk'} = \int_{-l/2}^{l/2} \psi_{k'}^*(\mathbf{x}) \psi_k(\mathbf{x}) \cdot x \, dx.$$

If the initial and final states have the same parity, the integrand is an odd function and $\langle x \rangle_{kk'}$ vanishes. Thus the general selection rule for electric dipole transition is that any such transition between states of the same parity is forbidden.

6046

Consider a particle in a one-dimensional infinite potential well. Let the origin be at the center as shown in Fig. 6.14.

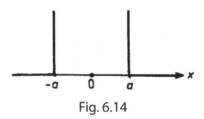

Fig. 6.14

a.   What are the allowed energies?

b.   What are the allowed wave functions?

c.   For what class of solutions will a perturbing potential $\Delta V(x) = kx$ have no first order effect on the energy?

d.   If transitions between states can occur by dipole radiation, what are the selection rules?

(Wisconsin)

Sol:

a.   The allowed energies are

$$E_n = \frac{\hbar^2 \pi^2 n^2}{8ma^2}. \qquad (n = 1, 2\ldots)$$

b.   There are two classes of allowed wave functions, one of even parity,

$$\psi_n^+(x) = \sqrt{\frac{1}{a}} \cos \frac{n\pi x}{2a},$$

where n is an odd integer, and one of odd parity,

$$\psi_n^-(x) = \sqrt{\frac{1}{a}} \sin \frac{n\pi x}{2a},$$

where n is an even integer.

c. First order perturbation gives the energy as

$$E_n = E_n^{(0)} + \langle \Delta V \rangle_{nn} = E_n^{(0)} + \langle kx \rangle_{nn}.$$

As $\Delta V = kx$ is an odd function, the diagonal matrix elements are all zero. This means that, as long as the wave function has a definite parity (whatever it is), there is no energy correction of the first order. Only for states of mixed parities will there be energy correction arising from first order perturbation.

d. Electric dipole transitions are determined by $\langle x \rangle_{kk'}$. Since

$$\langle x \rangle_{kk'} = \langle k' | x | k \rangle \propto \langle k' | a^+ + a | k \rangle$$
$$\approx \sqrt{k+1}\, \delta_{k', k+1} + \sqrt{k}\, \delta_{k', k-1},$$

the selection rule is $\Delta k = \pm 1$.

6047

A particle of charge q moving in one dimension is initially bound to a delta function potential at the origin. From time $t = 0$ to $t = \tau$ it is exposed to a constant electric field ε_0 in the x direction as shown in Fig. 6.15. The object of this problem is to find the probability that for $t > \tau$ the particle will be found in an unbound state with energy between E_k and $E_k + dE_k$.

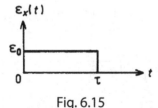

Fig. 6.15

a. Find the normalized bound-state energy eigenfunction corresponding to the delta function potential $V(x) = -A\delta(x)$.

b. Assume that the unbound states may be approximated by free particle states with periodic boundary conditions in a box of length L. Find the normalized wave function of wave vector k, $\psi_k(x)$, the density of states as a function of k, $D(k)$, and the density of states as a function of free-particle energy, $D(E_k)$.

c. Assume that the electric field may be treated as a perturbation. Write down the perturbation term in the Hamiltonian, $\hat{H}_1$, and find the matrix element of $\hat{H}_1$ between the initial state and the final state, $\langle 0|\hat{H}_1|k\rangle$.

d. The probability of a transition between an initially occupied state $|I\rangle$ and a final state $|F\rangle$ due to a weak perturbation Hamiltonian $\hat{H}_1(t)$ is given by

$$P_{I\to F}(t)=\frac{1}{\hbar^2}\left|\int_{-\infty}^{t}\left\langle F\left|\hat{H}_1(t')\right|I\right\rangle e^{i\omega_{FI}t'}\,dt'\right|^2,$$

where $\omega_{FI}=(E_F-E_I)/\hbar$. Find an expression for the probability $P(E_k)dE_k$ that the particle will be in an unbound state with energy between E_k and E_k+dE_k for $t>\tau$.

(MIT)

Sol:

a. The energy eigenfunction satisfies the Schrödinger equation

$$-\frac{\hbar^2}{2m}\frac{d^2\psi}{dx^2}-A\delta(x)\psi=E\psi,$$

where $E<0$, or

$$\frac{d^2\psi}{dx^2}-k^2\psi+A_0\delta(x)\psi=0,$$

with

$$k=\sqrt{\frac{2m|E|}{\hbar^2}},\qquad A_0=\frac{2mA}{\hbar^2}.$$

Integrating this equation from $-\varepsilon$ to $+\varepsilon$, ε being an arbitrary small positive number and then letting $\varepsilon\to0$, we get

$$\psi'(+0)-\psi'(-0)=-A_0\psi(0).$$

We also have from the continuity of the wave function at $x=0$

$$\psi(+0)=\psi(-0)=\psi(0).$$

Thus

$$\frac{\psi'(+0)}{\psi(+0)} - \frac{\psi'(-0)}{\psi(-0)} = -A_0.$$

Solving the Schrödinger equation for $x \neq 0$ we get $\psi(x) = Ce^{-k|x|}$. Then as $\psi(x) = Ce^{-kx}$ for $x > 0$ and $\psi(x) = Ce^{kx}$ for $x < 0$ we have

$$\frac{\psi'(+0)}{\psi(+0)} - \frac{\psi'(-0)}{\psi(-0)} = -2k.$$

Hence

$$k = \frac{A_0}{2} = \frac{mA}{\hbar^2}.$$

The energy level for the bound state is

$$E = -\frac{\hbar^2 k^2}{2m} = -\frac{mA^2}{2\hbar^2},$$

and the corresponding normalized eigenfunction is

$$\psi(\mathbf{x}) = \sqrt{\frac{mA}{\hbar^2}} \exp\left(-\frac{mA}{\hbar^2}|x|\right).$$

b. If the unbound states can be approximately represented by a plane wave e^{ikx} in a one-dimensional box of length L with periodic boundary conditions, we have

$$\exp\left[ik\left(-\frac{L}{2}\right)\right] = \exp\left(ik\frac{L}{2}\right),$$

which gives $e^{ikL} = 1$, or

$$kL = 2n\pi. \qquad (n = 0, \pm 1, \pm 2, \ldots)$$

Hence

$$k = \frac{2n\pi}{L} = k_n, \qquad \text{say.}$$

Thus the normalized plane wave function for wave vector k is

$$\psi_k(\mathbf{x}) = \frac{1}{\sqrt{L}} e^{ikx} = \frac{1}{\sqrt{L}} \exp\left[i\left(\frac{2n\pi}{L} x\right)\right].$$

Note that the state of energy E_k is two-fold degenerate when $k \neq 0$, so the number of states with momenta between p and $p + dp$ is

$$\frac{L dp}{2\pi\hbar} = D(k)dk = \frac{1}{2} D(E_k) dE_k.$$

As k, p and E_k are related by

$$k = \frac{p}{\hbar}, \quad E_k = \frac{p^2}{2m} = \frac{\hbar^2 k^2}{2m},$$

we have

$$D(k) = \frac{L}{2\pi}, \qquad D(E_k) = \frac{L}{\pi\hbar}\sqrt{\frac{m}{2E_k}}.$$

c. Treating ε_0 as a perturbation, the perturbation Hamiltonian is $\hat{H}_1 = -q\varepsilon_0 x$
Its matrix element between the initial and final states is

$$\langle k|\hat{H}_1|0\rangle = \int_{-\infty}^{\infty}\psi_k^*(-q\varepsilon_0 x)\psi\,dx$$

$$= -\frac{q\varepsilon_0}{\sqrt{L}}\sqrt{\frac{mA}{\hbar^2}}\int_{-\infty}^{+\infty}x\exp\left(-ikx - k_0|x|\right)dx$$

$$= -\frac{q\varepsilon_0}{\sqrt{L}}\sqrt{\frac{mA}{\hbar^2}}\,i\frac{d}{dk}\int_{-\infty}^{+\infty}\exp\left(-ikx - k_0|x|\right)dx$$

$$= -\frac{q\varepsilon_0}{\sqrt{L}}\sqrt{\frac{mA}{\hbar^2}}\,i\frac{d}{dk}\left[\int_{-\infty}^{0}\exp(-ikx + k_0 x)dx\right.$$

$$\left. + \int_{0}^{\infty}\exp(-ikx - k_0 x)dx\right]$$

$$= -\frac{q\varepsilon_0}{\sqrt{L}}\sqrt{\frac{mA}{\hbar^2}}(-4ikk_0)\frac{1}{[k^2 + k_0^2]^2}$$

$$= \frac{4iq\varepsilon_0}{\sqrt{L}}\left(\frac{mA}{\hbar^2}\right)^{3/2}\frac{k}{\left[k^2 + \left(\frac{mA}{\hbar^2}\right)^2\right]^2}.$$

d. The perturbation Hamiltonian is

$$\hat{H}_1 = \begin{cases} 0, & (-\infty < t < 0) \\ -q\varepsilon_0 x, & (0 \le t \le \tau) \\ 0. & (\tau < t < +\infty) \end{cases}$$

The transition probability at $t > \tau$ is

$$P_{I\to F}(t) = \frac{1}{\hbar^2}\left|\langle k|\hat{H}_1|0\rangle\right|^2\left|\int_0^\tau dt'\exp(iw_{FI}t')\right|^2$$

$$= \frac{1}{\hbar^2}\left|\langle k|\hat{H}_1|0\rangle\right|^2\frac{\sin^2\left(\omega_{FI}\tau/2\right)}{\left(\omega_{FI}/2\right)^2}.$$

As

$$E_F = \frac{\hbar^2 k^2}{2m}, \qquad E_I = -\frac{mA^2}{2\hbar^2}, \qquad \omega_{FI} = \frac{1}{\hbar}(E_F - E_I),$$

we have

$$\frac{\sin^2(\omega_{FI}\tau/2)}{(\omega_{FI}/2)^2} = \frac{\sin^2\left\{\frac{\hbar\tau}{4m}\left[k^2+\left(\frac{mA}{\hbar^2}\right)^2\right]\right\}}{\left\{\frac{\hbar}{4m}\left[k^2+\left(\frac{mA}{\hbar^2}\right)^2\right]\right\}^2} \cdot$$

Hence the probability required is

$$P(E_k)dE_k = P_{I\to F}(t)D(E_k)dE_k$$

$$= \frac{(16q\varepsilon_0)^2 m^3 k_0^3}{\pi\hbar^7\left(k_0^2+\frac{2mE_k}{\hbar^2}\right)^6}\sqrt{2mE_k}$$

$$\times \sin^2\left[\frac{\hbar\tau}{4m}\left(k_0^2+\frac{2mE_k}{\hbar^2}\right)\right]dE_k$$

$$= \frac{(16q\varepsilon_0)^2 m^3 \left(\frac{mA}{\hbar^2}\right)^3}{\pi\hbar^7\left[\left(\frac{mA}{\hbar^2}\right)^2+\frac{2mE_k}{\hbar^2}\right]^6}\sqrt{2mE_k}$$

$$\times \sin^2\left\{\frac{\hbar\tau}{4m}\left[\left(\frac{mA}{\hbar^2}\right)^2+\frac{2mE_k}{\hbar^2}\right]\right\}dE_k.$$

6048

Consider a two-level atom with internal states $|1\rangle$ and $|2\rangle$ of energy separation $E_2-E_1=\hbar\omega_{21}$. It is initially in its ground state $|1\rangle$ and is exposed to electromagnetic radiation described by $\mathbf{E}=\mathbf{E}_m(e^{iwt}+e^{-iwt})$.

a. If $\omega=\omega_{12}$, calculate the probability that the atom will be in the state $|2\rangle$ at a later time t.

b. If ω is only approximately equal to ω_{12}, what qualitative difference will this make? Calculate the same probability for this case as you did in part (a).

(Buffalo)

Sol:

Let

$$H_0|1\rangle = E_1|1\rangle = \hbar\omega_1|1\rangle,$$
$$H_0|2\rangle = E_2|2\rangle = \hbar\omega_2|2\rangle.$$

The Hamiltonian can be written as

$$H = H_0 + H',$$

where

$$H' = -e\mathbf{x} \cdot \mathbf{E}_m (e^{i\omega t} + e^{-i\omega t})$$

is the perturbation arising from the presence of the electromagnetic radiation. The time-dependent Schrödinger equation

$$i\hbar \frac{\partial}{\partial t} |t\rangle = H |t\rangle$$

is to be solved with the initial condition $|t = 0\rangle = |1\rangle$. Suppose the solution is

$$|t\rangle = c_1(t)e^{-i\omega_1 t}|1\rangle + c_2(t)e^{-i\omega_2 k}|2\rangle$$

with $c_1(0) = 1$, $c_2(0) = 0$. Substitution in the Schrödinger equation gives

$$\begin{aligned}
i\hbar(\dot{c}_1 e^{-i\omega_1 t}|1\rangle &- ic_1\omega_1 e^{-i\omega_1 t}|1\rangle \\
&+ \dot{c}_2 e^{-i\omega_2 t}|2\rangle - ic_2\omega_2 e^{-i\omega_2 t}|2\rangle) \\
&= c_1 e^{-i\omega_1 t}\hbar\omega_1|1\rangle + c_2 e^{-i\omega_2 t}\hbar\omega_2|2\rangle \\
&+ c_1 e^{-i\omega_1 t} H'|1\rangle + c_2 e^{-i\omega_2 t} H'|2\rangle.
\end{aligned}$$

As

$$i\hbar \frac{\partial}{\partial t}(e^{-i\omega_1 t}|1\rangle) = \hbar\omega_1 e^{-i\omega_1 t}|1\rangle, \qquad\qquad \text{etc.},$$

the above simplifies to

$$\begin{aligned}
i\hbar\dot{c}_1 e^{-i\omega_1 t}|1\rangle &+ i\hbar\dot{c}_2 e^{-i\omega_2 t}|2\rangle \\
&= c_1 e^{-i\omega_1 t} H'|1\rangle + c_2 e^{-i\omega_2 t} H'|2\rangle.
\end{aligned}$$

Multiplying the above equation by $\langle 1|$ and by $\langle 2|$, we obtain

$$i\hbar\dot{c}_1 = c_1 \langle 1|H'|1\rangle + c_2 e^{-i(\omega_2 - \omega_1)t} \langle 1|H'|2\rangle,$$

$$i\hbar\dot{c}_2 = c_1 e^{i(\omega_2 - \omega_1)t} \langle 2|H'|1\rangle + c_2 \langle 2|H'|2\rangle.$$

Writing

$$\langle i| - e\mathbf{x} \cdot \mathbf{E}_m |j\rangle = \hbar a_{ij},$$

or

$$\langle i|H'|j\rangle = (e^{i\omega t} + e^{-i\omega t})\hbar a_{ij},$$

the above equations become

$$i\hbar\dot{c}_1 = c_1(e^{i\omega t} + e^{-i\omega t})\hbar a_{11} + c_2 e^{-i\omega_{21} t}(e^{i\omega t} + e^{-i\omega t})\hbar a_{12},$$

$$i\hbar\dot{c}_2 = c_1 e^{i\omega_{21} t}(e^{i\omega t} + e^{-i\omega t})\hbar a_{21} + c_2(e^{i\omega t} + e^{-i\omega t})\hbar a_{22},$$

where $\omega_{21} = \omega_2 - \omega_1$. If E_m is small, the fast-oscillation terms can be neglected and the equations written as

$$\dot{c}_1 = -ia_{12}c_2 e^{i(\omega - \omega_{21})t},$$
$$\dot{c}_2 = -ia_{21}c_1 e^{i(\omega_{21} - \omega)t}.$$

Eliminating c_1 from the above we find

$$\ddot{c}_2 - i(\omega_{21} - \omega)\dot{c}_2 + a_{12}a_{21}c_2 = 0.$$

As $|t = 0\rangle = |1\rangle$, we have the initial conditions

$$c_1(0) = 1, \ c_2(0) = 0,$$
$$\dot{c}_2(0) = -ia_{21}c_1(0) = -ia_{21}.$$

a. For $\omega = \omega_{21}$ the above becomes

$$\ddot{c}_2 + \Omega^2 c_2 = 0,$$

where $\Omega^2 = a_{12}a_{21} = |a_{12}|^2$.

The solution is

$$c_2 = Ae^{i\Omega t} + Be^{-i\Omega t}.$$

The boundary conditions for c_2 give

$$A = -B = -\frac{a_{21}}{2\Omega}.$$

Hence

$$c_2(t) = -i\frac{a_{21}}{\Omega}\sin\Omega t,$$

and the probability that the atom will be in the state $|2\rangle$ at time t is

$$|c_2(t)|^2 = \sin^2\Omega t.$$

b. For $\omega \approx \omega_{21}$, try a solution $c_2 \sim e^{i\lambda t}$. Substitution gives

$$c_2 = Ae^{i\lambda_+ t} + Be^{i\lambda_- t}$$

where

$$\lambda_\pm = \frac{1}{2}\left[(\omega_{21} - \omega) \pm \Lambda\right]$$

with $\Lambda = [(\omega_{21} - \omega)^2 + 4\Omega^2]^{1/2}$. The boundary conditions for c_2 thus give

$$c_2(t) = -\frac{2ia_{21}}{\Lambda}\exp\left[\frac{i}{2}(\omega_{21} - \omega)\,t\right]\sin\left(\frac{\Lambda}{2}t\right).$$

Hence the probability is

$$|c_2(t)|^2 = \frac{4|a_{21}|^2}{\Lambda^2}\sin^2\left(\frac{\Lambda}{2}t\right).$$

6049

In HCl a number of absorption lines with wave numbers (in cm^{-1}) 83.03, 103.73, 124.30, 145.03, 165.51 and 185.86 have been observed. Are these vibrational or rotational transition? If the former, what is the characteristic frequency? If the latter, what J values do they correspond to, and what is the moment of inertia of HCl?

In that case, estimate the separation between the nuclei.

(Chicago)

Sol: Diatomic molecular energy levels are given to a first approximation by

$$W_{nJ} = -U_0 + \left(n + \frac{1}{2}\right)\hbar\omega + \frac{J(J+1)\hbar^2}{2MR^2},$$

where n is an integer, M and R are the reduced mass and separation of the two atoms. On the right-hand side, the terms are energies associated with, respectively, the electronic structure, nuclear vibrational motion, and rotation of the molecule. As

$$W_{n+1,J} - W_{n,J} = \hbar\omega,$$

the vibrational lines have only one frequency, and so the lines are not due to vibrational transitions. The rotational energy levels

$$E_J = \frac{\hbar^2}{2I}J(J+1),$$

where $I = MR^2$, and the selection rule $|\Delta J| = 1$ give the wave number for the line arising from $J+1 \rightarrow J$ as

$$\tilde{\nu} \equiv \frac{1}{\lambda} = \frac{1}{hc}\Delta E_J = \frac{\hbar^2(J+1)}{hcI}.$$

Hence

$$I = \frac{\hbar^2}{hc\Delta\tilde{\nu}}.$$

For $J \rightarrow J-1$, the energy of the spectral line is

$$hc\tilde{\nu} = \frac{\hbar^2 J}{I},$$

which is proportional to J. The spacing of the neighboring lines $\Delta\tilde{\nu} = \frac{\hbar^2}{I}\Delta J = \frac{\hbar^2}{I}$

is a constant. For the given lines we have

$\bar{\nu}=\frac{1}{\lambda}(\text{cm}^{-1})$	Transition $J \to J-1$	$\Delta\left(\frac{1}{\lambda}\right)(\text{cm}^{-1})$
83.03	$4 \to 3$	20.70
103.73	$5 \to 4$	20.57
124.30	$6 \to 5$	20.73
145.03	$7 \to 6$	20.48
165.51	$8 \to 7$	20.35
185.86	$9 \to 8$	

$$\overline{\left(\Delta\frac{1}{\lambda}\right)}=20.57(\text{cm}^{-1}).$$

The moment of inertia of the HCl molecule is therefore

$$I = \frac{\hbar^2}{hc\Delta\left(\frac{1}{\lambda}\right)}$$

$$= \frac{6.63\times 10^{-34}}{(2\pi)^2 \times 3.0\times 10^8 \times 20.57\times 10^2}$$

$$= 2.72\times 10^{-47} \text{ kg}\cdot\text{m}^2.$$

As $M^{-1} = m_{\text{H}}^{-1} + m_{\text{Cl}}^{-1}$, the nuclear separation is

$$R = \left[\left(\frac{m_{\text{H}}+m_{\text{Cl}}}{m_{\text{H}}m_{\text{Cl}}}\right)I\right]^{1/2}$$

$$= \left[\frac{(1+35)}{1\times 35\times 1.67\times 10^{-27}}\times 2.72\times 10^{-47}\right]^{1/2}$$

$$= 1.29\times 10^{-10}\text{m} = 1.29 \text{ Å}.$$

6050

An arbitrary quantum mechanical system is initially in the ground state $|0\rangle$. At $t=0$, a perturbation of the form $H'(t)=H_0 e^{-t/T}$ is applied. Show that at large times the probability that the system is in state $|1\rangle$ is given by

$$\frac{\left|\langle 0|H_0|1\rangle\right|^2}{\left(\frac{\hbar}{T}\right)^2 + (\Delta\varepsilon)^2},$$

where $\Delta\varepsilon$ is the difference in energy of states $|0\rangle$ and $|1\rangle$. Be specific about what assumption, if any, were made arriving at your conclusion.

(Columbia)

Sol: In the first order perturbation method the transition probability amplitude is given by

$$c_{k'k}(t)=\frac{1}{i\hbar}\int_0^t H'_{k'k}e^{i\omega_{k'k}t'}\,dt',$$

where $H'_{k'k}=\langle k'|H'|k\rangle$. Then

$$c_{10}(t)=\frac{1}{i\hbar}\int_0^\infty e^{i\omega_{10}t}e^{-t/T}\langle 1|H_0|0\rangle dt$$

$$=\frac{1}{i\hbar}\frac{1}{\left(\frac{1}{T}-i\omega_{10}\right)}<1|H_0|0\rangle$$

$$=\frac{1}{i(\hbar/T)+\Delta\varepsilon}\langle 1|H_0|0\rangle,$$

where $\Delta\varepsilon=\omega_{10}\hbar$ is the energy difference between the $|0\rangle$ and $|1\rangle$ states. Hence the probability that the system is in state $|1\rangle$ at large times is

$$P_{10}=|c_{10}(t)|^2=\frac{|\langle 0|H_0|1\rangle|^2}{\hbar^2/T^2+(\Delta\varepsilon)^2}.$$

It has been assumed in the above that H_0 is very small so that first order perturbation method is applicable.

6051

A particle of charge e is confined to a three-dimensional cubical box of side $2b$. An electric field **E** given by

$$\mathbf{E}=\begin{cases}0, & t<0,\\ \mathbf{E}_0e^{-\alpha t}, & t>0,\end{cases}\qquad(\alpha=\text{a positive constant}),$$

is applied to the system. The vector $\mathbf{E}_0$ is perpendicular to one of the surfaces of the box. Calculate to lowest order in E_0 the probability that the charged particle, in the ground state at $t=0$, is excited to the first excited state by the ime $t = \infty$. (If you wish, you can leave your result in terms of properly defined

dimensionless definite integrals.)

(*Berkeley*)

Sol: Replace the cubical box by the potential

$$V(x,y,z) = \begin{cases} 0, & 0 < x < 2b, \ 0 < y < 2b, \ 0 < z < 2b, \\ \infty, & \text{otherwise.} \end{cases}$$

The zero order wave function of the particle in the box is

$$\psi_{lmn}(x,y,z) = \sqrt{\frac{1}{b^3}} \sin\frac{l\pi x}{2b} \sin\frac{m\pi y}{2b} \sin\frac{n\pi z}{2b} \equiv |lmn\rangle.$$

Then the ground state is $|111\rangle$, the first excited states are $|211\rangle, |121\rangle, |112\rangle$. Let E be in the x direction, i.e., $\mathbf{E}_0 = E_0\mathbf{e}_x$. Then $H' = -eE_0 x e^{-\alpha t}$. Calculate the matrix elements for the transitions between the ground and the first excited states:

$$\langle 111|x|211\rangle = \frac{1}{b}\int_0^{2b} x \sin\frac{\pi x}{2b} \sin\frac{\pi x}{b} \, dx = -\frac{32b}{9\pi^2}.$$

$$\langle 111|x|121\rangle = \langle 111|x|112\rangle = 0.$$

Hence the transition probability is

$$P = \frac{1}{\hbar^2} \left| \int_0^\infty \langle 211|H'|111\rangle \, \exp\left(\frac{i\Delta E t}{\hbar}\right) dt \right|^2,$$

where

$$\Delta E = \frac{\pi^2 \hbar^2}{8mb^2}(2^2 + 1^2 + 1^2 - 1^2 - 1^2 - 1^2) = \frac{3\pi^2 \hbar^2}{8mb^2}.$$

Thus

$$P = \left(\frac{32beE_0}{9\hbar\pi^2}\right)^2 \left| \int_0^\infty \exp\left(-\alpha t + i\frac{\Delta E t}{\hbar}\right) dt \right|^2$$

$$= \left(\frac{32beE_0}{9\hbar\pi^2}\right)^2 \frac{\hbar^2}{\alpha^2\hbar^2 + (\Delta E)^2}.$$

6052

An ^{27}Al nucleus is bound in a one-dimensional harmonic oscillator potential with natural frequency ω. Label the states as $\psi_{m\alpha} = \psi_m(x)\phi_\alpha$, where $\psi_m(x)$, $m = 0, 1, 2, \ldots$, an eigenstate of the harmonic oscillator potential, describes center-of-mass motion and $\phi_\alpha(x)$, $\alpha = 0, 1, 2, \ldots$, is the wave function specifying the intrinsic nuclear state. Suppose the nucleus is initially in the state $\psi_0(x)\phi_1$ and then decays to the ground state ϕ_0 by emitting a photon in the $-x$ direction. Assume that the nuclear excitation energy is much greater than the harmonic oscillator excitation energy: $E^* = \left(E_{\alpha=1} - E_{\alpha=0}\right) \gg \hbar\omega$.

a. What is the wave function for the nucleus after the photon has been emitted?

b. Write an expression for the relative probability P_1/P_0, where P_n is the probability for the nucleus to be left in the state $\psi_{n0} = \psi_n(x)\phi_0$.

c. Estimate numerically P_1/P_0 with $E^* = 840$ keV and $\hbar\omega = 1$ keV.

(MIT)

Sol:

a. The Galilean transformation
$$x' = x - \upsilon t, \qquad t' = t$$
transforms a wave function $\psi(x,t)$ by
$$\psi(x,\ t) = \exp\left(i\frac{M\upsilon^2}{2\hbar}t' + i\frac{M\upsilon}{\hbar}x'\right)\psi(x',\ t'),$$
where υ is the velocity of frame L' with respect to frame L and is taken to be in the x direction, and M is the mass of the particle.

By emitting a photon of energy E^* in the $-x$ direction the nucleus acquires a velocity $\upsilon = \frac{E^*}{Mc}$ in the x direction. At the same time it decays to the ground state ϕ_0. Thus initially $(t' = t = 0)$ the nucleus has a velocity υ and is, in its own frame of reference L', in the ground state ϕ_0. Hence after emitting the photon the nucleus is initially in the state ψ given by $(t = 0, x = x')$
$$\psi(x,\ 0) = \exp\left(i\frac{M\upsilon}{\hbar}x\right)\psi_0(x)\phi_0$$
in the observer's frame L.

b. The probability that the nucleus is in the state $\psi_{n0} = \psi_n(x)\phi_0$ is

$$P_n = \left| \langle \psi_{n0} | \psi(x,\,0) \rangle \right|^2$$

$$= \left| \langle \psi_n(x)\phi_0 | \exp\left(i\frac{Mv}{h}x \right) | \phi_0 \psi_0(x) \rangle \right|^2$$

$$= \left| \langle n | \exp\left(i\frac{Mv}{h}x \right) | 0 \rangle \right|^2,$$

where $|n\rangle = |\psi_n(x)\rangle$. Using the creation and destructron operators a^+, a, we can express

$$x = \sqrt{\frac{\hbar}{2M\omega}}\,(a^+ + a),$$

and as $e^{A+B} = e^A e^B e^{-[A,B]/2}$ ($[A,\,B]$ commutes with $A,\,B$) we have

$$P_n = \left| \langle n | \ \exp\left(i\frac{Mv}{\hbar}\sqrt{\frac{\hbar}{2M\omega}}(a^+ + a) \right) | 0 \rangle \right|^2$$

$$= \left| \langle n | \exp\left(i\frac{Mv}{\hbar}\sqrt{\frac{\hbar}{2M\omega}}a^+ \right) \exp\left(i\frac{Mv}{\hbar}\sqrt{\frac{\hbar}{2M\omega}}a \right) \right.$$

$$\left. \times \exp\left(-\frac{Mv^2}{4\hbar\omega} \right) | 0 \rangle \right|^2$$

$$= \exp\left(-\frac{Mv^2}{2\hbar\omega} \right) \left| \sum_{m,l=0}^{\infty} \frac{\left(i\sqrt{\frac{Mv^2}{2\hbar\omega}} \right)^m}{m!} \frac{\left(i\sqrt{\frac{Mv^2}{2\hbar\omega}} \right)^l}{l!} \langle n | (a^+)^m a^l | 0 \rangle \right|^2$$

$$= \exp\left(-\frac{Mv^2}{2\hbar\omega} \right) \left| \frac{\left(i\sqrt{\frac{Mv^2}{2\hbar\omega}} \right)^n}{n!} \sqrt{n!} \right|^2$$

$$= \frac{1}{n!}\left(\frac{\frac{1}{2}Mv^2}{\hbar\omega} \right)^n \exp\left(-\frac{\frac{1}{2}Mv^2}{\hbar\omega} \right).$$

Then

$$\frac{P_1}{P_0} = \frac{Mv^2}{2\hbar\omega} \approx \frac{E^{*2}}{2Mc^2\hbar\omega} = \frac{0.84^2}{2 \times 27 \times 931.5 \times 10^{-3}}$$

$$\approx 1.4 \times 10^{-2}.$$

6053

Consider the situation which arises when a negative muon is captured by an aluminum atom (atomic number $Z = 13$). After the muon gets inside the "electron cloud" it forms a hydrogen-like muonic atom with the aluminum nucleus. The mass of the muon is 105.7 MeV .

- **a.** Compute the wavelength (in Å) of the photon emitted when this muonic atom decays from the 3d state. (Slide rule accuracy; neglect nuclear motion)

- **b.** Compute the mean life of the above muonic atom in the 3d state, taking into account the fact that the mean life of a hydrogen atom in the 3d state is 1.6×10^{-8} sec.

(Berkeley)

Sol:

- **a.** For spontaneous transitions from the 3d state, the largest probability is for $3d \to 2p$. In nonrelativistic approximation, the photon energy is given by

$$
h\upsilon = \frac{m_\mu Z^2 e^4}{2\hbar^2}\left(\frac{1}{2^2} - \frac{1}{3^2}\right)
$$

$$
= \frac{m_\mu c^2 Z^2}{2}\left(\frac{e^2}{\hbar c}\right)^2\left(\frac{5}{36}\right)
$$

$$
= \frac{105.7 \times 13^2}{2}\left(\frac{1}{137}\right)^2\left(\frac{5}{36}\right) = 6.61 \times 10^{-2} \text{ MeV.}
$$

The corresponding wavelength is

$$
\lambda = \frac{c}{\upsilon} = \frac{hc}{h\upsilon} = \frac{4.135 \times 10^{-15} \times 3 \times 10^{10}}{6.61 \times 10^4} = 1.88 \times 10^{-9}\text{cm .}
$$

- **b.** The transition probability per unit time is

$$
A \propto \omega^3 |\mathbf{r}_{kk'}|^2 .
$$

For hydrogen-like atoms, as

$$
|\mathbf{r}_{kk'}| \propto \frac{1}{Z}, \qquad\qquad \omega \propto mZ^2, \qquad\qquad \text{and so} \qquad\qquad A \propto m^3 Z^4,
$$

the mean life of the muonic atom in the $3d$ state is

$$T_\mu = \left(\frac{A_0}{A}\right) T_0 = \left(\frac{m_0}{m_\mu}\right)^3 \frac{T_0}{Z^4}$$

$$= \left(\frac{0.51}{105.7}\right)^3 \times \frac{1}{13^4} \times 1.6 \times 10^{-8} = 6.3 \times 10^{-20}\, s.$$

6054

A particle of mass M, charge e, and spin zero moves in an attractive potential $k\left(x^2 + y^2 + z^2\right)$. Neglect relativistic effects.

a. Find the three lowest energy levels E_0, E_1, E_2; in each case state the degeneracy.

b. Suppose the particle is perturbed by a small constant magnetic field of magnitude B in the z direction. Considering only states with unperturbed energy E_2, find the perturbations to the energy.

c. Suppose a small perturbing potential $Ax \cos \omega t$ causes transitions among the various states in (a). Using a convenient basis for degenerate states, specify in detail the allowed transitions, neglecting effects proportional to A^2 or higher powers of A.

d. In (c), suppose the particle is in the ground state at time $t = 0$. Find the probability the energy is E_1 at time t.

e. For the unperturbed Hamiltonian, what are the constants of the motion?

(*Berkeley*)

Sol:

a. The Schrödinger equation for the particle in a rectangular coordinate system,

$$\left[-\frac{\hbar^2}{2m}\left(\frac{\partial^2}{\partial x^2} + \frac{\partial^2}{\partial y^2} + \frac{\partial^2}{\partial z^2}\right) + k\left(x^2 + y^2 + z^2\right)\right] \psi(x, y, z)$$

$$= E\psi(x, y, z),$$

can be reduced to three equations of the harmonic oscillator type and the energy of the particle can be written as a sum

$$E_N = E_l + E_m + E_n = \frac{3}{2}\hbar\omega + (l+m+n)\,\hbar\omega,$$

where

$$\omega = \sqrt{2k/M}, \qquad\qquad N = l+m+n = 0,\,1,\,2,\,\dots\,.$$

Therefore,

$$E_0 = \frac{3}{2}\hbar\omega = \frac{3}{2}\hbar\sqrt{2k/M},$$

no degeneracy;

$$E_1 = \frac{5}{2}\hbar\omega = \frac{5}{2}\hbar\sqrt{2k/M},$$

three-fold degeneracy $\left(\psi_{100},\ \psi_{010},\ \psi_{001}\right)$;

$$E_2 = \frac{7}{2}\hbar\omega = \frac{7}{2}\hbar\sqrt{2k/M},$$

six-fold degeneracy $\left(\psi_{200},\ \psi_{020},\ \psi_{002},\ \psi_{110},\ \psi_{101},\ \psi_{011}\right)$.

In spherical coordinates the wave function is

$$\psi_{nlm}(r,\ \theta,\ \varphi) = R_{n,l}(r)Y_{lm}(\theta,\ \varphi),$$

where

$$R_{n,l}(r) = \alpha^{3/2}\left[\frac{2^{l+2-n_r}(2l+2n_r+1)!!}{\sqrt{\pi}n_r!\left[(2l+1)!!\right]^2}\right]^{1/2}(\alpha r)^l e^{-\alpha^2 r^2/2}$$

$$\times F\left(-n_r,\ l+3/2,\ \alpha^2 r^2\right),$$

$$l = N - 2n_r = \begin{cases} 0,\,2,\dots\,N, & (N = \text{even}), \\ 1,\,3,\dots\,N, & (N = \text{odd}), \end{cases}$$

N being related to the energy by

$$E_N = \left(N + \frac{3}{2}\right)\hbar\omega, \quad N = 0,\,1,\,2,\,\dots\,,$$

and the degeneracy is $f_N = \frac{1}{2}(N+1)(N+2)$.

b. For a weak magnetic field B in the z direction, the perturbation Hamiltonian is

$$H' = -\frac{eB}{2Mc}\hat{L}_z.$$

Then in spherical coordinates we have

$$E_{nlm} = E_{nl} - \frac{eB}{2Mc}m\hbar,$$

where $m\hbar$ is the eigenvalue of $\hat{L}_z$. Thus the different degenerate states of E_2 have perturbed energies:

$$E_{200} = E_{20}$$

$$E_{222} = E_{22} - \frac{eB}{Mc}\hbar,$$

$$E_{221} = E_{22} - \frac{eB}{2Mc}\hbar,$$

$$E_{220} = E_{22},$$

$$E_{22-1} = E_{22} + \frac{eB}{2Mc}\hbar,$$

$$E_{22-2} = E_{22} + \frac{eB}{Mc}\hbar.$$

It is seen that the degeneracy is partially destroyed.

c. At time $t = 0$, $H' = Ax\cos(\omega t)$. Consider the three-dimensional harmonic oscillator in the rectangular coordinate system. The first order perturbation gives, with l being the quantum number for the component oscillator along the x-axis,

$$\langle l'm'n'| H'(x, t)| lmn\rangle = \delta_{m'm}\delta_{n'n}\langle l'| H'(x, t)| l\rangle$$

$$= A\cos(\omega t)\delta_{m'm}\delta_{n'n}\langle l'| x| l\rangle$$

$$= A\alpha^{-1}\cos(\omega t)\left[\sqrt{\frac{l+1}{2}}\delta_{l',l+1} + \sqrt{\frac{l}{2}}\delta_{l',l-1}\right]$$

$$\times \delta_{m'm}\delta_{n'n},$$

where $\alpha = \sqrt{\dfrac{M\sqrt{2k/M}}{\hbar}} = (2kM/\hbar^2)^{1/4}$. Thus the allowed transitions are those between states for which

$$\Delta m = \Delta n = 0, \qquad \Delta l = \pm 1.$$

d. Between the states E_0 and E_1, the selection rules allow only the transition $\psi_{000} \to \psi_{100}$. The probability is

$$P_{10} = \frac{1}{\hbar^2}\left|\int_0^t H'_{10}e^{i\omega't'}\,dt'\right|^2 = \frac{A^2}{2\alpha^2\hbar^2}\left|\int_0^t \cos(\omega t')e^{x\omega't'}\,dt'\right|^2,$$

where

$$\omega' = \sqrt{\frac{2k}{M}}, \qquad H'_{10} = \langle 100| H' | 000 \rangle,$$

$$\int_0^t \cos(\omega t') e^{i\omega' t'} \, dt' = \frac{1}{2} \int_0^t (e^{i\omega t'} + e^{-i\omega t'}) e^{i\omega' t'} \, dt'$$

$$= \frac{1}{2i} \left[\frac{e^{i(\omega'+\omega)t} - 1}{\omega' + \omega} + \frac{e^{i(\omega'-\omega)t} - 1}{\omega - \omega'} \right].$$

In the microscope world, ω and ω' are usually very large. Only when $\omega \sim \omega'$ will the above integral make a significant contribution to the integral. Hence

$$P_{10} \approx \frac{A^2}{8\alpha^2 h^2} \frac{\sin^2\left[(\omega'-\omega)t/2\right]}{\left[(\omega'-\omega)/2\right]^2},$$

or, when t is large enough,

$$P_{10} \approx \frac{A^2 \pi t}{4\alpha^2 \hbar^2} \delta(\omega' - \omega).$$

Note that when t is large,

$$\left[\frac{\sin(x'-x)t/2}{(x'-x)/2} \right]^2 \approx 2\pi t \delta(x'-x).$$

e. The energy, angular momentum, third component of the angular momentum, and parity are constants of the motion.

6055

a. Suppose the state of a certain harmonic oscillator with angular frequency ω is given by the wave function

$$\psi = N \sum_{n=0}^{\infty} \frac{\alpha^n}{\sqrt{n!}} \psi_n(x) e^{-n\omega t} \qquad \alpha = x_0 \sqrt{\frac{m\omega}{2\hbar}} e^{i\phi}, \qquad N = e^{-|\alpha|^2/2}.$$

Calculate the average position of the oscillator, $\langle x \rangle$, in this state and show that the time dependence of $\langle x \rangle$ is that of a classical oscillator with amplitude x_0 and phase ϕ.

b. The Hamiltonian for a one-dimensional harmonic oscillator in a laser electromagnetic field is given by

$$H = \frac{p^2}{2m} + \frac{ep}{2m\omega} E_0 \sin\omega t - \frac{1}{2} eE_0 x \cos\omega t + \frac{1}{2} m\omega_0^2 x^2,$$

where ω_0, m and e are the angular frequency, mass and charge of the oscillator, and ω is the angular frequency of the radiation.

Assume the laser is turned on at $t=0$ with the oscillator in its ground state ψ_0. Treat the electromagnetic interaction as a perturbation in first order, and find the probability for any time $t>0$ that the oscillator will be found in one of its excited states ψ_n.

Useful information: The normalized oscillator wave functions $\psi_n(x)$ have the property that

$$\left(x + \frac{\hbar}{m\omega} \frac{d}{dx} \right) \psi_n = \sqrt{\frac{2\hbar n}{m\omega}} \psi_{n-1},$$

$$\left(x - \frac{\hbar}{m\omega} \frac{d}{dx} \right) \psi_n = \sqrt{\frac{2\hbar(n+1)}{m\omega}} \psi_{n+1}.$$

(Wisconsin)

Sol:

a. Adding up the two equations for ψ_n given in the question, we have

$$2x\psi_n = \sqrt{\frac{2\hbar}{m\omega}} \left(\sqrt{n}\,\psi_{n-1} + \sqrt{n+1}\,\psi_{n+1} \right),$$

or

$$x|n\rangle = \sqrt{\hbar/2m\omega} \left(\sqrt{n}\,|n-1\rangle + \sqrt{n+1}\,|n+1\rangle \right).$$

Hence

$$\langle x \rangle = \langle \psi | x | \psi \rangle = N^2 \sum_{n=0}^{\infty} \frac{\alpha^{*n}}{\sqrt{n!}} \sum_{k=0}^{\infty} \frac{\alpha^k}{\sqrt{k!}} e^{i(n-k)\omega t} \langle n|x|k \rangle$$

$$= N^2 \sum_{n=0}^{\infty} \frac{\alpha^{*n}}{\sqrt{n!}} \sum_{k=0}^{\infty} \frac{\alpha^k}{\sqrt{k!}} e^{i(n-k)\omega t} \sqrt{\frac{\hbar}{2m\omega}} \langle n | [\sqrt{k}|k-1\rangle$$

$$+ \sqrt{k+1}|k+1\rangle]$$

$$= N^2 \sum_{n=0}^{\infty} \frac{\alpha^{*n}}{\sqrt{n!}} \sum_{k=0}^{\infty} \frac{\alpha^k}{\sqrt{k!}} e^{i(n-k)\omega t} \left(\sqrt{k} \delta_{k,n+1} + \sqrt{k+1} \delta_{k,n-1} \right)$$

$$= N^2 \sum_{n=0}^{\infty} \sqrt{\frac{\hbar}{2m\omega}} \left[\frac{\alpha^{*n}}{\sqrt{n!}} \cdot \frac{\alpha^{n+1}}{\sqrt{(n+1)!}} \sqrt{n+1} e^{-i\omega t} \right.$$

$$\left. + \frac{\alpha^{*(n+1)}}{\sqrt{(n+1)!}} \cdot \frac{\alpha^n}{\sqrt{n!}} \sqrt{n+1} e^{i\omega t} \right]$$

$$= N^2 \sqrt{\frac{\hbar}{2m\omega}} \sum_{n=0}^{\infty} \frac{|\alpha|^{2n}}{n!} \left(\alpha e^{-i\omega t} + \alpha^* e^{i\omega t} \right)$$

$$= N^2 e^{|\alpha|^2} \sqrt{\frac{\hbar}{2m\omega}} \left(\alpha e^{-i\omega t} + \alpha^* e^{i\omega t} \right)$$

$$= \frac{1}{2} x_0 \left[e^{i(\phi - \omega t)} + e^{-i(\phi - \omega t)} \right]$$

$$= x_0 \cos(\phi - \omega t).$$

Thus $\langle x \rangle$ is the same as that for a classical oscillator of amplitude x_0 and initial phase ϕ.

b.   Initially the oscillator is in the state $\psi_0 = |0\rangle$. Writing ω_0 for ω, the given equations for ψ_n give

$$x|n\rangle = \sqrt{h/2m\omega_0} \left(\sqrt{n}|n-1\rangle + \sqrt{n+1}|n+1\rangle \right),$$
$$\hat{p}|n\rangle = i\sqrt{hm\omega_0/2} \left(\sqrt{n+1}|n+1\rangle - \sqrt{n}|n-1\rangle \right),$$

where $\hat{p} = -i\hbar \frac{d}{dx}$. It follows that

$$x|0\rangle = \sqrt{h/2m\omega_0}|1\rangle,$$
$$\hat{p}|0\rangle = i\sqrt{hm\omega_0/2}|1\rangle.$$

The perturbation Hamiltonian is

$$\hat{H}' = \frac{ep}{2m\omega} E_0 \sin \omega t - \frac{1}{2} eE_0 x \cos \omega t.$$

As $\langle n|1\rangle 1 = \delta_{n,1}$, $H'_{n0} = 0$ for $n \neq 1$. Hence $P_{n0} = 0$ for $n > 1$. Consider H'_{10}. We have

$$H'_{10} = \left\langle 1 \left| \frac{e\hat{p}}{2m\omega} E_0 \sin \omega t - \frac{1}{2} eE_0 x \cos \omega t \right| 0 \right\rangle$$

$$= \frac{eE_0}{2} \left(\frac{i}{m\omega} \sqrt{\frac{\hbar m \omega_0}{2}} \sin \omega t - \sqrt{\frac{\hbar}{2m\omega_0}} \cos \omega t \right)$$

$$= \frac{eE_0}{2} \sqrt{\frac{\hbar}{2m\omega_0}} \left(i\frac{\omega_0}{\omega} \sin \omega t - \cos \omega t \right),$$

and hence the probability that the oscillator transits to state ψ_1 at time t

$$
\begin{aligned}
P_{10} &= \frac{1}{\hbar^2} \left| \int_0^t \frac{eE_0}{2} \sqrt{\frac{\hbar}{2m\omega_0}} \left(i\frac{\omega_0}{\omega} \sin\omega t' - \cos\omega t' \right) e^{i\omega t'} \, dt' \right|^2 \\
&= \frac{e^2 E_0^2}{8m\omega_0 \hbar} \left| \int_0^t \left[\frac{\omega_0}{2\omega} \left(e^{i\omega t'} - e^{-i\omega t'} \right) - \frac{1}{2} \left(e^{i\omega t'} + e^{-i\omega t'} \right) \right] e^{i\omega t'} \, dt' \right|^2 \\
&= \frac{e^2 E_0^2}{8m\omega_0 \hbar} \left[\frac{1}{2\omega^2} + \frac{8\omega_0^2}{(\omega_0^2 - \omega^2)^2} - \frac{1}{2\omega^2} \cos 2\omega t \right] \\
&\quad + \frac{e^2 E_0^2}{4m\omega \hbar} \left[\frac{\cos(\omega_0 + \omega)t}{(\omega_0 + \omega)^2} - \frac{\cos(\omega_0 - \omega)t}{(\omega_0 - \omega)^2} \right].
\end{aligned}
$$

6056

Suppose that, because of a small parity-violating force, the $2^2 S_{1/2}$ level of the hydrogen atom has a small P-wave admixture

$$
\begin{aligned}
\psi\left(n=2, j=\frac{1}{2} \right) &= \psi_s\left(n=2, j=\frac{1}{2}, l=0 \right) \\
&\quad + \varepsilon \psi_p\left(n=2, j=\frac{1}{2}, l=1 \right).
\end{aligned}
$$

What first-order radiative decay will de-excite this state? What is the form of the decay matrix element? What does it become if $\varepsilon \to 0$, and why?

(Wisconsin)

Sol: The first-order radiative decay is related to electric dipole transition. It causes the state to decay to $\psi\left(n=1, j=\frac{1}{2} \right)$, which is two-fold degenerate, corresponding to $m_j = \pm\frac{1}{2}$, $l=0$. The matrix element for such an electric dipole transition is given by

$$
\begin{aligned}
H_{12}' &= \left\langle \psi\left(n=1, j=\frac{1}{2} \right) \middle| -e\mathbf{r} \middle| \psi\left(n=2, j=\frac{1}{2} \right) \right\rangle \\
&= \varepsilon \left\langle \psi\left(n=1, j=\frac{1}{2} \right) \middle| -e\mathbf{r} \middle| \psi_p\left(n=2, j=\frac{1}{2}, l=1 \right) \right\rangle
\end{aligned}
$$

because of the selection rule $\Delta l = \pm 1$ for the transition. Using non-coupling representation basic vectors to denote the relevant coupling representation basic vectors we have the following final and initial states:

$$\psi\left(n=1, j=\frac{1}{2}, m_j=\frac{1}{2}\right)=|100\rangle\begin{pmatrix}1\\0\end{pmatrix},$$

$$\psi\left(n=1, j=\frac{1}{2}, m_j=-\frac{1}{2}\right)=|100\rangle\begin{pmatrix}0\\1\end{pmatrix},$$

$$\psi_p\left(n=2, j=\frac{1}{2}, l=1, m_j=\frac{1}{2}\right)=-\sqrt{\frac{1}{3}}|210\rangle\begin{pmatrix}1\\0\end{pmatrix}$$

$$+\sqrt{\frac{2}{3}}|211\rangle\begin{pmatrix}0\\1\end{pmatrix},$$

$$\psi_p\left(n=2, j=\frac{1}{2}, l=1, m_j=-\frac{1}{2}\right)=-\sqrt{\frac{2}{3}}|21,-1\rangle\begin{pmatrix}1\\0\end{pmatrix}$$

$$+\sqrt{\frac{1}{3}}|210\rangle\begin{pmatrix}0\\1\end{pmatrix}.$$

Hence the non-vanishing matrix elements of H'_{12} are

$$\varepsilon\left\langle\psi\left(n=1, j=m_j=\frac{1}{2}\right)\middle|-e\mathbf{r}\middle|\psi_p\left(m_j=\frac{1}{2}\right)\right\rangle$$

$$=\sqrt{\frac{1}{3}}e\varepsilon\langle100|\mathbf{r}|210\rangle=\sqrt{\frac{1}{3}}e\varepsilon\langle100|z|210\rangle\mathbf{e}_z$$

$$=\frac{e\varepsilon}{3}\langle100|r|200\rangle\mathbf{e}_z=\frac{e\varepsilon A}{3}\mathbf{e}_z,$$

where $A=\langle100|r|200\rangle$, $\mathbf{e}_z$ is a unit vector along the z direction,

$$\varepsilon\left\langle\psi\left(n=1, j=\frac{1}{2}, m_j=-\frac{1}{2}\right)\middle|-e\mathbf{r}\middle|\psi_p\left(m_j=\frac{1}{2}\right)\right\rangle$$

$$=-\sqrt{\frac{2}{3}}e\varepsilon\langle100|\mathbf{r}|211\rangle$$

$$=-\sqrt{\frac{2}{3}}e\varepsilon\langle100|x\mathbf{e}_x+y\mathbf{e}_y|211\rangle=-\frac{e\varepsilon A}{3}\left(\mathbf{e}_x+i\mathbf{e}_y\right),$$

$$\varepsilon\left\langle\psi\left(n=1, j=m_j=\frac{1}{2}\right)\middle|-e\mathbf{r}\middle|\psi_p\left(m_j=-\frac{1}{2}\right)\right\rangle$$

$$=-\frac{e\varepsilon A}{3}\left(\mathbf{e}_x-i\mathbf{e}_y\right),$$

$$\varepsilon\left\langle\psi\left(n=1, j=\frac{1}{2}, m_j=-\frac{1}{2}\right)\middle|-e\mathbf{r}\middle|\psi_p\left(m_j=-\frac{1}{2}\right)\right\rangle$$

$$=-\sqrt{\frac{1}{3}}e\varepsilon\langle100|r|210\rangle=\frac{e\varepsilon A}{3}\mathbf{e}_x.$$

In the above we have used the relations

$$(1\ 0)\begin{pmatrix}1\\0\end{pmatrix}=1, \qquad (1\ 0)\begin{pmatrix}0\\1\end{pmatrix}=0, \qquad \text{etc.,}$$

$$\mathbf{r}=xe_x+ye_y+ze_z$$
$$=r\ \sin\theta\cos\varphi e_x+r\ \sin\theta\sin\varphi e_y+r\ \cos\theta e_z,$$

$$\langle100|z|210\rangle=\langle100|r|200\rangle\langle l=0, m=0|\cos\theta|l=1, m=0\rangle, \quad \text{etc.,}$$

and the selection rules

$\Delta m=0$ for the z-component of $e\mathbf{r}$,

$\Delta m=\pm1$ for the x-, y-components of $e\mathbf{r}$.

Thus if the parity of the $2^2S_{1/2}$ state is destroyed by the inclusion of the ε term the electric dipole radiation would cause transition from the $2^2S_{1/2}$ state to the ground state $1^2S_{1/2}$, the probability of such de-excitation being $\propto \varepsilon^2$. If $\varepsilon=0$ electric dipole radiation does not allow de-excitation from the state

$$\psi_s\left(n=2, j=\frac{1}{2}, l=0\right) \text{ to the state } \psi\left(n=1, j=\frac{1}{2}, l=0\right)$$

because the perturbation H' is a polar vector whose the matrix elements are nonzero only when $\Delta l=\pm1$.

6057

a. The part of the Hamiltonian describing the hyperfine interaction between the electron and proton in atomic hydrogen is given by

$$H'=-\frac{8\pi}{3}\mu_e\cdot\mu_p\delta^3(\mathbf{r}),$$

where $\mu_i=\frac{e_i g_i}{2m_ic}\mathbf{S}_i$ is the magnetic moment and $\mathbf{s}_i=\frac{1}{2}\mathbf{r}_i$ is the spin of particle i (the σ's are Pauli matrices). Calculate the hyperfine splitting between the $1s\,^3S_1$ and $1s\,^1S_0$ states of atomic hydrogen. Which state has the lower energy? Explain why physically.

b. The vector potential of the radiation field emitted in a transition between the states in part (a) has the general form that, as $r\to\infty$,

$$A = \left[-i\frac{\omega}{c}\langle \mathbf{x}\rangle + i\frac{\omega}{c}\hat{\mathbf{n}}\times\frac{e}{2m_ec}\langle \mathbf{L}\rangle + i\frac{\omega e}{2m_ec^2}\hat{\mathbf{n}}\times\langle \sigma_e\rangle + \ldots \right]$$
$$\times \frac{e^{i\frac{\omega}{c}r - i\omega t}}{r},$$

where $\hat{\mathbf{n}}$ is a unit vector along the direction of propagation of the radiation and $\langle \cdot \rangle$ denotes the matrix element for this transition. Show explicitly for each of the three terms whether or not $\langle \cdot \rangle$ is nonzero. What is the character of the radiation emitted in the transition?

(Wisconsin)

Sol: Let the spatial wave function of ls state be $\psi_0(\mathbf{r})$, the spin singlet state be χ_{00}, and the spin triplet state be $\chi_{1M}(M = 0, \pm 1)$.

a.   The perturbation method for degenerate states is to be used, the perturbation Hamiltonian being

$$H' = \frac{2\pi}{3}\cdot\frac{e^2 g_e g_p}{m_e m_p c^2}\mathbf{S}_e\cdot\mathbf{S}_p\delta^3(\mathbf{r}) = \frac{B}{2}\left[(\mathbf{S}_e + \mathbf{S}_p)^2\right.$$

$$\left. -\mathbf{S}_e^2 - \mathbf{S}_p^2\right]\delta^3(\mathbf{r}) = \frac{B}{2}\left[S^2 - \frac{3}{2}\hbar^2\right]\delta^3(\mathbf{r}),$$

when $B = \frac{2\pi}{3}\frac{e^2 g_e g_p}{m_e m_p c^2}$ and $\mathbf{S} = \mathbf{S}_e + \mathbf{S}_p$, with $S_e^2 = S_p^2 = \frac{1}{2}\cdot\frac{3}{2}\hbar$. If $\psi_0(\mathbf{r})\chi_{00}$ and $\psi_0(\mathbf{r})\chi_{1M}$ are chosen to be the basis vectors, then H' is a diagonal matrix. For the ls $^1S_0$ energy level, $S = 0$ and we have

$$\Delta E_1 = \langle \psi_0 \chi_{00}|H'|\psi_0 \chi_{00}\rangle$$
$$= -\frac{3}{4}B\hbar^2|\psi_0(\mathbf{r} = 0)|^2.$$

For the ls $^3S_1$ energy level, $S^2 = 1(1+1)\hbar^2$ and we have

$$\Delta E_2 = \langle \psi_0 \chi_{1M}|H'|\psi_0 \chi_{1M}\rangle = \frac{1}{4}B\hbar^2|\psi_0(\mathbf{r} = 0)|^2.$$

As

$$\psi_0(\mathbf{r}) = \frac{1}{\sqrt{4\pi}}\frac{2}{a^{3/2}}e^{-r/a}, \qquad |\psi_0(\mathbf{r} = 0)|^2 = \frac{1}{\pi a^3}.$$

Hence the hyperfine splitting is

$$\Delta E = \Delta E_2 - \Delta E_1 = \frac{1}{\pi a^3} B\hbar^2.$$

The above calculation shows that the singlet state $(^1S_0)$ has lower energy. The reason is as follows. The intensity of the field produced by a magnetic dipole decreases rapidly with increasing distance. So for the magnetic dipole interaction between the electron and proton we have to consider the case when they are very close. When μ_e is parallel to μ_p, the energy of the magnetic interaction is lower $\left(\text{as } E = -\mu \cdot \mathbf{B} \right)$ than when they are antiparallel. Since when μ_e and μ_p are parallel, S_e and S_p are antiparallel. The gap in singlet state has the lower energy.

b. For the transition from the triplet state to the singlet state, due to the vector potential **A**, as the terms for $\hat{x}$ and $\hat{\mathbf{L}}$ do not contain spin operators, we have

$$\langle \mathbf{x} \rangle = \langle \psi_0 \chi_{00} | \hat{x} | \psi_0 \chi_{1M} \rangle = \langle \psi_0 | \hat{x} | \psi_0 \rangle \langle \chi_{00} | \chi_{1M} \rangle = 0,$$

$$\langle \mathbf{L} \rangle = \langle \psi_0 \chi_{00} | \hat{\mathbf{L}} | \psi_0 \chi_{1M} \rangle = 0,$$

$$\langle \sigma_e \rangle = \langle \psi_0 \chi_{00} | \sigma_e | \psi_0 \chi_{1M} \rangle = \langle \chi_{00} | \sigma_e | \chi_{1M} \rangle.$$

As

$$\chi_{00} = \frac{1}{\sqrt{2}} \left[\begin{pmatrix} 1 \\ 0 \end{pmatrix}_e \begin{pmatrix} 0 \\ 1 \end{pmatrix}_p - \begin{pmatrix} 1 \\ 0 \end{pmatrix}_p \begin{pmatrix} 0 \\ 1 \end{pmatrix}_e \right],$$

$$\chi_{11} = \begin{pmatrix} 1 \\ 0 \end{pmatrix}_e \begin{pmatrix} 1 \\ 0 \end{pmatrix}_p,$$

$$\chi_{10} = \frac{1}{\sqrt{2}} \left[\begin{pmatrix} 1 \\ 0 \end{pmatrix}_e \begin{pmatrix} 0 \\ 1 \end{pmatrix}_p + \begin{pmatrix} 1 \\ 0 \end{pmatrix}_p \begin{pmatrix} 0 \\ 1 \end{pmatrix}_e \right],$$

$$\chi_{1,-1} = \begin{pmatrix} 0 \\ 1 \end{pmatrix}_e \begin{pmatrix} 0 \\ 1 \end{pmatrix}_p,$$

and if we take the z-axis parallel to **n** the z component of $\langle \sigma_e \rangle$ will contribute nothing to $\mathbf{n} \times \langle \sigma_e \rangle$, thus we have effectively

$$\langle \chi_{00} | \sigma_e | \chi_{11} \rangle = -\frac{1}{\sqrt{2}} (0 \ 1) \sigma_e \begin{pmatrix} 1 \\ 0 \end{pmatrix} = -\frac{1}{\sqrt{2}} e_x - \frac{i}{\sqrt{2}} e_y,$$

$$\langle \chi_{00} | \sigma_e | \chi_{1,-1} \rangle = \frac{1}{\sqrt{2}} (1 \ 0) \sigma_e \begin{pmatrix} 0 \\ 1 \end{pmatrix} = \frac{1}{\sqrt{2}} e_x - \frac{i}{\sqrt{2}} e_y,$$

$$\langle \chi_{00} | \sigma_e | \chi_{10} \rangle = \frac{1}{2} (1 \ 0) \sigma_e \begin{pmatrix} 1 \\ 0 \end{pmatrix} - \frac{1}{2} (0 \ 1) \sigma_e \begin{pmatrix} 0 \\ 1 \end{pmatrix} = e_z,$$

where $\mathbf{e}_x, \mathbf{e}_y, \mathbf{e}_z$ are unit vectors along the x-, y-, z-axis respectively. Hence $\langle \sigma_e \rangle \neq 0$. Note that the direction of **A** is parallel to $\mathbf{n} \times \langle \sigma_e \rangle$. It is similar to the vector potential of magnetic dipole radiation, so the radiation emitted in the transition has the character of magnetic dipole radiation.

6058

Protons (magnetic moment μ) are in a magnetic field of the form

$$B_x = B_0 \cos \omega t, \qquad\qquad B_y = B_0 \sin \omega t,$$
$$B_z = \text{constant}, \qquad\qquad B_0 \ll B_z.$$

At $t = 0$ all the protons are polarized in the $+z$ direction.

a. What value of ω gives resonant transitions?

b. What is the probability for a proton at time t to have spin in the $-z$ direction? (Assume $B_0 \ll B_z$)

(*Princeton*)

Sol:

a. As $B_0 \ll B_z$, $\mathbf{B}' = B_x\hat{\mathbf{x}} + B_y\hat{\mathbf{y}}$ may be considered as a perturbation. Then the unperturbed Hamiltonian (spin part) $H_0 = -\mu B_z \sigma_z$ gives the energy difference of the two states $\begin{pmatrix} 1 \\ 0 \end{pmatrix}$ and $\begin{pmatrix} 0 \\ 1 \end{pmatrix}$ with spins along $+z$ and $-z$ directions as $2\mu B_z$. Hence resonant transition occurs at angular frequency $\omega = 2\mu B_z / \hbar$.

b. As

$$H = -\mu\boldsymbol{\sigma}\cdot\mathbf{B}$$
$$= -\mu\left(\sigma_x B_x + \sigma_y B_y + \sigma_z B_z\right)$$
$$= -\mu\begin{pmatrix} B_z & B_x - iB_y \\ B_x + iB_y & -B_z \end{pmatrix},$$

the Schrödinger equation can be written as

$$i\hbar\frac{\partial}{\partial t}\begin{pmatrix} a \\ b \end{pmatrix} = -\mu\begin{pmatrix} B_z & B_0 e^{-i\omega t} \\ B_0 e^{i\omega t} & -B_z \end{pmatrix}\begin{pmatrix} a \\ b \end{pmatrix},$$

where a and b are the probability amplitudes of the electron with its spin oriented along $+z$ and $-z$ directions respectively. Letting

$$a = e^{-i\frac{\omega}{2}t} f, \qquad b = e^{i\frac{\omega}{2}t} g,$$

one obtains the equations for f and g:

$$\hbar \frac{\omega}{2} f + i\hbar \frac{\partial f}{\partial t} + \mu B_z f + \mu B_0 g = 0, \tag{1}$$

$$\mu B_0 f + i\hbar \frac{\partial g}{\partial t} - \hbar \frac{\omega}{2} g - \mu B_z g = 0. \tag{2}$$

Taking the time derivative of Eq. (2) and substituting in the expressions of $\frac{\partial f}{\partial t}, \frac{\partial g}{\partial t}$ from (1) and (2), we obtain

$$\frac{\partial^2 g}{\partial t^2} + \Omega^2 g = 0, \tag{3}$$

where

$$\Omega^2 = \frac{1}{\hbar^2} \left[\mu^2 B_0^2 + \left(\frac{\hbar \omega}{2} + \mu B_z \right)^2 \right].$$

Initially the protons are polarized in the $+z$ direction. Hence $|f| = 1$, $g = 0$ at $t = 0$. Then the solution of (3) is $g = A \sin \Omega t$, where A is a constant. Assume $f = B \sin \Omega t + C \cos \Omega t$ and substitute these in (1). Supposing $f = i$ at $t = 0$, we have

$$C = i, \qquad B = -\frac{1}{\hbar\Omega} \left(\frac{\hbar\omega}{2} + \mu B_z \right), \qquad A = -\frac{\mu B_0}{\hbar\omega}.$$

Hence

$$f = -\frac{1}{\hbar\Omega} \left(\hbar \frac{\omega}{2} + \mu B_z \right) \sin \Omega t + i \cos \Omega t,$$

$$g = \frac{-\mu B_0}{\hbar\Omega} \sin \Omega t.$$

Thus the probability for the protons to have spin in the $-z$ direction at time t is

$$P = |b|^2 = |g|^2 = \left(\frac{\mu B_0}{\hbar\Omega} \right)^2 \sin^2 \Omega t$$

with

$$\Omega = \frac{1}{\hbar} \sqrt{\mu^2 B_0^2 + \left(\frac{\hbar\omega}{2} + \mu B_z \right)^2} \approx \frac{1}{\hbar} \left(\frac{\hbar\omega}{2} + \mu B_z \right) \qquad \text{as} \qquad B_z \gg B_0.$$

6059

A piece of paraffin is placed in a uniform magnetic field $\mathbf{H}_0$. The sample contains many hydrogen nuclei. The spins of these nuclei are relatively free from interaction with their environment and, to first approximation, interact only with the applied magnetic field.

a. Give an expression for the numbers of protons in the various magnetic substates at a temperature T.

b. A radio-frequency coil is to be introduced in order to observe resonance absorption produced by an oscillating magnetic field. What should be the direction of the oscillating field relative to the steady magnetic field H_0 and why?

c. At what frequency will resonance absorption be observed? Give the units of all quantities appearing in your expression so that the frequency will be given in megacycles per second.

d. In terms of the transition mechanism of the proton spins, explain why the absorption of energy from the radio-frequency field does not disappear after an initial pulse, but in fact continues at a steady rate. What happens to the absorption rate as the strength of the oscillating field is increased to very large values? Explain.

(CUS)

Sol:

a. As the spins of the hydrogen nuclei are assumed to interact only with the external field, the interaction Hamiltonian is

$$\hat{H} = -\mathbf{m} \cdot \mathbf{H}_0 = -g\mu_N \hat{s}_z H_0,$$

taking the z-axis in the direction of H_0. Then there are two states $\left| s_z = \tfrac{1}{2} \right\rangle$ and $\left| s_z = -\tfrac{1}{2} \right\rangle$ with respective energies

$$E_{1/2} = -\frac{1}{2} g\mu_N H_0, \quad E_{-1/2} = \frac{1}{2} g\mu_N H_0,$$

where $g = 5.6$ is a constant and μ_N the nuclear magneton $\mu_N = e\hbar / 2m_p c$. The condition for statistical equilibrium at temperature T gives the probabilities for a nucleus to be in the two states as for $\left| s_z = \tfrac{1}{2} \right\rangle$:

$$P = \exp\left(\frac{1}{2} g\mu_N H_0 / kT \right) \Bigg/ \left[\exp\left(\frac{1}{2} g\mu_N H_0 / kT \right) \right.$$
$$\left. + \exp\left(-\frac{1}{2} g\mu_N H_0 / kT \right) \right],$$

for $\left| s_z = -\tfrac{1}{2} \right\rangle$:

$$P = \exp\left(-\frac{1}{2} g\mu_N H_0 / kT \right) \Bigg/ \left[\exp\left(\frac{1}{2} g\mu_N H_0 / kT \right) \right.$$
$$\left. + \exp\left(-\frac{1}{2} g\mu_N H_0 / kT \right) \right],$$

which are also the proportions of protons in the two states.

b. The oscillating magnetic field $\hat{H}_1$ must be perpendicular to $\hat{H}_0$, say along the x direction. This is because only if the spin part of the Hamiltonian has the form

$$\hat{H} = -\mathbf{m} \cdot \mathbf{H} = -g\mu_N s_z H_0 - g\mu_N s_x H_1$$

will the matrix elements $\left\langle s_z = \tfrac{1}{2} \middle| \hat{H} \middle| s_z = -\tfrac{1}{2} \right\rangle$ and $\left\langle s_z = -\tfrac{1}{2} \middle| \hat{H} \middle| s_z = \tfrac{1}{2} \right\rangle$ be non-vanishing and transitions between the spin states occur, since

$$\left\langle \tfrac{1}{2} \middle| s_z \middle| -\tfrac{1}{2} \right\rangle = \frac{1}{2}(1 \ 0) \begin{pmatrix} 1 & 0 \\ 0 & -1 \end{pmatrix} \begin{pmatrix} 0 \\ 1 \end{pmatrix} = 0,$$

$$\left\langle \tfrac{1}{2} \middle| s_x \middle| -\tfrac{1}{2} \right\rangle = \frac{1}{2}(1 \ 0) \begin{pmatrix} 0 & 1 \\ 1 & 0 \end{pmatrix} \begin{pmatrix} 0 \\ 1 \end{pmatrix} = \frac{1}{2}.$$

c. Resonance absorption occurs only when the oscillating frequency satisfies the condition

$$\hbar\omega = E_{-1/2} - E_{1/2},$$

or

$$\omega = g\mu_N H_0 / \hbar .$$

With $g = 5.6$, $\hbar = 1.054 \times 10^{-34}$ Js,

$$\mu_N = \frac{e\hbar}{2m_p c} = \frac{e\hbar}{2m_e c} \frac{m_e}{m_p} = \frac{9.274 \times 10^{-28}}{1836} \text{J Gs}^{-1},$$

and ω in megacycles per second is given by

$$\omega = \frac{5.6}{1836} \times \frac{9.274 \times 10^{-28}}{1.054 \times 10^{-34}} \times 10^{-6} H_0 = 2.7 \times 10^{-2} H_0 .$$

where H_0 is in gauss.

d. Spin interactions between the protons tend to maintain a thermal equilibrium, so that even if the external field vanishes the magnetic interaction between a proton and the magnetic field caused by other protons still exists and the transitions take place. When the external magnetic field is very strong, the absorption rate saturates.

6060

An electron is bound in the ground state by a potential

$$V = \begin{cases} -\dfrac{\beta}{x}, & x > 0, \\ \infty, & x < 0, \end{cases}$$

which is independent of y and z. Find the cross section for absorption of plane-wave light of energy ωh, direction $\mathbf{k}$, polarization ε. Specify the final state of the electron. Take

$$\frac{\beta^2 m}{\hbar^2} \ll \hbar\omega \ll mc^2 .$$

(*Wisconsin*)

Sol:

As initially the electron moves freely in the y, z directions, its initial state is given by

$$\psi_i(\mathbf{r}) = \varphi(\mathbf{x}) \exp\left(\frac{i\left(p_y y + p_z z\right)}{\hbar} \right),$$

with

$$\begin{cases} -\dfrac{\hbar^2}{2m}\dfrac{d^2\varphi}{dx^2} - \dfrac{\beta}{x}\varphi = E\varphi, & x > 0, \\ \varphi = 0, & x < 0. \end{cases}$$

The equation for φ is the same as the radial equation of a hydrogen atom with $l = 0$. Hence the energy states are

$$E_n = -\frac{m\beta^2}{2\hbar^2}\frac{1}{n^2}, \quad n = 1, 2, \dots .$$

Thus the ground (initial) state for x motion has energy and wave function

$$E_1 = -\frac{m\beta^2}{2\hbar^2},$$

$$\varphi_1(\mathbf{x}) = \frac{2x}{a^{3/2}} e^{-x/a}, \quad x > 0,$$

where

$$a = \frac{\hbar^2}{m\beta}.$$

The condition $\hbar\omega \gg \frac{m\beta^2}{\hbar^2} \approx |E_1|$ means that the photon energy is much higher than the mean binding energy of the electron and can therefore liberate it from the potential well. However, as $\hbar\omega \ll mc^2$, this energy is much lower than the electron rest mass and cannot produce electron pairs. So the electron initial state is

$$\psi_i(\mathbf{r}) \equiv \langle \mathbf{r} | i \rangle = C\varphi_1(x) \exp\left[i\left(k_y^{(e)} y + k_z^{(e)} z \right) \right],$$

where $k_y^{(e)} = p_y/\hbar$, $k_z^{(e)} = p_z/\hbar$ are the wave numbers of the electron in the y, z directions respectively, $C = \left(\frac{1}{\sqrt{L}}\right)^2 = \frac{1}{L}$ if the initial-state wave function is normalized to one electron in a 2-dimensional box of sides L in the y – z plane. The final state of the electron is that of a free electron moving in direction $k_f^{(e)}$ (direction of observation):

$$\psi_f(\mathbf{r}) \equiv \langle \mathbf{r} | f \rangle = \left(\frac{1}{L}\right)^{3/2} \exp\left(i\mathbf{k}_f^{(e)} \cdot \mathbf{r} \right),$$

where L^3 is the 3-dimensional box used for normalization. The perturbation Hamiltonian is

$$H' = H - H_0 = \left\{ \frac{1}{2m} (\hat{\mathbf{p}} + \frac{e}{c}\mathbf{A})^2 + V \right\} - \left\{ \frac{1}{2m} \hat{\mathbf{p}}^2 + V \right\}$$

$$\approx \frac{e}{mc} \mathbf{A} \cdot \hat{\mathbf{p}}.$$

where $\mathbf{A}$ is the vector potential of the photon field and the charge of the electron is $-e$. In the above we have chosen the gauge $\nabla \cdot \mathbf{A} = 0$ and omitted terms of orders higher than A^2. Let the corresponding initial electric field be

$$\mathbf{E} = E\varepsilon \, \sin\left(\omega t - \mathbf{k}_i \cdot \mathbf{r} + \delta_0 \right),$$

where $\mathbf{k}_i$ is the wave vector of the incident photon and $\varepsilon = \left\{ \varepsilon_x, \, \varepsilon_y, \, \varepsilon_z, \right\}$ is a unit vector in the direction of $\mathbf{E}$. The vector potential can be taken to be

$$\mathbf{A} = \frac{cE}{\omega} \varepsilon \cos\left(\omega t - \mathbf{k}_i \cdot \mathbf{r} + \delta_0 \right)$$

as $\mathbf{E} = -\frac{1}{c}\frac{\partial \mathbf{A}}{\partial t}$, and the perturbation Hamiltonian is then

$$H' \simeq \frac{e}{mc}\mathbf{A} \cdot \hat{\mathbf{P}} = \frac{-i\hbar e}{2m\omega}\left\{\exp\left[i(\omega t - \mathbf{k}_i \cdot \mathbf{r} + \delta_0)\right]\right.$$
$$\left. + \exp\left[-i(\omega t - \mathbf{k}_i \cdot \mathbf{r} + \delta_0)\right]\right\}E\varepsilon\nabla$$

In photon absorption, $E_f > E_i$ and we need to consider only the second term (the first term is for photo-emission); so the perturbation Hamiltonian is

$$H' = \frac{-i\hbar e}{2m\omega}\exp\left[-i(\omega t - \mathbf{k}_i \cdot \mathbf{r} + \delta_0)\right]E\varepsilon\nabla.$$

For a plane electromagnetic wave,

$$\mathbf{H} = \frac{1}{k}\mathbf{k} \times \mathbf{E}$$

and so the Poynting vector is

$$\mathbf{S} = \frac{c}{4\pi}(\mathbf{E} \times \mathbf{H}) = \frac{c}{4\pi k}E^2\mathbf{k}.$$

Averaging over time we have

$$\bar{S} = \frac{cE^2}{8\pi}.$$

Hence the number of incident photons crossing unit area per unit time is

$$n = \frac{\bar{S}}{\hbar\omega} = \frac{cE^2}{8\pi\hbar\omega}.$$

The differential absorption cross section for photoelectric effect is given by

$$\frac{d\sigma}{d\Omega_f} = \frac{\omega_{i\to f}}{n},$$

where $\omega_{i\to j}$ is the number of electrons in solid angle $d\Omega_f$ which transit from the initial state to final states f near energy E_f in unit time. First order perturbation theory gives the transition probability per unit time as

$$\omega_{i\to f}d\Omega_f = \frac{2\pi}{\hbar}\rho(E_f)\left|W_{fi}\right|^2 d\Omega_f,$$

where $\rho(E_f)$ is the density of the final states per unit energy range.

For nonrelativistic electrons,

$$\rho = \frac{mk_f^{(e)}L^3}{8\pi^3\hbar^2},$$

where $k_f^{(e)}$ is the wave number of the electrons in the final states of energies near E_f,

$$W_{fi} = \langle f|H'|i\rangle = \left\langle f\left|\frac{-i\hbar e}{2m\omega}\exp\left[-i(-\mathbf{k}_i\cdot\mathbf{r}+\delta_0)\right]E\varepsilon\cdot\nabla\right|i\right\rangle$$

$$= \frac{-i\hbar e}{2m\omega}e^{-i\delta_0}\int_0^{+\infty}dx\int_{-\infty}^{+\infty}\int dy\,dz\cdot L^{-3/2}$$

$$\times \exp\left[-ik_f^{(e)}\cdot\mathbf{r}+ik_i\cdot\mathbf{r}\right](E\varepsilon\cdot\nabla)C\varphi_1(x)$$

$$\times \exp\left[i\left(k_y^{(e)}y+k_z^{(e)}z\right)\right]$$

$$= \frac{-i\hbar eCEe^{-i\delta_0}}{2m\omega}L^{-3/2}\int dx\,dy\,dz\left\{\varepsilon_x\left(\frac{1}{x}-\frac{1}{a}\right)\right.$$

$$\left. + i\varepsilon_y k_y^{(e)}+i\varepsilon_z k_z^{(e)}\right\}\varphi_1(x)\exp\left[-ik_f^{(e)}\cdot\mathbf{r}+ik_i\cdot\mathbf{r}\right]$$

$$\times \exp\left[i\left(k_y^{(e)}y+k_z^{(e)}z\right)\right]$$

$$= \frac{4\pi^2\sqrt{a}\hbar eEe^{-i\delta_0}}{m\omega L^{5/2}\left[1+ia\left(k_x^{(f)}-k_x^{(i)}\right)\right]^2}\left\{\varepsilon_x\left(k_x^{(f)}-k_x^{(i)}\right)\right.$$

$$\left. -\varepsilon_y k_y^{(e)}-\varepsilon_z k_z^{(e)}\right\}\delta(-k_y^{(f)}+k_y^{(i)}$$

$$\left. +k_y^{(e)})\delta\left(-k_z^{(f)}+k_z^{(i)}+k_z^{(e)}\right).\right.$$

Hence the differential absorption cross section is

$$\frac{d\sigma}{d\Omega_f} = \frac{8\pi a k_f e^2}{m\omega c(1+a^2\Delta^2)^2}\left[\varepsilon_x\Delta-\varepsilon_y k_y^{(e)}-\varepsilon_z k_z^{(e)}\right]^2$$

$$\times \delta\left(k_y^{(i)}+k_y^{(e)}-k_y^{(f)}\right)\delta\left(k_z^{(i)}+k_z^{(e)}-k_z^{(f)}\right).$$

In above calculation, we have made the change of symbols

$$\mathbf{k}_f^{(e)}\to\mathbf{k}^f\equiv\left\{k_x^{(f)},\,k_y^{(f)},\,k_z^{(f)}\right\},\quad \mathbf{k}_i\to\mathbf{k}^{(i)}\equiv\left\{k_x^{(i)},\,k_y^{(i)},\,k_z^{(i)}\right\},$$

$$k_x^{(f)}-k_x^{(i)}=\Delta$$

and used

$$\left[\delta\left(k_y^{(i)}+k_y^{(e)}-k_y^{(f)}\right)\right]^2 = \frac{1}{2\pi}\int_{-L/2}^{L/2}\delta\left(k_y^{(i)}+k_y^{(e)}-k_y^{(f)}\right)$$

$$\times \exp\left[iy\left(k_y^{(i)}+k_y^{(e)}-k_y^{(f)}\right)\right]dy$$

$$= \frac{L}{2\pi}\delta\left(k_y^{(i)}+k_y^{(e)}-k_y^{(f)}\right),$$

and

$$\left[\delta\left(k_z^{(i)}+k_z^{(e)}-k_z^{(f)}\right)\right]^2=\frac{L}{2\pi}\delta\left(k_z^{(i)}+k_z^{(e)}-k_z^{(f)}\right).$$

Note that the two δ-functions express momentum conservation in the directions of y and z. Also, energy conservation requires

$$\frac{\hbar^2k_f^2}{2m}=E_1+\frac{\hbar^2\left(k_y^{(e)^2}+k_z^{(e)^2}\right)}{2m}+\hbar\omega$$

$$=-|E_1|+\frac{\hbar^2\left(k_y^{(e)^2}+k_z^{(e)^2}\right)}{2m}+\hbar\omega.$$

The δ functions mean that the y and z components of $\mathbf{k}_f$ are fixed, and so is the x component of $\mathbf{k}_f$. The physical reason why the δ-functions appear in the expression for the differential absorption cross section is that when an electron which has definite momenta in the y and z directions collide with an incident photon which also has a definite momentum, energy and momentum conservation laws require the scattering direction of the electron in the final state to be fixed.

To find the total absorption cross section, we note that

$$\delta(\alpha x)=\frac{1}{\alpha}\delta(x)$$

and so

$$\begin{cases}\delta\left(k_y^{(f)}-k_y^{(i)}-k_y^{(e)}\right)=\frac{1}{k_f}\delta\left(\sin\theta_f\sin\varphi_f-\frac{k_y^{(i)}+k_y^{(e)}}{k_f}\right),\\[4mm]\delta\left(k_z^{(f)}-k_z^{(i)}-k_z^{(e)}\right)=\frac{1}{k_f}\delta\left(\cos\theta_f-\frac{k_z^{(i)}+k_z^{(e)}}{k_f}\right).\end{cases}$$

Then the total absorption cross section is

$$\sigma_a=\int\frac{d\sigma}{d\Omega_f}d\Omega_f$$

$$=\frac{8\pi ae^2k_f}{m\omega c}\int\frac{1}{k_f^2}\left[\frac{\varepsilon_z\left(k_f\sin\theta_f\cos\varphi_f-k_x^{(i)}\right)-\varepsilon_yk_y^{(e)}-\varepsilon_zk_z^{(e)}}{1+a^2\left(k_f\sin\theta_f\cos\varphi_f-k_x^{(i)}\right)^2}\right]^2$$

$$\times\delta\left(\sin\theta_f\sin\varphi_f-\frac{k_y^{(i)}+k_y^{(e)}}{k_f}\right)\delta\left(\cos\theta_f-\frac{k_z^{(i)}+k_z^{(e)}}{k_f}\right)$$

$$\times\sin\theta_f d\theta_f d\varphi_f$$

$$= \frac{8\pi a e^2}{m\omega c k_f} \int_0^{2\pi} \left[\frac{\varepsilon_x \left(k_f \sin\theta_f \cos\varphi_f - k_x^{(i)} \right) - \varepsilon_y k_y^{(e)} - \varepsilon_z k_z^{(e)}}{1 + a^2 \left(k_f \sin\theta_f \cos\varphi_f - k_x^{(i)} \right)^2} \right]^2$$

$$\times \frac{\delta \left(\sin\varphi_f - \dfrac{k_y^{(i)} + k_y^{(e)}}{k_f \sin\theta_f} \right) \dfrac{d\left(\sin\varphi_f \right)}{\cos\varphi_f}}{\sin\theta_f}$$

$$= \frac{8\pi a e^2}{m\omega c k_f} \cdot \left[\frac{\varepsilon_x \left(k_f \sin\theta_f \cos\varphi_f - k_x^{(i)} \right) - \varepsilon_y k_y^{(e)} - \varepsilon_z k_z^{(e)}}{1 + a^2 \left(k_f \sin\theta_f \cos\varphi_f - k_x^{(i)} \right)^2} \right]^2 \cdot$$

$$\times \frac{1}{\sin\theta_f \cdot \cos\varphi f},$$

where

$$\cos\theta_f = \frac{k_z^{(i)} + k_z^{(e)}}{k_f}, \qquad \sin\theta_f = \frac{\sqrt{k_f^2 - \left(k_z^{(i)} + k_z^{(e)} \right)^2}}{k_f},$$

$$\cos\varphi f = \frac{\sqrt{k_f^2 \sin^2\theta_f - \left(k_y^{(i)} + k_y^{(e)} \right)^2}}{k_f \sin\theta_f},$$

$$\sin\varphi_f = \frac{k_y^{(i)} + k_y^{(e)}}{k_f \sin\theta_f}.$$

Finally we get

$$\sigma_a = \frac{8\pi a e^2}{m\omega c} \frac{1}{\sqrt{k_f^2 - \left(k_y^{(i)} + k_y^{(e)} \right)^2 - \left(k_z^{(i)} + k_z^{(e)} \right)^2}}$$

$$\times \varepsilon_x \left[\sqrt{k_f^2 - \left(k_y^{(i)} + k_y^{(e)} \right)^2 - \left(k_z^{(i)} + k_z^{(e)} \right)^2} - k_x^{(i)} \right]$$

$$\times \left\{ \frac{\varepsilon_y k_y^{(e)} + \varepsilon_z k_z^{(e)}}{1 + a^2 \left[\sqrt{k_f^2 - \left(k_y^{(i)} + k_y^{(e)} \right)^2 - \left(k_z^{(i)} + k_z^{(e)} \right)^2} - k_x^{(i)} \right]^2} \right\}^2.$$

6061

A system of two distinguishable spin-$\frac{1}{2}$ particles is described by the Hamiltonian

$$H = -\frac{\hbar^2}{2m_1}\nabla_1^2 - \frac{\hbar^2}{2m_2}\nabla_2^2 + \frac{1}{2}\frac{m_1 m_2}{m_1 + m_2}\omega^2(\mathbf{r}_1 - \mathbf{r}_2)^2 + g\sigma_1 \cdot \sigma_2$$

with $g \ll \hbar\omega$.

a. What are the energy levels of the system? Give the explicit form of the wave functions for the two lowest energy levels (you need not specify the normalization).

b. The system is in its ground state at time $t \to -\infty$. An external time-dependent potential is applied which has the form

$$V(t) = \left[V_1 + V_2 \frac{z_1 - z_2}{L} \right] f(t) s_1 \cdot \hat{\mathbf{x}}$$

with $f(t) = 0$ as $|t| \to \infty$. Derive a set of coupled equations for the probability amplitudes $C_n(t) = \langle n | \psi(t) \rangle$, where $|n\rangle$ denotes an eigenstate of H_0 and $\psi(t)$ is the time-dependent wave function.

c. Calculate $C_n(\infty)$ for the case

$$f(t) = \begin{cases} 0, & t < 0 \\ 1, & 0 < t < \tau \end{cases} \qquad \text{and} \qquad t > \tau,$$

with $\frac{g\tau}{\hbar} \ll 1$ and V_2 very small. Work through first order in V_2 and specify clearly the quantum numbers for the states.

(*MIT*)

Sol:

a. Let

$$\mathbf{r} = \mathbf{r}_1 - \mathbf{r}_2,$$

$$\mathbf{R} = \frac{m_1 \mathbf{r}_1 + m_2 \mathbf{r}_2}{m_1 + m_2},$$

$$\mathbf{S} = s_1 + s_2 = \frac{1}{2}(s_1 + s_2).$$

Then the Hamiltonian of the system can be reduced to

$$H = -\frac{\hbar^2}{2M}\nabla_R^2 - \frac{\hbar^2}{2\mu}\nabla_T^2 + \frac{\mu}{2}\omega^2 r^2 + 2g\left[S(S+1) - \frac{3}{2} \right],$$

where $M = m_1 + m_2$, $\mu = \frac{m_1 m_2}{m_1 + m_2}$, S is the total spin.

Note that in the expression for H, the first term is due to the motion of the system as a whole, the second and third terms together are the Hamiltonian of a spinless isotropic harmonic oscillator, and the last term is due to the spins of the particles. Hence the energy of the system is

$$E_{nS} = \frac{P^2}{2M} + \left(n + \frac{3}{2}\right)\hbar\omega + 2g\left[S(S+1) - \frac{3}{2}\right],$$

$$n = 0, 1, 2, \ldots .$$

For energy levels of the internal motion, we shall omit the first term on the right-hand side of the above. For the ground state of the internal motion, $n = 0$, $S = 0$ and

$$E_{00} = \frac{3}{2}\hbar\omega - 3g .$$

Write the wave function as $\psi_0 = |0\rangle\alpha_{00}$. For the first excited state, we similarly have

$$E_{01} = \frac{3}{2}\hbar\omega + g, \quad \psi_1 = |0\rangle\alpha_{10}, \quad \alpha_{1,\pm}.$$

Note that $|0\rangle$ is the wave function of the harmonic oscillator ground state and α_{SM} is the coupled spin wave function.

b. Let

$$H_0 = -\frac{\hbar^2}{2\mu}\nabla_r^2 + \frac{\mu}{2}\omega^2 r^2 + 2g\left[S(S+1) - \frac{3}{2}\right],$$

$$H_0|n\rangle = E_n|n\rangle ,$$

$$[H_0 + V(t)]\psi(t) = i\hbar\frac{\partial\psi}{\partial t}.$$

Expanding $\psi(t)$:

$$|\psi(t)\rangle = \sum_n C_n(t)\exp(-iE_n t/\hbar)|n\rangle,$$

and substituting it in the last equation we obtain

$$[H_0 + V(t)]\sum_n C_n e^{-i\frac{E_n t}{\hbar}}|n\rangle = i\hbar\sum_n \dot{C}_n e^{-i\frac{E_n t}{\hbar}}|n\rangle$$

$$+ i\hbar\sum_n C_n\left(\frac{-iE_n}{\hbar}\right)e^{-i\frac{E_n t}{\hbar}}|n\rangle.$$

Multiplying both sides by $\langle m | e^{i\frac{E_m t}{\hbar}}$ and summing over m we get

$$\sum_m C_m E_m + \sum_m \sum_n C_n e^{i\frac{(E_m - E_n)t}{\hbar}} \langle m | V(t) | n \rangle$$

$$= i\hbar \sum_m \dot{C}_m + i\hbar \sum_m C_m \left(\frac{-iE_m}{\hbar} \right),$$

or

$$i\hbar \dot{C}_n(t) = \sum_m \langle n | V(t) | m \rangle \exp\left[-i\left(E_m - E_n \right) t / \hbar \right] C_m(t),$$

which is the required set of coupled equations.

c. Denote the initial state by $|000\alpha_{00}\rangle$, the final state by $|nlm\alpha_{SM}\rangle$. As

$$\sigma_1 \cdot \hat{\mathbf{x}} = \sin\theta\cos\varphi\sigma_{1x} + \sin\theta\sin\varphi\sigma_{1y} + \cos\theta\sigma_{1z};$$

$$\alpha_{00} = \frac{1}{\sqrt{2}}\left(\alpha_1\beta_2 - \beta_1\alpha_2 \right),$$

$$\alpha_{11} = \alpha_1\alpha_2, \qquad \alpha_{1,-1} = \beta_1\beta_2,$$

$$\alpha_{10} = \frac{1}{\sqrt{2}}\left(\alpha_1\beta_2 + \alpha_2\beta_1 \right);$$

$$\sigma_{1x}\alpha_1 = \beta_1, \qquad \sigma_{1y}\alpha_1 = i\beta_1, \qquad \sigma_{1z}\alpha_1 = \alpha_1; \qquad \sigma_{1x}\beta_1 = \alpha_1,$$

$$\sigma_{1y}\beta_1 = -i\alpha_1, \qquad \sigma_{1z}\beta_1 = -\beta_1;$$

we have

$$\sigma_1 \cdot \hat{\mathbf{x}}\alpha_{00} = \frac{1}{\sqrt{2}}\left(\sin\theta e^{i\varphi}\alpha_{1,-1} - \sin\theta e^{-i\varphi}\alpha_{11} + \sqrt{2}\cos\theta\alpha_{10} \right).$$

Therefore, as $Y_{11} = \sqrt{\frac{3}{8\pi}}\sin\theta e^{i\varphi}$, $Y_{1,-1} = \sqrt{\frac{3}{8\pi}}\sin\theta e^{-i\varphi}$, $Y_{10} = \sqrt{\frac{3}{4\pi}}\cos\theta$, we have

$$\langle nlm\alpha SM | \sigma_1 \times \hat{\mathbf{x}} | \alpha_{00} 000 \rangle = \langle nlm\alpha_{SM} | \sqrt{\frac{4\pi}{3}} Y_{11}\alpha_{1,-1}$$

$$- \sqrt{\frac{4\pi}{3}} Y_{1,-1}\alpha_{11} + \sqrt{\frac{4\pi}{3}} Y_{10}\alpha_{10} | 000 \rangle$$

$$= \sqrt{\frac{1}{3}}\delta_{n0}\delta_{l1}\delta_{S1}\left(\delta_{m0}\delta_{M0} - \delta_{m1}\delta_{M,-1} \right)$$

$$- \delta_{m1}\delta_{M1} \bigg) \equiv 0,$$

since $\ell = 0$ for $n = 0$. Thus to first order perturbation the first term $V_1 f(t)\sigma_1 \cdot \hat{\mathbf{x}}$ in $V(t)$ makes no contribution to the transition. Consider next

$$\langle \text{nlm}\alpha_{sM} | r \cos\theta\sigma_1 \times \hat{x} | \alpha_{00}000\rangle$$

$$= \langle \text{nlm}\alpha_{SM} | r\left(-\sqrt{\frac{4\pi}{15}}Y_{21}\alpha_{1,-1} - \sqrt{\frac{4\pi}{15}}Y_{2,-1}\alpha_{11}\right.$$

$$\left. + \frac{1}{3}\sqrt{\frac{16\pi}{5}}Y_{20}\alpha_{10} + \frac{1}{3}\alpha_{10}\right)|000\rangle$$

$$= \lambda_1\delta_{l2}\delta_{S1}\left[\frac{2}{3\sqrt{5}}\delta_{m0}\delta_{M0} - \frac{1}{\sqrt{15}}\delta_{m1}\delta_{M,-1}\right.$$

$$\left. - \frac{1}{\sqrt{15}}\delta_{m,-1}\delta_{M1}\right] + \frac{1}{3}\lambda_1\delta_{l0}\delta_{S1}\delta_{m0}\delta_{M0},$$

where

$$\lambda_1 = \int_0^\infty R_{nl}\cdot R_{00}\cdot r^3\, dr = \int_0^\infty R_{n2}R_{00}r^3\, dr\,.$$

For the three-dimensional harmonic oscillator, $n = l + 2n_r = 2(1+n_r) = \text{even}$.
We have

$$C_n(\infty) = \frac{1}{i\hbar}\int_0^\infty e^{i\omega_{n0}t}\langle \text{nlm}\alpha_{SM} | V(t) | 000\alpha_{00}\rangle dt$$

$$= \frac{1}{i\hbar}\int_0^\tau e^{in\omega t}\langle \text{nlm}\alpha_{SM} | \left(V_1 + V_2\frac{z}{L}\right)\sigma_1\cdot\hat{x} | 000\alpha_{00}\rangle$$

$$= \frac{1}{n\omega\hbar}\left(e^{in\omega\tau} - 1\right)\langle \text{nlm}\alpha_{SM} | \frac{V_2}{L}r \cos\theta\sigma_1\cdot\hat{x} | \alpha_{00}000\rangle,$$

and

$$C_{2k-1}(\infty) = 0,$$

$$C_{2k}(\infty) = \frac{1}{2k\hbar\omega}\left(e^{i2k\omega\tau} - 1\right)\frac{V_2}{L}\lambda_1$$

$$\times \delta_{S1}\left[\frac{2}{3\sqrt{5}}\delta_{m0}\delta_{M0} - \frac{1}{\sqrt{15}}\delta_{m,1}\delta_{M,-1}\right.$$

$$\left. - \frac{1}{\sqrt{15}}\delta_{m,-1}\delta_{M1}\right],$$

where $k = 1, 2, \ldots, l = 2$, $\lambda_1 = \int_0^\infty r^3 R_{(2k)2}R_{00}\, dr$.

Part VII
Many-Particle Systems

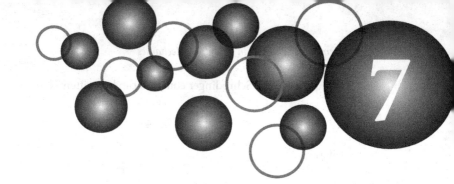

7001

In one dimension, a particle of mass m is attracted to the origin by a linear force $-kx$. Its Schrödinger equation has eigenfunctions

$$\Psi_n(\xi) = H_n(\xi)\exp\left(-\frac{1}{2}\xi^2\right),$$

where

$$\xi = \left(\frac{mk}{\hbar^2}\right)^{1/4} x$$

and H_n is the Hermite polynomial of order n. The eigenvalues are

$$E_n = \left(n+\frac{1}{2}\right)\hbar\omega, \quad \text{where } \omega = \left(\frac{k}{m}\right)^{1/2}.$$

Consider two non-interacting distinguishable particles ($i = 1, 2$), each of mass m, each attracted to the origin by a force $-kx_i$. Write down expressions for the eigenfunctions, eigenvalues, and degeneracies for the two-particle system using each of the following coordinate systems:

a. single-particle coordinates x_1 and x_2,

b. relative $(x = x_2 - x_1)$ and center-of-mass $\left(X = \frac{x_1 + x_2}{2}\right)$ coordinates.

(MIT)

Sol:

    a. For the single-particle system, the Hamiltonian is

$$H = -\frac{\hbar^2}{2m}\frac{d^2}{dx^2} + \frac{1}{2}kx^2$$

$$= -\frac{\hbar^2}{2m}\left(\frac{d^2}{dx^2} - \alpha^4 x^2\right)$$

with $\alpha^4 = \frac{mk}{\hbar^2}$. The Schrödinger equation can be written as

$$\frac{d^2\psi}{d\xi^2} + (\lambda - \xi^2)\psi = 0,$$

where

$$\xi = \alpha x, \quad \lambda = \frac{2}{\hbar}\sqrt{\frac{m}{k}}E.$$

The eigenfunctions are

$$\psi_n(\xi) = H_n(\xi)\exp\left(-\frac{1}{2}\xi^2\right),$$

with eigenvalues

$$E_n = \left(n + \frac{1}{2}\right)\hbar\omega,$$

where

$$\omega = \sqrt{\frac{k}{m}}.$$

Using the single-particle coordinates x_1 and x_2, we can write the Hamiltonian for the two-particle system as

$$H = -\frac{\hbar^2}{2m}\frac{\partial^2}{\partial x_1^2} - \frac{\hbar^2}{2m}\frac{\partial^2}{\partial x_2^2} + \frac{1}{2}m\omega^2 x_1^2 + \frac{1}{2}m\omega^2 x_2^2$$
$$= H_1 + H_2.$$

The energy eigenfunctions can be obtained as the common eigenfunctions of $\{H_1, H_2\}$, i.e., $\psi(x_1, x_2) = \psi(x_1)\psi(x_2)$, the corresponding energy being $E = E_1 + E_2$. Thus

$$\psi_{nm}(x_1, x_2) = H_n(\alpha x_1)H_m(\alpha x_2)\exp\left[-\frac{1}{2}\alpha^2(x_1^2 + x_2^2)\right],$$

$$E_{nm}^{(N)} = (n + m + 1)\hbar\omega = (N + 1)\hbar\omega,$$

where

$$\alpha = \left(\frac{mk}{\hbar^2}\right)^{1/4}, \quad \omega = \left(\frac{k}{m}\right)^{1/2}, \quad N = n + m.$$

The degeneracy of the energy level $E_{nm}^{(N)}$ is equal to the number of non-negative integer pairs (n, m) which satisfy the condition $n + m = N$, i.e.,

$$f^{(N)} = N + 1.$$

b. Using the relative and center-of-mass coordinates $x = x_2 - x_1$ and $X = \frac{x_1 + x_2}{2}$, we can write the Hamiltonian for the system as

$$H = -\frac{\hbar^2}{2m}\frac{\partial^2}{\partial X^2} - \frac{\hbar^2}{2\mu}\frac{\partial^2}{\partial x^2} + \frac{1}{2}M\omega^2 X^2 + \frac{1}{2}\mu\omega^2 x^2,$$

where $M = 2m, \mu = \frac{1}{2}m, \omega = \left(\frac{k}{m}\right)^{1/2}$. As in (a) we have

$$\psi_{nm}(X, x) = H_n(\alpha X)H_m(\beta x)\exp\left[-\frac{1}{2}(\alpha^2 X^2 + \beta^2 x^2)\right],$$

$$E_{nm}^{(N)} = (n + m + 1)\hbar\omega = (N + 1)\hbar\omega,$$

$$f^{(N)} = N + 1,$$

where

$$\alpha = \left(\frac{M\omega}{\hbar}\right)^{1/2}, \qquad\qquad \beta = \left(\frac{\mu\omega}{\hbar}\right)^{1/2}.$$

7002

Consider three identical noninteracting particles confined to an infinite square well of width L. Find the energy of the ground state, first excited state, and second exited state in terms of ε_1, where $\varepsilon_1 = \dfrac{\pi^2\hbar^2}{2mL^2}$ if the particles are as follows:

i. Bosons

ii. Spin-1/2 fermions

Sol:

i. For bosons, there is no restriction, so any number of particles can be placed in a single state.

The energy-level diagram for bosons are shown in the following figure:

For the first excited state, the energy-level diagram is given by

For the second excited state, the energy-level diagram is given by

—————————— $9\,\varepsilon_1$

——— ○ ○ ——— $4\,\varepsilon_1$ Second excited state energy is $2 \times 4\,\varepsilon_1 + 1 \times \varepsilon_1 = 9\,\varepsilon_1$

——— ○ ——— ε_1

ii. For spin-1/2 fermions, there is a restriction concerning the Pauli's exclusion principle; therefore, the maximum number of fermions in a state is given by $2s + 1$, where s is the spin of the fermion.

For spin-1/2 fermions, the maximum number is $2(1/2) + 1 = 2$.

Therefore, for a state, only two fermions can be placed.

The ground state of the fermion is given by

—————————— $9\,\varepsilon_1$

——— ○ ——— $4\,\varepsilon_1$ Ground state energy is $2 \times \varepsilon_1 + 4 \times \varepsilon_1 = 6\,\varepsilon_1$

——— ○ ○ ——— ε_1

The first excited state is given by

—————————— $9\,\varepsilon_1$

——— ○ ○ ——— $4\,\varepsilon_1$ First excited state energy is $2 \times 4\,\varepsilon_1 + 1 \times \varepsilon_1 = 9\,\varepsilon_1$

——— ○ ——— ε_1

The second excited state is given by

——— ○ ——— $9\,\varepsilon_1$

—————————— $4\,\varepsilon_1$ Second excited state energy is $1 \times 9\,\varepsilon_1 + 2 \times \varepsilon_1 = 11\,\varepsilon_1$

——— ○ ○ ——— ε_1

7003

a. Write down the Hamiltonian and Schrödinger equation for a one-dimensional harmonic oscillator.

b. If xe^{-vx^2} is a solution find v and then give the energy E_1 and the expectation values for $\langle x \rangle, \langle x^2 \rangle, \langle p^2 \rangle, \langle px \rangle$.

c. Show that for two equal particles in a single one-dimensional harmonic oscillator well the ground state may be written either as $\phi_0(m, x_1) \times \phi_0(m, x_2)$ or

$$\phi_0\left(2m, \frac{x_1 + x_2}{2}\right)\phi_0\left(\frac{m}{2}, (x_1 - x_2)\right),$$

where $\phi_0(m, x)$ is the ground state solution for a single particle of mass m.

(Wisconsin)

ol:

a. The Hamiltonian for a one-dimensional harmonic oscillator is

$$H = \frac{p^2}{2m} + \frac{1}{2}m\omega^2 x^2.$$

The stationary Schrödinger equation is

$$\left(-\frac{\hbar^2}{2m}\frac{d^2}{dx^2} + \frac{1}{2}m\omega^2 x^2\right)\psi(x) = E\psi(x).$$

b. Substitution of $\psi = xe^{-vx^2}$ in the Schrödinger equation gives

$$\left[-\frac{\hbar^2}{2m}(-2v)(3 - 2vx^2) + \frac{1}{2}m\omega^2 x^2\right]xe^{-vx^2}$$

$$= \left[\frac{3\hbar^2}{m}v + \left(\frac{1}{2}m\omega^2 - \frac{2\hbar^2 v^2}{m}\right)x^2\right]xe^{-vx^2}$$

$$= E_1 xe^{-vx^2}.$$

Equating the coefficients gives

$$\begin{cases} \dfrac{1}{2}m\omega^2 - \dfrac{2\hbar^2 v^2}{m} = 0, \\[2mm] E_1 = \dfrac{3\hbar^2}{m}v, \end{cases}$$

whose solution is

$$\begin{cases} v = \dfrac{m\omega}{2\hbar}, \\[2mm] E_1 = \dfrac{3}{2}\hbar\omega. \end{cases}$$

From the symmetry we know $\langle x \rangle = 0$. The Virial theorem gives

$$\frac{1}{2m}\langle p^2 \rangle = \frac{1}{2}m\omega^2\langle x^2 \rangle = \frac{1}{2}\cdot\frac{3}{2}\hbar\omega.$$

Therefore

$$\langle p^2 \rangle = \frac{3}{2}m\hbar\omega, \qquad \langle x^2 \rangle = \frac{3}{2}\frac{\hbar}{m\omega}.$$

To find $\langle px \rangle$ we first normalize the wave function:

$$A^2\int_{-\infty}^{\infty} x^2 e^{-2vx^2}\,dx = 1,$$

giving

$$A^2 = 4v\sqrt{\frac{2v}{\pi}}.$$

Then consider

$$\begin{aligned}
\langle px \rangle &= -\frac{\hbar}{i}\int_{-\infty}^{\infty}\psi\frac{d}{dx}(x\psi)\,dx \\
&= \frac{\hbar}{i}A^2\int_{-\infty}^{\infty} xe^{-vx^2}\frac{d}{dx}(x^2 e^{-vx^2})\,dx \\
&= \frac{\hbar}{i}A^2\int_{-\infty}^{\infty} 2(x^2 - vx^4)e^{-2vx^2}\,dx \\
&= -\frac{i\hbar}{2}.
\end{aligned}$$

c.   The Schrödinger equation for the two-particle system is

$$\left[-\frac{\hbar^2}{2m}(\nabla_1^2 + \nabla_2^2) + \frac{1}{2}m\omega^2(x_1^2 + x_2^2)\right]\psi(x_1, x_2) = E\psi(x_1, x_2).$$

Suppose $\psi(x_1, x_2) = \phi(x_1)\phi(x_2)$. The variables can be separated:

$$\left(-\frac{\hbar^2}{2m}\nabla_i^2 + \frac{1}{2}m\omega^2 x_i^2\right)\phi(x_i) = E_i\phi(x_i), \qquad i = 1, 2,$$

with $E = E_1 + E_2$.

Hence the system can be considered as consisting of two identical harmonic oscillators without coupling. The ground state is then

$$\psi_0(x_1, x_2) = \phi_0(m, x_1)\phi_0(m, x_2).$$

On the other hand, introducing the Jacobi coordinates

$$R = \frac{1}{2}(x_1 + x_2), \qquad r = x_1 - x_2,$$

the Schrödinger equation becomes

$$\left[-\frac{\hbar^2}{2m}\left(\frac{1}{2}\nabla_R^2 + 2\nabla_r^2\right) + \frac{1}{2}m\omega^2\left(2R^2 + \frac{1}{2}r^2\right)\right]\psi(R,r) = E\psi(R,r).$$

Writing $\psi(R, r) = \phi(R)\varphi(r)$, it can also be separated in the variables to

$$\left(-\frac{\hbar^2}{4m}\nabla_R^2 + m\omega^2 R^2\right)\phi(R) = E_R\phi(R),$$

$$\left(-\frac{\hbar^2}{m}\nabla_r^2 + \frac{1}{4}m\omega^2 r^2\right)\varphi(r) = E_r\varphi(r),$$

where $E_R + E_r = E$, with E_R, E_r respectively describing the motion of the center of mass and the relative motion.

Therefore the wave function of the ground state can be written as

$$\psi_0(x_1, x_2) = \phi_0(2m, R)\phi_0\left(\frac{m}{2}, r\right)$$

$$= \phi_0\left(2m, \frac{x_1 + x_2}{2}\right)\phi_0\left(\frac{m}{2}, x_1 - x_2\right).$$

7004

Consider two particles of masses $m_1 \neq m_2$ interacting via the Hamiltonian

$$H = \frac{p_1^2}{2m_1} + \frac{p_2^2}{2m_2} + \frac{1}{2}m_1\omega^2 x_1^2 + \frac{1}{2}m_2\omega^2 x_2^2 + \frac{1}{2}K(x_1 - x_2)^2.$$

a. Find the exact solutions.

b. Sketch the spectrum in the weak coupling limit $K \ll \mu\omega^2$, where μ is the reduced mass.

(Berkeley)

Sol:

a. Let

$$R = (m_1 x_1 + m_2 x_2)/(m_1 + m_2), \qquad r = x_1 - x_2.$$

We have

$$\frac{d}{dx_1} = \frac{\partial R}{\partial x_1}\frac{d}{dR} + \frac{\partial r}{\partial x_1}\frac{d}{dr}$$

$$= \frac{m_1}{m_1 + m_2}\frac{d}{dR} + \frac{d}{dr},$$

and so

$$\frac{d^2}{dx_1^2} = \left(\frac{m_1}{m_1+m_2}\right)^2 \frac{d^2}{dR^2} + 2\frac{m_1}{m_1+m_2}\frac{d^2}{drdR} + \frac{d^2}{dr^2},$$

and similarly

$$\frac{d^2}{dx_2^2} = \left(\frac{m_2}{m_1+m_2}\right)^2 \frac{d^2}{dR^2} - 2\frac{m_2}{m_1+m_2}\frac{d^2}{dRdr} + \frac{d^2}{dr^2}.$$

We also have

$$x_1^2 = R^2 + 2\frac{m_2}{m_1+m_2}Rr + \frac{m_2^2}{(m_1+m_2)^2}r^2,$$

$$x_2^2 = R^2 - 2\frac{m_1}{m_1+m_2}Rr + \frac{m_1^2}{(m_1+m_2)^2}r^2.$$

Hence

$$H = -\frac{\hbar^2}{2(m_1+m_2)}\frac{d^2}{dR^2} - \frac{\hbar^2}{2}\frac{m_1+m_2}{m_1 m_2}\frac{d^2}{dr^2}$$
$$+ \frac{1}{2}(m_1+m_2)\omega^2 R^2 + \frac{1}{2}\frac{m_1 m_2}{m_1+m_2}\omega^2 r^2 + \frac{1}{2}Kr^2.$$

Let

$$M = m_1 + m_2, \qquad \mu = \frac{m_1 m_2}{m_1+m_2}.$$

The equation of motion becomes

$$\left[\frac{-\hbar^2}{2M}\frac{d^2}{dR^2} + \frac{1}{2}M\omega^2 R^2 - \frac{\hbar^2}{2\mu}\frac{d^2}{dr^2} + \frac{\mu}{2}\left(1 + \frac{K}{\mu\omega^2}\right)\omega^2 r^2\right]\psi(R,r)$$
$$= E\psi(R,r).$$

It can be reduced to two independent oscillator equations with energy states and wave functions

$$E = E_{lm} = E_l + E_m = \left(l + \frac{1}{2}\right)\hbar\omega + \left(m + \frac{1}{2}\right)\hbar\omega\sqrt{1 + \frac{K}{\mu\omega^2}},$$

$$\psi_{lm}(R,r) = \psi_l(R)\psi_m(r) = N_l N_m \exp\left[-\frac{1}{2}\alpha_1^2 R^2\right]H_l(\alpha_1 R)$$
$$\times \exp\left[-\frac{1}{2}\alpha_2^2 r^2\right]H_m(\alpha_2 r),$$

where

$$N_l = \left(\frac{\alpha_1}{\sqrt{\pi}2^l l!}\right)^{1/2}, \qquad \alpha_1 = \left(\frac{M\omega}{\hbar}\right)^{1/2},$$

$$N_m = \left(\frac{\alpha_2}{\sqrt{\pi}\, 2^m m!} \right)^{1/2}, \qquad \alpha_2 = \left(\frac{\mu\omega}{\hbar} \right)^{1/2} \left(1 + \frac{K}{\mu\omega^2} \right)^{1/4},$$

and H_m are Hermite polynomials.

b. For $K \ll \mu\omega^2$, if we can take

$$\left(1 + \frac{K}{\mu\omega^2} \right)^{1/2} \approx 1,$$

we have

$$E_{lm} \approx (l+m+1)\hbar\omega = (N+1)\hbar\omega, \qquad N \equiv l+m = 0,\ 1,\ 2, \dots .$$

This means that the degeneracy of the Nth energy level is $N+1$:

$N \dots$	$\dots$			
$N=3$	$l=3, m=0;$	$l=2, m=1;$	$l=1, m=2;$	$l=0, m=3.$
$N=2$	$l=2, m=0;$	$l=1, m=1;$	$l=0, m=2.$	
$N=1$	$l=1, m=0;$	$l=0, m=1.$		
$N=0$	$l=m=0.$			

If K is not so small and we have to take

$$\sqrt{1 + \frac{K}{\mu\omega^2}} = 1 + \frac{K}{2\mu\omega^2} + \cdots$$

then the energy levels with the same m will move upward by

$$\left(m + \frac{1}{2} \right) \hbar\omega \left(\frac{K}{2\mu\omega^2} + \cdots \right),$$

and so the degeneracy is destroyed.

7005

Consider two identical spin-1/2 fermions confined to an infinite square well of width L. What is the spatial wave function of the system such that the total energy $\varepsilon = \dfrac{5\pi^2\hbar^2}{mL^2}$ if the spin state is singlet?

Sol: The total wave function of the fermionic system is antisymmetric, so the spatial part of wave function must be symmetric in nature.

$$\psi_{\text{total}}^{\text{fermion}} = \psi_{\text{spin}}^{\text{fermion}} \times \psi_{\text{space}}^{\text{fermion}}$$

The given energy is $10\varepsilon_1$, where ε_1 is the ground state energy of infinite square well.

$$\varepsilon = \frac{10\pi^2\hbar^2}{2mL^2} = 10\varepsilon_1$$

Distributing the two fermions in the energy level,

$$\psi_{\text{space}}^{\text{sym}} = \frac{1}{\sqrt{2}}\left[\psi_1(1)\psi_3(2) + \psi_3(1)\psi_1(2)\right]$$

$$\psi_{\text{space}}^{\text{sym}} = \frac{1}{\sqrt{2}}\frac{2}{L}\left[\sin\left(\frac{\pi x_1}{L}\right)\sin\left(\frac{3\pi x_2}{L}\right) + \sin\left(\frac{\pi x_2}{L}\right)\sin\left(\frac{3\pi x_1}{L}\right)\right].$$

7006

Consider a system defined by the Schrödinger eigenvalue equation

$$\left\{-\frac{\hbar^2}{2m}(\nabla_1^2 + \nabla_2^2) + \frac{k}{2}|\mathbf{r}_1 - \mathbf{r}_2|^2\right\}\psi(\mathbf{r}_1, \mathbf{r}_2) = E\psi(\mathbf{r}_1, \mathbf{r}_2).$$

a. List all symmetries of this Schrödinger equation.

b. Indicate all constants of motion.

c. Indicate the form of the ground-state wave function.
You may assume that the ground-state wave function of one-dimensional harmonic oscillator is a Gaussian.

(Berkeley)

Sol:

a. The Schrödinger equation has symmetries with respect to time translation, space inversion, translation of the whole system, and the exchange of $\mathbf{r}_1$ and $\mathbf{r}_2$, as well as symmetry with respect to the Galilean transformation.

b. Let $\mathbf{r} = \mathbf{r}_1 - \mathbf{r}_2$, $\mathbf{R} = \frac{1}{2}(\mathbf{r}_1 + \mathbf{r}_2)$. The Schrödinger equation can then be written as

$$\left\{-\frac{\hbar^2}{4m}\nabla_R^2 - \frac{\hbar^2}{m}\nabla_r^2 + \frac{k}{2}r^2\right\}\psi(\mathbf{R}, \mathbf{r}) = E\psi(\mathbf{R}, \mathbf{r}).$$

This equation can be separated into two, one for the motion of a particle of mass $2m$ at the center of mass and one for the motion of a harmonic oscillator of mass $m/2$ relative to the second particle. The motion of the center of mass is a free motion, so that $P_R^2, P_x, P_y, P_z, E_R, L_R^2, L_x, L_y, L_z$ are all constants of the motion. Of the relative motion, E_r, L_r^2, L_z, as well as the parity of the wave function, are constants of the motion.

c. The ground-state wave function has the form $\psi(\mathbf{R}, \mathbf{r}) = \phi(\mathbf{R})\varphi(\mathbf{r})$. $\varphi(\mathbf{r})$ is the wave function of a harmonic oscillator of mass $\frac{m}{2}$:

$$\varphi(\mathbf{r}) \sim \exp\left(-\frac{1}{2}\alpha^2 r^2\right)$$

with

$$\alpha^2 = \sqrt{\frac{mk}{2\hbar^2}}.$$

$\phi(\mathbf{R})$ is the wave function of a free particle of mass $2m$:

$$\phi(\mathbf{R}) \sim \exp(-i\mathbf{p}\cdot\mathbf{R}/\hbar)$$

with

$$|\mathbf{p}| = \sqrt{4mE_R}, \qquad E_R = E - \frac{1}{2}\hbar\sqrt{\frac{2k}{m}}.$$

7007

Two identical bosons, each of mass m, move in the one-dimensional harmonic oscillator potential $V = \frac{1}{2}m\omega^2 x^2$. They also interact with each other via the potential

$$V_{int}(x_1, x_2) = \alpha e^{-\beta(x_1-x_2)^2},$$

where β is a positive parameter. Compute the ground state energy of the system to first order in the interaction strength parameter α.

(Berkeley)

Sol: Being bosons the two particles can simultaneously stay in the ground state. Taking V_{int} as perturbation, the unperturbed wave function of the ground state for the system is

$$\psi_0(x_1, x_2) = \phi_0(x_1)\phi_0(x_2) = \frac{\alpha_0}{\sqrt{\pi}}\exp\left[-\frac{1}{2}\alpha_0^2(x_1^2 + x_2^2)\right], \quad \alpha_0 = \sqrt{\frac{m\omega}{\hbar}}.$$

The perturbation energy to first order in α is then

$$\Delta E = \int\int_{-\infty}^{\infty} \psi_0^*(x_1,x_2)V_{\text{int}}(x_1,x_2)\psi_0(x_1,x_2)dx_1\,dx_2$$

$$= \frac{\alpha_0^2\alpha}{\pi}\int\int_{-\infty}^{\infty}\exp[-\alpha_0^2(x_1^2+x_2^2)-\beta(x_1-x_2)^2]dx_1\,dx_2$$

$$= \frac{\alpha_0\alpha}{(\alpha_0^2+2\beta)^{1/2}},$$

where the integration has been facilitated by the transformation

$$\frac{x_1+x_2}{2} = y_1, \qquad \frac{x_1-x_2}{2} = y_2.$$

The ground state energy of the system is therefore

$$E = \hbar\omega + \frac{\alpha_0\alpha}{(\alpha_0^2+2\beta)^{1/2}} \quad \text{with} \quad \alpha_0 = \left(\frac{m\omega}{\hbar}\right)^{1/2}.$$

7008

A one-dimensional square well of infinite depth and 1 Å width contains 3 electrons. The potential well is described by $V = 0$ for $0 \le x \le 1\,\text{Å}$ and $V = +\infty$ for $x < 0$ and $x > 1\,\text{Å}$. For a temperature of $T = 0\,\text{K}$, the average energy of the 3 electrons is $E = 12.4\,\text{eV}$ in the approximation that one neglects the Coulomb interaction between electrons. In the same approximation and for $T = 0\,\text{K}$, what is the average energy for 4 electrons in this potential well?

(Wisconsin)

Sol: For a one-dimensional potential well the energy levels are given by

$$E_n = E_1 n^2,$$

where E_1 is the ground state energy and $n = 1, 2,\ldots$ Pauli's exclusion principle and the lowest energy principle require that two of the three electrons are in the energy level E_1 and the third one is in the energy level E_2. Thus $12\cdot4\times3 = 2E_1 + 4E_1$, giving $E_1 = 6\cdot2$ eV. For the case of four electrons, two are in E_1 and the other two in E_2, and so the average energy is

$$E = \frac{1}{4}(2E_1 + 2E_2) = \frac{5}{2}E_1 = 15.5\text{ eV}.$$

(Note: the correct value of E_1 is

$$\frac{\pi^2 \hbar^2}{2ma^2} = \frac{1}{2mc^2} \left(\frac{\pi \hbar c}{a} \right)^2 = \frac{1}{1.02 \times 10^6} \left(\frac{\pi \times 6.58 \times 10^{-16} \times 3 \times 10^{10}}{10^{-8}} \right)^2$$

$$= 37.7 \text{ eV.})$$

7009

Consider two electrons moving in a central potential well in which there are only three single-particle states ψ_1, ψ_2 and ψ_3.

a. Write down all of the wave functions $\psi(\mathbf{r}_1, \mathbf{r}_2)$ for the two-electron system.

b. If now the electrons interact with a Hamiltonian $\delta H = V'(\mathbf{r}_1, \mathbf{r}_2) = V'(\mathbf{r}_2, \mathbf{r}_1)$, show that the following expression for the matrix element is correct:

$$\langle \psi_{13} | \delta H | \psi_{12} \rangle = \langle \psi_3(\mathbf{r}_1)\psi_1(\mathbf{r}_2) | V'(\mathbf{r}_1, \mathbf{r}_2) | \psi_2(\mathbf{r}_1)\psi_1(\mathbf{r}_2) \rangle$$
$$- \langle \psi_1(\mathbf{r}_1)\psi_3(\mathbf{r}_2) | V'(\mathbf{r}_1, \mathbf{r}_2) | \psi_2(\mathbf{r}_1)\psi_1(\mathbf{r}_2) \rangle.$$

(Buffalo)

Sol:

a. The wave functions for a fermion system are antisymmetric for interchange of particles, so the possible wave functions for the system are

$$\psi_{12} = \frac{1}{\sqrt{2}} (\psi_1(\mathbf{r}_1)\psi_2(\mathbf{r}_2) - \psi_1(\mathbf{r}_2)\psi_2(\mathbf{r}_1)),$$

$$\psi_{13} = \frac{1}{\sqrt{2}} (\psi_1(\mathbf{r}_1)\psi_3(\mathbf{r}_2) - \psi_1(\mathbf{r}_2)\psi_3(\mathbf{r}_1)),$$

$$\psi_{23} = \frac{1}{\sqrt{2}} (\psi_2(\mathbf{r}_1)\psi_3(\mathbf{r}_2) - \psi_2(\mathbf{r}_2)\psi_3(\mathbf{r}_1)).$$

b. We can write

$$\langle \psi_{13} | \delta H | \psi_{12} \rangle = \frac{1}{2} \langle \psi_1(\mathbf{r}_1)\psi_3(\mathbf{r}_2) | V'(\mathbf{r}_1, \mathbf{r}_2) | \psi_1(\mathbf{r}_1)\psi_2(\mathbf{r}_2) \rangle$$

$$- \frac{1}{2} \langle \psi_1(\mathbf{r}_1)\psi_3(\mathbf{r}_2) | V'(\mathbf{r}_1, \mathbf{r}_2) | \psi_2(\mathbf{r}_1)\psi_1(\mathbf{r}_2) \rangle$$

$$- \frac{1}{2} \langle \psi_1(\mathbf{r}_2)\psi_3(\mathbf{r}_1) | V'(\mathbf{r}_1, \mathbf{r}_2) | \psi_1(\mathbf{r}_1)\psi_2(\mathbf{r}_2) \rangle$$

$$+ \frac{1}{2} \langle \psi_1(\mathbf{r}_2)\psi_3(\mathbf{r}_1) | V'(\mathbf{r}_1, \mathbf{r}_2) | \psi_2(\mathbf{r}_1)\psi_1(\mathbf{r}_2) \rangle.$$

Since the particles are identical, r_1 and r_2 may be interchanged in each term. Do this and as $V'(\mathbf{r}_1, \mathbf{r}_2) = V'(\mathbf{r}_2, \mathbf{r}_1)$, we again obtain the same expression, showing its correctness.

7010

Two identical nonrelativistic fermions of mass m, spin 1/2 are in a one-dimensional square well of length L, with V infinitely large and repulsive outside the well. The fermions are subject to a repulsive inter-particle potential $V(x_1 - x_2)$, which may be treated as a perturbation. Classify the three lowest-energy states in terms of the states of the individual particles and state the spin of each. Calculate (in first-order perturbation theory) the energies of second- and third-lowest states; leave your result in the form of an integral. Neglect spin-dependent forces throughout.

(Berkeley)

Sol: For the unperturbed potential

$$V(x) = \begin{cases} 0, & x \in [0, L], \\ \infty, & \text{otherwise,} \end{cases}$$

the single-particle spatial wave functions are

$$\psi_n(x) = \begin{cases} \sqrt{\dfrac{2}{L}} \sin \dfrac{n\pi x}{L}, & x \in [0, L] \\ 0, & \text{otherwise,} \end{cases}$$

where n is an integer.

The spin wave function of a single particle has the form $\begin{pmatrix} a \\ b \end{pmatrix}$.

As we do not consider spin-dependent forces, the wave function of the two particles can be written as a product of a space part and a spin part. The spin part $\chi_J(M) \equiv \chi_{JM}$ is chosen as the eigenstate of $S = s_1 + s_2$ and $S_z = s_{1z} + s_{2z}$, i.e.,

$$S^2 \chi_{JM} = J(J+1)\chi_{JM},$$
$$S_J \chi_{JM} = M \chi_{JM}.$$

$J = 0$ gives the spin singlet state, which is antisymmetric for interchange of the two particles. $J = 1$ gives the spin triplet state, which is symmetric for interchange of the two particles. Symmetrizing and antisymmetrizing the space wave functions of the two-particle system we have

$$\psi_{nm}^A(x_1,x_2)=\frac{1}{\sqrt{2}}[\psi_n(x_1)\psi_m(x_2)-\psi_n(x_2)\psi_m(x_1)],$$

$$\psi_{nm}^s(x_1,x_2)=\begin{cases}\frac{1}{\sqrt{2}}[\psi_n(x_1)\psi_m(x_2)+\psi_n(x_2)\psi_m(x_1)], & n\neq m,\\ \psi_n(x_1)\psi_n(x_2), & n=m.\end{cases}$$

The corresponding energies are

$$E=\frac{\pi^2\hbar^2}{2mL^2}(n^2+m^2),\quad n,m=1,2,\ldots.$$

The total wave functions, which are antisymmetric, can be written as

$$\psi_{nm}^A(x_1,x_2)\chi_{JM}^s,$$
$$\psi_{nm}^s(x_1,x_2)\chi_{JM}^A.$$

The three lowest energy states are the following.

i. Ground state, $n=m=1$. The space wave function is symmetric, so the spin state must be singlet. Thus

$$\psi_0=\psi_{11}^s(x_1,x_2)\chi_{00}.$$

ii. First excited states, $n=1,m=2$.

$$\psi_1=\begin{cases}\psi_{12}^A(x_1,x_2)\chi_{1M}, & M=0,\pm1,\\ \psi_{12}^s(x_1,x_2)\chi_{00}.\end{cases}$$

The degeneracy is 4.

iii. Second excited state, $n=2,m=2$. The space wave function is symmetric, the spin state is singlet:

$$\psi_2=\psi_{22}^s(x_1,x_2)\chi_{00},$$

which is nondegenerate. Because the perturbation Hamiltonian is independent of spin, the calculation of perturbation of the first excited state can be treated as a nondegenerate case. The perturbation energies of second and third lowest energy states are given by

$$\Delta E_1^A=\int dx_1 dx_2\,|\psi_{12}^A(x_1,x_2)|^2\,V(x_1-x_2),$$
$$\Delta E_1^s=\int dx_1 dx_2\,|\psi_{12}^s(x_1,x_2)|^2\,V(x_1-x_2),$$
$$\Delta E_2=\int dx_1 dx_2\,|\psi_{22}^s(x_1,x_2)|^2\,V(x_1-x_2).$$

7011

Consider two particles of masses m_1 and m_2 interacting via the Hamiltonian

$$H = \frac{p_1^2}{2m_1} + \frac{p_2^2}{2m_2} + \frac{1}{2}m_1\omega x_1^2 + \frac{1}{2}m_2\omega x_2^2 + \frac{1}{2}k(x_1 - x_2)^2.$$ Find the exact solutions to the energy and the wave function?

Sol:

We let $R = \dfrac{m_1 x_1 + m_2 x_2}{m_1 + m_2}$, $r = x_1 - x_2$

We then have $\dfrac{d}{dx_1} = \dfrac{\partial R}{\partial x_1}\dfrac{d}{dR} + \dfrac{\partial r}{\partial x_1}\dfrac{d}{dr} = \dfrac{m_1}{m_1 + m_2}\dfrac{d}{dR} + \dfrac{d}{dr}$

$$\left(\frac{d}{dx_1}\right)^2 = \frac{d^2}{dR^2}\left(\frac{m_1}{m_1+m_2}\right)^2 + \frac{2m_1}{m_1+m_2}\frac{d^2}{drdR} + \frac{d^2}{dr^2}$$

And similarly, $\left(\dfrac{d}{dx_2}\right)^2 = \dfrac{d^2}{dR^2}\left(\dfrac{m_2}{m_1+m_2}\right)^2 - \dfrac{2m_2}{m_1+m_2}\dfrac{d^2}{drdR} + \dfrac{d^2}{dr^2}$

In addition, $x_1^2 = R^2 + \left(\dfrac{m_1}{m_1+m_2}\right)^2 r^2 + \dfrac{2m_2}{m_1+m_2}rR$

$$x_2^2 = R^2 + \left(\frac{m_1}{m_1+m_2}\right)^2 r^2 - \frac{2m_1}{m_1+m_2}rR$$

Therefore, the Hamiltonian becomes

$$H = \left(-\frac{\hbar^2}{2M}\frac{d^2}{dR^2} + \frac{1}{2}M\omega R^2\right) + \left(-\frac{\hbar^2}{2\mu}\frac{d^2}{dr^2} + \frac{1}{2}\mu\left(1 + \frac{K}{\mu\omega^2}\right)\omega^2 r^2\right)$$

where $M = m_1 + m_2$ and $\mu = \dfrac{m_1 m_2}{m_1 + m_2}$

We thus have two independent harmonic oscillators and we can write

$$E_{pq} = \left(p + \frac{1}{2}\right)\hbar\omega + \left(q + \frac{1}{2}\right)\hbar\omega\sqrt{1 + \frac{K}{\mu\omega^2}}, \text{ where } p, q = 0, 1, 2, 3 \ldots$$

and $\psi_{pq}(R,r)=\psi_p(R)\psi_q(r)=N_pN_q e^{-\alpha_1^2R^2/2}H_p(\alpha_1 R)e^{-\alpha_2^2r^2/2}H_q(\alpha_2 r),$

where $N_p = \left(\dfrac{\alpha_1}{\sqrt{\pi}2^p p!}\right)^{1/2}, N_q = \left(\dfrac{\alpha_2}{\sqrt{\pi}2^q q!}\right)^{1/2}$ and $\alpha_1 = \sqrt{\dfrac{M\omega}{\hbar}}, \alpha_2 = \sqrt{\dfrac{\mu\omega}{\hbar}}\left(1+\dfrac{K}{\mu\omega^2}\right)^{1/4}.$

7012

Consider a system of two spin half particles in a state with total spin quantum number $S = 0$. Find the eigenvalue of the spin Hamiltonian $H = AS_1 \times S_2$, where A is a positive constant in this state.

Sol:

The total spin angular momentum S of this two spin-half system is

$$S=S_1+S_2$$

$$S^2 = S_1^2+S_2^2+2S_1.S_2$$

Hence, $H = \dfrac{A}{2}(S^2 - S_1^2 - S_2^2)$

Let the simultaneous eigenkets of $S^2, S_z, S_1^2,$ and S_2^2 be $|sm_s\rangle$. Then,

$$H|sm_s\rangle = \frac{A}{2}(S^2 - S_1^2 - S_2^2)|sm_s\rangle$$

$$= \frac{A(0-3/4-3/4)}{2}\hbar^2 = \frac{-3}{4}A\hbar^2$$

The eigenvalue of the spin Hamiltonian H is $\dfrac{-3}{4}A\hbar^2.$

7013

a. A 2-fermion system has a wave function $\psi(1, 2)$. What condition must it satisfy if the particles are identical?

b. How does this imply the elementary statement of the Pauli exclusion principle that no two electrons in an atom can have identical quantum numbers?

c. The first excited state of Mg has the configuration ($3s$, $3p$) for the valence electrons. In the LS-coupling limit, which values of L and S are possible? What is the form of the spatial part of their wave functions using the single-particle functions $\psi_s(\mathbf{r})$ and $\phi_p(\mathbf{r})$? Which will have the lowest energy, and Why?

(Berkeley)

Sol:

a. $\psi(1,2)$ must satisfy the condition of antisymmetry for interchange of the two particles:

$$\hat{P}_{12}\psi(1,2)=\psi(2,1)=-\psi(1,2).$$

b. In an atom, if there are two electrons having identical quantum numbers then $\psi(1,2)=\psi(2,1)$. The antisymmetry condition above then gives $\psi(1,2)=0$, implying that such a state does not exist.

c. The electron configuration ($3s$, $3p$) correspond to

$$l_1=0,\quad l_2=1,$$
$$s_1=s_2=1/2.$$

Hence

$$L=1,\quad S=0,1.$$
$$\psi_S^L(1,2)=\phi_S^L(1,2)\chi_S(1,2),$$

where

$$\begin{cases}\phi_0^1(1,2)=\dfrac{1}{\sqrt{2}}(\phi_s(\mathbf{r}_1)\phi_p(\mathbf{r}_2)+\phi_s(\mathbf{r}_2)\phi_p(\mathbf{r}_1))\\[2mm] \qquad\quad=\dfrac{1}{\sqrt{2}}(1+\hat{P}_{12})\phi_s(\mathbf{r}_1)\phi_p(\mathbf{r}_2),\\[2mm]\phi_1^1(1,2)=\dfrac{1}{\sqrt{2}}(1-\hat{P}_{12})\phi_s(\mathbf{r}_1)\phi_p(\mathbf{r}2).\end{cases}$$

The lowest energy state is $\Psi_1^1(1,2)$, i.e. the state of $S=1$. Because the spatial part of the state $S=1$ is antisymmetric for the interchange $1\leftrightarrow2$, the probability that the two electrons get close together is small, and so the Coulomb repulsive energy is small, resulting in a lower total energy.

7014

Two particles, each of mass M, are bound in a one-dimensional harmonic oscillator potential $V = \frac{1}{2}kx^2$ and interact with each other through an attractive harmonic force $F_{12} = -K(x_1 - x_2)$. You may take K to be small.

a. What are the energies of the three lowest states of this system?

b. If the particles are identical and spinless, which of the states of (a) are allowed?

c. If the particles are identical and have spin 1/2, what is the total spin of each of the states of (a)?

(Wisconsin)

Sol: The Hamiltonian of the system is

$$\hat{H} = \frac{-\hbar^2}{2M}\left(\frac{\partial^2}{\partial x_1^2} + \frac{\partial^2}{\partial x_2^2}\right) + \frac{1}{2}k(x_1^2 + x_2^2) + \frac{K}{2}(x_1 - x_2)^2.$$

Let $\xi = \frac{1}{\sqrt{2}}(x_1 + x_2)$, $\eta = \frac{1}{\sqrt{2}}(x_1 - x_2)$ and write $\hat{H}$ as

$$\hat{H} = -\frac{\hbar^2}{2M}\left(\frac{\partial^2}{\partial \xi^2} + \frac{\partial^2}{\partial \eta^2}\right) + \frac{1}{2}k(\xi^2 + \eta^2) + K\eta^2$$

$$= -\frac{\hbar^2}{2M}\left(\frac{\partial^2}{\partial \xi^2} + \frac{\partial^2}{\partial \eta^2}\right) + \frac{1}{2}k\xi^2 + \frac{1}{2}(k+2K)\eta^2.$$

The system can be considered as consisting of two independent harmonic oscillators of angular frequencies ω_1 and ω_2 given by

$$\omega_1 = \sqrt{\frac{k}{M}}, \qquad \omega_2 = \sqrt{\frac{k+2K}{M}}.$$

The total energy is therefore

$$E_{nm} = \left(n + \frac{1}{2}\right)\hbar\omega_1 + \left(m + \frac{1}{2}\right)\hbar\omega_2,$$

and the corresponding eigenstate is

$$|nm\rangle = \psi_{nm} = \varphi_n^{(k)}(\xi)\varphi_m^{(k+2K)}(\eta),$$

where $n,m = 0, 1, 2, ...$, and $\varphi_n^{(k)}$ is the nth eigenstate of a harmonic oscillator of spring constant k.

a. The energies of the three lowest states of the system are

$$E_{00} = \frac{1}{2}\hbar(\omega_1 + \omega_2),$$

$$E_{10} = \frac{1}{2}\hbar(\omega_1 + \omega_2) + \hbar\omega_1,$$

$$E_{01} = \frac{1}{2}\hbar(\omega_1 + \omega_2) + \hbar\omega_2.$$

b. If the particles are identical and spinless, the wave function must be symmetric with respect to the interchange of the two particles. Thus the states $|00\rangle, |10\rangle$ are allowed, while the state $|01\rangle$ is not allowed.

c. If the particles are identical with spin 1/2, the total wave function, including both spatial and spin, must be antisymmetric with respect to an interchange of the two particles. As the spin function for total spin $S = 0$ is antisymmetric and that for $S = 1$ is symmetric, we have

$$S = 0 \quad \text{for } |00\rangle,$$
$$S = 0 \quad \text{for } |10\rangle,$$
$$S = 1 \quad \text{for } |01\rangle.$$

7015

A particular one-dimensional potential well has the following bound-state single-particle energy eigenfunctions:

$$\psi_a(x), \psi_b(x), \psi_c(x)\cdots, \quad \text{where} \quad E_a < E_b < E_c \cdots.$$

Two non-interacting particles are placed in the well. For each of the cases (a), (b), (c) listed below write down:

The two lowest total energies available to the two-particle system, the degeneracy of each of the two energy levels, the possible two-particle wave functions associated with each of the levels.

(Use ψ to express the spatial part and a ket $|S, m_s\rangle$ to express the spin part. S is the total spin.)

a. Two distinguishable spin-$\frac{1}{2}$ particles.

b. Two identical spin-$\frac{1}{2}$ particles.

c. Two identical spin-0 particles.

(MIT)

Sol: As the two particles, each of mass M, have no mutual interaction, their spatial wave functions separately satisfy the Schrödinger equation:

$$\left[-\frac{\hbar^2}{2M}\frac{\partial^2}{\partial x_1^2}+V(x_1)\right]\psi_i(x_1)=E_i\psi_i(x_1),$$

$$\left[-\frac{\hbar^2}{2M}\frac{\partial^2}{\partial x_2^2}+V(x_2)\right]\psi_j(x_2)=E_j\psi_j(x_2).$$

$$(i,j=a,b,c...)$$

The equations may also be combined to give

$$\left[-\frac{\hbar^2}{2M}\frac{\partial^2}{\partial x_1^2}-\frac{\hbar^2}{2M}\frac{\partial^2}{\partial x_2^2}+V(x_1)+V(x_2)\right]\psi_i(x_1)\psi_j(x_2)$$

$$=(E_i+E_j)\psi_i(x_1)\psi_j(x_2).$$

Consider the two lowest energy levels (i) and (ii).

a. Two distinguishable spin-$\frac{1}{2}$ particles.

    i. Total energy $=E_a+E_a$, degeneracy $=4$. The wave functions are

$$\begin{cases}\psi_a(x_1)\psi_a(x_2)|0,0\rangle,\\ \psi_a(x_1)\psi_a(x_2)|1,m\rangle.\ (m=0,\pm1)\end{cases}$$

    ii. Total energy $=E_a+E_b$, degeneracy $=8$. The wave functions are

$$\begin{cases}\psi_a(x_1)\psi_a(x_2)|0,0\rangle,\\ \psi_a(x_1)\psi_a(x_2)|1,m\rangle,\end{cases}\quad\begin{cases}\psi_a(x_1)\psi_b(x_2)|0,0\rangle,\\ \psi_a(x_1)\psi_b(x_2)|1,m\rangle.\end{cases}\quad(m=0,\pm1)$$

b. Two identical spin-1/2 particles.

    i. total energy $=E_a+E_a$, degeneracy $=1$. The wave function is $\psi_a(x_1)\psi_a(x_2)|0,0\rangle$.

    ii. total energy $=E_a+E_b$, degeneracy $=4$. The wave functions are

$$\begin{cases}\dfrac{1}{\sqrt{2}}[\psi_a(x_1)\psi_b(x_2)+\psi_b(x_1)\psi_a(x_2)]|0,0\rangle,\\ \dfrac{1}{\sqrt{2}}[\psi_a(x_1)\psi_b(x_2)-\psi_b(x_1)\psi_a(x_2)]|1,m\rangle.\ (m=0,\pm1)\end{cases}$$

c. Two identical spin-0 particles.

    i. Total energy $=E_a+E_a$, degeneracy $=1$. The wave function is $\psi_a(x_1)\psi_a(x_2)|0,0\rangle$.

   ii. Total energy $= E_a + E_a$, degeneracy $= 1$. The wave function is

$$\frac{1}{\sqrt{2}}[\psi_a(x_1)\psi_b(x_2)+\psi_b(x_1)\psi_a(x_2)|0,0\rangle.$$

7016

Two electrons move in a central field. Consider the electrostatic interaction $e^2/|\mathbf{r}_1 - \mathbf{r}_2|$ between the electrons as a perturbation.

a. Find the first order energy shifts for the states (terms) of the $(1s)(2s)$ configuration. (Express your answers in terms of unperturbed quantities and matrix elements of the interaction $e^2/|\mathbf{r}_1-\mathbf{r}_2|$)

b. Discuss the symmetry of the two-particle wave function for the states in part (a).

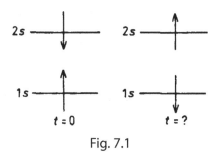

Fig. 7.1

c. Suppose that, at time $t=0$, one electron is found to be in the $1s$ unperturbed state with spin up and the other electron in the $2s$ unperturbed state with spin down as shown in Fig. 7.1. At what time t will the occupation of the states be reversed?

(Berkeley)

Sol:

a. The zero order wave function of the two electrons has the forms

$$\phi_+(\mathbf{r}_1,\mathbf{r}_2)\chi_{00}(s_{1z},s_{2z}),$$
$$\phi_-(\mathbf{r}_1,\mathbf{r}_2)\chi_{1M_s}(s_{1z},s_{2z}),$$

where

$$\phi_\varepsilon = \frac{1}{\sqrt{2}}[u_{1s}(1)v_{2s}(2) + \varepsilon u_{1s}(2)v_{2s}(1)], \quad \varepsilon = \pm 1,$$

being the normalized symmetric (+) and antisymmetric (−) wave functions, χ_0 and χ_1 denote the singlet and triplet spin states respectively. Denoting $u_{1s}(1)v_{2s}(2)$ by $|1,2\rangle$ and $u_{1s}(2)v_{2s}(1)$ by $|2,1\rangle$, we can write the above as

$$|\phi_\varepsilon\rangle = (|1,2\rangle + \varepsilon|2,1\rangle)/\sqrt{2}.$$

Because the perturbation Hamiltonian is independent of spin, we need not consider χ. Thus

$$\Delta E_\varepsilon = \int d^3 r_1 d^3 r_2 \phi_\varepsilon^* \frac{e^2}{|\mathbf{r}_1 - \mathbf{r}_2|} \phi_\varepsilon$$

$$= \frac{1}{2} (\langle 1,2| + \varepsilon\langle 2,1|) \frac{e^2}{|\mathbf{r}_1 - \mathbf{r}_2|} (|1,2\rangle + \varepsilon|2,1\rangle)$$

$$= \frac{1}{2}[\langle 1,2|A|1,2\rangle + \langle 2,1|A|2,1\rangle$$

$$+ \varepsilon\langle 1,2|A|2,1\rangle + \varepsilon\langle 2,1|A|1,2\rangle] = K + \varepsilon J,$$

where $A = e^2/|\mathbf{r}_1 - \mathbf{r}_2|, K = \langle 1,2|A|1,2\rangle = \langle 2,1|A|2,1\rangle$ is the direct integral, $J = \langle 1,2|A|2,1\rangle = \langle 2,1|A|1,2\rangle$ is the exchange integral.

b. The singlet state χ_0 is antisymmetric for interchange of spins. The triplet state χ_1 is symmetric for interchange of spins. Similarly, ϕ_+ is symmetric for the interchange of $\mathbf{r}_1$ and $\mathbf{r}_2$, and ϕ_- is antisymmetric for the interchange. Hence the total wave function is always antisymmetric for interchange of the electrons.

c. The initial state of the system is

$$\psi(t=0) = \frac{1}{\sqrt{2}}[|1s2s\rangle|\uparrow\downarrow\rangle - |2s1s\rangle|\downarrow\uparrow\rangle]$$

$$= \frac{1}{2\sqrt{2}}[(|1,2\rangle + |2,1\rangle)(|\uparrow\downarrow\rangle - |\uparrow\downarrow\rangle)$$

$$+ (|1,2\rangle - |2,1\rangle)(|\uparrow\downarrow\rangle + |\uparrow\downarrow\rangle)]$$

$$= \frac{1}{\sqrt{2}}(\psi_+\chi_{00} + \psi_-\chi_{10}),$$

and so the wave function at t is

$$\psi(t) = \frac{1}{\sqrt{2}}(\psi_+\chi_{00}e^{-iE_+t/\hbar} + \psi_-\chi_{10}e^{-iE_-t/\hbar}).$$

When $e^{-iE_-t/\hbar}/e^{-iE_+t/\hbar} = -1$, the wave function becomes

$$\psi(t_n) = e^{-iE_+t_n/\hbar} \cdot \frac{1}{\sqrt{2}}(\psi + \chi_{00} - \psi - \chi_{10})$$

$$= e^{-E+t_n/\hbar} \cdot \frac{1}{\sqrt{2}}\left[|2,1\rangle|\downarrow\uparrow\rangle - |1,2\rangle|\downarrow\uparrow\rangle\right],$$

which shows that at this time the 1s electron has spin down and the 2s electron has spin up i.e., the spins are reversed. As $-1 = e^{i(2n+1)\pi}$, $n = 0, 1, 2\cdots$, this happens at times

$$t = (2n+1)\pi \cdot \frac{\hbar}{E_+ - E_-} = \frac{(2n+1)\pi\hbar}{2J}.$$

7017

a. Show that the parity operator commutes with the orbital angular momentum operator. What is the parity quantum number of the spherical harmonic $Y_{lm}(\theta,\varphi)$?

b. Show for a one-dimensional harmonic oscillator in state $E_n = (n+\frac{1}{2})\hbar\omega$ that $\langle \Delta x^2 \rangle_n \langle \Delta p^2 \rangle_n = (n+\frac{1}{2})^2 \hbar^2$.

c. Consider the rotation of a hydrogen molecule H_2. What are its rotational energy levels? How does the identity of the two nuclei affect this spectrum? What type of radiative transition occurs between these levels? Remember that the proton is a fermion.

d. Show that $(\mathbf{n}\cdot\sigma)^2 = 1$, where $\mathbf{n}$ is a unit vector in an arbitrary direction and σ are the Pauli spin matrices.

(Berkeley)

Sol:

a. Applying the parity operator $\hat{p}$ and the orbital angular momentum operator

$$\mathbf{L} = \mathbf{r} \times \mathbf{p}$$

to an arbitrary wave function $f(\mathbf{r})$, we have

$$\hat{P}\mathbf{L}f(\mathbf{r}) = \hat{P}(\mathbf{r}\times\mathbf{p})f(\mathbf{r}) = (-\mathbf{r})\times(-\mathbf{p})f(-\mathbf{r})$$
$$= \mathbf{r}\times\mathbf{p}f(-\mathbf{r}) = \mathbf{L}\hat{P}f(\mathbf{r}).$$

Hence $\hat{P}$ and **L** commute, i.e.,

$$[\hat{P}, \mathbf{L}] = 0.$$

As

$$\hat{P} Y_{lm}(\theta, \phi) = Y_{lm}(\pi - \theta, \pi + \phi) = (-1)^l Y_{lm}(\theta, \phi),$$

the parity quantum number of $Y_{lm}(\theta, \varphi)$ is $(-1)^l$.

b. For a one-dimensional harmonic oscillator, we can use the Fock representation

$$x = \sqrt{\frac{\hbar}{2m\omega}}(a + a^+), \qquad p = i\sqrt{\frac{m\hbar\omega}{2}}(a^+ - a),$$

where a, a^+ are annihilation and creation operators having the properties

$$a|n\rangle = \sqrt{n}|n-1\rangle,$$
$$a^+|n\rangle = \sqrt{n+1}|n+1\rangle.$$

Using these operators we have

$$\langle n|x|n \rangle = \sqrt{\frac{\hbar}{2m\omega}}\langle n|a + a^+|n\rangle$$

$$= \sqrt{\frac{\hbar}{2m\omega}}\left(\sqrt{n}|n-1\rangle\right) + \sqrt{n+1}\langle n|n+1\rangle = 0,$$

$$\langle n|x^2|n\rangle = \frac{\hbar}{2m\omega}\langle n|(a + a^+)(a + a^+)|n\rangle$$

$$= \frac{\hbar}{2m\omega}\left(\sqrt{n}\langle n|a + a^+|n-1\rangle + \sqrt{n+1}\langle n|a + a^+|n+1\rangle\right)$$

$$= \frac{\hbar}{2m\omega}\left[\sqrt{n(n-1)}\langle n|n-2\rangle + n\langle n|n\rangle\right.$$

$$\left. + (n+1)\langle n|n\rangle + \sqrt{(n+1)(n+2)}\langle n|n+2\rangle\right]$$

$$= \frac{\hbar}{2m\omega}(2n+1),$$

and similarly

$$\langle n|p|n\rangle = 0,$$

$$\langle n|p^2|n\rangle = \frac{m\hbar\omega}{2}(2n+1).$$

As

$$\langle \Delta x^2 \rangle_n = \langle (x - \langle x \rangle)^2 \rangle_n = \langle x^2 \rangle_n - \langle x \rangle_n^2 = \frac{\hbar}{2m\omega}(2n+1),$$

$$\langle \Delta p^2 \rangle_n = \langle p^2 \rangle_n - \langle p \rangle_n^2 = \frac{m\hbar\omega}{2}(2n+1),$$

we find

$$\left\langle \Delta x^2 \right\rangle_n \cdot \left\langle \Delta p^2 \right\rangle_n = \hbar^2 \left(n + \frac{1}{2} \right)^2.$$

c.   The rotational energy levels of a hydrogen molecule are given by

$$E_r = \hbar^2 K(K+1)/2I_0,$$

where $I_0 = MR_0^2$ is the moment of inertia of the molecule about the rotating axis, which is perpendicular to the line connecting the two nuclei, K is the angular momentum quantum number, $K = 0, 1, 2,....$ Since the spin of a proton is $\hbar/2$, the total wave function of the molecule is antisymmetric for interchange of the two protons. When the two protons are interchanged, the wave function for the motion of the center of mass and the wave function for the atomic vibration are not changed; only the wave function for rotation is altered:

$$Y_{KM_K}(\theta, \varphi) \rightarrow Y_{KM_K}(\pi - \theta, \pi + \varphi) = (-1)^K Y_{KM_K}(\theta, \varphi).$$

If K is even, $(-1)^K Y_{KM_K}(\theta, \varphi) = Y_{KM_K}(\theta, \varphi)$ and the spin wave function χ_0 must be antisymmetric, i.e., χ_0 is a spin singlet state; If K is odd, $(-1)^K Y_{KM_K}(\theta, \varphi) = -Y_{KM_K}(\theta, \varphi)$ and the spin wave function χ_1 must be symmetric, i.e., χ_1 is a spin triplet state. The hydrogen molecule in the former case is called a para-hydrogen, and in the latter case is called an ortho-hydrogen. There is no inter-conversion between para-hydrogen and ortho-hydrogen. Transitions can take place between rotational energy levels with $\Delta K = 2,4,6,...$ within each type. Electric quadruple transitions may also occur between these levels.

d.

$$(\mathbf{n} \cdot \boldsymbol{\sigma})^2 = \left(\sum_i n_i \sigma_i \right)^2 = \sum_{ij} n_i n_j \sigma_i \sigma_j$$

$$= \frac{1}{2} \sum_{i,j} n_i n_j \{\sigma_i, \sigma_j\} = \sum_{i,j} n_i n_j \delta_{ij} = \sum_i n_i n_i = 1.$$

In the above i, j refer to x, y, z, and $\{\sigma_i, \sigma_j\} \equiv \sigma_i \sigma_j + \sigma_j \sigma_i = 2\delta_{ij}$.

7018

Two particles of mass m are placed in a rectangular box of sides $a > b > c$ in the lowest energy state of the system compatible with the conditions below. The particles interact with each other according to the potential $V = A\delta(\mathbf{r}_1 - \mathbf{r}_2)$. Using first order perturbation theory to calculate the energy of the system under the following conditions:

a. Particles not identical.

b. Identical particles of spin zero.

c. Identical particles of spin one-half with spins parallel.

(Berkeley)

Sol:

a. The unperturbed system can be treated as consisting of two separate single-particle systems and the wave function as a product of two single-particle wave functions:

$$\psi(\mathbf{r}_1,\mathbf{r}_2)=\psi(\mathbf{r}_1)\psi(\mathbf{r}_2).$$

The lowest energy state wave function is thus

$$\psi_0(\mathbf{r}_1,\mathbf{r}_2)=\begin{cases}\dfrac{8}{abc}\sin\dfrac{\pi x_1}{a}\sin\dfrac{\pi x_2}{a}\sin\dfrac{\pi y_1}{b}\sin\dfrac{\pi y_2}{b}\sin\dfrac{\pi z_1}{c}\sin\dfrac{\pi z_2}{c},\\ \qquad\text{for } 0<x_i<a,\ 0<y_i<b,\ 0<z_i<c,\ (i=1,2)\\ 0,\qquad\text{otherwise,}\end{cases}$$

corresponding to an energy

$$E_0=\frac{\hbar^2\pi^2}{m}\left(\frac{1}{a^2}+\frac{1}{b^2}+\frac{1}{c^2}\right).$$

First order perturbation theory gives an energy correction

$$\Delta E=\int\psi_0^*(\mathbf{r}_1,\mathbf{r}_2)A\delta(\mathbf{r}_1-\mathbf{r}_2)\psi_0(\mathbf{r}_1,\mathbf{r}_2)d^3r_1 d^3r_2$$

$$=\int A|\psi_0(\mathbf{r}_1,\mathbf{r}_1)|^2\,d^3r_1=\int_0^c\int_0^b\int_0^a A\left(\frac{8}{abc}\right)^2$$

$$\times\left(\sin\frac{\pi x_1}{a}\sin\frac{\pi y_1}{b}\sin\frac{\pi z_1}{c}\right)^4 dx_1 dy_1 dz_1=\frac{27A}{8abc},$$

and hence

$$E'=\frac{\hbar^2\pi^2}{m}\left(\frac{1}{a^2}+\frac{1}{b^2}+\frac{1}{c^2}\right)+\frac{27A}{8abc}.$$

b. For a system of spin-0 particles, the total wave function must be symmetric for interchange of a pair of particles. Hence the lowest energy state is

$$\psi_s(\mathbf{r}_1,\mathbf{r}_2)=\psi_0(\mathbf{r}_1,\mathbf{r}_2),$$

which is the same as that in (a). The energy to first order perturbation is also

$$E'_s=\frac{\hbar^2\pi^2}{m}\left(\frac{1}{a^2}+\frac{1}{b^2}+\frac{1}{c^2}\right)+\frac{27A}{8abc}.$$

c. For a system of spin-$\frac{1}{2}$ particles the total wave function must be antisymmetric. As the spins are parallel, the spin wave function is symmetric and so the spatial wave function must be antisymmetric. As $\frac{1}{a^2}<\frac{1}{b^2}<\frac{1}{c^2}$, the lowest energy state is

$$\psi_A(\mathbf{r}_1,\mathbf{r}_2)=\frac{1}{\sqrt{2}}[\psi_{211}(\mathbf{r}_1)\psi_{111}(\mathbf{r}_2)-\psi_{211}(\mathbf{r}_2)\psi_{111}(\mathbf{r}_1)],$$

where $\psi_{111}(\mathbf{r})$ and $\psi_{211}(\mathbf{r})$ are the ground and first excited single-particle states respectively. The unperturbed energy is

$$E_{A0}=\frac{\hbar\pi^2}{m}\left(\frac{5}{2a^2}+\frac{1}{b^2}+\frac{1}{c^2}\right).$$

First order perturbation theory gives

$$\Delta E=\int\psi_A^*(\mathbf{r}_1,\mathbf{r}_2)A\delta(\mathbf{r}_1-\mathbf{r}_2)\psi_A(\mathbf{r}_1,\mathbf{r}_2)d^3r_1d^3r_2=0.$$

Therefore

$$E_A'=\frac{\hbar^2\pi^2}{m}\left(\frac{5}{2a^2}+\frac{1}{b^2}+\frac{1}{c^2}\right).$$

7019

A porphyrin ring is a molecule which is present in chlorophyll, hemoglobin, and other important compounds. Some aspects of the physics of its molecular properties can be described by representing it as a one-dimensional circular path of radius $r=4$ Å, along which 18 electrons are constrained to move.

a. Write down the normalized one-particle energy eigenfunctions of the system, assuming that the electrons do not interact with each other.

b. How many electrons are there in each level in the ground state of the molecule?

c. What is the lowest electronic excitation energy of the molecule? What is the corresponding wavelength (give a numerical value) at which the molecule can absorb radiation?

(Berkeley)

Sol:

a. Denote the angular coordinate of an electron by θ. The Schrödinger equation of an electron

$$-\frac{\hbar^2}{2mr^2}\frac{\partial^2}{\partial\theta^2}\psi(\theta)=E\psi(\theta)$$

has solution

$$\psi(\theta)=\frac{1}{\sqrt{2\pi}}e^{ik\theta}.$$

The single-valuedness of $\psi(\theta)$, $\psi(\theta)=\psi(\theta+2\pi)$, requires $k=0,\ \pm1,\ \pm2....$
The energy levels are given by

$$E=\frac{\hbar^2}{2mr^2}k^2\ .$$

b. Let 0,1,2, ... denote the energy levels $E_0,E_1,E_2,...$ respectively. In accordance with Pauli's exclusion principle, E_0 can accommodate two electrons of opposite spins, while E_k, $k\neq0$, which is 2-fold degenerate with respect to $\pm|k|$, can accommodate four electrons. Thus the electron configuration of the ground state of the system is

$$0^21^42^43^44^4.$$

c. The electron configuration of the first excited state is

$$0^21^42^43^44^35^1.$$

The energy difference between E_4 and E_5 is

$$\Delta E=E_5-E_4=\frac{\hbar^2}{2mr^2}(5^2-4^2)=\frac{9\hbar^2}{2mr^2}$$

and the corresponding absorption wavelength is

$$\lambda=\frac{ch}{\Delta E}=\frac{8\pi^2}{9}\left(\frac{mc}{h}\right)r^2=\frac{8\pi^2}{9}\times\frac{4^2}{0.0242}=5800\ \text{Å},$$

where $\frac{h}{mc}=0.0242$ Å is the Compton wavelength of electron.

7020

A large number N of spinless fermions of mass m are placed in a one-dimensional oscillator well, with a repulsive δ-function potential between each pair:

$$V=\frac{k}{2}\sum_{i=1}^{N}x_i^2+\frac{\lambda}{2}\sum_{i\neq j}\delta(x_i-x_j),\qquad k,\lambda>0.$$

a. In terms of normalized single-particle harmonic oscillator functions $\psi_n(x)$, obtain the normalized wave functions and energies for the three lowest energies. What are the degeneracies of these levels?

b. Compute the expectation value of $\sum_{i=1}^{N} x_i^2$ for each of these states. For partial credit, you may do parts (a) and (b) with $\lambda = 0$.

(Berkeley)

Sol:

a. Treat the δ-function potential as perturbation. As for a system of fermions, the total wave function is antisymmetric, the zero-order wave function for the system can be written as the determinant

$$\psi_{n_1\ldots n_N}(x_1\ldots x_N) = \frac{1}{\sqrt{N!}} \begin{vmatrix} \psi_{n_1}(x_1) & \psi_{n_1}(x_2) & \cdots & \psi_{n_1}(x_N) \\ \psi_{n_2}(x_1) & \psi_{n_2}(x_2) & \cdots & \psi_{n_2}(x_N) \\ \vdots & \vdots & & \vdots \\ \psi_{n_N}(x_1) & \psi_{n_N}(x_2) & \cdots & \psi_{n_N}(x_N) \end{vmatrix}$$

$$= \frac{1}{\sqrt{N!}} \sum_P \delta_P P[\psi_{n_1}(x_1)\cdots\psi_{n_N}(x_N)],$$

where n_i label the states from ground state up, P denotes permutation of x and $\delta_P = +1, -1$ for even and odd permutations respectively. On account of the δ-function, the matrix elements of the perturbation Hamiltonian are all zero. Thus the energy levels are

$$E_{(n_1, n_2, \cdots, n_N)} = \langle n_1 \cdots n_N | H | n_1 \ldots n_N \rangle = \hbar\omega \left(\frac{N}{2} + \sum_{i=1}^{N} n_i \right),$$

where $\omega = \sqrt{k/m}$.

i. For the ground state: $n_1 \cdots n_N$ are respectively $0, 1, \ldots, N-1$, the energy is

$$E_{(0,\ldots,N-1)} = \hbar\omega \left[\frac{N}{2} + \frac{N(N-1)}{2} \right] = \frac{\hbar\omega}{2} N^2,$$

and the wave function is

$$\psi_{0,1,\ldots,N-1}(x_1\cdots x_N) = \frac{1}{N!} \sum_P \delta_P P[\psi_0(x_1)\cdots\psi_{N-1}(x_N)].$$

ii. For the first excited state: $n_1 \cdots n_N$ are respectively $0, 1, \ldots, N-2, N$, the energy is

$$E_{(0,\ldots,N-2,N)} = \frac{1}{2}\hbar\omega(N^2 + 2),$$

and the wave function is

$$\psi_{0,1\cdots N-2,N}(x_1\cdots x_N)=\frac{1}{\sqrt{N!}}\sum_P \delta_P P[\psi_0(x_1)\cdots\psi_{N-2}(x_{N-1})\psi_N(x_N)].$$

iii. For the second excited state: $n_1\cdots n_N$ are respectively either $0, 1, \ldots N-2, N+1$, or $0, 1, \ldots N-3, N-1, N$. The energy is

$$E_{(0,\ldots N-2,N+1)}=E_{(0,1,\ldots N-3,N-1,N)}=\frac{\hbar\omega}{2}(N^2+4),$$

and the corresponding wave functions are

$$\psi_{0,\ldots N-2,N+1}(x_1\cdots x_N)=\frac{1}{\sqrt{N!}}\sum_P \delta_P P[\psi_0(x_1)\cdots$$
$$\psi_{N-2}(x_{N-1})\psi_{N+1}(x_N)],$$
$$\psi_{0,\ldots N-3,N-1,N}(x_1\cdots x_N)=\frac{1}{\sqrt{N!}}\sum_P \delta_P P[\psi_0(x_1)\cdots$$
$$\psi_{N-3}(x_{N-2})\psi_{N-1}(x_{N-1})\psi_N(x_N)].$$

It can be seen that the ground and first excited states are nondegenerate, where the second excited state is two-fold degenerate.

b. For stationary states,

$$2\langle T\rangle=\left\langle\sum_i x_i\,\partial_i V(x_1\cdots x_N)\right\rangle.$$

As

$$\left\langle\sum_k x_k\partial_k\sum_{i\neq j}\frac{\lambda}{2}\delta(x_i-x_j)\right\rangle=0,$$
$$\left\langle\sum_i x_i\frac{\partial}{\partial x_i}\left(\frac{k}{2}\sum_j x_j^2\right)\right\rangle=k\left\langle\sum_i x_i^2\right\rangle,$$

we have

$$2\langle T\rangle=k\left\langle\sum_i^N x_i^2\right\rangle.$$

The virial theorem

$$\langle T\rangle=\langle V(x_1\cdots x_N)\rangle,$$

or

$$E=2\langle T\rangle,$$

then gives

$$\left\langle\sum_{i=1}^N x_i^2\right\rangle=\frac{1}{m\omega^2}E.$$

Hence

$$\left\langle 0 \left| \sum_{i=1}^{N} x_i^2 \right| 0 \right\rangle = \frac{\hbar}{2m\omega} N^2,$$

$$\left\langle 1 \left| \sum_{i=1}^{N} x_i^2 \right| 1 \right\rangle = \frac{\hbar}{2m\omega} (N^2 + 2),$$

$$\left\langle 2 \left| \sum_{i=1}^{N} x_i^2 \right| 2 \right\rangle = \left\langle 2' \left| \sum_{i=1}^{N} x_i^2 \right| 2' \right\rangle = \frac{\hbar}{2m\omega} (N^2 + 4),$$

where $|0\rangle, |1\rangle, |2\rangle$, and $|2'\rangle$ are the ground state, the first excited state and the two second excited states respectively.

7021

What is the energy difference in eV between the two lowest rotational levels of the HD molecule? The HD (D is a deuteron) distance is 0.75 Å.

(Berkeley)

Sol: The rotational energy levels are given by

$$E_J = \frac{\hbar^2}{2I} J(J+1).$$

Thus for the two lowest levels,

$$\Delta E_{10} = \frac{\hbar^2}{2I} J(J+1) \bigg|_{J=1} - \frac{\hbar^2}{2I} J(J+1) \bigg|_{J=0} = \frac{\hbar^2}{I}.$$

As the mass m_D of the deuteron is approximately twice that of the hydrogen nucleus m_p, we have

$$I = \mu r^2 = \frac{2m_p \cdot m_p}{2m_p + m_p} r^2 = \frac{2}{3} m_p r^2,$$

and hence

$$\Delta E_{10} = \frac{\hbar^2}{\frac{2}{3} m_p r^2} = \frac{3}{2} \frac{(\hbar c)^2}{m_p c^2} \frac{1}{r^2} = \frac{1.5}{938 \times 10^6}$$

$$\times \left(\frac{6.582 \times 10^{-16} \times 3 \times 10^{10}}{0.75 \times 10^{-8}} \right)^2 = 1.11 \times 10^{-2} \text{ eV}.$$

7022

Consider the (homonuclear) molecule N_2^{14}. Use the fact that a nitrogen nucleus has spin $I = 1$ in order to derive the result that the ratio of intensities of adjacent rotational lines in the molecule's spectrum is 2:1.

(Chicago)

Sol: In the adiabatic approximation, the wave function of N_2 molecule whose center of mass is at rest can be expressed as the product of the electron wave function ψ_e, the total nuclear spin wave function ψ_s, the vibrational wave function ψ_0, and the rotational wave function ψ_I; that is, $\psi = \psi_e \psi_s \psi_0 \psi_I$. For the molecular rotational spectrum, the wave functions of the energy states involved in the transition have the same ψ_e, ψ_0, but different ψ_s, ψ_I. For interchange of the nitrogen nuclei, we have $\psi_e \psi_0 \to \psi_e \psi_0$ or $-\psi_e \psi_0$.

The N nucleus is a boson as its spin is 1, so the total nuclear spin of the N_2 molecule can only be 0, 1 or 2, making it a boson also. For the exchange operator $\hat{P}$ between the N nuclei, we have

$$\hat{P}\psi_s = \begin{cases} +\psi_s & \text{for } S = 0, 2, \\ -\psi_s & \text{for } S = 1, \end{cases} \qquad \hat{P}\psi_I = \begin{cases} \psi_I & \text{for } I = \text{even integer}, \\ -\psi_I & \text{for } I = \text{odd integer}. \end{cases}$$

As N_2 obeys the Bose-Einstein statistics, the total wave function does not change on interchange of the two nitrogen nuclei. So for adjacent rotational energy levels with $\Delta I = 1$, one must have $S = 0$ or 2, the other $S = 1$, and the ratio of their degeneracies is $[2 \times 2 + 1 + 2 \times 0 + 1] : (2 \times 1 + 1) = 2 : 1$.

For the molecular rotational spectrum, the transition rule is $\Delta J = 2$. As S usually remains unchanged in optical transitions, two adjacent lines are formed by transitions from $I = $ even to even and $I = $ odd to odd. Since the energy difference between two adjacent rotational energy levels is very small compared with kT at room temperature, we can neglect the effect of any heat distribution. Therefore, the ratio of intensities of adjacent spectral lines is equal to the ratio of the degeneracy of $I = $ even rotational energy level to that of the adjacent $I = $ odd rotational energy level, which has been given above as 2:1.

7023

a. Assuming that two protons of H_2^+ molecule are fixed at their normal separation of 1.06 Å, sketch the potential energy of the electron along an axis passing through the protons.

b. Sketch the electron wave functions for the two lowest states in H_2^+, indicating roughly how they are related to hydrogenic wave functions. Which wave function corresponds to the ground state, and why?

c. What happens to the two lowest energy levels of H_2^+ in the limit that the protons are moved far apart?

(Wisconsin)

Sol:

a. As the protons are fixed, the potential energy of the system is that of the electron, apart from the potential energy due to the Coulomb interaction between the nuclei $\frac{e^2}{R}$. Thus

$$V = -\frac{e^2}{|\mathbf{r}_1|} - \frac{e^2}{|\mathbf{r}_2|},$$

where $\mathbf{r}_1$, $\mathbf{r}_2$ are as shown in Fig. 7.2. When the electron is on the line connecting the two protons, the potential energy is

$$V = -\frac{e^2}{|x|} - \frac{e^2}{|R-x|},$$

where x is the distance from the proton on the left. V is shown in Fig. 7.3 as a function of x. The wave function must be symmetrical or antisymmetrical with respect to the interchange of the protons. Thus the wave functions of the two lowest states are

$$\psi_\pm = \frac{1}{\sqrt{2}}[\phi(\mathbf{r}_1) \pm \phi(\mathbf{r}_2)],$$

where $\phi(\mathbf{r})$ has the form of the wave function of the ground state hydrogen atom.

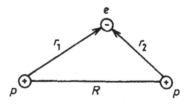

Fig. 7.2

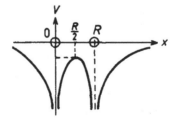

Fig. 7.3

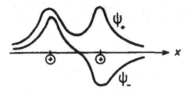

Fig. 7.4

$$\phi(r)=\frac{1}{\sqrt{\pi}}\left(\frac{\lambda}{a}\right)^{3/2}e^{-\lambda r/a}$$

where a is the Bohr radius and λ is a constant. The shape of the two wave functions along the x-axis are sketched in Fig. 7.4. It can be seen that the probability that the electron is near the two nuclei is larger for ψ_+. Hence ψ_+ corresponds to a lower V and is therefore the ground state wave function. The fact that $E_+ < E_-$ can also be seen from

$$E_{\pm}=\langle\psi_{\pm}|H|\psi_{\pm}\rangle$$
$$=\langle\phi(\mathbf{r}_1)|H|\phi(\mathbf{r}_1)\rangle\pm\langle\phi(\mathbf{r}_1)|H|\phi(\mathbf{r}_2)\rangle,$$

since

$$\langle\phi(\mathbf{r}_1)|H|\phi(\mathbf{r}_1)\rangle=\langle\phi(\mathbf{r}_2)|H|\phi(\mathbf{r}_2)\rangle,$$
$$\langle\phi(\mathbf{r}_1)|H|\phi(\mathbf{r}_2)\rangle=\langle\phi(\mathbf{r}_2)|H|\phi(\mathbf{r}_1)\rangle.$$

and all the integrals are negative.

c. As the protons are being moved apart, the overlap of the two bound states $\phi(\mathbf{r}_1)$ and $\phi(\mathbf{r}_2)$ becomes less and less, and so $\langle\phi(\mathbf{r}_1)|H|\phi(\mathbf{r}_2)\rangle$ and $\langle\phi(\mathbf{r}_2)|H|\phi(\mathbf{r}_1)\rangle\to 0$. In other words, the two lowest energy levels will become the same, same as the ground state energy of a hydrogen atom.

7024

Write the Schrödinger equation for atomic helium, treating the nucleus as an infinitely heavy point charge.

(Berkeley)

Sol: Treating the nucleus as an infinitely heavy point charge, we can neglect its motion, as well as the interaction between the nucleons inside the nucleus and the distribution of the nuclear charge.

The Schrödinger equation is then

$$\left(\frac{\mathbf{p}_1^2}{2m_e}+\frac{\mathbf{p}_2^2}{2m_e}-\frac{2e^2}{R_1}-\frac{2e^2}{R_2}+\frac{e^2}{|\mathbf{R}_1-\mathbf{R}_2|}\right)\psi(\mathbf{R}_1,\mathbf{R}_2)=E\psi(\mathbf{R}_1,\mathbf{R}_2),$$

where $\mathbf{R}_1$, $\mathbf{R}_2$ are as shown in Fig. 7.5.

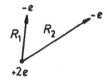

Fig. 7.5

On the left side of the equation, the first and second terms are the kinetic energies of the electrons, the third and fourth terms correspond to the attractive potentials between the nucleus and the electrons, and the last term is the repulsive potential between the two electrons.

7025

The excited electronic configuration $(1s)^1(2s)^1$ of the helium atom can exist as either a singlet or a triplet state. Tell which state has the lower energy and explain why. Give an expression which represents the energy separation between the singlet and triplet states in terms of the one-electron orbitals $\psi_{1s}(\mathbf{r})$ and $\psi_{2s}(\mathbf{r})$.

(MIT)

Sol: Electrons being fermions, the total wave function of a system of electrons must be anti-symmetric for the interchange of any two electrons. As the spin triplet state of helium atom is symmetric, its spatial wave function must be antisymmetric. In this state the electrons have parallel spins so the probability for them to get close is small (Pauli's principle), and consequently the repulsive energy, which is positive, is small. Whereas for the spin singlet state the reverse is true, i.e., the probability for the two electrons to get close is larger, so is the repulsive energy. Hence the triplet state has the lower energy.

Consider the interaction between the electrons as perturbation. The perturbation Hamiltonian is

$$H' = \frac{e^2}{r_{12}},$$

where $r_{12} = |\mathbf{r}_1 - \mathbf{r}_2|$. For the singlet state, using the one-electron wave functions ψ_{1s}, ψ_{2s}, we have

$$^1\psi = \frac{1}{\sqrt{2}}[\psi_{1s}(\mathbf{r}_1)\psi_{2s}(\mathbf{r}_2) + \psi_{1s}(\mathbf{r}_2)\psi_{2s}(\mathbf{r}_1)]\chi_{00},$$

and for the triplet state

$$^3\psi = \frac{1}{\sqrt{2}}[\psi_{1s}(\mathbf{r}_1)\psi_{2s}(\mathbf{r}_2) - \psi_{1s}(\mathbf{r}_2)\psi_{2s}(\mathbf{r}_1)]\chi_{1m}$$

with $m = 1, 0, -1$. The energy separation between the states is

$$\Delta E = \left\langle {}^1\psi \mid H' \mid {}^1\psi \right\rangle - \left\langle {}^3\psi \mid H' \mid {}^3\psi \right\rangle.$$

With $\psi_{ns}^* = \psi_{ns}$, we have

$$\Delta E = 2 \int \frac{e^2}{r_{12}} [\psi_{1s}(\mathbf{r}_1)\psi_{2s}(\mathbf{r}_1)\psi_{1s}(\mathbf{r}_2)\psi_{2s}(\mathbf{r}_2)] d\mathbf{r}_1 d\mathbf{r}_2.$$

7026

a. Suppose you have solved the Schrödinger equation for the singly-ionized helium atom and found a set of eigenfunctions $\psi_N(\mathbf{r})$.

    1. How do the $\phi_N(\mathbf{r})$ compare with the hydrogen atom wave functions?

    2. If we include a spin part σ^+ (or σ^-) for spin up (or spin down), how do you combine the ϕ's and σ's to form an eigenfunction of definite spin?

b. Now consider the helium atom to have two electrons, but ignore the electromagnetic interactions between them.

    1. Write down a typical two-electron wave function, in terms of the ϕ's and σ's, of definite spin. Do not choose the ground state.

    2. What is the total spin in your example?

    3. Demonstrate that your example is consistent with the Pauli exclusion principle.

    4. Demonstrate that your example is antisymmetric with respect to electron interchange.

(Buffalo)

Sol:

    a.

        1. The Schrödinger equation for singly-charged He atom is the same as that for H atom with $e^2 \rightarrow Ze^2$, where Z is the charge of the He nucleus. Hence the wave functions for hydrogen-like ion are the same as those for H atom with the Bohr radius replaced:

$$r_0 = \frac{\hbar^2}{\mu e^2} \rightarrow a = \frac{\hbar^2}{\mu Z e^2},$$

        μ being the reduced mass of the system. For helium $Z = 2$.

2. As ϕ_N and $\sigma^{\pm}$ belong to different spaces we can simply multiply them to form an eigenfunction of a definite spin.

b.

1. and 2. A He atom, which has two electrons, may be represented by a wave function

$$\frac{1}{\sqrt{2}}\phi_N(1)\phi_N(2)[\sigma^+(1)\sigma^-(2)-\sigma^-(1)\sigma^+(2)]$$

if the total spin is zero, and by

$$\frac{1}{\sqrt{2}}[\phi_{N1}(1)\phi_{N2}(2)-\phi_{N2}(1)\phi_{N1}(2)]\sigma^+(1)\sigma^+(2)$$

if the total spin is 1.

3. If $\sigma^+=\sigma^-, \phi_{N1}=\phi_{N2}$, the wave functions vanish, in agreement with the Pauli exclusion principle.

4. Denote the wave functions by $\psi(1, 2)$. Interchanging particles 1 and 2 we have

$$\psi(2, 1)=-\psi(1, 2).$$

7027

Ignoring electron spin, the Hamiltonian for the two electrons of helium atom, whose positions relative to the nucleus are $r_i (i=1, 2)$, can be written as

$$H=\sum_{i=1}^{2}\left(\frac{p_i^2}{2m}-\frac{2e^2}{|r_i|}\right)+V, \quad V=\frac{e^2}{|r_1-r_2|}.$$

a. Show that there are 8 orbital wave functions that are eigenfunctions of $H-V$ with one electron in the hydrogenic ground state and the others in the first excited state.

b. Using symmetry arguments show that all the matrix elements of V among these 8 states can be expressed in terms of four of them. [Hint: It may be helpful to use linear combinations of $l=1$ spherical harmonics proportional to

$$\frac{x}{|r|}, \frac{y}{|r|} \text{ and } \frac{z}{|r|}.]$$

c. Show that the variational principle leads to a determinantal equation for the energies of the 8 excited states if a linear combination of the

8 eigenfunctions of $H-V$ is used as a trial function. Express the energy splitting in terms of the four independent matrix elements of V.

d. Discuss the degeneracies of the levels due to the Pauli principle.

(Buffalo)

Sol: Treating V as perturbation, the zero-order wave function is a product of two eigenfunctions $|n, l, m\rangle$ of a hydrogen-like atom. Thus the 8 eigenfunctions for $H_0 = H - V$ with one electron in the hydrogen ground state can be written as

$$|lm\pm\rangle = \frac{1}{\sqrt{2}}\left[|(100)_1(2lm)_2\rangle \pm |(2lm)_1(100)_2\rangle\right]$$

with $l = 0, 1$, $m = -l, \ldots, l$, where the subscripts 1 and 2 refer to the two electrons. The corresponding energies are

$$E_b = E_1 + E_2 = -\frac{\mu(2e^2)^2}{2\hbar^2}\left(1 + \frac{1}{2^2}\right) = -\frac{5\mu e^4}{2\hbar^2}.$$

To take account of the perturbation we have to calculate the matrix elements

$$\langle l'm'\pm|V|lm\pm\rangle.$$

As V is rotation-invariant and symmetric in the two electrons and $|lm\pm\rangle$ are spatial rotation eigenstates, we have

$$\langle(100)_1(2l'm')_2|V|(100)_1(2lm)_2\rangle$$
$$= \langle(2l'm')_1(100)_2|V|(2lm)_1(100)_2\rangle$$
$$= \delta_{ll'}\delta_{mm'}A_l,$$
$$\langle(100)_1(2l'm')_2|V|(2lm)_1(100)_2\rangle$$
$$= \langle(2l'm')_1(100)_2|V|(100)_1(2lm)_2\rangle$$
$$= \delta_{ll'}\delta_{mm'}B_l,$$

and hence

$$\langle l'm'+|V|lm+\rangle = \delta_{ll'}\delta_{mm'}(A_l + B_l),$$
$$\langle l'm'+|V|lm-\rangle = 0,$$
$$\langle l'm'-|V|lm+\rangle = 0,$$
$$\langle l'm'-|V|lm-\rangle = \delta_{ll'}\delta_{mm'}(A_l - B_l).$$

Because the wave functions were formed taking into account the symmetry with respect to the interchange of the two electrons, the perturbation matrix is diagonal, whence the four discrete energy levels follow:

The first levels $|1m+\rangle$ have energy $E_b + A_1 + B_1$, second levels $|1m-\rangle$ have energy $E_b + A_1 - B_1$, the third level $|00+\rangle$ has energy $E_b + A_0 + B_0$, the fourth level

$|00-\rangle$ has energy $E_b + A_0 - B_0$. Note that the levels $|1m+\rangle$ and $|1m-\rangle$ are each three-fold degenerate ($m = \pm1, 0$).

According to Pauli's principle, we must also consider the spin wave functions. Neglecting spin-orbit coupling, the total spin wave functions are χ_0, antisymmetric, a singlet state; χ_{1s_z}, symmetric, a triplet state.

Since the total electron wave function must be antisymmetric for inter-change of the electrons, we must take combinations as follows,

$$|lm+\rangle\chi_0,$$

$$|lm-\rangle\chi_{1s_z}.$$

Hence, the degeneracies of the energy levels are

$$E_b + A_0 - B_0 \;:\; 1\times3=3$$
$$E_b + A_0 + B_0 \;:\; 1\times1=1$$
$$E_b + A_1 - B_1 \;:\; 3\times3=9$$
$$E_b + A_1 + B_1 \;:\; 3\times1=3.$$

7028

Describe approximate wave functions and energy levels of the lowest set of P-states ($L=1$) of the neutral helium atom, using as a starting basis the known wave functions for the hydrogen atom of nuclear charge Z:

$$\psi_{1s} = \pi^{-1/2} a^{-3/2} e^{-r/a}, a = a_0/Z,$$
$$\psi_{2p,m_j=0} = (32\pi)^{-1/2} a^{-5/2} re^{-r/2a}\cos\theta, \text{ etc.}$$

a. There are a total of 12 states (2 spin components × 2 spin components × 3 orbital components) which you should classify according to the Russell-Saunders coupling scheme, giving all the appropriate quantum numbers. Be sure that the states are properly antisymmetrized.

b. Give an estimate (to the nearest integer) for the values of "Z" to use for each of the two orbital wave functions. What energy above the ground state results? What mathematical process could be used to calculate the optimum Z values?

c. Write down an integral which gives the separation between two subsets of these 12 states due to the Coulomb repulsion between the two electrons. Which states are lower in energy?

d. Which of these *P*-states, if any, can decay to the atomic ground state by the emission of a single photon. (Electric dipole only)

e. Do there exist any other excited states with $L=1$ which can decay to any one of the *P*-states discussed above by emission of a single photon by electric dipole interaction? If so, give an example of such a state in the usual scheme of spectroscopic notation.

(Berkeley)

Sol:

a. Since $\mathbf{L}=\mathbf{l}_1+\mathbf{l}_2$, $L_z=l_{1z}+l_{2z}$, $L=1$ means that $l_1, l_2=0,1$ or $1, 0$, i.e. one electron is in 1s state, the other in 2p state. For convenience, Dirac's bra-ket notation is used to represent the states. The symmetrized and antisymmetrized spatial wave functions are

$$|\psi_1\rangle=\frac{1}{\sqrt{2}}\big(|1s\rangle|2p,m_l=1\rangle+|2p,m_l=1\rangle|1s\rangle\big),$$

$$|\psi_2\rangle=\frac{1}{\sqrt{2}}\big(|1s\rangle|2p,m_l=1\rangle-|2p,m_l=1\rangle|1s\rangle\big),$$

$$|\psi_3\rangle=\frac{1}{\sqrt{2}}\big(|1s\rangle|2p,m_l=0\rangle+|2p,m_l=0\rangle|1s\rangle\big),$$

$$|\psi_4\rangle=\frac{1}{\sqrt{2}}\big(|1s\rangle|2p,m_l=0\rangle-|2p,m_l=0\rangle|1s\rangle\big),$$

$$|\psi_5\rangle=\frac{1}{\sqrt{2}}\big(|1s\rangle|2p,m_l=-1\rangle+|2p,m_l=-1\rangle|1s\rangle\big),$$

$$|\psi_6\rangle=\frac{1}{\sqrt{2}}\big(|1s\rangle|2p,m_l=-1\rangle-|2p,m_l=-1\rangle|1s\rangle\big).$$

For the total wave functions to be antisymmetric, we must choose the products of the spin singlet state χ_{00} and the symmetric space wave functions $|\psi_1\rangle,|\psi_3\rangle,|\psi_5\rangle$, forming three singlet states $|\psi_i\rangle\chi_{00}$ ($i=1,3,5$); and the products of the spin triplet states χ_{11} and the antisymmetric space wave functions $|\psi_2\rangle,|\psi_4\rangle,|\psi_6\rangle$, forming nine triplet states $|\psi_i\rangle\chi_{11}$ ($i=2,4,6,m=0,\pm1$). To denote the twelve states in the coupling representation, we must combine the above antisymmetrized wave functions: The wave functions of the three singlet states are

$$^1P_1: \qquad |m_j = 1\rangle = |\psi_1\rangle\chi_{00},$$

$$|m_j = 0\rangle = |\psi_3\rangle\chi_{00},$$

$$|m_j = -1\rangle = |\psi_5\rangle\chi_{00}.$$

The wave functions of the nine triplet states are

$$^3P_2: \qquad |m_j = 2\rangle = |\psi_2\rangle\chi_{11},$$

$$|m_j = 1\rangle = \frac{1}{\sqrt{2}}\big(|\psi_2\rangle\chi_{10} + |\psi_4\rangle\chi_{11}\big),$$

$$|m_j = 0\rangle = \sqrt{\frac{1}{6}}|\psi_2\rangle\chi_{1,-1} + \sqrt{\frac{2}{3}}|\psi_4\rangle\chi_{10} + \sqrt{\frac{1}{6}}|\psi_6\rangle\chi_{11},$$

$$|m_j = -1\rangle = \frac{1}{\sqrt{2}}\big(|\psi_4\rangle\chi_{1,-1} + |\psi_6\rangle\chi_{10}\big),$$

$$|m_j = -2\rangle = |\psi_6\rangle\chi_{1,-1}.$$

$$^3P_1: |m_j = 1\rangle = \frac{1}{\sqrt{2}}\big(|\psi_2\rangle\chi_{10} - |\psi_4\rangle\chi_{11}\big),$$

$$|m_j = 0\rangle = \frac{1}{\sqrt{2}}\big(|\psi_2\rangle\chi_{1,-1} - |\psi_6\rangle\chi_{11}\big),$$

$$|m_j = -1\rangle = \frac{1}{\sqrt{2}}|\psi_4\rangle\chi_{1,-1} - |\psi_6\rangle\chi_{10}).$$

$$^3P_0: |m_j = 0\rangle = \frac{1}{\sqrt{3}}\big(|\psi_2\rangle\chi_{1,-1} - |\psi_4\rangle\chi_{10} + |\psi_6\rangle\chi_{11}\big).$$

b. As the electron cloud of the $2p$ orbits is mainly outside the electron cloud of the $1s$ orbit, the value of Z of the $|1s\rangle$ wave function is 2 and that of the $|2p\rangle$ wave function is 1. The energy levels of a hydrogen-like atom is given by

$$E = -\frac{mZ^2e^4}{2\hbar^2 n^2}.$$

Hence the energy of the $2p$ states above the ground state is

$$\Delta E = -\frac{mc^2}{2}\left(\frac{e^2}{\hbar c}\right)^2\left(\frac{1}{2^2} - \frac{2^2}{1}\right)$$

$$= \frac{0.51\times10^6}{2}\times\left(\frac{1}{137}\right)^2\times\frac{15}{4}$$

$$= 51 \text{ eV}.$$

The optimum Z can be obtained from shielding effect calculations using the given wave functions.

c. Denote the two subsets of symmetric and antisymmetric spatial wave functions with a parameter $\varepsilon = \pm 1$ and write

$$|\psi_\varepsilon\rangle = \frac{1}{\sqrt{2}}\left(|1s\rangle|2p\rangle + \varepsilon|2p\rangle|1s\rangle\right).$$

The repulsive interaction between the electrons,

$$H' = \frac{e^2}{|\mathbf{r}_1 - \mathbf{r}_2|},$$

results in a splitting of the energy levels of the two sets of wave functions. As

$$\langle\psi_\varepsilon|H'|\psi_\varepsilon\rangle = \frac{1}{2}\left(\langle 1s|\langle 2p| + \varepsilon\langle 2p|\langle 1s|\right)H'\left(|1s\rangle|2p\rangle + \varepsilon|2p\rangle|1s\rangle\right)$$

$$= \langle 1s2p|H'|1s2p\rangle + \varepsilon\langle 1s2p|H'|2p1s\rangle,$$

the splitting is equal to twice the exchange integral in the second term of the right-hand side, i.e.,

$$K = \int \psi_{1s}^*(\mathbf{r}_1)\psi_{1s}(\mathbf{r}_2)\psi_{2p}^*(\mathbf{r}_2)\psi_{2p}(\mathbf{r}_1)\frac{e^2}{|\mathbf{r}_1 - \mathbf{r}_2|}d\mathbf{r}_1 d\mathbf{r}_2.$$

As $K > 0$, the energy of the triplet state ($\varepsilon = -1$) is lower than that of the singlet state. (This is to be expected since when the space wave function is antisymmetric, the two electrons having parallel spins tend to avoid each other.)

d. The selection rules for electric dipole radiation transition are $\Delta L = 0, \pm 1$; $\Delta S = 0$; $\Delta J = 0, \pm 1$ and a change of parity. Hence the state that can transit to the ground state $^1S_0$ is the $^1P_1$ state.

e. Such excited states do exist. For example, the $^3P_1$ state of the electronic configuration $2p3p$ can transit to any of the above $^3P_{2,1,0}$ states through electric dipole interaction.

7029

ustify, as well as you can, the following statement: "In the system of two ground state H atoms, there are three repulsive states and one attractive bound) state."

(Wisconsin)

Sol: In the adiabatic approximation, when discussing the motion of the two electrons in the two **H** atoms we can treat the distance between the nuclei as fixed and consider only the wave functions of the motion of the two electrons. For total spin $S=1$, the total spin wave function is symmetric for interchange of the two electrons and so the total space wave function is antisymmetric. Pauli's principle requires the electrons, which in this case have parallel spins, to get away from each other as far as possible. This means that the probability for the two electrons to come near each other is small and the states are repulsive states. As $S=1$ there are three such states. For total spin $S=0$, the space wave function is symmetric. The probability of the electrons being close together is rather large and so the state is an attractive one. As $S=0$, there is only one such state.

7030

In a simplified model for a deuteron the potential energy part of the Hamiltonian is $V = V_a(r) + V_b(r)\mathbf{s}_n \cdot \mathbf{s}_p$. The spin operators for the two spin-1/2 particles are $\mathbf{s}_n$ and $\mathbf{s}_p$; the masses are m_n and m_p; V_a and V_b are functions of the particle separation r.

a. The energy eigenvalue problem can be reduced to a one-dimensional problem in the one variable r. Write out this one-dimensional equation.

b. Given that V_a and V_b both are negative or zero, state (and explain) whether the ground state is singlet or triplet.

(Princeton)

Sol:

a.   In units where $\hbar = 1$, we have for the singlet state $(S=0)$ of the deuteron,

$$\mathbf{s}_n \cdot \mathbf{s}_p = \frac{1}{2}(\mathbf{s}_n + \mathbf{s}_p)^2 - \frac{1}{2}\mathbf{s}_n^2 - \frac{1}{2}\mathbf{s}_p^2 = -\frac{1}{2}\left(\frac{1}{2} \times \frac{3}{2} \times 2\right) = -\frac{3}{4}$$

and the potential energy

$$V_{singlet} = V_a(r) - \frac{3}{4}V_b(r).$$

The Hamiltonian is then

$$H = -\frac{1}{2m_n}\nabla_n^2 - \frac{1}{2m_p}\nabla_p^2 + V_a(r) - \frac{3}{4}V_b(r),$$

whence the Hamiltonian describing the relative motion

$$H_r = -\frac{1}{2m}\nabla_r^2 + V_a(r) - \frac{3}{4}V_b(r),$$

where ∇_r^2 is the Laplacian with respect to the relative position coordinate $r = |\mathbf{r}_p - \mathbf{r}_n|$, $m = \frac{m_n m_p}{m_n + m_p}$ is the reduced mass of the two particles.

After separating out the angular variables from the Schrödinger equation, the energy eigenvalues are obtained from the one-dimensional equation satisfied by the radial wave function $R(r)$:

$$-\frac{1}{2mr}\frac{d^2}{dr^2}(rR) + \left[\frac{l(l+1)}{2mr^2} + V_a(r) - \frac{3}{4}V_b(r)\right](rR) = E(rR).$$

Similarly, for the triplet state $(S = 1)$ we have

$$V_{\text{triplet}} = V_a(r) + \frac{1}{4}V_b(r),$$

and the corresponding one-dimensional equation

$$-\frac{1}{2mr}\frac{d^2}{dr^2}(rR) + \left[\frac{l(l+1)}{2mr^2} + V_a(r) + \frac{1}{4}V_b(r)\right](rR) = E(Rr).$$

b. We shall make use of the lemma: For a one-dimensional problem of energy eigenvalues, if the conditions are all the same except that two potential energies satisfy the inequality

$$V'(x) \geq V(x), \quad (-\infty < x < \infty),$$

then the corresponding energy levels satisfy the inequality $E'_n \geq E_n$. For the ground state, $l = 0$. As $V_b \leq 0$ for a stable deuteron, $V_{\text{singlet}} \geq V_{\text{triplet}}$ and so the triplet state is the ground state.

7031

a. The ground state of the hydrogen atom is split by the hyperfine interaction. Indicate the level diagram and show from first principles which state lies higher in energy.

b. The ground state of the hydrogen molecule is split into total nuclear spin triplet and singlet states. Show from first principles which state lies higher in energy.

(Chicago)

ol:

a. The hyperfine interaction is one between the intrinsic magnetic moment μ_p of the proton nucleus and the magnetic field $\mathbf{B}_e$ arising from the external electron structure, and is represented by the Hamiltonian $H_{hf} = -1\mu_p \cdot \mathbf{B}_e$. For the ground state, the probability density for the electron is spherically symmetric and so $\mathbf{B}_e$ can be considered to be in the same direction as μ_e, the intrinsic magnetic moment of the electron. Then as

$$\mu_e = -\frac{e}{m_e c}\mathbf{s}_e, \ \mu_p = \frac{eg_p}{2m_p c}\mathbf{s}_p, \ (g_p > 0)$$

$\mathbf{B}_e$ is antiparallel to $\mathbf{s}_e$ and $-\langle \mu_p \cdot \mathbf{B}_e \rangle$ has the same sign as $\langle \mathbf{s}_e \cdot \mathbf{s}_p \rangle$.

Let $\mathbf{S} = \mathbf{s}_e + \mathbf{s}_p$ and consider the eigenstates of $\mathbf{S}^2$ and S_z. We have

$$\langle \mathbf{s}_e \cdot \mathbf{s}_p \rangle = \frac{1}{2}\langle \mathbf{S}^2 - \mathbf{s}_e^2 - \mathbf{s}_p^2 \rangle$$

$$= \frac{1}{2}\left[S(S+1)\hbar^2 - \frac{3}{4}\hbar^2 - \frac{3}{4}\hbar^2 \right]$$

$$= \frac{1}{4}[2S(S+1) - 3]\hbar^2.$$

As the spins of electron and proton are both $\frac{1}{2}\hbar$, we can have

$$S = \begin{cases} 0, & \text{singlet state,} \\ 1, & \text{triplet state,} \end{cases}$$

and correspondingly

$$\langle \mathbf{s}_e \cdot \mathbf{s}_p \rangle = \begin{cases} -\dfrac{3}{4}\hbar^2 < 0, & \text{singlet state,} \\ \dfrac{1}{4}\hbar^2 > 0, & \text{triplet state.} \end{cases}$$

The hyperfine interaction causes the ground state to split into two states, $S = 0$ and $S = 1$ (respectively the singlet and triplet total spin states). As H_{n_l} has the same sign as $\langle \mathbf{s}_e \cdot \mathbf{s}_p \rangle$, the energy of the triplet states is higher. The diagram of the energy levels of the ground state is shown in Fig. 7.6.

Physically, hyperfine splitting is caused by the interaction of the intrinsic magnetic moments of the electron and the proton. For the electron the intrinsic magnetic moment is antiparallel to its spin; while for the proton

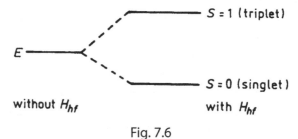

Fig. 7.6

the magnetic moment is parallel to its spin. For the spin triplet, the spins of the electron and the proton are parallel, and so their magnetic moments are antiparallel. For the spin singlet, the reverse is true. If the spatial wave functions are same, the Coulomb energy between the election and proton is higher for the triplet state.

b. For the hydrogen molecule H_2, as protons are fermions, the total wave function must be antisymmetric for interchange of the two protons. Then for the nuclear spin singlet, the rotation quantum number can only be $L = 0, 2, 4 \ldots$, where $L = 0$ has the lowest energy; for the spin triplet, the rotation quantum number can only be $L = 1, 3, 5, \ldots$, where $L = 1$ has the lowest energy. As the energy difference caused by difference of L is larger than that caused by difference of nuclear spins, the energy of the state $L = 1$ (total nuclear spin $S = 1$) is higher than that of the state $L = 0$ (total nuclear spin $S = 0$). So, for the ground state splitting of H_2, the nuclear spin triplet ($S = 1$) has the higher energy.

Because the spatial wave functions of $L = 1$ and $L = 0$ states are antisymmetric and symmetric respectively, the probability for the protons to come close is larger in the latter case than in the former, and so the Coulomb interaction energy is higher (for the same principal quantum number n). However, the difference between the energies of $L = 1$ and $L = 0$ is larger for the rotational energy levels than for the Coulomb energy levels. So for the ground state splitting of hydrogen atom, the nuclear spin triplet ($S = 1$) has the higher energy.

7032

The wave function for a system of two hydrogen atoms can be described approximately in terms of hydrogenic wave functions.

a. Give the complete wave functions for the lowest states of the system for singlet and triplet spin configurations. Sketch the spatial part of each wave function along a line through the two atoms.

b. Sketch the effective potential energy for the atoms in the two cases as functions of the internuclear separation. (Neglect rotation of the system.) Explain the physical origin of the main features of the curves, and of any differences between them.

(Wisconsin)

Sol: The Hamiltonian of the system of two hydrogen atoms can be written as $H = H_n + H_e$, and correspondingly the total wave function is $\psi = \psi_n \phi$, consisting of a nuclear part ψ_n and an electron part ϕ, with

$$\psi_n = \begin{cases} R_v(r)Y_{Im}(\theta, \varphi)\chi_0, & \text{for} \quad I = \text{even, (para--hydrogen)}, \\ R_v(r)Y_{Im}(\theta, \varphi)\chi_1, & \text{for} \quad I = \text{odd, (ortho--hydrogen)}, \end{cases}$$

where v denotes vibration, I denotes rotation quantum numbers, and χ_0, χ_1 are nuclear spin singlet and triplet wave functions.

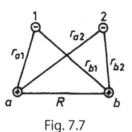

Fig. 7.7

a. The configuration of the system is shown in Fig. 7.7. The wave function of a single electron is taken to be approximately

$$\varphi(r) = \frac{1}{\sqrt{\pi}} \left(\frac{\lambda}{a} \right)^{3/2} e^{-\lambda r/a}.$$

Note that when $\lambda = 1$, $\varphi(r)$ is the wave function of an electron in the ground state of a hydrogen atom. For two electrons, the lowest singlet state wave function is

$$\phi_s = \frac{1}{\sqrt{2}}[\varphi(r_{a1})\varphi(r_{b2}) + \varphi(r_{a2})\varphi(r_{b1})]\chi_{0e},$$

and the lowest triplet state wave function is

$$\phi_t = \frac{1}{\sqrt{2}}[\varphi(r_{a1})\varphi(r_{b2}) - \varphi(r_{a2})\varphi(r_{b1})]\chi_{1e},$$

where χ_{0e} and χ_{1e} are electron spin singlet and triplet wave functions. Taking the x-axis along ab with the origin at a, we can express the spatial parts of ϕ_s and ϕ_t by

$$\phi_s = b(e^{-k|x_1|}e^{-k|R-x_2|} + e^{-k|x_2|}e^{-k|R-x_1|}),$$
$$\phi_t = b(e^{-k|x_1|}e^{-k|R-x_2|} - e^{-k|x_2|}e^{-k|R-x_1|}).$$

Keeping one variable (say x_2) fixed, we sketch the variation of the spatial wave functions with the other variable in Fig. 7.8. It is seen that if one electron gets close to a nucleus, the probability is large for the other electron to be close to the other nucleus.

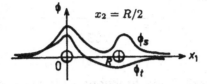

Fig. 7.8

b.

$$V = -\left(\frac{1}{r_{a1}} + \frac{1}{r_{a2}} + \frac{1}{r_{b1}} + \frac{1}{r_{b2}}\right)e^2$$

$$+ \frac{e^2}{r_{12}} + \frac{e^2}{R} + V_0.$$

The effective potential energy, $\bar{V} \equiv \langle \phi | V | \phi \rangle$, for the ground state as a function of R/a is shown in Fig. 7.9. It is seen that the potential energy vanishes when the neutral atoms are infinitely far apart: $R \to \infty$, $\bar{V} \to 0$. When $R \to 0$, the potential energy between the two hydrogen nuclei becomes infinitely large while that between the electrons and the nuclei is finite, similar to the electron potential energy of a He atom. Hence $R \to 0$,

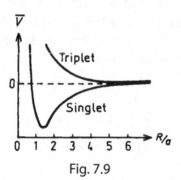

Fig. 7.9

$\bar{V} \to +\infty$. As R decreases from a large value, the repulsive potential between the nuclei increases, at the same time the attractive potential between the electrons and the nuclei increases also, competing against each other. For the singlet state, the probability that the electrons are close to the adjacent nuclei is large, and so the potential has a minimum value. For the triplet state, the probability that the electrons are close to the nuclei is small, and so the decrease of the potential energy due to the attractive force between the electrons and

nuclei, which is negative, has a small value, and the repulsive potential between
the nuclei, which is positive, is the main part of the total potential. Therefore
$\bar{V} > 0$ and no minimum occurs.

7033

a. Using hydrogen atom ground state wave functions (including the
electron spin), write wave functions for the hydrogen molecule which
satisfy the Pauli exclusion principle. Omit terms which place both electrons on the same nucleus. Classify the wave functions in terms of their
total spin.

b. Assuming that the only potential energy terms in the Hamiltonian arise
from Coulomb forces, discuss qualitatively the energies of the above
states at the normal internuclear separation in the molecule and in the
limit of very large internuclear separation.

c. What is meant by an "exchange force"?

(Wisconsin)

Sol:

a. The configuration of a hydrogen molecule is as shown in Fig. 7.7. Denote the
ground state wave function of hydrogen atom by $|100\rangle$ and let $\varphi(\mathbf{r}) = (|100\rangle)^\lambda$,
where λ is a parameter to be determined. Then the singlet state $(S = 0)$ wave
function of hydrogen molecule is

$$\psi_1 = \frac{1}{\sqrt{2}}[\varphi(\mathbf{r}_{a1})\varphi(\mathbf{r}_{b2}) + \varphi(\mathbf{r}_{a2})\varphi(\mathbf{r}_{b1})]\chi_{00},$$

and the triplet state $(S = 1)$ wave functions are

$$\psi_3 = \frac{1}{\sqrt{2}}[\varphi(\mathbf{r}_{a1})\varphi(\mathbf{r}_{b2}) - \varphi(\mathbf{r}_{a2})\varphi(\mathbf{r}_{b1})]\chi_{1M}$$

with $M = -1, 0, 1$.

b. The energy of a hydrogen atom is

$$E_{H_2} = -\frac{me^4}{2\hbar^2}\frac{1}{n^2} = -\frac{mc^2}{2}\left(\frac{e^2}{\hbar c}\right)^2\frac{1}{n^2}$$

$$= -\frac{0.511 \times 10^6}{2} \times \left(\frac{1}{137}\right)^2\frac{1}{n^2}$$

$$= -\frac{13.6}{n^2}\text{ eV.}$$

Thus the sum of the energies of two separate ground-state hydrogen atoms is $-2 \times 13.6 = -27.2$ eV. On the other hand, for the He atom which also contains two protons and two electrons, the ground state energy is

$$E_{He} = -2 \times \frac{mZ'^2 e^4}{2\hbar^2} = -2 \times 13.6 \times \left(2 - \frac{5}{16}\right)^2$$
$$= -77.5 \text{ eV}.$$

where the factor 2 is for the two electrons of He atom and $Z' = 2 - \frac{5}{16}$ is the effective charge number of the He nucleus.

i.   For the singlet state, the probability for the two electrons to be close to each other is rather large (on account of the Pauli principle), which enhances the repulsive exchange potential energy between them. The probability that the two electrons are near to the two nuclei is also large which tends to increase the attractive exchange potential. Taking both into account the exchange interaction potential lowers the energy. It is easily seen that for the singlet state, -77.5 eV $< E_1 < -27.2$ eV. For the triplet state, the spins are parallel and so the spatial wave function is antisymmetric. In this case the potential energy is increased by the exchange interaction and so $E_3 > -27.2$ eV, which makes it difficult to form a bound state.

ii. When the distance between the nuclei $\to \infty$, H_2 reduces to two separate hydrogen atoms. Hence the energy $\to -27.2$ eV.

c.   The symmetrization or antisymmetrization of the wave function causes a mean shift of the potential energy by

$$\Delta V = \iint \varphi(\mathbf{r}_{a1}) \varphi(\mathbf{r}_{b2}) V \varphi(\mathbf{r}_{a2}) \varphi(\mathbf{r}_{b1}) d\tau_1 d\tau_2.$$

This is said to be caused by an "exchange force".

7034

Describe the low-lying states of the H_2 molecule. Give a rough value for their excitation energies. Characterize the radiative transitions of the first two excited states to the ground state.

(Wisconsin)

Sol: In an approximate treatment of hydrogen atom, the zero order wave function is taken to be the product of two ground state hydrogen-like wave functions, which have the form

$$\varphi(\mathbf{r}) = \frac{1}{\sqrt{\pi}} \left(\frac{\lambda}{a_0} \right) e^{-\lambda r / a_0},$$

where a_0 is the Bohr radius, λ is a parameter to be determined. The spin part of the electron wave function of the H_2 molecule ground state $(S = 0)$ is antisymmetric which requires the spatial part to be symmetric. As the spins of the two electrons are antiparallel, they can get quite close to each other (Pauli's principle). This means that the density of "electron cloud" is rather large in the region of space between the two nuclei. In this region, the attractive potential between the two electrons and the two nuclei is quite large and thus can form a bound state, with wave function

$$\psi = \frac{1}{\sqrt{2}} [\varphi(\mathbf{r}_{a1})\varphi(\mathbf{r}_{b2}) + \varphi(\mathbf{r}_{a2})\varphi(\mathbf{r}_{b1})]\chi_{00},$$

where the variables are as shown in Fig. 7.7.

If the spins of the electrons are parallel $(S = 1)$, then the spatial wave function must be antisymmetric, the probability that the two electrons getting close is small, and no bound state occurs.

Of the energy levels related to the electronic, vibrational and rotational motions of H_2, the rotational levels have the smallest spacing between two adjacent levels. For simplicity, we shall only consider rotational energy levels with the electrons in the ground state initially and in the absence of vibration between the nuclei. With no loss of generality, we can take the molecule's energy to be zero when there is no rotation. The rotational levels are given by

$$E = \frac{\hbar^2}{2I} J(J+1),$$

where I is the moment of inertia and J is the total angular momentum of the two-nuclei system. When $J = $ even, the total spin of the two protons in H_2 is $S = 0$ and para-hydrogen results; when $J = $ odd, the total spin of the two protons is $S = 1$ and ortho-hydrogen results. Suppose the distance between the two protons is $R = 1.5 \times 0.53 = 0.80$ Å (Fig. 7.9). As

$$\frac{\hbar^2}{2I} = \frac{\hbar^2}{\mu R^2} = \frac{1}{\mu c^2} \left(\frac{\hbar c}{R} \right)^2$$

$$= \frac{2}{938 \times 10^6} \left(\frac{6.582 \times 10^{-16} \times 3 \times 10^{10}}{0.8 \times 10^{-8}} \right)^2$$

$$= 1.3 \times 10^{-2} \text{ eV},$$

the energies of the low-lying states are as follows.

Para-hydrogen :	J	0	2	4
	$E(10^{-2}\text{eV})$	0	7.8	26.0

Ortho-hydrogen :	J	1	3	5
	$E(10^{-2}\,eV)$	2.6	15.6	39.0

As the interactions between two atoms are spin-independent, para-hydrogen and ortho-hydrogen cannot transform to each other, hence the selection rule $\Delta J = even$.

In nature the ratio of the number of molecules of ortho-hydrogen to that of para-hydrogen is 3:1. This means that the spectral line for $J = 2 \rightarrow J = 0$ is weaker than for $J = 3 \rightarrow J = 1$.

7035

The density matrix for a collection of atoms of spin J is ρ. If these spins are subject to a randomly fluctuating magnetic field, it is found that the density matrix relaxes according to the following equation:

$$\frac{\partial \rho}{\partial t} = \frac{1}{T}[\mathbf{J}_{op} \cdot \rho \mathbf{J}_{op} - J(J+1)\rho].$$

Prove that the relaxation equation implies the following:

a.

$$\frac{\partial}{\partial t}\langle J_z \rangle = \frac{\partial}{\partial t}\mathrm{Tr}(J_{op}\rho) = -\frac{1}{T}\langle J_z \rangle,$$

b.

$$\frac{\partial}{\partial t}\langle J_z^2 \rangle = \frac{\partial}{\partial t}\mathrm{Tr}(J_z^2\rho) = -\frac{3}{T}\langle J_z^2 \rangle + \frac{J(J+1)}{T}.$$

[Hint: Raising and lowering operators are useful here]

(Columbia)

Sol: From definition, $\langle J_z \rangle = \mathrm{Tr}(\rho J_z)$. Thus the following:

$$\frac{\partial}{\partial t}\langle J_z \rangle = \mathrm{Tr}\left(\frac{\partial \rho}{\partial t}J_z\right) = \frac{1}{T}\mathrm{Tr}(\mathbf{J}_{op} \cdot \rho \mathbf{J}_{op}J_z - J(J+1)\rho J_z]$$

$$= \frac{1}{T}\mathrm{Tr}[J_x\rho J_x J_z + J_y\rho J_y J_z + J_z\rho J_z^2 - J(J+1)\rho J_z].$$

As

$$\mathrm{Tr}\,AB = \mathrm{Tr}\,BA,$$

$$J_x J_y - J_y J_x = iJ_z,$$
$$J_y J_z - J_z J_y = iJ_x,$$
$$J_z J_x - J_x J_z = iJ_y,$$
$$\mathbf{J}^2 J_z = J(J+1)J_z,$$

(using units in which $\hbar = 1$) we have

$$\frac{\partial}{\partial t}\langle J_z \rangle = \frac{1}{T}\mathrm{Tr}[\rho J_x J_z J_x + \rho J_y J_z J_y + \rho J_z^3 - pJ(J+1)J_z]$$

$$= \frac{1}{T}\mathrm{Tr}\{\rho[(J_x^2 + J_y^2 + J_z^2)J_z + iJ_x J_y - iJ_y J_x - J(J+1)J_z]\}$$

$$= -\frac{1}{T}\mathrm{Tr}(\rho J_z) = -\frac{1}{T}\langle J_z \rangle.$$

b.

$$\frac{\partial}{\partial t}\langle J_z^2 \rangle = \frac{1}{T}\mathrm{Tr}\left(\frac{\partial \rho}{\partial t}J_z^2\right)$$

$$= \frac{1}{T}\mathrm{Tr}[J_x \rho J_x J_z^2 + J_y \rho J_y J_z^2 + J_z \rho J_z^3 - J(J+1)\rho J_z^2]$$

$$= \frac{1}{T}\mathrm{Tr}[\rho J_x J_z^2 J_x + \rho J_y J_z^2 J_y + \rho J_z^4 - \rho J(J+1)J_z^2]$$

$$= \frac{1}{T}\mathrm{Tr}[p(J_x J_z J_x J_z + J_y J_z J_y J_z + iJ_x J_z J_y - iJ_y J_z J_x + J_z^4)$$
$$- J(J+1)\rho J_z^2]$$

$$= \frac{1}{T}\mathrm{Tr}\{\rho[J_x^2 J_z^2 + J_y^2 J_z^2 + J_z^4 + iJ_x J_y J_z - iJ_y J_x J_z + iJ_x J_z J_y$$
$$- iJ_y J_z J_x - J(J+1)J_z^2]\}$$

$$= \frac{1}{T}\mathrm{Tr}\{\rho[i(iJ_z)J_z + iJ_x(J_y J_z) + iJ_x(-iJ_x) - iJ_y(J_x J_z)$$
$$- iJ_y(iJ_y)]\}$$

$$= \frac{1}{T}\mathrm{Tr}\{\rho[-J_z^2 + J_x^2 + J_y^2 + i(iJ_z)J_z]\}$$

$$= \frac{1}{T}\mathrm{Tr}\{\rho[-3J_z^2 + (J_x^2 + J_y^2 + J_z^2)]\}$$

$$= -\frac{3}{T}\langle J_z^2 \rangle + \frac{J(J+1)}{T},$$

since

$$\frac{1}{T}\mathrm{Tr}[\rho J(J+1)] = \langle J(J+1) \rangle = J(J+1).$$

7036

A molecule is made up of three identical atoms at the corners of an equilateral triangle as shown in Fig. 7.10. We consider its ion to be made by adding one electron with some amplitude on each site. Suppose the matrix element of the Hamiltonian for the electron on two adjacent sites i, j is $\langle i|H|j\rangle = -a$ for $i \neq j$.

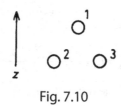

Fig. 7.10

a. Calculate the energy splittings.

b. Suppose an electric field in the z direction is applied, so that the potential energy for the electron on top is lowered by b with $|b| \ll |a|$. Now calculate the levels.

c. Suppose the electron is in the ground state. Suddenly the field is rotated by $120°$ and points toward site 2. Calculate the probability for the electron to remain in the ground state.

(Princeton)

Sol:

a. Denote the basis vectors by $|1\rangle$, $|2\rangle$, $|3\rangle$ and let $\langle i|H|i\rangle = E_0$, $i = 1, 2, 3$. Then

$$H = \begin{pmatrix} E_0 & -a & -a \\ -a & E_0 & -a \\ -a & -a & E_0 \end{pmatrix}.$$

To diagonalize H, solve

$$\begin{vmatrix} E_0 - \lambda & -a & -a \\ -a & E_0 - \lambda & -a \\ -a & -a & E_0 - \lambda \end{vmatrix} = 0.$$

The solution gives energy levels $E_{1,2} = E_0 + a$ (two-fold degenerate) and $E_3 = E_0 - 2a$.

b. The H matrix is now

$$H = \begin{pmatrix} E_0 - b & -a & -a \\ -a & E_0 & -a \\ -a & -a & E_0 \end{pmatrix}.$$

Its diagonalization gives energy levels

$$E_1 = E_0 + a,$$

$$E_2 = E_0 - \frac{a+b+\sqrt{(a-b)^2 + 8a^2}}{2},$$

$$E_3 = E_0 - \frac{a+b-\sqrt{(a-b)^2 + 8a^2}}{2}.$$

E_2 has the lowest energy and thus corresponds to the ground state, with wave function

$$\psi_0 = \frac{1}{\sqrt{(E_0 - E_2 - a)^2 + 2a^2}} [(E_0 - E_2 - a)|1\rangle + a|2\rangle + a|3\rangle].$$

c. After the rotation of the field the system has the same configuration as before but the sites are renamed:

$$1 \to 2,\ 2 \to 3,\ 3 \to 1.$$

Hence the new ground state is

$$\psi_0' = \frac{1}{\sqrt{(E_0 - E_2 - a)^2 + 2a^2}} [a|1\rangle + (E_0 - E_2 - a)|2\rangle + a|3\rangle].$$

Hence the probability for the electron to remain in the ground state is

$$|\langle \psi_0' | \psi_0 \rangle|^2 = \left[\frac{2a(E_0 - E_2 - a) + a^2}{(E_0 - E_2 - a)^2 + 2a^2} \right]^2.$$

7037

Consider three particles, each of mass m, moving in one dimension and bound to each other by harmonic forces, i.e.,

$$V = \frac{1}{2}[(x_1 - x_2)^2 + (x_2 - x_3)^2 + (x_3 - x_1)^2].$$

a. Write the Schrödinger equation for the system.

b. Transform to a center-of-mass coordinate system in which it is apparent that the wave functions and eigenenergies may be solved for exactly.

c. Using (b) find the ground state energy if the particles are identical bosons.

d. What is the ground state energy if the particles are identical spin $-1/2$ fermions?

(Wisconsin)

ol:

a. As

$$E = \sum_{i=1}^{3} \frac{p_i^2}{2m} + V,$$

The Schrödinger equation is

$$i\hbar \frac{\partial \psi}{\partial t} = -\frac{\hbar^2}{2m} \left(\frac{\partial^2}{\partial x_1^2} + \frac{\partial^2}{\partial x_2^2} + \frac{\partial^2}{\partial x_3^2} \right) \psi$$
$$+ \frac{k}{2}[(x_1 - x_2)^2 + (x_2 - x_3)^2 + (x_3 - x_1)^2]\psi.$$

b. Using the Jacobi coordinates

$$\begin{cases} y_1 = x_1 - x_2, \\ y_2 = \dfrac{x_1 + x_2}{2} - x_3, \\ y_3 = \dfrac{x_1 + x_2 + x_3}{3}, \end{cases}$$

or

$$\begin{cases} x_1 = y_3 + \dfrac{y_1}{2} + \dfrac{y_2}{3}, \\ x_2 = y_3 - \dfrac{y_1}{2} + \dfrac{y_2}{3}, \\ x_3 = y_3 - \dfrac{2}{3}y_2, \end{cases}$$

we have

$$V = \frac{k}{2}\left(\frac{3}{2}y_1^2 + 2y_2^2 \right),$$

$$\sum_{i=1}^{3} \frac{p_i^2}{2m} = -\frac{\hbar^2}{2m}\left(\frac{1}{3}\frac{\partial^2}{\partial y_3^2} + 2\frac{\partial^2}{\partial y_1^2} + \frac{3}{2}\frac{\partial^2}{\partial y_2^2} \right),$$

and hence the stationary eigenequation

$$E_T \psi = -\frac{\hbar^2}{6m}\frac{\partial^2 \psi}{\partial y_3^2} - \frac{\hbar^2}{2m}\left\{2\frac{\partial^2}{\partial y_1^2} + \frac{3}{2}\frac{\partial^2}{\partial y_2^2}\right\}\psi$$

$$+ \frac{k}{2}\left\{\frac{3}{2}y_1^2 + 2y_2^2\right\}\psi.$$

Try

$$\psi = Y(y_3)\phi(y_1, y_2).$$

The equation is separated into two equations:

$$\begin{cases} -\dfrac{\hbar^2}{6m}\dfrac{\partial^2 Y}{\partial y_3^2} = E_c Y, \\[2ex] -\dfrac{\hbar^2}{2m}\left(2\dfrac{\partial^2}{\partial y_1^2} + \dfrac{3}{2}\dfrac{\partial^2}{\partial y_2^2}\right)\phi + \dfrac{k}{2}\left(\dfrac{3}{2}y_1^2 + 2y_2^2\right)\phi = E\phi, \end{cases}$$

where $E_c = E - E_T$ is the energy due to the motion of the center of mass. The first equation gives

$$Y = \frac{1}{\sqrt{2\pi}}e^{i\sqrt{6mE_c}\,y_3/\hbar}.$$

With

$$\phi = \phi_1(y_1)\phi_2(y_2)$$

the second equation is separated into two equations

$$\begin{cases} -\dfrac{\hbar^2}{m}\dfrac{\partial^2 \phi_1}{\partial y_1^2} + \dfrac{3}{4}ky_1^2\phi_1 = E_1\phi_1, \\[2ex] -\dfrac{3\hbar^2}{4m}\dfrac{\partial^2 \phi_2}{\partial y_2^2} + ky_2^2\phi_2 = E_2\phi_2. \end{cases}$$

where $E = E_1 + E_2$.

These are equations for harmonic oscillators of masses $\frac{m}{2}, \frac{2m}{3}$ and force constants $2k$ and $3k$ respectively, both having the same angular frequency $\omega = \sqrt{\frac{3k}{m}}$. Hence the total energy is

$$E = E_1 + E_2 = \left(n + \frac{1}{2}\right)\hbar\sqrt{\frac{3k}{m}} + \left(l + \frac{1}{2}\right)\hbar\sqrt{\frac{3k}{m}},$$

with $n, l = 0, 1, 2, 3, \ldots$.

c. Let $\alpha^2 = \frac{mw}{\hbar}$. The ground state wave functions of ϕ_1, ϕ_2 are

$$\phi_{10}(y_1) = \left(\frac{1}{2\pi}\right)^{1/4} \sqrt{\alpha} \exp\left(-\frac{1}{4}\alpha^2 y_1^2\right),$$

$$\phi_{20}(y_2) = \left(\frac{2}{3\pi}\right)^{1/4} \sqrt{\alpha} \exp\left(-\frac{1}{3}\alpha^2 y_2^2\right),$$

and so

$$\phi_0(y_1, y_2) = \phi_{10}(y_1)\phi_{20}(y_2) = \left(\frac{1}{3\pi^2}\right)^{1/4} \alpha \exp\left[\frac{-\alpha^2}{12}\left(3y_1^2 + 4y_2^2\right)\right],$$

where

$$3y_1^2 + 4y_2^2 = 3(x_1 - x_2)^2 + (x_1 + x_2 - 2x_3)^2$$
$$= 4(x_1^2 + x_2^2 + x_3^2 - x_1 x_2 - x_2 x_3 - x_3 x_1).$$

As $y_3 = \frac{1}{3}(x_1 + x_2 + x_3)$ it is obvious that the spatial wave function ψ_0 is symmetric for the interchange of any two of the particles, which is required as the bosons are identical. The ground state energy of the three bosons, excluding the translational energy of the center of mass, is

$$E_0 = \frac{1}{2}\hbar\sqrt{\frac{3k}{m}} + \frac{1}{2}\hbar\sqrt{\frac{3k}{m}} = \hbar\sqrt{\frac{3k}{m}}.$$

d. If the particles are identical spin-1/2 fermions, as spin is not involved in the expression for the Hamiltonian, the eigenfunction is a product of the spatial wave function and the spin wave function, and must be antisymmetric for interchange of particles.

For the coordinate transformation in (b), we could have used

$$\begin{cases} y_1' = x_2 - x_3, \\ y_2' = \dfrac{x_2 + x_3}{2} - x_1, \\ y_3' = \dfrac{x_1 + x_2 + x_3}{3} \end{cases}$$

and still obtain the same result. In this case the spatial eigenfunction is

$$\psi(y_1', y_2', y_3') = \phi_{1n}(y_1')\phi_{2l}(y_2')Y(y_3')$$

and the energy is

$$E = (n + l + 1)\hbar\sqrt{\frac{3k}{m}}.$$

Since $\phi_{10}(y_1)\phi_{20}(y_2) = \phi_{10}(y_1')\phi_{20}(y_2')$, the spatial wave function is symmetric for the interchange of two particles. However, for three spin-1/2 fermions it is not possible to construct a spin wave function which is antisymmetric. Hence this state cannot be formed for three spin-1/2 fermions and higher states are to be considered.

Looking at the wave functions of a harmonic oscillator, we see that the exponential part of $\phi_{1n}(y_1)\phi_{2l}(y_2)$ is the same as that of $\phi_{10}(y_1)\phi_{20}(y_2)$ and is symmetric. Let

$$\Phi_1 = \phi_{11}(y_1)\phi_{20}(y_2),$$
$$\Phi_2 = \phi_{11}(y_1')\phi_{20}(y_2').$$

and construct the total wave function

$$\Phi = \Phi_1 \begin{pmatrix} 1 \\ 0 \end{pmatrix}_1 \begin{pmatrix} 1 \\ 0 \end{pmatrix}_2 \begin{pmatrix} 0 \\ 1 \end{pmatrix}_3 + \Phi_2 \begin{pmatrix} 0 \\ 1 \end{pmatrix}_1 \begin{pmatrix} 1 \\ 0 \end{pmatrix}_2 \begin{pmatrix} 1 \\ 0 \end{pmatrix}_3$$

$$- (\Phi_2 + \Phi_1) \begin{pmatrix} 1 \\ 0 \end{pmatrix}_1 \begin{pmatrix} 0 \\ 1 \end{pmatrix}_2 \begin{pmatrix} 1 \\ 0 \end{pmatrix}_3 .$$

As $\Phi_1 = C(x_1 - x_2)$, $\dot{\Phi_2} = C(x_2 - x_3)$, $\Phi_1 + \Phi_2 = C(x_1 - x_3)$, where C is symmetric for interchange of the particles, Φ is antisymmetric as required for a system of identical fermions. Hence the ground state energy of the system, excluding the translational energy of the center of mass, is

$$E_0 = 2\hbar\sqrt{\frac{3k}{m}} .$$

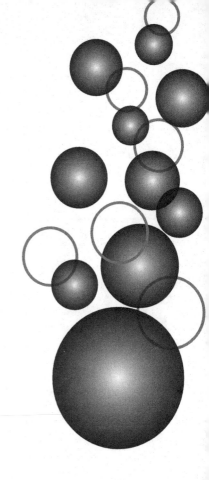

Part VIII
Miscellaneous Topics

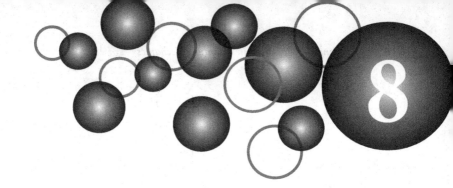

xpress $e^{\begin{pmatrix} 0 & a \\ -a & 0 \end{pmatrix}}$ as a 2×2 matrix; a is a positive constant.

(Berkeley)

ol 1: Let

$$S(a) = e^{\begin{pmatrix} 0 & a \\ -a & 0 \end{pmatrix}} = e^{a\begin{pmatrix} 0 & 1 \\ -1 & 0 \end{pmatrix}} = e^{aA}$$

with

$$A \equiv \begin{pmatrix} 0 & 1 \\ -1 & 0 \end{pmatrix}.$$

As

$$A^2 = \begin{pmatrix} 0 & 1 \\ -1 & 0 \end{pmatrix}\begin{pmatrix} 0 & 1 \\ -1 & 0 \end{pmatrix} = -\begin{pmatrix} 1 & 0 \\ 0 & 1 \end{pmatrix} = -I,$$

I being the unit matrix, we have

$$\frac{d}{da}S(a) = AS(a),$$

$$\frac{d^2}{da^2}S(a) = A^2 S(a) = -S(a),$$

and thus

$$S''(a) + S(a) = 0.$$

The general solution is

$$S(a) = c_1 e^{ia} + c_2 e^{-ia},$$

subject to boundary conditions $S(0) = I, S'(0) = A$

Hence

$$\begin{cases} c_1 + c_2 = I, \\ c_1 - c_2 = -iA, \end{cases}$$

giving

$$\begin{cases} c_1 = \dfrac{I - iA}{2}, \\ c_2 = \dfrac{I + iA}{2}. \end{cases}$$

Therefore

$$S(a) = \frac{I - iA}{2} e^{ia} + \frac{I + iA}{2} e^{-ia}$$

$$= \frac{I}{2}\left(e^{ia} + e^{-ia}\right) + \frac{iA}{2}\left(e^{-ia} - e^{ia}\right) = I \cos a + A \sin a$$

$$= \begin{pmatrix} \cos a & \sin a \\ -\sin a & \cos a \end{pmatrix}.$$

Sol 2: Let $A = \begin{pmatrix} 0 & 1 \\ -1 & 0 \end{pmatrix}$. As $A^2 = -\begin{pmatrix} 1 & 0 \\ 0 & 1 \end{pmatrix} = -I$, $A^3 = -A$, $A^4 = I, \ldots$

$$e^{aA} = \sum_{n=0}^{\infty} \frac{a^n A^n}{n!} = \sum_{k=0}^{\infty} \frac{a^{2k}(-1)^k}{(2k)!} I + \sum_{k=0}^{\infty} \frac{a^{2k+1}(-1)^k}{(2k+1)!} A$$

$$= \cos a I + \sin a A = \begin{pmatrix} \cos a & \sin a \\ -\sin a & \cos a \end{pmatrix}.$$

8002

a. Sum the series $y = 1 + 2x + 3x^2 + 4x^3 + \cdots, |x| < 1$.

b. If $f(x) = xe^{-x/\lambda}$ over the interval $0 < x < \infty$, find the mean and most probable values of x. $f(x)$ is the probability density of x.

c. Evaluate $I = \int_0^{\infty} \dfrac{dx}{4 + x^4}$.

d. Find the eigenvalues and normalized eigenvectors of the matrix

$$\begin{pmatrix} 1 & 2 & 4 \\ 2 & 3 & 0 \\ 5 & 0 & 3 \end{pmatrix}.$$

Are the eigenvectors orthogonal? Comment on this.

(Chicago)

ol:

a. As $|x| < 1$,

$$y - xy = 1 + x + x^2 + x^3 + \cdots = \frac{1}{(1-x)},$$

or

$$y = \frac{1}{(1-x)^2}.$$

b. The mean value of x is

$$\bar{x} = \int_0^\infty x f(x)\,dx \Big/ \int_0^\infty f(x)\,dx$$

$$= \int_0^\infty x \cdot x e^{-x/\lambda}\,dx \Big/ \int_0^\infty x e^{-x/\lambda}\,dx$$

$$= \lambda \frac{\Gamma(3)}{\Gamma(2)} = 2\lambda.$$

The probability density is an extremum when

$$f'(x) = e^{-x/\lambda} - \frac{1}{\lambda} x e^{-x/\lambda} = 0,$$

i.e. at $x = \lambda$ or $x \to \infty$. Note that $\lambda > 0$ if $f(x)$ is to be finite in $0 < x < \infty$. As

$$f''(\lambda) = -\frac{1}{\lambda} e^{-1} < 0, f(\lambda) = \lambda e^{-1} > \lim_{x \to \infty} f(x) = 0,$$

the probability density is maximum at $x = \lambda$. Hence the most probable value of x is λ

c. Consider the complex integral

$$\int_c \frac{dz}{4+z^4} = \int_{c_1} \frac{dz}{4+z^4} + \int_{c_2} \frac{dz}{4+z^4}$$

along the contour $c = c_1 + c_2$ as shown in Fig. 8.1.

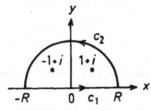

Fig. 8.1

The integrand has singular points $-1+i$, $1+i$ inside the closed contour c. Hence the residue theorem gives

$$\oint_c \frac{dz}{4+z^4} = 2\pi i [\text{Res}(1+i) + \text{Res}(-1+i)]$$

$$= 2\pi i \left(-\frac{1+i}{16} - \frac{-1+i}{16} \right) = \frac{\pi}{4}.$$

Now let $R \to \infty$, we have

$$\int_{c_2} \frac{dz}{4+z^2} \to 0.$$

Then as

$$\int_{c_1} \frac{dz}{4+z^4} = \int_{-\infty}^0 \frac{dx}{4+x^4} + \int_0^\infty \frac{dx}{4+x^4} = 2\int_0^\infty \frac{dx}{4+x^4},$$

we have

$$\int_0^\infty \frac{dx}{4+x^4} = \frac{\pi}{8}.$$

d. Let the eigenvalue be E and the eigenvector be

$$X = \begin{pmatrix} x_1 \\ x_2 \\ x_3 \end{pmatrix}.$$

Then

$$\begin{pmatrix} 1 & 2 & 4 \\ 2 & 3 & 0 \\ 5 & 0 & 3 \end{pmatrix} \begin{pmatrix} x_1 \\ x_2 \\ x_3 \end{pmatrix} = E \begin{pmatrix} x_1 \\ x_2 \\ x_3 \end{pmatrix}.$$

For non-vanishing X, we require

$$\begin{vmatrix} E-1 & -2 & -4 \\ -2 & E-3 & 0 \\ -5 & 0 & E-3 \end{vmatrix} = 0.$$

The solution is

$$E_1 = 3,\, E_2 = -3,\, E_3 = 7.$$

Substitution in the matrix equation gives the eigenvectors, which, after normalization, are

$$X_1 = \frac{1}{\sqrt{5}} \begin{pmatrix} 0 \\ 2 \\ -1 \end{pmatrix}$$

for $E = E_1$ and

$$X_2 = \frac{1}{\sqrt{65}}\begin{pmatrix} -6 \\ 2 \\ 5 \end{pmatrix},$$

$$X_3 = \frac{1}{3\sqrt{5}}\begin{pmatrix} 4 \\ 2 \\ 5 \end{pmatrix}$$

for $E = E_2, E_3$. Note that these eigenvectors are not orthogonal. Generally, only for a Hermition matrix are the eigenvectors corresponding to different eigenvalues orthogonal.

8003

Please indicate briefly (in one sentence) what contributions to physics are associated with the following pairs of names. (Where applicable write an appropriate equation.)

a. Franck–Hertz
b. Davisson–Germer
c. Breit–Wigner
d. Hartree–Fock
e. Lee–Yang
f. Dulong–Petit
g. Cockroft–Walton
h. Hahn–Strassmann
i. Ramsauer–Townsend
j. Thomas–Fermi

(Berkeley)

Sol:

a. Franck and Hertz verified experimentally the existence of discrete energy levels of an atom.

b. Davisson and Germer verified the wave properties of electrons by demonstrating their diffraction in a crystal.

c. Breit and Wigner discovered the Breit–Wigner resonance formula in nuclear physics.

d. Hartree and Fock developed a self-consistent field method for obtaining approximate many-electron wave functions.

e. Lee and Yang proposed the non-conservation of parity in weak interactions.

f. Dulong and Petit discovered that atomic heat is the same for all solids at high temperatures, being equal to $3R$, R being the ideal gas constant.

g. Cockroft and Walton effected the first artificial disintegration of an atomic nucleus.

h. Hahn and Strassmann first demonstrated the fission of uranium by neutrons.

i. Ramsauer and Townsend first observed the resonant transmission of low energy electrons through rare-gas atoms.

j. Thomas and Fermi proposed an approximate statistical model for the structure of metals.

8004

Give estimates of magnitude order for the following quantities.

a. The kinetic energy of a nucleon in a typical nucleus.

b. The magnetic field in gauss required to give a Zeeman splitting in atomic hydrogen comparable to the Coulomb binding energy of the ground state.

c. The occupation number n of the harmonic oscillator energy eigenstate that contributes most to the wave function of a classical onedimensional oscillator with mass $m = 1$ gram, period $T = 1$ sec, amplitude $x_0 = 1$ cm.

d. The ratio of the hyperfine structure splitting to the binding energy in the 1s state of atomic hydrogen, expressed in terms of the fine structure constant α, the electron mass m_e, and the proton mass m_p.

(Berkeley)

ol:

a. The kinetic energy $T = \dfrac{p^2}{2m}$ of a nucleon in a nucleus can be estimated using the approximation $p \sim \Delta p$ and the uncertainty principle $\Delta x \Delta p \sim h$. As $\Delta x \sim 10^{-12}\,\text{cm}$, $\Delta p \sim \dfrac{h}{\Delta x}$,

$$T \sim \frac{h^2}{2m}\left(\frac{1}{\Delta x}\right)^2 = \frac{1}{2mc^2}\left(\frac{hc}{\Delta x}\right)^2$$

$$= \frac{1}{2\times 938\times 10^6}\left(\frac{4.1\times 10^{-15}\times 3\times 10^{10}}{10^{-12}}\right)^2$$

$$\sim \frac{150}{2000}\times 10^8 \sim 10^7\,\text{eV}.$$

b. The Zeeman splitting is given by $\Delta E \sim \mu_B \cdot B$, μ_B being the Bohr magneton, and the Coulomb binding energy of a hydrogen atom is 13.6 eV. For the two to be comparable we require

$$B \approx \frac{13.6\times 1.6\times 10^{-19}}{9.3\times 10^{-32}} = \frac{13.6\times 1.6}{9.3}\times 10^{13}\,\text{wbm}^{-2} \sim 10^9\,\text{Gs}.$$

c. The energy of a classical one-dimensional oscillator is

$$E = \frac{m}{2}\left(\omega x_0\right)^2 = 2\pi^2 mx_0^2 / T^2 = 2\pi^2\,\text{erg}.$$

For

$$n\hbar\omega = E,$$

we require

$$n = \frac{E}{\hbar\omega} = \frac{2\pi^2}{\hbar\omega} = \frac{\pi T}{\hbar} = \frac{\pi\times 1}{1.054\times 10^{-27}} = 3\times 10^{27}.$$

d. The energy shift due to hyperfine-structure splitting of a hydrogen atom in the ground state (in units where $c = \hbar = 1$) is

$$\Delta E \sim m_e^2 \alpha^4 / m_p$$

where α is the fine-structure constant. The binding energy of the electron in the ground state is $E_e = m_e \alpha^2/2$. Hence

$$\Delta E / E = 2\alpha^2\left(\frac{m_e}{m_p}\right)$$

8005

Consider the *density matrix* $\rho(t)=|\psi(t)\rangle\langle\psi(t)|$ corresponding to the state $\|\psi(t)\rangle$ of the system. Show that it satisfies the time-evolution equation

$$\frac{d\rho}{dt}=\frac{i}{\hbar}[H,\rho].$$

Sol:

Using the Schrödinger equation, $H|\psi\rangle=i\hbar\frac{\partial}{\partial t}|\psi\rangle$ and its conjugate, we express the density matrix as follows:

$$\frac{d\rho}{dt}=|\frac{\partial}{\partial t}\psi(t)\rangle\langle\psi(t)|+|\psi(t)\rangle\langle\frac{\partial}{\partial t}\psi(t)|$$

$$\frac{d\rho}{dt}=-\frac{i}{\hbar}\left(|H\psi(t)\rangle\langle\psi(t)|-|\psi(t)\rangle\langle\psi(t)H|\right)$$

$$\frac{d\rho}{dt}=-\frac{i}{\hbar}\left(H\rho-\rho H\right)$$

$$\frac{d\rho}{dt}=-\frac{i}{\hbar}[H,\rho].$$

8006

Answer each of the following questions with a brief and, where possible, quantitative statement. Give your reasoning.

a. A beam of neutral atoms passes through a Stern–Gerlach apparatus. Five equally-spaced lines are observed. What is the total angular momentum of the atom?

b. What is the magnetic moment of an atom in the state $^3 P_0$? (Disregard nuclear effects)

c. Why are the noble gases chemically inert?

d. Estimate the energy density of black body radiation in this room in erg cm^{-3}. Assume the walls are black.

e. In a hydrogen gas discharge both the spectral lines corresponding to the transitions $2^2P_{1/2} \to 1^2S_{1/2}$ and $2^2P_{3/2} \to 1^2S_{1/2}$ are observed. Estimate the ratios of their intensities.

f. What is the cause for the existence of two independent term level schemes, the singlet and triplet systems, in atomic helium?

(Chicago)

ol:

a. When unpolarized neutral atoms of total angular momentum J pass through the Stern-Gerlach apparatus, the incident beam will split into $2J+1$ lines. Thus $2J+1 = 5$, giving $J = 2$.

b. An atom in the state $^3P_0$ has total angular momentum $J = 0$. Hence its magnetic moment is equal to zero, if nuclear spin is neglected.

c. The molecules of noble gases consist of atoms with full-shell structures, which makes it very difficult for the atoms to gain or lose electrons. Hence noble gases are chemically inert.

d. The energy density of black body radiation at room temperature $T \approx 300\,K$ is

$$\mu = \frac{4}{c}\sigma T^4$$

$$= \frac{4}{3\times10^{10}}\times 5.7\times10^{-5}\times 300^4$$

$$= 6\times10^{-5}\,\mathrm{erg/cm^3}.$$

e.

$$\frac{I\left(2^2P_{1/2} \to 1^2S_{1/2}\right)}{I\left(2^2P_{3/2} \to 1^2S_{1/2}\right)} \approx \frac{2J_1+1}{2J_2+1}$$

$$= \frac{2\times1/2+1}{2\times3/2+1} = \frac{1}{2}.$$

f. The helium atom contains two spin-1/2 electrons, whose total spin $S = s_1 + s_2$ can have two values $S = 1$ (triplet) and $S = 0$ (singlet). Transition between the two states is forbidden by the selection rule $\Delta S = 0$. As a result we have two independent term level schemes in atomic helium.

8007

a. Derive the conditions for the validity of the WKB approximation for the one-dimensional time-independent Schrödinger equation, and show that the approximation must fail in the immediate neighborhood of a classical turning point.

b. Explain, using perturbation theory, why the ground state energy of an atom always decreases when the atom is placed in an external electric field.

(Berkeley)

Sol:

a. The WKB method starts from the Schrödinger equation

$$\left[-\frac{\hbar^2}{2m}\frac{d^2}{dx^2}+V(x)\right]\psi(x)=E\psi(x),$$

where it is assumed

$$\psi(x)=e^{is(x)/\hbar}.$$

Substitution in the Schrödinger equation gives

$$\frac{1}{2m}\left(\frac{ds}{dx}\right)^2+\frac{\hbar}{i}\frac{1}{2m}\frac{d^2s}{dx^2}=E-V(x). \tag{1}$$

Expanding s as a series in powers of $\hbar/i$

$$s=s_0+\frac{\hbar}{i}s_1+\left(\frac{\hbar}{i}\right)^2 s_2+\cdots,$$

and substituting it in Eq. (1), we obtain

$$\frac{1}{2m}s_0'^2+\frac{\hbar}{i}\frac{1}{2m}\left(s_0''+2s_0's_1'\right)+\left(\frac{\hbar}{i}\right)^2\left(s_1'^2+2s_0's_2'+s_1''\right)$$
$$+\cdots=E-V(x). \tag{2}$$

If we impose the conditions

$$\left|\hbar s_0''\right|\ll\left|s_0'^2\right|, \tag{3}$$

$$\left|2\hbar s_0's_1'\right|\ll\left|s_0'^2\right|, \tag{4}$$

Eq. (2) can be approximated by

$$\frac{1}{2m}s_0'^2=E-V(x), \tag{5}$$

which is equivalent to setting

$$2s_0's_1' + s_0'' = 0,$$
$$2s_0's_2' + s_1'^2 + s_1'' = 0,$$

$$\cdots\cdots$$

(3) and (4) are the conditions for the validity of the WKB method. Integration of Eq. (5) gives

$$s_0(x) = \pm\int^x \sqrt{2m(E - V(x))}dx = \pm\int^x p\,dx,$$

so that (3) can be written as

$$\left|\frac{\hbar}{p^2}\frac{dp}{dx}\right| \ll 1, \qquad (6)$$

i.e.,

$$\left|\hbar\frac{d}{dx}\left(\frac{1}{p}\right)\right| \ll 1,$$

or

$$\left|\frac{d\lambda}{dx}\right| \ll 1,$$

where

$$\lambda = \frac{\hbar}{p} = \frac{\hbar}{\sqrt{2m(E - V(x))}}.$$

Near a turning point $V(x) \sim E, p \to 0$ and (6) is not satisfied. Hence the WKB method cannot be used near classical turning points.

b. Consider an atom in an external electric field ε in the z direction. The perturbation Hamiltonian is

$$H' = -e\varepsilon z,$$

where $z = \sum_i z_i$ is the sum of the z coordinates of the electrons of the atom, and the energy correction is

$$\Delta E_0 = H_{00}' + \sum_{n\neq 0}|H_{on}'|^2 / (E_0 - E_n).$$

As z is an odd operator and the parity of the ground state is definite, $H_{00}' = 0$. Futhermore $E_0 - E_n < 0$. Hence $\Delta E_0 < 0$. This means that the energy of the ground state decreases when the atom is placed in an electric field.

8008

Consider a particle of mass m in the ground state of an infinitely high box extending $x = 0$ to $x = a$. It is suddenly expanded from $x = a$ to $x = 2a$. What is the probability of that particle in the ground state after this sudden expansion?

Sol:

$$\psi_i = \sqrt{\frac{2}{a}}\sin\frac{\pi x}{a} \qquad \psi_f = \sqrt{\frac{2}{2a}}\sin\frac{\pi x}{2a}$$

Probability $= \left|\langle\psi_i|\psi_f\rangle\right|^2$

Probability amplitude $= \langle\psi_i|\psi_f\rangle$

$$= \int_0^a \sqrt{\frac{2}{a}}\sin\frac{\pi x}{a}\sqrt{\frac{2}{2a}}\sin\frac{\pi x}{2a}\,dx$$

$$= \frac{2}{a}\times\frac{1}{\sqrt{2}}\int_0^a \sin\frac{\pi x}{a}\sin\frac{\pi x}{2a}\,dx$$

$$= \frac{\sqrt{2}}{a}\left[\frac{\sin\frac{\pi}{a}\left(1-\frac{1}{2}\right)x}{2\frac{\pi}{a}\left(1-\frac{1}{2}\right)} - \frac{\sin\frac{\pi}{a}\left(1+\frac{1}{2}\right)x}{2\frac{\pi}{a}\left(1+\frac{1}{2}\right)}\right]$$

$$= \frac{\sqrt{2}}{a}\left[\sin\frac{\pi x}{2a}\times\frac{a}{\pi} - \sin\frac{3\pi x}{a}\times\frac{a}{3\pi}\right]_0^a$$

$$= \frac{\sqrt{2}}{a}\left\{\frac{a}{\pi}[1-0] - \frac{a}{3\pi}[-1-0]\right\}$$

$$= \frac{\sqrt{2}}{a}\frac{a}{\pi}\left[1+\frac{1}{3}\right] = \frac{\sqrt{2}}{\pi}\times\frac{4}{3}$$

Probability $= \left(\frac{\sqrt{2}}{\pi}\times\frac{4}{3}\right)^2 = \frac{16\times 2}{9\pi^2} = \frac{32}{9\pi^2}$.

8009

Set up the relevant equations with estimates of all missing parameters. The molecular bond (spring constant) of HCl is about 470 N/m. The moment of inertia is 2.3×10^{-47} kg·m^2.

a. At 300 K what is the probability that the molecule is in its lowest excited vibrational state?

b. Of the molecules in the vibrational ground state what is the ratio of the number in the ground rotational state to the number in the first excited rotational state?

(*Wisconsin*)

Sol:

a. The Hamiltonian for the vibrational motion of the system is

$$\hat{H}_v = \frac{\hat{p}^2}{2\mu} + \frac{1}{2}\mu\omega^2\hat{x}^2,$$

and the vibrational states are

$$E_v^{(n)} = \left(n+\frac{1}{2}\right)\hbar\omega, \quad n = 0, 1, 2, \ldots,$$

with $\omega = \sqrt{K/\mu}$, K being the force constant and μ the reduced mass of the oscillating atoms.

Statistically, the number of molecules in state $E^{(n)}$ is proportional to $\exp(-nx)$, where $x = \dfrac{\hbar\omega}{kT}$, k being Boltzmann's constant and T is the absolute temperature. Thus the probability that the molecule is in the first excited state is

$$P_1 = \frac{e^{-x}}{1+e^{-x}+e^{-2x}+\ldots} = e^{-x}\left(1-e^{-x}\right),$$

As

$$\mu = \left(\frac{1\times 35}{1+35}\right)m_p \approx m_p = 1.67\times 10^{-27} \text{ kg,}$$

$$x = \frac{\hbar\omega}{kT} = \frac{1.054\times 10^{-34}\times\left(470/1.67\times 10^{27}\right)^{1/2}}{1.38\times 10^{-23}\times 300} = 13.5,$$

we have $P_1 \approx e^{-13.5} = 1.37\times 10^{-6}$.

b. The Hamiltonian for rotation is

$$\hat{H}_r = \frac{1}{2I}\hat{J}^2,$$

and the energy states are

$$E_r^{(J)} = \frac{\hbar^2}{2I}J(J+1), \quad J = 0, 1, 2, \ldots.$$

Since the number of molecules in rotational state J is proportional to$(2J+1)\exp\left(-\dfrac{E_r^{(J)}}{kT}\right)$ as the J state is $(2J+1)$-times degenerate $\left(m_J=-J,-J+1,\dots J\right)$, we have

$$\frac{N(J=0)}{N(J=1)}=\frac{1}{3}\exp\left(\frac{\hbar^2}{IkT}\right).$$

As

$$\frac{\hbar^2}{IkT}=\frac{\left(1.054\times10^{-34}\right)^2}{2.3\times10^{-47}\times1.38\times10^{-23}\times300}=0.117,$$

$$\frac{N(J=0)}{N(J=1)}=e^{0.117}/3=0.37.$$

8010

The potential curves for the ground electronic state (A) and an excited electronic state (B) of a diatomic molecule are shown in Fig. 8.2. Each electronic state has a series of vibrational levels which are labelled by the quantum number v.

a. The energy differences between the two lowest vibrational levels are designated as Δ_A and Δ_B for the electronic states A and B respectively. Is Δ_A larger or smaller than Δ_B? Why?

b. Some molecules were initially at the lowest vibrational level of the electronic state B, followed by subsequent transitions to the various vibrational levels of the electronic state A through spontaneous emission of radiation. Which vibrational level of the electronic state A would be most favorably populated by these transitions? Explain your reasoning.

(*Wisconsin*)

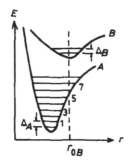

Fig. 8.2

ol:

a. The force constant is $K = \left(\dfrac{\partial^2 V}{\partial r^2}\right)_{r=r_0}$ where r_0 is the equilibrium position. It

can be seen from Fig. 8.2 that $K_A > K_B$. The vibrational energy levels are given by

$$E^{(n)} = \left(n+\frac{1}{2}\right)\hbar\omega, \quad \omega = \sqrt{\frac{K}{\mu}}.$$

Hence.

$$\Delta_A \approx \hbar\sqrt{\frac{K_A}{\mu}}, \Delta_B \approx \hbar\sqrt{\frac{K_B}{\mu}},$$

and so $\Delta_A > \Delta_B$.

b. Electrons move much faster than nuclei in vibration. When an electron transits to another state, the distance between the vibrating nuclei remains practically unchanged. Hence the probability of an electron to transit to the various levels is determined by the electrons' initial distribution probability. As the molecules are initially on the ground state of vibrational levels, the probability that the electrons are at the equilibrium position $r = r_{0B}$ is largest. Then from Fig. 8.2 we see that the vibrational level $v \approx 5$ of A is most favorably occupied.

8011

inglet positronium decays by emitting two photons which are polarized t right angles with respect to each other. An experiment is performed with photon detectors behind polarization analyzers, as shown in Fig. 8.3. Each analyzer has a preferred axis such that light polarized in that direction is transmitted perfectly, while light polarized in the perpendicular direction is absorbed completely. The analyzer axes are at right angles with respect to each other. When many events are observed, what is the ratio of the number of events in which both detectors record a photon to the number in which only one detector records a photon?

(MIT)

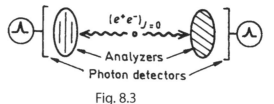

Fig. 8.3

Sol: Suppose the positronium is initially at rest. Then the two photons will move in opposite directions to conserve momentum, and will reach the respective analyzers at the same time. Assume further that the detector solid angle is very smaller. Then the directions of those photons that reach the analyzers must be almost perpendicular to the latter. Hence the directions of polarization of these photons are parallel to the analyzers.

Denote by θ the angle between one photon's direction of polarizationn and the direction of transmission of the analyzer reached by it. The probability that it can pass through the analyzer is $\cos^2\theta$. Consider the second photon produced in the same decay. As it is polarized at right angles with respect to the first one, the angle between its direction of polarization and the direction of transmission of the second analyzer, which is oriented at right angles to that of the first analyzer, is also θ. Hence the probability that, of the two detectors, only one records the passage of a photon is

$$P_1 \propto \Omega\left[\frac{1}{2\pi}\int_0^{2\pi}\cos^2\theta\left(1-\cos^2\theta\right)d\theta + \frac{1}{2\pi}\int_0^{2\pi}\left(1-\cos^2\theta\right)\cos^2\theta\,d\theta\right]$$

$$= \frac{\Omega}{4},$$

where Ω is the solid angle subtended by the detector, and the probability that both detectors record the passage of photons is

$$P_2 \propto \Omega\left[\frac{1}{2\pi}\int_0^{2\pi}\cos^2\theta\cos^2\theta\,d\theta\right] = \frac{3\Omega}{8}.$$

Hence the ratio of the number of events of both detectors recording to that of only one detector recording in a given time is

$$\frac{P_2}{P_1} = \frac{3}{8} \div \frac{1}{4} = \frac{3}{2}.$$

8012

A point source Q emits coherent light isotropically at two frequencies ω and $\omega + \Delta\omega$ with equal power I joules/sec at each frequency. Two detectors A and each with a (small) sensitive area s, capable of responding to individual photons are located at distances l_A and l_B from Q as shown in Fig. 8.4. In the following take $\Delta\omega/\omega \ll 1$ and assume the experiment is carried out in vacuum.

a. Calculate the individual photon counting rates (photons/sec) at A and B as functions of time. Consider time scales $\gg 1/\omega$.

b. If now the output pulses from A and B are put into a coincidence circuit of resolving time τ, what is the time-averaged coincidence counting rate? Assume that $\tau \ll 1/\Delta\omega$ and recall that a coincidence circuit will produce an output pulse if the two input pulses arrive within a time τ of each other.

(CUS)

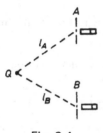

Fig. 8.4

ol:

a. The wave function of a photon at A is

$$\psi_A(l_A, t) = C_1 \left[e^{i\omega\left(\frac{l_A}{c} - t\right)} + e^{i(\omega + \Delta\omega)\left(\frac{l_A}{c} - t\right)} \right],$$

where C_1 is real and, hence, the probability of finding a proton at A in unit time is

$$P_A = \psi_1^* \psi_1$$

$$= C_1^2 \left\{ 2 + 2\cos\left[\Delta\omega\left(\frac{l_A}{c} - t\right) \right] \right\}$$

$$= 4C_1^2 \cos^2\left[\frac{\Delta\omega}{2}(l_A/c - t) \right].$$

If there is only a single frequency, $P_A = C_1^2$. As each photon has energy $\hbar\omega$, the number of photons arriving at A per second is

$$\frac{s}{4\pi l_A^2} \cdot \frac{I}{\hbar\omega}.$$

Hence

$$C^2 = \frac{Is}{4\pi l_A^2 \hbar \omega},$$

and

$$P_A = \frac{Is}{\pi l_A^2 \hbar \omega} \cos^2\left[\frac{\Delta\omega}{2}(l_A/c - t)\right].$$

Similarly we have

$$P_B = 4C_2^2 \cos^2\left[\frac{\Delta\omega}{2}(l_A/c - t)\right],$$

where

$$C_2^2 = \frac{Is}{4\pi l_B^2 \hbar \omega}.$$

b. In a coincidence of resolving time τ, the time-averaged coincidence counting rate

$$
\begin{aligned}
P &= \lim_{T\to\infty} \frac{1}{2T} \int_{-T}^{T} dt \int_{-\tau}^{\tau} P_A(t)P_B(t+x)dx \\
&= \lim_{T\to\infty} \frac{1}{2T} \int_{-T}^{T} dt \int_{-\tau}^{\tau} 4C^4 \left[1 + \cos\Delta\omega(l_A/c - t)\right] \\
&\quad \times \left[1 + \cos\Delta\omega\left(\frac{l_B}{c} - t - x\right)\right] dx \\
&= \lim_{T\to\infty} \frac{1}{2T} \int_{-T}^{T} 4C^4 \cdot \left\{1 + \cos\left[(l_A/c - t)\Delta\omega\right]\right\} \\
&\quad \times \left\{2\tau + 2\tau\cos\left[\Delta\omega\left(\frac{l_B}{c} - t\right)\right]\right\} dt,
\end{aligned}
$$

where

$$C^4 = \frac{I^2 s^2}{16\pi^2 l_B^2 l_A^2 \hbar^2 \omega^2} = C_1^2 C_2^2,$$

as

$$
\begin{aligned}
\int_{-\tau}^{\tau} \cos\left[\Delta\omega(l_B/c - t - x)\right] dx &= \frac{1}{\Delta\omega} 2\sin(\tau\Delta\omega)\cos\left[\left(\frac{l_B}{c} - t\right)\Delta\omega\right] \\
&\approx 2\tau\cos\left[\Delta\omega(l_B/c - t)\right].
\end{aligned}
$$

Hence

$$P = 8\tau C^4 \lim_{T\to\infty} \frac{1}{2T}\int_{-T}^{T}\left\{1+\cos\left[(l_A/c-t)\Delta\omega\right]\right\}$$

$$\times\left\{1+\cos\left[\left(\frac{l_B}{c}-t\right)\Delta\omega\right]\right\}dt$$

$$= 8\tau C^4 \lim_{T\to\infty} \frac{1}{2T}\int_{-T}^{T}\left\{1+\cos\left[\left(\frac{l_B}{c}-t\right)\Delta\omega\right]\right.$$

$$\left.+\cos\left[\left(\frac{l_A}{c}-t\right)\Delta\omega\right]+\cos\left[\left(\frac{l_A}{c}-t\right)\Delta\omega\right]\cos\left[\left(\frac{l_B}{c}-t\right)\Delta\omega\right]\right\}dt$$

$$= 8\tau C^4 \lim_{T\to\infty} \frac{1}{2T}\int_{-T}^{T}\left\{1+\frac{1}{2}\cos\left[\frac{\Delta\omega}{c}(l_A-l_B)\right]\right.$$

$$\left.+\cos\left[\left(\frac{l_A+l_B}{2c}-t\right)\Delta\omega\right]\cos\left[\left(\frac{l_A-l_B}{2c}\right)\Delta\omega\right]\right.$$

$$\left.+\frac{1}{2}\cos\left[\left(\frac{l_A+l_B}{c}-2t\right)\Delta\omega\right]\right\}dt$$

$$= 8\tau C^4 \lim_{T\to\infty} \frac{1}{2T}\left\{2T+\frac{2T}{2}\cos\left[\left(\frac{l_A-l_B}{c}\right)\Delta\omega\right]\right\}$$

$$= 8\tau C^4\left\{1+\frac{1}{2}\cos\left[\frac{\Delta\omega(l_A-l_B)}{c}\right]\right\}$$

$$= \frac{\tau I^2 s^2}{2\pi^2 l_B^2 l_A^2 \hbar^2 \omega^2}\left\{1+\frac{1}{2}\cos\left[\frac{\Delta\omega}{c}(l_A-l_B)\right]\right\}.$$

8013

The state of an infinite potential well of width a is $\psi = i\psi_0 +(1+i)\psi_1+(2-i)\psi_2$. Normalize the wave function and find the expectation value of energy.

Sum of squares of coefficients = 1

Sol:

$$1 = A^2(1+2+5) = A^2 8 \qquad A = \frac{1}{\sqrt{8}}$$

$$\varepsilon_1 = \varepsilon$$
$$\varepsilon_2 = 4\varepsilon$$
$$\varepsilon_3 = 9\varepsilon$$
$$\varepsilon = \frac{\pi^2\hbar^2}{2ma^2}$$

$$\langle \varepsilon \rangle = \varepsilon_s P_s$$

$$= \frac{1}{8}\varepsilon + \frac{2}{8}\times 4\varepsilon + \frac{5}{8}\times 9\varepsilon = \frac{\overset{27}{\cancel{54}}\varepsilon}{\cancel{8}_4}$$

$$= \frac{27\varepsilon}{4} = \frac{27\pi^2\hbar^2}{8ma^2}.$$

8014

A spinless particle of mass m and charge q is constrained to move in a circle of radius R as shown in Fig. 8.5. Find its allowed energy levels (up to a common additive constant) for each of the following cases:

  a. The motion of the particle is nonrelativistic.

  b. There is a uniform magnetic field **B** perpendicular to the plane of the circle.

  c. The same magnetic flux which passed through the circle is now contained into a solenoid of radius $b(b<R)$.

  d. There is a very strong electric field **F** in the plane of the circle $\left(q|\mathbf{F}|\gg \hbar^2/mR^2\right)$.

  e. **F** and **B** are zero, but the electron's motion around the circle is extremely relativistic.

  f. The circle is replaced by an ellipse with the same perimeter but half the area.

                                                    (CUS)

Sol:

  a. Let the momentum of the particle be p. The quantization condition

$$p\cdot 2\pi R = nh$$

gives

$$p = \frac{n\hbar}{R}$$

and hence

$$E = \frac{p^2}{2m} = \frac{1}{2m}\left(\frac{n\hbar}{R}\right)^2 = \frac{\hbar^2}{2mR^2}n^2,$$

where

$$n = 0, \pm 1, \pm 2, \ldots.$$

(a)

(b)

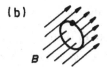

B

(c)

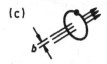

(d)

Fig. 8.5

b. Take coordinates with origin at the center of the circle and the z-axis along the direction of **B**. Then the vector potential at a point on the circle is

$$\mathbf{A} = \frac{1}{2} B R e_\varphi.$$

The Schrödinger equation

$$\hat{H}\psi = E\psi,$$

where

$$\hat{H} = \frac{1}{2m}\left(\mathbf{p} - \frac{q}{c}\mathbf{A}\right)^2,$$

can be written as

$$\frac{1}{2m}\left(-\frac{i\hbar}{R}\frac{\partial}{\partial\varphi} - \frac{q}{c}\cdot\frac{1}{2}BR\right)^2 \psi(\varphi) = E\psi(\varphi).$$

Its solution is $\psi(\varphi) = Ce^{in\varphi}$, The single-valuedness condition $\psi(\varphi) = \psi(\varphi + 2\pi)$ demands $n = 0, \pm 1, \pm 2, \ldots$. Substituting the solution in the equation gives

$$E = \frac{1}{2m}\left(\frac{n\hbar}{R} - \frac{q}{2c}BR\right)^2.$$

c. When the magnetic flux is confined to the inside of a solonoid of radius b enclosed by the circle, magnetic field is zero on the circular path. As $\nabla \times \mathbf{A} = \mathbf{B} = 0, A$ can be taken to be a constant which is equal to $\frac{1}{2} BR$ when $b \to R$. Then

$$A = \frac{1}{2} \frac{B \pi R^2}{\pi R} = \frac{\phi}{2 \pi R}.$$

As ϕ remains the same, the energy levels are the same as in (b).

d. Take the x-axis parallel to **F**. Then

$$\mathbf{F} = F(\cos \varphi, -\sin \varphi), d\mathbf{r} = (0, Rd\varphi),$$

and hence

$$V = -\int q \mathbf{F} \cdot d\mathbf{r} = qFR \int \sin \varphi d\varphi = -qFR \cos \varphi.$$

Thus the Hamiltonian is

$$\hat{H} = \frac{-\hbar^2}{2m} \frac{1}{R^2} \frac{d^2}{d\varphi^2} - qFR \cos \varphi.$$

Because the electric field **F** is very strong, the probability that the particle moves near $\varphi \sim 0$ is large. Hence we can make the approximation

$$\cos \varphi = 1 - \frac{1}{2} \varphi^2 + O(\varphi^4) \approx 1 - \frac{\varphi^2}{2}$$

and obtain

$$\hat{H} = -\frac{\hbar^2}{2mR^2} \frac{d^2}{d\varphi^2} - qFR \left(1 - \frac{1}{2} \varphi^2 \right),$$

or

$$\hat{H} + qFR = -\frac{\hbar^2}{2mR^2} \frac{d^2}{d\varphi^2} + \frac{1}{2} qFR \varphi^2,$$

which has the form of the Hamiltonian of a harmonic oscillator of mass $M = mR^2$ and angular frequency ω given by $M\omega^2 = qFR$, whose eigenvalues are

$$E_n + qFR = \left(n + \frac{1}{2} \right) \hbar \omega,$$

or

$$E_n = \left(n + \frac{1}{2} \right) \hbar \omega - qFR,$$

with

$$\omega = \sqrt{\frac{qFR}{M}} = \sqrt{\frac{qF}{mR}}, \quad n = 0, 1, 2, \ldots.$$

Therefore

$$E_n = -qFR + \left(n+\frac{1}{2}\right)\hbar\sqrt{\frac{qF}{mR}}.$$

e. The quantization condition gives

$$p \cdot 2\pi R = nh,$$

or $p = n\hbar / R$.

If the particle is highly relativistic,

$$E = pc = \frac{n\hbar c}{R}, \quad n = 0, 1, 2, \ldots .$$

f. The quantization condition gives

$$p = n\hbar / R,$$

and hence

$$E = pc = \frac{n\hbar c}{R},$$

same as for a circular orbit.

8015

Consider the *density matrix* $\rho(t) = |\psi(t)\rangle\langle\psi(t)|$ corresponding to the state $|\psi(t)\rangle$ of the system. Show that it satisfies the time-evolution equation

$$\frac{d\rho}{dt} = -\frac{i}{\hbar}[H, \rho]$$

Sol:

Using the Schrodinger equation, $H|\psi\rangle = i\hbar\frac{\partial}{\partial t}|\psi\rangle$ and its conjugate, we express the density matrix as follows:

$$\frac{d\rho}{dt} = |\frac{\partial}{\partial t}\psi(t)\rangle\langle\psi(t)| + |\psi(t)\rangle\langle\frac{\partial}{\partial t}\psi(t)|$$

$$\frac{d\rho}{dt} = -\frac{i}{\hbar}\left(|H\psi(t)\rangle\langle\psi(t)| - |\psi(t)\rangle\langle\psi(t)H|\right)$$

$$\frac{d\rho}{dt} = -\frac{i}{\hbar}\left(H\rho - \rho H\right)$$

$$\frac{d\rho}{dt} = -\frac{i}{\hbar}[H, \rho]$$

8016

To find approximate eigenfunctions of the Hamiltonian H we can use trial functions of the form $\psi = \sum_{k=1}^{n} a_k \phi_k$ in the variational method (where the ϕ_k are given functions, and the a_k are parameters to be varied). Show that one gets n solutions ψ_α with energies $\varepsilon_\alpha = \langle \psi_\alpha | H | \psi_\alpha \rangle / \langle \psi_\alpha | \psi_\alpha \rangle$, where H is the Hamiltonian. We will order them so that $\varepsilon_1 \le \varepsilon_2 \le \varepsilon_3 \dots$ Show from the Hermitian properties of the Hamiltonian that the ψ_α either automatically have or can be chosen to have the properties $\langle \psi_\alpha | \psi_\beta \rangle = \delta_{\alpha\beta}$, $\langle \psi_\alpha | H | \psi_\beta \rangle = \varepsilon_\alpha \delta_{\alpha\beta}$. From the fact that one can certainly find a linear combination of ψ_1 and ψ_2 which is orthogonal to ψ_1, the exact ground state of H with eigenvalue E_1, prove that $\varepsilon_2 \ge E_2$, where E_2 is the exact energy of the first excited state.

(Wisconsin)

Sol: Suppose $\{\phi_k\}$ is the set of linearly independent functions. We may assume that $\langle \phi_i | \phi_j \rangle = \delta_{ij}$, first using Schmidt's orthogonalization process if necessary. Then

$$\bar{H} = \frac{\langle \psi | H | \psi \rangle}{\langle \psi | \psi \rangle} = \frac{\sum_{i,j}^{n} a_i^* a_j \lambda_{ij}}{\sum_{i,j}^{n} a_i^* a_j \delta_{ij}} = \sum_{i,j} x_i^* \lambda_{ij} x_j = X^+ \hat{\lambda} X,$$

where

$$x_i = \frac{a_i}{\sqrt{\sum_j |a_j|^2}}, \quad \lambda_{ij} = \langle \phi_i | H | \phi_j \rangle = \lambda_{ji}^*.$$

Note that

$$\sum_{i=1}^{n} |x_i|^2 = 1.$$

As $\hat{\lambda}$ is Hermitian, we can choose a rotational transformation $X = \hat{p} Y$, such that $\hat{\Lambda} = \hat{p}^+ \hat{\lambda} \hat{p} = \hat{p}^{-1} \hat{\lambda} \hat{p}$ is a diagonal matrix with diagonal elements $\Lambda_{11} \le \Lambda_{22} \le \Lambda_{33}$. Then

$$\bar{H} = \sum_{i=1}^{n} \Lambda_{ii} |y_i|^2,$$

where y_i satisfy $\sum_{i=1}^{n}|y_i|^2 = 1$.

Applying the variational principle

$$0 = \delta\left[\bar{H} - \alpha\left(\sum_i |y_i|^2 - 1\right)\right] = \delta\left[\sum_i (\Lambda_{ii} - \alpha)|y_i|^2 + \alpha\right],$$

where α is the Lagrange multiplier, we get

$$\sum_i (\Lambda_{ii} - \alpha)|y_i|\,\delta|y_i| = 0,$$

or

$$(\Lambda_{ii} - \alpha)|y_i| = 0, \quad (i = 1, 2, \cdots)$$

i.e., $\alpha = \Lambda_{ii}$ or $|y_i| = 0$.

Hence the solutions of the variational equations are

$$\alpha = \Lambda_{ii}, \quad y_j^{(i)} = \delta_j^i = \delta_{ij}, \quad (i = 1, 2, \cdots, n)$$

Thus we get n solutions ψ_α, the αth solution $y_i^{(\alpha)} = \delta_i^{(\alpha)}$ corresponding to energy

$$\varepsilon_\alpha = \frac{\langle \psi_\alpha | H | \psi_\alpha \rangle}{\langle \psi_\alpha | \psi_\alpha \rangle} = \sum_i \Lambda_{ii} |y_i^{(\alpha)}|^2 = \Lambda_{\alpha\alpha}$$

with $\varepsilon_1 \le \varepsilon_2 \le \varepsilon_3 \dots$.

For $\psi_\alpha = \psi_\alpha[X(Y)]$, we have

$$\langle \psi_\alpha | \psi_\beta \rangle = \sum_i a_i^{(\alpha)*} a_i^{(\beta)}$$

$$= \sqrt{\sum_j |a_j^{(\alpha)}|^2 \sum_j |a_j^{(\beta)}|^2} \cdot \sum_i x_i^{(\alpha)*} x_i^{(\beta)}$$

$$= \sqrt{\sum_j |a_j^{(\alpha)}|^2 \sum_j |a_j^{(\beta)}|^2} \cdot \sum_i y_i^{(\alpha)*} y_i^{(\beta)} = \left[\sum_j |a_j^{(\alpha)}|^2\right] \delta_{\alpha\beta},$$

$$\langle \psi_\alpha | H | \psi_\beta \rangle = \sum_{i,j} a_i^{(\alpha)*} \lambda_{ij} a_j^{(\beta)}$$

$$= \sqrt{\sum_j |a_j^{(\alpha)}|^2 \sum_j |a_j^{(\beta)}|^2} \cdot \sum_{i,j} x_i^{(\alpha)*} \lambda_{ij} x_j^{(\beta)}$$

$$= \sqrt{\sum_j |a_j^{(\alpha)}|^2 \sum_j |a_j^{(\beta)}|^2} \cdot \sum_{i,j} y_i^{(\alpha)*} \Lambda_{ij} y_j^{(\beta)}$$

$$= \left[\sum_j |a_j^{(\alpha)}|^2\right] \varepsilon_\alpha \delta_{\alpha\beta}.$$

Then, by setting $\Psi_\alpha = \psi_\alpha / \sqrt{\sum_j |a_j^{(\alpha)}|^2}$, we have

$$\langle \Psi_\alpha | \Psi_\beta \rangle = \delta_{\alpha\beta}, \langle \Psi_\alpha | H | \Psi_\beta \rangle = \varepsilon_\alpha \delta_{\alpha\beta}.$$

Let the exact wave functions of the ground state and the first excited state of H be Φ_1 and Φ_2, their exact energies be E_1 and E_2 respectively. Then there must exist two numbers μ_1 and μ_2 such that $\Phi_1 = \mu_1 \Psi_1 + \mu_2 \Psi_2, |\mu_1|^2 + |\mu_2|^2 = 1$. From the orthogonality of Φ_1 and Φ_2, we have $\Phi_2 = \mu_2^* \Psi_1 - \mu_1^* \Psi_2$, and hence

$$E_1 = \varepsilon_1 |\mu_1|^2 + \varepsilon_2 |\mu_2|^2,$$

$$E_2 = \varepsilon_1 |\mu_2|^2 + \varepsilon_2 |\mu_1|^2 = (\varepsilon_1 - \varepsilon_2)|\mu_2|^2 + \varepsilon_2 \le \varepsilon_2.$$

8017

Find the value of the parameter λ in the trial function $\phi(x) = Ae^{-\lambda^2 x^2}$, where A is a normalization constant, which would lead to the best approximation for the energy of the ground state of the one-particle Hamiltonian $H = -\dfrac{\hbar^2}{2m}\dfrac{d^2}{dx^2} + bx^4$, where b is a constant. The following integrals may be useful:

$$\int_{-\infty}^{\infty} e^{-ax^2}dx = \sqrt{\frac{\pi}{a}}, \int_{-\infty}^{\infty} x^2 e^{-ax^2}dx = \frac{1}{2}\sqrt{\frac{\pi}{a^3}},$$

$$\int_{-\infty}^{\infty} x^4 e^{-ax^2}dx = \frac{3}{4}\sqrt{\frac{\pi}{a^5}}.$$

(Wisconsin)

Sol: Using the trial function $\phi = Ae^{-\lambda^2 x^2}$, consider the integrals

$$\int_{-\infty}^{\infty} \phi^*(x)\phi(x)dx = \int_{-\infty}^{\infty} A^2 e^{-2\lambda^2 x^2}dx = A^2 \sqrt{\frac{\pi}{2\lambda^2}} = 1,$$

$$\int_{-\infty}^{\infty} \phi^*(x)H\phi(x)dx = \int_{-\infty}^{\infty} A^2 e^{-\lambda^2 x^2}\left(-\frac{\hbar^2}{2m}\frac{d^2}{dx^2} + bx^4\right)e^{-\lambda^2 x^2}dx$$

$$= A^2 \int_{-\infty}^{\infty}\left[-\frac{\hbar^2}{m}(2\lambda^4 x^2 - \lambda^2) + bx^4\right]e^{-2\lambda^2 x^2}dx$$

$$= A^2\left[-\frac{\hbar^2}{m}\left(2\lambda^4 \cdot \frac{1}{2}\sqrt{\frac{\pi}{(2\lambda^2)^3}} - \lambda^2\sqrt{\frac{\pi}{2\lambda^2}}\right) + b\frac{3}{4}\sqrt{\frac{\pi}{(2\lambda^2)^5}}\right]$$

$$= A^2\left[\frac{1}{2}\frac{\hbar^2}{m}\sqrt{\frac{\pi}{2}}\lambda + b\sqrt{\frac{\pi}{2}}\frac{3}{16\lambda^5}\right],$$

and obtain

$$\langle H \rangle = \frac{\int \phi^* H \phi \, dx}{\int \phi^* \phi \, dx} = \frac{1}{2}\left(\frac{\hbar^2}{m}\lambda^2 + b\frac{3}{8\lambda^4} \right).$$

As $\frac{1}{3}(a+b+c) \geq (abc)^{1/3}$ for positive numbers a, b, c, we have

$$\langle H \rangle = \frac{1}{2}\left(\frac{\hbar^2\lambda^2}{2m} + \frac{\hbar^2\lambda^2}{2m} + \frac{3b}{8\lambda^4} \right) \geq \frac{3}{2}\left(\frac{\hbar^4}{4m^2}\cdot\frac{3b}{8} \right)^{\frac{1}{3}}.$$

Hence the best approximation for the energy of the ground state is

$$\langle H \rangle_{\min} = \frac{3}{4}\left(\frac{3}{4} \right)^{\frac{1}{3}}\left(\frac{b\hbar^4}{m^2} \right)^{\frac{1}{3}}.$$

8018

Two electrons in a 10-potential well with a state $\psi = \frac{1}{\sqrt{2}}\psi_G + \frac{1}{\sqrt{2}}\psi_{FE}$. Find the state and expectation value of energy.

Ground state can be either ψ_{12} or ψ_{21}.

First excited state can be either ψ_{13} or ψ_{31}.

Sol:

$$\Rightarrow |\psi\rangle = \frac{1}{\sqrt{2}}\left[\frac{2}{a}\sin\frac{\pi x_1}{a}\sin\frac{2\pi x_2}{a} - \frac{2}{a}\sin\frac{2\pi x_1}{a}\sin\frac{\pi x_2}{a} \right] +$$
$$\frac{1}{\sqrt{2}}\left[\frac{2}{a}\sin\frac{\pi x_1}{a}\sin\frac{3\pi x_2}{a} - \frac{2}{a}\sin\frac{3\pi x_1}{a}\sin\frac{\pi x_2}{a} \right]$$

$$\langle \psi|\pi|\psi\rangle = \frac{1}{2}\cdot\frac{\pi^2\hbar^2}{2ma^2}(1+4) + \frac{1}{2}\frac{\pi^2\hbar^2}{2ma^2}(1+9)$$
$$= \frac{15\pi^2\hbar^2}{4ma^2}.$$

8019

(Use nonrelativistic methods to solve this problem.)

Most mesons can be described as bound quark-antiquark states $(q\bar{q})$.

Consider the case of a meson made of a $(q\bar{q})$ pair in an s-state. Let m_q be the quark mass.

Assume the potential binding the q to the $\bar{q}$ can be written as $V = \dfrac{A}{r} + Br$ with $A < 0$ and $B > 0$. You are asked to find a reasonable approximation to the ground state energy of this system in terms of A, B, m_q and $\hbar$. Unfortunately, for a class of trial functions appropriate to this problem, a cubic equation has to be solved. If this happens to you, and you do not want to spend your limited time trying to solve such a cubic equation, you may complete your solution for the case $A = 0$ (without loss of credit). Please express your final answer in terms of a numerical constant, which you should explicitly evaluate, multiplying a function of B, m_q and $\hbar$.

(Berkeley)

Sol:

Method I

Use for the trial function the wave function of a ground state hydrogen atom.

$$\psi(\mathbf{r}) = e^{-r/a}$$

and calculate

$$\bar{H} = \langle \psi \mid H \mid \psi \rangle / \langle \psi \mid \psi \rangle$$

$$= \int_0^\infty drr^2 e^{-r/a} \left[-\frac{\hbar^2}{2\mu} \frac{1}{r^2} \frac{\partial}{\partial r} \left(r^2 \frac{\partial}{\partial r} \right) \right.$$

$$\left. + A r^{-1} + B r \right] e^{-r/a} / \int_0^\infty drr^2 e^{-2r/a}$$

$$= \frac{3Ba}{2} + \frac{\hbar^2}{2\mu} \frac{1}{a^2} + \frac{A}{a},$$

where $\mu = \dfrac{m_q}{2}$ is the reduced mass of the $q\bar{q}$. system. Vary a to minimize $\bar{H}$ by letting $\dfrac{\delta \bar{H}}{\delta a} = 0$, which gives

$$\frac{3}{2} Ba^3 - Aa - \frac{\hbar^2}{\mu} = 0.$$

When $A = 0$, the solution is $a = \left(\dfrac{2\hbar^2}{3B\mu} \right)^{1/3}$. Hence the estimated ground state energy is

$$E_g = \bar{H} = \frac{3}{4} \left(\frac{36B^2\hbar^2}{m_q} \right)^{1/3} = 2.48 \left(\frac{B^2\hbar^2}{m_q} \right)^{1/3}.$$

Method II

Another estimate of the energy of the ground state can be obtained from the uncertainty principle. Consider

$$H = \frac{p^2}{2\mu} + \frac{A}{r} + Br.$$

As the principle requires

$$p_x x \geq \frac{\hbar}{2}, \; p_y y \geq \frac{\hbar}{2}, \; p_z z \geq \frac{\hbar}{2},$$

we take the equal sign for the ground state and obtain

$$H = \frac{\hbar^2}{8\mu x^2} + \frac{\hbar^2}{8\mu y^2} + \frac{\hbar^2}{8\mu z^2} + \frac{A}{r} + Br.$$

To minimize *H*, let

$$\frac{\partial H}{\partial x} = 0,$$

i.e.

$$\frac{-\hbar^2}{4\mu x^2} - \frac{Ax}{r^3} + \frac{Bx}{r} = 0.$$

As *H* is symmetric with respect to *x*, *y*, *z*, when it reaches the optimal value, we have $x = y = z$, or $r = \sqrt{3}x$, and the above equation becomes

$$-\frac{\hbar^2}{4\mu x^3} - \frac{A}{3\sqrt{3}x^2} + \frac{B}{\sqrt{3}} = 0.$$

Letting $A = 0$ we get

$$x = 3^{\frac{1}{6}}\left(\frac{\hbar^2}{4\mu B}\right)^{1/3}, \quad \text{or} \quad r = \left(\frac{9\hbar^2}{4\mu B}\right)^{1/3},$$

Hence

$$\bar{H} = \frac{3\hbar^2}{8\mu x^2} + Br = 2\left(\frac{9\,\hbar^2 B^2}{2\,m_q}\right)^{\frac{1}{3}} = 3.30\left(\frac{\hbar^2 B^2}{m_q}\right)^{1/3}.$$

8020

An attractive potential well in one dimension satisfies

$$V(x) < 0, \int_{-\infty}^{+\infty} V(x)dx \text{ finite}, \quad \int_{-\infty}^{+\infty} x^2 V(x)dx \text{ finite}.$$

a. Using trial wave functions of the form $e^{-\beta x^2/2}$, prove that the potential has at least one bound state.

b. Assuming further that the potential is quite weak $\left(\int_{-\infty}^{+\infty} V(x)dx,\right.$ $\int_{-\infty}^{+\infty} x^2 V(x)dx$ are both "small"), find the best upper bound (for the energy) for this class of trial functions.

c. In a dimensionless statement, state precisely what is meant by "small" in part (b).

(Berkeley)

Sol:

a. The given trial function is the ground state wave function of a one-dimensional harmonic oscillator. We shall use the normalized function

$$\psi(x)=\left(\frac{\beta}{\pi}\right)^{1/4} e^{-\beta x^2/2},$$

where $\beta = \dfrac{m\omega}{\hbar}$. The Hamiltonian can be written as

$$H = -\frac{\hbar^2}{2m}\frac{d^2}{dx^2} + \frac{1}{2}m\omega^2 x^2 + V(x) - \frac{1}{2}m\omega^2 x^2 = H_0 + V(x) - \frac{1}{2}m\omega^2 x^2.$$

As

$$\langle\psi|H_0|\psi\rangle = \frac{1}{2}\hbar\omega,$$

we have

$$\bar{H} = \langle\psi|H|\psi\rangle = \frac{1}{2}\hbar\omega + \left\langle\psi\left|V(x)-\frac{1}{2}m\omega^2 x^2\right|\psi\right\rangle$$

$$= \frac{1}{4}\hbar\omega + \langle\psi|V(x)|\psi\rangle = \frac{\hbar^2}{4m}\beta + \langle\psi|V(x)|\psi\rangle,$$

and

$$\frac{\delta\bar{H}}{\delta\beta} = \frac{\hbar^2}{4m} + \frac{1}{2\beta}\langle\psi|V(x)|\psi\rangle - \langle\psi|x^2 V(x)|\psi\rangle.$$

Since when $\beta \to 0$

$$\frac{1}{2\beta}\langle\psi|V|\psi\rangle = \frac{1}{2\beta}\sqrt{\frac{\beta}{\pi}}\int_{-\infty}^{\infty} V dx \to -\infty$$

as V is negative,

$$\langle\psi|x^2 V|\psi\rangle = \sqrt{\frac{\beta}{\pi}}\int_{-\infty}^{\infty} x^2 V dx \to 0,$$

we have $\dfrac{\delta \bar{H}}{\delta \beta} \to -\infty$ as $\beta \to 0$. When $\beta \to \infty$, $\dfrac{\delta \bar{H}}{\delta \beta} \to \dfrac{\hbar^2}{4m} > 0$. Therefore $\dfrac{\delta \bar{H}}{\delta \beta} = 0$ at least for a certain positive β, say β_0. Thus the trial function is suitable and the energy for the corresponding state is

$$E = \bar{H}(\beta_0) = \frac{\hbar^2 \beta_0}{4m} + 2\beta_0\left(-\frac{\hbar^2}{4m} + \langle \psi | x^2 V(x)| \psi \rangle\right)$$

$$= -\frac{\hbar^2 \beta_0}{4m} + 2\beta_0 \langle \psi | x^2 V(x)| \psi \rangle < 0.$$

Therefore the system has at least one bound state. Note that we have used the fact $\left(\dfrac{\delta \bar{H}}{\delta \beta}\right)_{\beta_0} = 0$, which gives, for $\beta = \beta_0$,

$$\langle \psi | V(x) | \psi \rangle = 2\beta_0\left(-\frac{\hbar^2}{4m} + \langle \psi | x^2 V(x)| \psi \rangle\right).$$

b. c. Let $\displaystyle\int_{-\infty}^{\infty} V(x)dx = A$, $\displaystyle\int_{-\infty}^{\infty} x^2 V(x)dx = B$. The requirement that A and B are small means that the potential $V(x)$ can have large values only in the region of small $|x|$. Furthermore, for large $|x|$, $V(x)$ must attenuate rapidly. This means that we can expand the integrals

$$A_1 = \int_{-\infty}^{\infty} e^{-\beta x^2} V(x)dx \simeq \int_{-\infty}^{\infty}\left(1 - \beta x^2\right)V(x)dx = A - \beta B,$$

$$B_1 = \int_{-\infty}^{\infty} x^2 e^{-\beta x^2} V(x)dx \simeq \int_{-\infty}^{\infty} x^2 V(x)dx = B,$$

Then the minimization condition $\dfrac{\delta \bar{H}}{\delta \beta} = 0$ gives

$$\frac{\hbar^2}{4m} + \frac{A_1}{2\sqrt{\pi\beta}} - \sqrt{\frac{\beta}{\pi}} B_1 = 0,$$

or

$$\frac{\hbar^2}{4m} + \frac{A}{2\sqrt{\pi\beta}} - \frac{3}{2}\sqrt{\frac{\beta}{\pi}} B = 0,$$

i.e.

$$\sqrt{\beta} = \frac{\sqrt{\pi}\hbar^2}{12mB} + \sqrt{\frac{\pi\hbar^4}{144m^2 B^2} + \frac{A}{3B}}.$$

Hence the bound state energy is estimated to be

$$\bar{E} = -\frac{\hbar^2 \beta}{4m} + 2\beta \sqrt{\frac{\beta}{\pi}} B$$

$$= \left(\frac{\sqrt{\pi}\hbar^2}{12mB} + \sqrt{\frac{\pi\hbar^4}{144m^2B^2} + \frac{A}{3B}} \right)^2 \left(\frac{-\hbar^2}{12m} - \sqrt{\frac{\hbar^4}{36m^2} + \frac{4AB}{3\pi}} \right).$$

As A and B are both negative, $\bar{E} < 0$. Hence

$$\bar{E} < \left(\frac{\sqrt{\pi}\hbar^2}{12mB} + \sqrt{\frac{\pi\hbar^4}{144m^2B^2} + \frac{A}{3B}} \right)^2 \left(-\frac{\hbar^2}{4m} \right).$$

Since for two arbitrary real numbers a and b, $(a+b)^2 \geq 4ab$, the upper bound of $\bar{E}$ is given by

$$\bar{E} < 4 \left(-\frac{\hbar^2}{4m} \right) \frac{\sqrt{\pi}\hbar^2}{12mB} \sqrt{\frac{\pi\hbar^4}{144m^2B^2} + \frac{A}{3B}} = \frac{-\pi\hbar^6}{144m^3B^2} \sqrt{1 + \frac{48m^2AB}{\pi\hbar^4}}.$$

8021

A particle moves in an attractive central potential $V(r) = -g^2/r^{3/2}$. Use the variational principle to find an upper bound to the lowest s-state energy. Use a hydrogenic wave function as your trial function.

(Chicago)

Sol: As the trial function we use the normalized ground state wave function of the hydrogen atom,

$$\psi = \left(\frac{k^3}{8\pi} \right)^{1/2} e^{-kr/2},$$

to calculate the energy. For an s-state, $l = 0$ and

$$\bar{H}(k) = \int \psi^* H\psi \, d\tau$$

$$= \int \psi^* \left[-\frac{\hbar^2}{2mr^2} \frac{\partial}{\partial r} \left(r^2 \frac{\partial}{\partial r} \right) + V(r) \right] \psi \, d\tau$$

$$= \frac{k^3}{8\pi} \cdot 4\pi \int_0^{+\infty} r^2 e^{-kr/2} \left[-\frac{\hbar^2}{2m} \frac{1}{r^2} \frac{\partial}{\partial r} \left(r^2 \frac{\partial}{\partial r} \right) - \frac{g^2}{r^{3/2}} \right] e^{-kr/2} dr$$

$$= \frac{k^3}{2} \int_0^{+\infty} e^{-kr} \left[-g^2 r^{1/2} + \frac{\hbar^2}{2m} kr - \frac{k^2\hbar^2}{8m} r^2 \right] dr$$

$$= \frac{k^3}{2}\left[\frac{\hbar^2}{4mk} - \frac{\sqrt{\pi}g^2}{2k^{3/2}}\right]$$

$$= \frac{\hbar^2}{8m}k^2 - \frac{\sqrt{\pi}g^2}{4}k^{3/2}.$$

For $\bar{H}$ to be a minimum, $\dfrac{\partial\bar{H}}{\partial k} = 0$, i.e. $\dfrac{\hbar^2}{4m}k - \dfrac{3\sqrt{\pi}}{8}g^2 k^{1/2} = 0$, giving two solutions

$k_1 = 0$,

$$k_2^{1/2} = \frac{3\sqrt{\pi}g^2 m}{2\hbar^2}.$$

The first solution implies $\psi = 0$ and is to be discarded. On the other hand, if

$k_2^{1/2} = \dfrac{3\sqrt{\pi}g^2 m}{2\hbar^2}$, reaches a minium $-\dfrac{27\pi^2 g^8 m^3}{128\hbar^6}$. This is the upper bound to the

lowest s-state energy.

8022

A system of spin-1 particles consists of an incoherent mixture of the following 3 pure spin states, each state being equally probable, i.e. one third of the particles are in state $\psi^{(1)}$, etc.

$$\psi^{(1)} = \begin{pmatrix}1\\0\\0\end{pmatrix}, \psi^{(2)} = \frac{1}{\sqrt{2}}\begin{pmatrix}0\\1\\0\end{pmatrix} + \frac{1}{\sqrt{2}}\begin{pmatrix}0\\0\\1\end{pmatrix}, \psi^{(3)} = \begin{pmatrix}0\\0\\1\end{pmatrix}.$$

a. Find the polarization vector for each of these 3 pure states.
b. Find the polarization vector per particle **P** for the above mixed state.
c. Calculate the density matrix P for the system and verify that $\text{Tr}\,\rho = 1$
d. Using ρ, find the polarization vector **P** and check against (b).
   Reminder: for $J = 1$,

$$J_x = \frac{1}{\sqrt{2}}\begin{pmatrix}0&1&0\\1&0&1\\0&1&0\end{pmatrix}, J_y = \frac{1}{\sqrt{2}}\begin{pmatrix}0&-i&0\\i&0&-i\\0&i&0\end{pmatrix}, J_z = \begin{pmatrix}1&0&0\\0&0&0\\0&0&-1\end{pmatrix}.$$

(*Chicago*)

Sol:

a.   The polarization vector for a state i is given by

$$\mathbf{P}^{(i)} = \left\langle \psi^{(i)} \,|\, \mathbf{J} \,|\, \psi^{(i)} \right\rangle.$$

Thus

$$P_x^{(1)} = \frac{1}{\sqrt{2}}(1, 0, 0)\begin{pmatrix} 0 & 1 & 0 \\ 1 & 0 & 1 \\ 0 & 1 & 0 \end{pmatrix}\begin{pmatrix} 1 \\ 0 \\ 0 \end{pmatrix} = 0,$$

$$P_y^{(1)} = \frac{1}{\sqrt{2}}(1, 0, 0)\begin{pmatrix} 0 & -i & 0 \\ i & 0 & -i \\ 0 & i & 0 \end{pmatrix}\begin{pmatrix} 1 \\ 0 \\ 0 \end{pmatrix} = 0,$$

$$P_z^{(1)} = (1, 0, 0)\begin{pmatrix} 1 & 0 & 0 \\ 0 & 0 & 0 \\ 0 & 0 & -1 \end{pmatrix}\begin{pmatrix} 1 \\ 0 \\ 0 \end{pmatrix} = 1.$$

and so $\mathbf{P}^{(1)} = (0, 0, 1)$.

Similarly we have

$$\mathbf{P}^{(2)} = \left(\frac{1}{\sqrt{2}}, 0, -\frac{1}{2} \right),$$

$$\mathbf{P}^{(3)} = (0, 0, -1).$$

b.   For the incoherent mixture, $\mathbf{P}$ is the sum of the polarization vectors:

$$\mathbf{P} = \frac{1}{3}\left[\mathbf{P}^{(1)} + \mathbf{P}^{(2)} + \mathbf{P}^{(3)}\right] = \frac{1}{6}(\sqrt{2}, 0, -1).$$

c.   In terms of the orthonormal vectors

$$|1\rangle = \begin{pmatrix} 1 \\ 0 \\ 0 \end{pmatrix}, |2\rangle = \begin{pmatrix} 0 \\ 1 \\ 0 \end{pmatrix}, |3\rangle = \begin{pmatrix} 0 \\ 0 \\ 1 \end{pmatrix},$$

we have

$$\left| \psi^{(1)} \right\rangle = \langle 1 |, \left| \psi^{(2)} \right\rangle = \frac{1}{\sqrt{2}}(|2\rangle + (|3\rangle),$$

$$\left| \psi^{(3)} \right\rangle = |3\rangle.$$

Generally a state can be expressed as

$$\left| \psi^{(i)} \right\rangle = \sum_{n=1}^{3} C_n^i |n\rangle,$$

where $i = 1, 2, 3$. The density matrix is defined as

$$\rho = \sum_i \omega^{(i)} \left| \psi^{(i)} \right\rangle \left\langle \psi^{(i)} \right|,$$

where $\omega^{(i)}$ is the probability that the system is in the pure state $\psi^{(i)}\rangle$, or

$$\rho_{mn} = \sum_i \omega^i C_n^{i*} C_m^i = \frac{1}{3} \sum_i C_n^{i*} C_m^i,$$

as $\omega^{(i)} = \dfrac{1}{3}$ for all i in the present case. The matrix of the coefficients is

$$C = \begin{pmatrix} 1 & 0 & 0 \\ 0 & \dfrac{1}{\sqrt{2}} & \dfrac{1}{\sqrt{2}} \\ 0 & 0 & 1 \end{pmatrix},$$

and so

$$\rho = \frac{1}{3} C^+ C = \begin{pmatrix} \dfrac{1}{3} & 0 & 0 \\ 0 & \dfrac{1}{6} & \dfrac{1}{6} \\ 0 & \dfrac{1}{6} & \dfrac{1}{2} \end{pmatrix}$$

and

$$Tr\rho = \frac{1}{3} + \frac{1}{6} + \frac{1}{2} = 1.$$

d. As $\mathbf{P} = \langle J \rangle = Tr(\rho J)$, we have

$$P_x = Tr(\rho J_x) = Tr \left\{ \frac{1}{\sqrt{2}} \begin{pmatrix} 0 & \dfrac{1}{3} & 0 \\ \dfrac{1}{6} & \dfrac{1}{6} & \dfrac{1}{6} \\ \dfrac{1}{6} & \dfrac{1}{2} & \dfrac{1}{6} \end{pmatrix} \right\} = \frac{1}{\sqrt{2}} \left(\frac{1}{6} + \frac{1}{6} \right) = \sqrt{2/6},$$

$$P_y = Tr(\rho J_y) = Tr \left\{ \frac{1}{\sqrt{2}} \begin{pmatrix} 0 & \dfrac{-i}{3} & 0 \\ \dfrac{i}{6} & \dfrac{i}{6} & \dfrac{-i}{6} \\ \dfrac{i}{6} & \dfrac{i}{2} & \dfrac{-i}{6} \end{pmatrix} \right\} = \frac{1}{\sqrt{2}} \left(\frac{i}{6} - \frac{i}{6} \right) = 0$$

$$P_z = Tr(\rho J_z) = Tr \begin{pmatrix} \dfrac{1}{3} & 0 & 0 \\[2mm] 0 & 0 & -\dfrac{1}{6} \\[2mm] 0 & 0 & -\dfrac{1}{2} \end{pmatrix} = \dfrac{1}{3} - \dfrac{1}{2} = -\dfrac{1}{6},$$

same as in (b)

8023

The deuteron is a bound state of a neutron and a proton in which the two spins are coupled with a resultant total angular momentum $S = 1$. By absorbing a gamma ray of more than 2.2MeV the deuteron may disintegrate into a free neutron and a free proton.

a. Write a wave function for the final state in the reaction $\gamma + D \rightarrow n + p$ using plane waves and being sure to include properly the spin coordinates for the two particles. Assume that the interaction with the gamma ray is via electric dipole coupling.

b. Suppose the neutron and the proton are to be detected far apart from each other after the disintegration of the deuteron. Looking at this in the center-of-mass system, what correlations will be found in *time* and *space*, and in *spin*? Assume that the target consists of unpolarized deuterons. (You may use the following definition of spin correlation: If a proton is detected with spin "up", what is the probability that the corresponding deuteron will also be detected with spin "up"?)

(*Berkeley*)

Sol:

a. The ground state deuteron $^3S_1$ has positive parity. The electric dipole transition requires a change of parity between the initial and final states. Hence the parity of the free (n, p) system must have parity -1. Assume that the wave function of (n, p) can be written as $\Psi(n, p) \sim \psi(r_n, r_p)\chi(n, p)$. For $\chi = \chi_1^3$, after the nucleons are interchanged the wave function becomes

$\Psi(p, n)=(-1)^l \Psi(n, p)$. For $\chi = \chi_0^1$, after the nucleons are interchanged, the wave function becomes $\Psi(p,n)=(-1)^{l+1} \Psi(n,p)$. A fermion system must be antisymmetric with respect to interchange of any two particles, which means that for the former case, $l=1,3,...,$ and for the latter case, $l=0, 2, 4, ...,$ and so the parities are -1 (l = odd) and $+1$ (l = even) respectively. Considering the requirement we see that only states with $\chi = \chi_1^3$, i.e. spin triplet states, are possible. Further, $S=1, L=1, 3, ...,$ and so $J=0, 1, 2, ...$ As the deuterons are unpolarized, its spin wave function has the same probability of being χ_{11}, χ_{10} or χ_{1-1}. Therefore, after the transition (n, p) can be represented by the product of a plane wave and the average spin wave function:

$$\Psi(n, p) \sim e^{i(k_n \cdot r_n + k_p \cdot r_p)} \cdot e^{-i(\omega_n t + \omega_p t)}(1/\sqrt{3})(\chi_{11} + \chi_{10} + \chi_{1-1}).$$

b. The correlation of time and space is manifested in conservation of energy and conservation of momentum. In the center-of-mass coordinates, if the energy of the proton is measured to be E_p, the energy of the neutron is $E_n = E_{cm} - E_p$; if the momentum of the proton is $\mathbf{p}$, the momentum of the neutron is $-\mathbf{p}$. Let α be the spin function for "up" spin, and β be that for "down" spin. Then $\chi_{11} = \alpha(n)\alpha(p)$, $\chi_{1-1} = \beta(n)\beta(p)$, $\chi_{10} = \frac{1}{\sqrt{2}}[\alpha(n)\beta(p)+\alpha(p)\beta(n)]$, and the spin wave function is

$$\chi(n, p) = \frac{1}{\sqrt{3}}\left[\alpha(n)+\frac{1}{\sqrt{2}}\beta(n)\right]\alpha(p)+\frac{1}{\sqrt{3}}\left[\frac{1}{\sqrt{2}}\alpha(n)+\beta(n)\right]\beta(p).$$

Thus, if the spin of p is detected to be up, we have

$$\chi = \frac{1}{\sqrt{3}}\alpha(n)\alpha(p)+\frac{1}{\sqrt{6}}\beta(n)\alpha(p).$$

Hence the probability that the spin state of n is also up is

$$\frac{\left(\frac{1}{\sqrt{3}}\right)^2}{\left(\frac{1}{\sqrt{3}}\right)^2+\left(\frac{1}{\sqrt{6}}\right)^2}=\frac{2}{3}.$$

8024

a. You are given a system of two identical particles which may occupy any of three energy levels $\varepsilon_n = n\varepsilon, n=0,1,2$. The lowest energy state, $\varepsilon_0 = 0$, is doubly degenerate. The system is in thermal equilibrium at temperature T.

For each of the following cases, determine the partition function and the energy and carefully enumerate the configurations.

1. The particles obey Fermi statistics.
2. The particles obey Bose statistics.
3. The (now distinguishable) particles obey Boltzmann statistics.

b. Discuss the conditions under which fermions or bosons might be treated as Boltzmann particles.

(Buffalo)

Sol: Denote the two states with $\varepsilon_0 = 0$ by A and B and the states with ε and 2ε by 1 and 2 respectively.

1. The system can have the following configurations if the particles obey fermi statistics:

   Configuration: (A, B) (A, 1) (B, 1) (A, 2) (B, 2) (1, 2)
   Energy: $0\varepsilon\ \varepsilon\ 2\varepsilon\ 2\varepsilon\ 3\varepsilon$
   Thus the partition function is $Z = 1 + 2e^{-\varepsilon} + 2e^{-2\varepsilon} + e^{-3\varepsilon}$,
   and the mean energy is $\bar{\varepsilon} = \left(2\varepsilon e^{-\varepsilon} + 4\varepsilon e^{-2\varepsilon} + 3\varepsilon e^{-3\varepsilon} \right) / Z$

2. If the particles obey Bose statistics, in addition to the above states, the following configurations are also possible:
   Configuration: (A, A) (B, B) (1, 1) (2, 2)
   Energy: $0\ 0\ 2\varepsilon\ 4\varepsilon$
   Hence the partition function and average energy are
   $$Z = 3 + 2e^{-\varepsilon} + 3e^{-2\varepsilon} + e^{-3\varepsilon} + e^{-4\varepsilon},$$
   $$\bar{\varepsilon} = \left(2\varepsilon e^{-\varepsilon} + 6\varepsilon e^{-2\varepsilon} + 3\varepsilon e^{-3\varepsilon} + 4\varepsilon e^{-4\varepsilon} \right) / Z.$$

3. for destinguisable particles obeying Boltzmann statistics, more configurations are possible. These are (B, A), (1, A), (1, B), (2, A), (2, B) and (2, 1). Thus we have
   $$Z = 4 + 4e^{-\varepsilon} + 5e^{-2\varepsilon} + 2e^{-3\varepsilon} + e^{-4\varepsilon},$$
   $$\bar{\varepsilon} = \left(4\varepsilon e^{-\varepsilon} + 10\varepsilon e^{-2\varepsilon} + 6\varepsilon e^{-3\varepsilon} + 4\varepsilon e^{-4\varepsilon} \right) / Z.$$

b. Fermions and bosons can be treated as Boltzmann particles when the number of particles is much less than the number of energy levels, for then the exchange effect can be neglected.

8025

Consider a free electron near a boundary surface.

a. If $\phi_k(x)$'s are the electron eigenfunctions, show that the function

$$u(x, t) = \sum_k \phi_k^*(x)\phi_k(0)\exp\left(\frac{-\varepsilon_k t}{\hbar}\right)$$

satisfies a diffusion-type equation. Identify the corresponding diffusion coefficient.

b. From the theory of diffusion how would you expect $u(0, t)$ to be influenced by the presence of a boundary at a distance l from the origin? Would the boundary be felt immediately or only after an elapse of time?

c. Examine the expression for $u(0, t)$ as a sum over k as given in (a). What is the range of ε_k which contribute significantly to $u(0, t)$ at the time when the influence of the boundary is felt by the electron?

(Buffalo)

Sol:

a. The wave function $\phi_k(x)$ satisfies the Schrödinger equation of a free particle

$$-\frac{\hbar^2}{2m}\nabla^2\phi_k(\mathbf{x}) = \varepsilon_k\phi_k(\mathbf{x}).$$

Thus

$$\nabla^2 u(\mathbf{x}, t) = -\frac{2m}{\hbar^2}\sum_k \varepsilon_k\phi_k^*(\mathbf{x})\phi_k(0)\exp\left(-\frac{\varepsilon_k t}{\hbar}\right).$$

Since

$$\frac{\partial}{\partial t}u(\mathbf{x}, t) = -\frac{1}{\hbar}\sum_k \varepsilon_k\phi_k^*(\mathbf{x})\phi_k(0)\exp\left(-\frac{\varepsilon_k t}{\hbar}\right),$$

$u(x, t)$ satisfies the following diffusion-type equation:

$$\frac{\partial}{\partial t}u(\mathbf{x}, t) = \frac{\hbar}{2m}\nabla^2 u(\mathbf{x}, t).$$

The corresponding diffusion coefficient is $\hbar/2m$.

b. Initially $u(x,0) = \delta(x)$. When $t > 0$, the function u starts diffusing to both sides. The boundary will not be felt by the electron before a lapse of time.

656 Problems and Solutions in Quantum Mechanics

c. Suppose the boundary is at $x = l$. The solution of the diffusion equation is

$$u(\mathbf{x}, t) = c \exp\left[-\frac{m}{2\hbar t}\left(y^2 + z^2\right)\right]$$
$$\times\left\{\exp\left[-\frac{m}{2\hbar t}x^2\right] - \exp\left[-\frac{m}{2\hbar t}(x-2l)^2\right]\right\}.$$

When there is no boundary (i.e., $l \to \infty$), the solution is

$$u(\mathbf{x}, t) = c \exp\left[-\frac{m}{2\hbar t}\left(y^2 + z^2\right)\right]\exp\left(-\frac{m}{2\hbar t}x^2\right).$$

From the above two expressions, we see that only when $\dfrac{m}{2\hbar t}(0 - 2l)^2 \sim 1$, i.e.,

at $t \sim 2ml^2 / \hbar$, will the electron start to feel the existence of the boundary.
Consider

$$u(0, t) = \sum_k |\phi_k(0)|^2 \exp\left(-\frac{\varepsilon_k t}{\hbar}\right).$$

Only states ϕ_k for which the energy ε_k is such that $\dfrac{\varepsilon_k t}{\hbar} \leq 1$ will contribute significantly to $u(0, t)$. At the time $t \sim \dfrac{2ml^2}{\hbar}$, we require $\varepsilon_k \leq \dfrac{\hbar^2}{2ml^2}$ for ϕ_k to make a significant contribution.

8026

Symmetrizing Maxwell's equations by postulating a magnetic monopole charge of strength g, Dirac derived a quantization condition

$$\frac{eg}{\hbar c} = n,$$

where n = an integer, e is the electronic charge, and g is the magnetic charge. In the spirit of the Bohr-Sommerfeld quantization procedure, derive semi-classically a similar quantization condition by quantizing the angular momentum of the field in the "mixed dipole" system shown in Fig. 8.6. Hint: How is the angular momentum of the field related to the Poynting vector?

(Columbia)

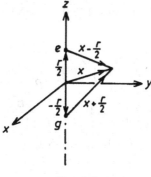

Fig. 8.6

ol: The electromagnetic field consists of two components

$$E = \frac{e\left(x - \dfrac{r}{2}\right)}{\left|x - \dfrac{r}{2}\right|^3},$$

$$B = \frac{g\left(x + \dfrac{r}{2}\right)}{\left|x + \dfrac{r}{2}\right|^3}.$$

In cylindrical coordinates (ρ, θ, z), we can write $\mathbf{r} = 2a\mathbf{e}_3$, where $a = |\mathbf{r}|/2$, and

$$\mathbf{x} = \rho\cos\theta\,\mathbf{e}_1 + \rho\sin\theta\,\mathbf{e}_2 + z\mathbf{e}_3,$$

The angular momentum of the electromagnetic field is

$$\mathbf{L}_{em} = \frac{1}{4\pi c}\int \mathbf{x}\times(\mathbf{E}\times\mathbf{B})d^3x.$$

As

$$\mathbf{E}\times\mathbf{B} = eg\frac{\left(\mathbf{x} - \frac{r}{2}\right)\times\left(\mathbf{x} + \frac{r}{2}\right)}{\left|\mathbf{x} - \frac{r}{2}\right|^3 \cdot \left|\mathbf{x} + \frac{r}{2}\right|^3}$$

$$= \frac{eg\,\mathbf{x}\times\mathbf{r}}{\left[\left(\mathbf{x} - \frac{r}{2}\right)^2\left(\mathbf{x} + \frac{r}{2}\right)^2\right]^{3/2}},$$

$$\mathbf{x}\times(\mathbf{E}\times\mathbf{B}) = \frac{eg\left[(\mathbf{x}\cdot\mathbf{r})\mathbf{x} - x^2\mathbf{r}\right]}{\left[\left(\rho^2 + z^2 + a^2 - 2az\right)\left(\rho^2 + z^2 + a^2 + 2az\right)\right]^{3/2}}$$

$$= \frac{2aeg\left(z\rho\cos\theta\,\mathbf{e}_1 + z\rho\sin\theta\,\mathbf{e}_2 - \rho^2\mathbf{e}_3\right)}{\left[\left(\rho^2 + z^2 + a^2\right) - 4a^2z^2\right]^{3/2}},$$

$$\int_0^{2\pi} \cos\theta \, d\theta = \int_0^{2\pi} \sin\theta \, d\theta = 0,$$

we have

$$\mathbf{L}_{em} = -\frac{aeg}{2\pi c}\mathbf{e}_3 \int_{-\infty}^{+\infty} dz \int_0^{2\pi} d\theta \int_0^{\infty} \frac{\rho^3 d\rho}{[(\rho^2 + z^2 + a^2)^2 - 4a^2 z^2]^{3/2}}$$

$$= -\mathbf{e}_3 \left(\frac{eg}{c}\right) \int_{-\infty}^{+\infty} dt \int_0^{\infty} \frac{s^3 ds}{[(s^2 + t^2 + 1)^2 - 4t^2]^{3/2}},$$

where $s = \rho/a$, $t = z/a$. It can be shown that

$$\int_{-\infty}^{+\infty} dt \int_0^{\infty} \frac{s^3 ds}{[(s^2 + t^2 + 1)^2 - 4t^2]^{3/2}} = 1.$$

Hence

$$\mathbf{L}_{em} = -\frac{eg}{c}\mathbf{e}_3.$$

The quantization condition is therefore

$$|\mathbf{L}_{emz}| = \frac{eg}{c} = n\hbar,$$

or

$$\frac{eg}{\hbar c} = n, \quad n = 0, \pm 1, \pm 2, \ldots.$$

8027

In a crude picture, a metal is viewed as a system of free electrons enclosed in a well of potential difference V_0. Due to thermal agitations, electrons with sufficiently high energies will escape from the well. Find and discuss the emission current density for this model.

(Buffalo)

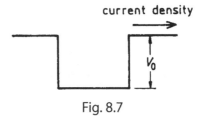

Fig. 8.7

Sol: The system of free electrons can be considered as an electron gas of volume V which obeys the Fermi statistics. At absolute temperature T the number density of electrons with momenta between P and $\mathbf{P} + d\mathbf{P}$, where $\mathbf{P} = (P_x, P_y, P_z)$, is

$$\frac{dN}{V} = \frac{1}{V} \cdot \frac{dP_x dP_y dP_z}{e^{(\varepsilon - \mu)/kT} + 1} \cdot \frac{2V}{h^3}$$

where the factor 2 is the degeneracy due to the electrons having two spin directions.

Consider the number of electrons, j_n, leaving V in the z direction per unit cross sectional area per unit time. Such electrons must have a speed

$$v_z = \frac{P_z}{m} \geq \frac{1}{m}\sqrt{2mV_0}.$$

Hence

$$j_n = \int v_z \frac{dN}{V} = \frac{2}{mh^3} \int_{\sqrt{2mV_0}}^{\infty} P_z dP_z$$

$$\times \int_{-\infty}^{\infty} \int_{-\infty}^{\infty} \frac{dP_x dP_y}{\exp\left\{\left[\frac{1}{2m}\left(P_x^2 + P_y^2 + P_z^2\right) - \mu\right]/kT\right\} + 1},$$

or, by setting $P_x^2 + P_y^2 = P_r^2$, $dP_x dP_y = 2\pi P_r dP_r$, and neglecting the number 1 in the denominator,

$$j_n \approx \frac{4\pi}{mh^3} \int_{\sqrt{2mV_0}}^{\infty} P_z dP_z \int_0^{\infty} \exp\left\{-\frac{1}{kT}\left[\frac{1}{2m}\left(P_z^2 + P_r^2\right) - \mu\right]\right\} P_r dP_r$$

$$= \frac{4\pi kT}{h^3} \int_{\sqrt{2mV_0}}^{\infty} \exp\left[-\frac{1}{kT}\left(\frac{P_z}{2m} - \mu\right)\right] P_z dP_z$$

$$= \frac{4\pi m k^2 T^2}{h^3} e^{-(V_0 - \mu)/kT}.$$

The electric current density is then

$$j_e = -e j_n = -\frac{4\pi m e k^2 T^2}{h^3} e^{-(V_0 - \mu)/kT}.$$

Note that in the above, to simplify the integration, we have assumed

$$kT \ll \frac{1}{2m} \cdot 2mV_0 - \mu = V_0 - \mu.$$

At $T = 0$ the electron number density is

$$n = \frac{2}{h^3} \cdot \frac{4\pi}{3} P_0^3 = \frac{1}{3\pi^2 \hbar^3}\left(2m\mu_0\right)^{3/2},$$

where P_0, μ_0, are the limiting momentum and energy. At ordinary temperatures we have

$$\mu \approx \mu_0 = \frac{\hbar^2}{2m}\left(3\pi^2 n\right)^{2/3}.$$

The quantity $V_0 - \mu$ is the work function of the metal and the emission of electrons from incandescent cathodes is known as Richardson's effect.

8028

It is generally recognized that there are at least three different kinds of neutrinos. They can be distinguished by the reactions in which the neutrinos are created or absorbed. Let us call these three types of neutrino v_e, v_μ and v_τ. It has been speculated that each of the neutrinos has a small but finite rest mass, possibly different for each type. Let us suppose, for this exam question, that there is a small perturbing interaction between these neutrino types, in the absence of which all three types have the same nonzero rest mass M_0. Let the matrix element of this perturbation have the same real value $\hbar\omega_1$ between each pair of neutino types. Let it have zero expectation value in each of the states v_e, v_μ and v_τ.

a. A neutrino of type v_e is produced at rest at time zero. What is the probability, as a function of time, that the neutrino will be in each of the other two states?

b. [Can be answered independently of (a)] An experiment to detect these "neutrino oscillations" is being performed. The flight path of the neutrinos is 2000 meters. Their energy is 100 GeV. The sensitivity is such that the presence of 1% of neutrinos of one type different from that produced at the start of the flight path can be measured with confidence. Take M_0 to be 20 electron volts. What is the smallest value of $\hbar\omega_1$ that can be detected? How does this depend on M_0?

(Berkeley)

a. In the representation of $|ve\rangle, |v\mu\rangle$ and $|v_\tau|$, the matrix of the Hamiltonian of the system is

$$H = \begin{pmatrix} M_0 & \hbar\omega_1 & \hbar\omega_1 \\ \hbar\omega_1 & M_0 & \hbar\omega_1 \\ \hbar\omega_1 & \hbar\omega_1 & M_0 \end{pmatrix}.$$

The Schrödinger equation

$$\frac{\hbar}{i}\frac{\partial\Psi}{\partial t} + H\Psi = 0,$$

where $\Psi = \begin{pmatrix} a_1 \\ a_2 \\ a_3 \end{pmatrix}$, a_i being the wave function for state v_i, has the matrix form

$$i\hbar\begin{pmatrix} \dot{a}_1 \\ \dot{a}_2 \\ \dot{a}_3 \end{pmatrix} = \begin{pmatrix} M_0 & \hbar\omega_1 & \hbar\omega_1 \\ \hbar\omega_1 & M_0 & \hbar\omega_1 \\ \hbar\omega_1 & \hbar\omega_1 & M_0 \end{pmatrix}\begin{pmatrix} a_1 \\ a_2 \\ a_3 \end{pmatrix}$$

with the initial condition

$$a_1(0) = 1, \ a_2(0) = a_3(0) = 0.$$

The solution is

$$\begin{cases} a_1(t) = e^{-iM_0 t/\hbar}\left(\frac{2}{3}e^{i\omega_1 t} + \frac{1}{3}e^{-i2\omega_1 t}\right), \\ a_2(t) = \frac{1}{3}e^{-iM_0 t/\hbar}\left(e^{-2\omega_1 t} - e^{i\omega_1 t}\right), \\ a_3(t) = \frac{1}{3}e^{-iM_0 t/\hbar}\left(e^{-2\omega_1 t} - e^{i\omega_1 t}\right). \end{cases}$$

Hence the probabilities of the neutrino being in states v_μ and v_τ are

$$\begin{cases} P(v_\mu) = |a_2(t)|^2 = \frac{2}{9}(1 - \cos 3\omega_1 t), \\ P(v_\tau) = |a_3(t)|^2 = \frac{2}{9}(1 - \cos 3\omega_1 t). \end{cases}$$

d. The time of flight of v_e is $\Delta t = \dfrac{l}{v}$ in the laboratory time, or $\Delta \tau = \Delta t \sqrt{1 - \left(\dfrac{v}{c}\right)^2}$

$\approx \dfrac{l}{c} \dfrac{M_0}{E}$, where E is the total energy, in the rest frame of v_e. For $P(v_\mu) \geq 1\%$,

i.e.,

$$\frac{2}{9}\left[1 - \cos(3\omega_1 \Delta \tau)\right] \gtrsim 0.01,$$

we require

$$\omega_1 \geq \frac{\cos^{-1} 0.955}{3\Delta\tau} = \frac{0.301}{3\Delta\tau} = \frac{0.1cE}{lM_0},$$

or

$$\hbar\omega_1 \geq \frac{0.1 \times 3 \times 10^8 \times 100 \times 10^9 \times 6.58 \times 10^{-16}}{2000 \times 20} = 0.05 \text{ eV}.$$

8029

To a good approximation, an electron in a crystal lattice experiences a periodic potential as shown in Fig. 8.8.:

It is a theorem (Floquet's), and a physical fact, that the spectrum of any such periodic potential sparates into continuous "bands" with forbidden "gaps". To construct a very crude model of (the lowest band of) this effect, imagine that the barriers are high, so that the set of "ground states" $|n\rangle (-\infty < n < +\infty)$ (one for each well) are approximate eigenstates. Call E_0 the energy of each $|n\rangle$. Now suppose $\varepsilon = |\varepsilon| e^{i\alpha}$ is the (small) amplitude for tunneling between any two nearest-neighbor wells (probability for $|n-1\rangle \leftarrow |n\rangle \rightarrow |n+1> is |\varepsilon|^2$). Set up a Hermitian Hamiltonian that describes this. Compute the energy $E(\theta)$ of the state(s)

$$|\theta\rangle = \sum_{n=-\infty}^{+\infty} e^{in\theta} |n\rangle$$

What is the width of your band?

(Berkeley)

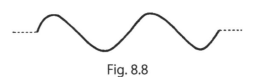

Fig. 8.8

Sol: We write the Hamiltonian as a matrix, choosing $|n\rangle$ as basis vectors. Supposing

$$H|n\rangle = E_0\left(1-\varepsilon-\varepsilon^*\right)|n\rangle + E_0\varepsilon|n+1\rangle + E_0\varepsilon^*|n-1\rangle,$$

we have

$$\langle m|H|n\rangle = \int \psi^*(x-ma)H\psi(x-na)dx$$

$$= E_0\int \psi^*(x-ma)\psi(x-na)dx$$

$$= \delta_{mn}E_0\left(1-\varepsilon-\varepsilon^*\right) + \delta_{m,n+1}\varepsilon E_0 + \delta_{m,n-1}\varepsilon^* E_0,$$

where we have used the assumption that tunneling occurs only between adjacent potential wells and the amplitude for tunneling to the right is $\varepsilon = |\varepsilon|e^{i\alpha}$, that to the left is $\varepsilon^* = |\varepsilon|e^{-i\alpha}$. Thus the matrix of H is

$$H = \begin{array}{c} \langle 1| \\ \langle 2| \\ \langle 3| \\ \vdots \\ \vdots \end{array} \left[\begin{array}{cccc} E_0(1-\varepsilon-\varepsilon^*) & \varepsilon E_0 & & 0 \\ \varepsilon^* E_0 & E_0(1-\varepsilon-\varepsilon^*) & \varepsilon E_0 & \\ & \varepsilon^* E_0 & E_0(1-\varepsilon-\varepsilon^*) & \\ & & & \\ 0 & & & \end{array} \right]$$
$$\qquad\qquad |1\rangle \qquad |2\rangle \qquad |3\rangle \ldots\ldots\ldots$$

and

$$H|\theta\rangle = E_0\sum_{n=-\infty}^{+\infty}e^{in\theta}|n\rangle\left(1-\varepsilon-\varepsilon^*\right) + E_0\sum_{n=-\infty}^{+\infty}e^{in\theta}\left(\varepsilon|n+1\rangle + \varepsilon^*|n-1\rangle\right)$$

$$= E_0(1-2|\varepsilon|\cos\alpha)|\theta\rangle + E_0\sum_{n=-\infty}^{+\infty}\left[e^{i(n-1)\theta}\varepsilon + e^{i(n+1)\theta}\varepsilon^*\right]|n\rangle$$

$$= E_0[1-2|\varepsilon|\cos\alpha + 2|\varepsilon|\cos(\theta-\alpha)]|\theta\rangle.$$

Hence the energy eigenvalue of $|\theta\rangle$ is

$$E_\theta = E_0[1-2|\varepsilon|(\cos\alpha-\cos(\theta-\alpha))]$$

$$= E_0\left[1-4|\varepsilon|\sin\frac{\theta}{2}\sin\left(\frac{\theta}{2}-\alpha\right)\right].$$

From these results it can be concluded as follows:

i. Since a continuous variation of θ results in a continuous variation of the energy, the energy levels become an energy band. Furthermore, when $\theta = \alpha$, $E_\theta = E_{max} = E_0\{1+2|\varepsilon|(1-\cos\alpha)\}$, and when $\theta = \pi+\alpha, E_0 = E_{min} = E_0\{1-2|\varepsilon|(1+\cos\alpha)\}$. So the width of the band is $E_{max}-E_{min} = 4|\varepsilon|E_0$.

ii. When α, which depends on the shape of the periodic potential well, is sufficiently small, tunneling between neighboring wells always results in a lowering of the ground state energy.

8030

Consider an idealized (point charge) Al atom ($Z = 13$, $A = 27$). If a negative lepton or meson is captured by this atom it rapidly cascades down to the lower n states which are inside the electron shells. In the case of μ-capture:

a. Compute the energy E_1 for the μ in the $n = 1$ orbit; estimate also a mean radius. Neglect relativistic effects and nuclear motion.

b. Now compute a correction to E_1 to take into account the nuclear motion.

c. Find a perturbation term to the Hamiltonian due to relativistic kinematics, ignoring spin. Estimate the resulting correction to E_1.

d. Define a nuclear radius. How does this radius for Al compare to the mean radius for the $n = 1$ orbit from (a)? Discuss qualitatively what happens to the μ^- when the μ^- atomic wave function overlaps the nucleus substantially. What happens to a π^- under the circumstances? Information that may be relevant:

$$M_\mu = 105 \, \text{MeV}/c^2, \quad \text{SPIN}(\mu) = 1/2,$$
$$M_\pi = 140 \, \text{MeV}/c^2, \quad \text{SPIN}(\pi) = 0.$$

(Berkeley)

Sol:

a. We shall neglect the effects of the electrons outside the nucleus and consider only the motion of the μ in the Coulomb field of the Al nucleus. The energy levels of μ in a hydrogen-like atom of nuclear charge Z (in the nonrelativistic approximation) are given by

$$E_n = -\frac{me^4}{2\hbar^2}\frac{Z^2}{n^2}.$$

Thus

$$E_1 = -\frac{m}{m_e}\frac{m_e e^4}{2\hbar^2}Z^2 = -\frac{105}{0.51}\times 13.6\times 13^2 \, \text{eV}$$
$$= -0.4732 \text{MeV},$$

$$a = \frac{\hbar^2}{Zme^2} = \frac{m_e}{Zm}\frac{\hbar^2}{m_e e^2} = \frac{0.5}{13\times105}\times 0.53\text{Å}$$

$$= 1.9\times10^{-4}\,\text{Å}.$$

b. To take into account the motion of the nucleus, we simply have to replace the mass m of the meson with its reduced mass $\mu = \dfrac{Mm}{M+m}$, M being the nuclear mass. Thus

$$E_1' = \frac{\mu}{m}E_1 = \frac{1}{1+\dfrac{m}{M}}E_1 = \frac{-0.4732}{1+\dfrac{105}{27\times938}}$$

$$= -0.471 \text{ MeV}.$$

c. Taking into account the relativistic effects the muon kinetic energy is

$$T = \sqrt{p^2c^2 + m^2c^4} - mc^2 = \frac{p^2}{2m} - \frac{p^4}{8m^3c^2} + \cdots.$$

The relativistic correction introduces a perturbation Hamiltonian

$$H = -\frac{p^4}{8m^3c^2}.$$

The energy correction ΔE for E_1 is then

$$\Delta E = \left\langle 100 \left| -\frac{p^4}{8m^3c^2} \right| 100 \right\rangle$$

$$= -\frac{1}{2mc^2}\left\langle 100 \left| \frac{p^2}{2m}\cdot\frac{p^2}{2m} \right| 100 \right\rangle$$

$$= -\frac{1}{2mc^2}\left\langle 100 \left| \left(H+\frac{Ze^2}{r}\right)\left(H+\frac{Ze^2}{r}\right) \right| 100 \right\rangle$$

$$= -\frac{1}{2mc^2}\left\langle 100 \left| \left(E_1+\frac{Ze^2}{r}\right)^2 \right| 100 \right\rangle.$$

For a rough estimate, take $r \approx a$. Then

$$\Delta E \approx -\frac{1}{2mc^2}\left(E_1 + \frac{Ze^2}{a}\right)^2 = -\frac{|E_1|^2}{2mc^2} = \frac{-0.47^2}{2\times105} = 1.06\times10^{-3}\,\text{MeV}$$

d. In the scattering of neutrons by a nucleus, an attractive strong nuclear force sets in when the distance becomes smaller than $r \sim r_0 A^{\frac{1}{3}}$, where $r_0 \sim 1.2\times10^{-13}$ cm and A is the atomic mass number of the nucleus. r is generally taken to be the radius of the nucleus, which for Al is

$$r = 1.2\times10^{-5}\times 27^{\frac{1}{3}} = 3.6\times10^{-5}\,\text{Å}.$$

The difference between the radius of the nucleus of Al and that of the first orbit of the μ-mesic atom is not very large, so that there is a considerable overlap of the wave functions of the nucleus and the muon. This effect, due to the finite volume of the nucleus, will give rise to a positive energy correction. At the same time there is also a large interaction between the muon and the magnetic moment of the nucleus.

Under similar circumstances, for the π-mesic atom there is also the volume effect, but no interaction with the magnetic moment of the nucleus as pions have zero spin.

8031

Low energy neutrons from a nuclear reactor have been used to test gravitationally induced quantum interference. In Fig. 8.9, neutrons incident from A can follow two paths of equal lengths, ABCEF and ABDEF, and interfere after they recombine at E. The three parallel slabs which diffract the neutrons are cut from one single crystal. To change the effects of the gravitational potential energy, the system can be rotated about the line ABD. Suppose ϕ is the angle of this rotation ($\phi = 0$ for the path ABCEF horizontal).

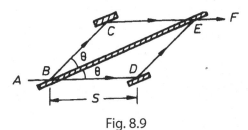

Fig. 8.9

a. Show that the phase difference at point E due to the effect of gravity can be expressed as $\beta = q\sin\phi$, where $q = K\lambda S^2 \sin 2\theta$, λ being the neutron wavelength and K an appropriate constant which depends on neutron mass m, gravitational acceleration g, Planck's constant $\hbar$, and numerical factors. Determine the constant K. Assume here that the gravitational potential energy differences are very small compared to the neutron kinetic energies.

Miscellaneous Topics **667**

b. The neutron wavelength used in the experiment is 1.45Å . What is the corresponding kinetic energy in electron volts?

c. If $S = 4$ cm, $\theta = 22.5°$, and $\lambda = 1.45$Å, how many maxima should be seen by a neutron counter at F as ϕ goes from $-90°$ to $+90°$?

Mass of neutron $= 939$ MeV $/ c^2$, $\hbar c = 1.97 \times 10^{-11}$ MeV $\cdot$ cm.

Sol:

a. The wave function of the incident neutrons can be taken as

$$\psi(\mathbf{r}, t) = c e^{(i\mathbf{p}\cdot\mathbf{r} - iEt)/\hbar},$$

where c is a constant. When they move along a certain orbit from $x = 0$ to $x = l$ it becomes

$$\psi(\mathbf{r}, t) = c \exp\left[\frac{i}{\hbar}\int_0^l \sqrt{2m(E - V)}\,dx - \frac{i}{\hbar}Et\right].$$

Thus the phase is

$$\varphi = \frac{1}{\hbar}\int_0^l \sqrt{2m(E - V)}\,dx - \frac{1}{\hbar}Et.$$

The neutrons are separated at point B into two beams 1 and 2, for which $\varphi_{B1} = \varphi_{B2}$.

The situations on lines BC and DE are same and so for the two neutron beams, $\Delta\varphi_{CB} = \varphi_{ED}$. On line BD, we can set the gravitational potential $V = 0$, $E = E_0$, and so

$$\Delta\varphi_{DB} = \frac{1}{\hbar}\int_0^S \sqrt{2mE_0}\,dx - \frac{1}{\hbar}E_0 \cdot \frac{S}{v_0}$$

$$= \frac{1}{\hbar}\left(\sqrt{2mE_0}\,S - \frac{1}{2}\sqrt{2mE_0}\,S\right) = \frac{S}{2\hbar}\sqrt{2mE_0},$$

where v_0 is the neutron velocity $\frac{1}{m}\sqrt{2mE_0}$.

On line CE, the gravitational potential is

$$V = mgh = mg \cdot \overline{BE} \sin\theta \sin\phi, \quad \text{with} \quad \overline{BE} = 2S\cos\theta,$$

i.e.,

$$V = mgS \sin 2\theta \sin\phi,$$

and

$$\Delta\varphi_{EC} = \frac{1}{\hbar}\int_0^S \sqrt{2m(E_0 - V)}\,dx - \frac{1}{\hbar}E_0 t$$

$$= \frac{S}{\hbar}\sqrt{2mE_0}\sqrt{1 - \frac{V}{E_0}} - \frac{1}{\hbar}E_0 \frac{S}{\sqrt{\dfrac{2(E_0 - V)}{m}}}$$

$$= S\frac{\sqrt{2mE_0}}{2\hbar}\left(2\sqrt{1 - \frac{V}{E_0}} - \frac{1}{\sqrt{1 - \dfrac{V}{E_0}}}\right).$$

Thus the phase difference of the two beams of neutrons at point F is

$$\beta = \Delta\varphi_{DB} - \Delta\varphi_{EC} = \frac{S}{2\hbar}\sqrt{2mE_0}\left(1 - 2\sqrt{1 - \frac{V}{E_0}} + \frac{1}{\sqrt{1 - \dfrac{V}{E_0}}}\right)$$

$$\approx \frac{S}{2\hbar}\sqrt{2mE_0}\left[1 - 2\left(1 - \frac{V}{2E_0}\right) + \left(1 + \frac{V}{2E_0}\right)\right]$$

$$= \frac{3VS}{4\hbar E_0}\sqrt{2mE_0}$$

as $V \ll E_0$. Thus

$$\beta = q\sin\phi,$$

where

$$q \approx \frac{3}{2}\frac{m^2 g}{\hbar\sqrt{2mE_0}}S^2\sin 2\theta$$

$$= K\lambda S^2 \sin 2\theta$$

with

$$\lambda = \frac{2\pi\hbar}{p} = \frac{2\pi\hbar}{\sqrt{2mE_0}},$$

$$K = \frac{3}{4}\frac{m^2 g}{\pi\hbar^2}.$$

b. The neutron has momentum

$$p = \frac{h}{\lambda} = \frac{2\pi\hbar}{\lambda}$$

and hence kinetic energy

$$E_k = \frac{p^2}{2m} = \frac{2}{mc^2}\left(\frac{\pi\hbar c}{\lambda}\right)^2$$

$$= \frac{2}{939\times10^6}\left(\frac{\pi\times1.97\times10^{-5}}{1.45\times10^{-8}}\right)^2$$

$$= 0.039\text{eV}$$

c. In the range $-1 < \sin\phi < 1$, the number of maxima seen by a neutron counter at F is

$$n = \frac{2q}{2\pi} = \frac{3}{4}\left(\frac{Smc^2}{\pi\hbar c^2}\right)^2 g\lambda\sin2\theta$$

$$= \frac{3}{4}\left(\frac{4\times939}{\pi\times1.97\times10^{-11}\times3\times10^{10}}\right)^2 \times980\times1.45\times10^{-8}\times\sin45°$$

$$= 30.$$

8032

Consider the Dirac equation in one dimension

$$H\psi = i\hbar\frac{\partial\psi}{\partial t},$$

where

$$H = c\alpha p_z + \beta mc^2 + V(z) = c\alpha\left(-i\hbar\frac{\partial}{\partial z}\right) + \beta mc^2 + V(z),$$

$$\alpha = \begin{pmatrix} 0 & \sigma_3 \\ \sigma_3 & 0 \end{pmatrix}, \quad \sigma_3 = \begin{pmatrix} 1 & 0 \\ 0 & -1 \end{pmatrix}, \quad \beta = \begin{pmatrix} I & 0 \\ 0 & -I \end{pmatrix},$$

I being the 2×2 unit matrix.

a. Show that $\sigma = \begin{pmatrix} \sigma_3 & 0 \\ 0 & \sigma_3 \end{pmatrix}$ commutes with H.

b. Use the results of (a) to show that the one-dimensional Dirac equation can be written as two coupled first order differential equations.

(Buffalo)

Sol:

a. As

$$[\sigma, \alpha] = \left[\begin{pmatrix} \sigma_3 & 0 \\ 0 & \sigma_3 \end{pmatrix}, \begin{pmatrix} 0 & \sigma_3 \\ \sigma_3 & 0 \end{pmatrix} \right] = 0,$$

$$[\sigma, \beta] = \left[\begin{pmatrix} \sigma_3 & 0 \\ 0 & \sigma_3 \end{pmatrix}, \begin{pmatrix} I & 0 \\ 0 & -I \end{pmatrix} \right] = 0,$$

we have

$$[\sigma, H] = \left[\sigma, c\alpha p_z + \beta mc^2 + V \right] = c[\sigma, \alpha]p_z + [\sigma, \beta]mc^2 = 0.$$

b. As $[\sigma, H] = 0$, σ and H have common eigenfunctions. σ is a diagonal matrix.

Let its eigenfunction be $\begin{pmatrix} \psi_1 \\ \psi_2 \\ \psi_3 \\ \psi_4 \end{pmatrix}$. As

$$\sigma \begin{pmatrix} \psi_1 \\ \psi_2 \\ \psi_3 \\ \psi_4 \end{pmatrix} = \begin{pmatrix} \psi_1 \\ -\psi_2 \\ \psi_3 \\ -\psi_4 \end{pmatrix} = \begin{pmatrix} \psi_1 \\ 0 \\ \psi_3 \\ 0 \end{pmatrix} - \begin{pmatrix} 0 \\ \psi_2 \\ 0 \\ \psi_4 \end{pmatrix},$$

σ has eigenfunctions $\begin{pmatrix} \psi_1 \\ 0 \\ \psi_3 \\ 0 \end{pmatrix}$ and $\begin{pmatrix} 0 \\ \psi_2 \\ 0 \\ \psi_4 \end{pmatrix}$ with eigenvalues $+1$ and -1

respectively.

Substituting these in the Dirac equation, we obtain

$$\left(-i\hbar c \frac{\partial}{\partial z} + V \right) \begin{pmatrix} \psi_3 \\ 0 \\ \psi_1 \\ 0 \end{pmatrix} + mc^2 \begin{pmatrix} \psi_1 \\ 0 \\ -\psi_3 \\ 0 \end{pmatrix} = i\hbar \frac{\partial}{\partial t} \begin{pmatrix} \psi_1 \\ 0 \\ \psi_3 \\ 0 \end{pmatrix},$$

$$\left(-i\hbar c \frac{\partial}{\partial z} + V \right) \begin{pmatrix} 0 \\ -\psi_4 \\ 0 \\ -\psi_2 \end{pmatrix} + mc^2 \begin{pmatrix} 0 \\ \psi_2 \\ 0 \\ -\psi_4 \end{pmatrix} = i\hbar \frac{\partial}{\partial t} \begin{pmatrix} 0 \\ \psi_2 \\ 0 \\ \psi_4 \end{pmatrix}.$$

Each of these represents two coupled differential equations. However, the two sets of equations become identical if we let $\psi_3 \rightarrow -\psi_4$, $\psi_1 \rightarrow \psi_2$. Thus the one-dimensional Dirac equation can be written as two coupled first order differential equations.

8033

a. Write down the Dirac equation in Hamiltonian form for a free particle, and give explicit forms for the Dirac matrices.

b. Show that the Hamiltonian H commutes with the operator $\sigma \cdot \mathbf{P}$ where $\mathbf{P}$ is the momentum operator and σ is the Pauli spin operator in the space of four component spinors.

c. Find plane wave solutions of the Dirac equation in the representation in which $\sigma \cdot \mathbf{P}$ is diagonal. Here $\mathbf{P}$ is the eigenvalue of the momentum operator.

(Buffalo)

Sol:

a.

$$\bar{H} = c\alpha \cdot \mathbf{P} + \beta mc^2 = c\alpha \cdot (-i\hbar\nabla) + \beta mc^2,$$

where

$$\alpha = \begin{pmatrix} 0 & \sigma \\ \sigma & 0 \end{pmatrix}, \beta = \begin{pmatrix} 1 & 0 \\ 0 & -1 \end{pmatrix}$$

are the Dirac matrices.

b. Write

$$\alpha \cdot \mathbf{P} = \begin{pmatrix} 0 & \sigma \\ \sigma & 0 \end{pmatrix} \cdot \mathbf{P} = \begin{pmatrix} 0 & \sigma \cdot \mathbf{P} \\ \sigma \cdot \mathbf{P} & 0 \end{pmatrix}.$$

As

$$\sigma \cdot \mathbf{P} = \sigma \cdot \mathbf{P}1 = \begin{pmatrix} \sigma \cdot \mathbf{P} & 0 \\ 0 & \sigma \cdot \mathbf{P} \end{pmatrix},$$

$$[\sigma \cdot \mathbf{P}, H] = \left[\begin{pmatrix} \sigma \cdot \mathbf{P} & 0 \\ 0 & \sigma \cdot \mathbf{P} \end{pmatrix}, c \begin{pmatrix} 0 & \sigma \cdot \mathbf{P} \\ \sigma \cdot \mathbf{P} & 0 \end{pmatrix} + mc^2 \begin{pmatrix} 1 & 0 \\ 0 & -1 \end{pmatrix} \right]$$

$$= c \left[\begin{pmatrix} \sigma \cdot \mathbf{P} & 0 \\ 0 & \sigma \cdot \mathbf{P} \end{pmatrix}, \begin{pmatrix} 0 & \sigma \cdot \mathbf{P} \\ \sigma \cdot \mathbf{P} & 0 \end{pmatrix} \right]$$

$$+ mc^2 \left[\begin{pmatrix} \sigma \cdot \mathbf{P} & 0 \\ 0 & \sigma \cdot \mathbf{P} \end{pmatrix}, \begin{pmatrix} 1 & 0 \\ 0 & -1 \end{pmatrix} \right] = 0 + 0 = 0.$$

c. Let $\mathbf{P}$ be along the z direction. Then as $\sigma_z = \begin{pmatrix} 1 & 0 \\ 0 & -1 \end{pmatrix}$, we have

$$\sigma \cdot \mathbf{P} = \begin{pmatrix} \sigma_z & 0 \\ 0 & \sigma_z \end{pmatrix} P_z = \begin{pmatrix} 1 & & & \\ & -1 & & \\ & & 1 & \\ & & & -1 \end{pmatrix} P_z,$$

where the unspecified elements are all zeros, which is diagonal. Then as shown in **Problem 8032** the plane wave solutions of the Dirac equation in

this representation are $\begin{pmatrix} \alpha \\ 0 \\ \gamma \\ 0 \end{pmatrix} e^{iP_z z/\hbar}$ and $\begin{pmatrix} 0 \\ \beta \\ 0 \\ \delta \end{pmatrix} e^{iP_z z/\hbar}$, where α and γ, β and

δ take two sets of different values. Substituting the eigenfunctions in the Schrödinger equation

$$\hat{H}\psi = i\hbar \frac{\partial \psi}{\partial t} = E\psi$$

we have

$$\begin{cases} cP_z \gamma + mc^2 \alpha = E\alpha, \\ cP_z \alpha - mc^2 \gamma = E\gamma, \end{cases}$$

giving

$$E_\pm = \pm \sqrt{m^2 c^4 + P_z^2 c^2},$$

$$\psi = \begin{pmatrix} 1 \\ 0 \\ \dfrac{E_\pm - mc^2}{cP_z} \\ 0 \end{pmatrix} e^{i(P_z z - E_\pm t)/\hbar};$$

and

$$\begin{cases} -cP_z\delta + mc^2\beta = E\beta, \\ -cP_z\beta - mc^2\delta = E\delta, \end{cases}$$

giving

$$E_\pm = \pm\sqrt{m^2c^4 + P_z^2 c^2},$$

$$\psi = \begin{pmatrix} 0 \\ 1 \\ 0 \\ \dfrac{mc^2 - E_\pm}{cP_z} \end{pmatrix} e^{i(P_z z - E_\pm t)/\hbar}.$$

8034

Consider a free real scalar field $\phi(x_\mu)$, where $x_\mu = x, y, z$ for $\mu = 1, 2, 3$ and $x_4 = ict$, satisfying the Klein-Gordon equation.

a. Write down the Lagrangian density for the system.

b. Using Euler's equations of motion, verify that ϕ does satisfy the Klein-Gordon equation.

c. Derive the Hamiltonian density for the system. Write down Hamilton's equations and show that they are consistent with the equation derived in (b)

(Buffalo)

Sol:

a. The Lagrangian density is

$$\mathcal{L}(x) = -\frac{1}{2}\partial_\mu\phi(x)\partial_\mu\phi(x) - \frac{m^2}{2}\phi(x)\phi(x).$$

b. Using the above expression for $\mathcal{L}$ in Euler's equations of motion

$$\partial_\mu\left[\frac{\partial\mathcal{L}(x)}{\partial(\partial_\mu\phi)}\right] - \frac{\partial\mathcal{L}(x)}{\partial\phi(x)} = 0$$

we obtain

$$\partial_\mu\partial_\mu\phi(x) - m^2\phi(x) = 0,$$

which is just the Klein-Gordon equation.

c.   The Hamiltonian density of the system is

$$\mathcal{H}(x)=\frac{\partial \mathcal{L}}{\partial\left(\partial_{\mu}\phi\right)}\partial_{\mu}\phi-\mathcal{L}=-\frac{1}{2}\partial_{\mu}\phi\partial_{\mu}\phi+\frac{m^{2}}{2}\phi^{2}.$$

Hamilton's canonical equations

$$\frac{\partial \mathcal{H}}{\partial\phi}=-\partial_{\mu}P_{\mu}, \quad \frac{\partial \mathcal{H}}{\partial P_{\mu}}=-\partial_{\mu}\phi,$$

where

$$P_{\mu}=\frac{\partial \mathcal{L}}{\partial\left(\partial_{\mu}\phi\right)}=-\partial_{\mu}\phi,$$

then give

$$-\partial_{\mu}P_{\mu}=m^{2}\phi,$$

i.e.,

$$\partial_{\mu}\partial_{\mu}\phi-m^{2}\phi=0,$$

same as obtained in (b).

8035

It can be shown that the probability for an on-shell charged particle with initial momentum P to emit a virtual photon with momentum q is proportional to the covariant tensor

$$W_{\mu\nu}=Ag_{\mu\nu}+BP_{\mu}P_{\nu}+Cq_{\mu}q_{\nu}+D\left(q_{\mu}P_{\nu}+P_{\mu}q_{\nu}\right),$$

where A, B, C and D are real Lorentz-invariant scalar functions of q^{2}, $q\cdot P$ and $P^{2}=m^{2}$.

a.   Use current conservation to show that $W_{\mu\nu}$ has the form

$$W_{\mu\nu}=W_{1}\left(g_{\mu\nu}-\frac{q_{\mu}q_{\nu}}{q^{2}}\right)+W_{2}\left(P_{\mu}-q_{\mu}\frac{q\cdot P}{q^{2}}\right)\cdot\left(P_{\nu}-q_{\nu}\frac{q\cdot P}{q^{2}}\right),$$

i.e., only two of A, B, C and D are independent.

b.   Compute W_{1} and W_{2} for a Dirac particle of mass m for which

$$W_{\mu\nu}=\text{Tr}\left[(\not{P}-\not{q}+m)\gamma_{\mu}(\not{P}+m)\gamma_{\nu}\right].$$

(Buffalo)

Sol:

a. Current conservation requires $q^\mu W_{\mu\nu} = 0$, i.e.,

$$Aq_\nu + B(q\cdot P)P_\nu + Cq^2 q_\nu + D\left[q^2 P_\nu + (q\cdot P)q_\nu\right] = 0,$$

where $q\cdot P = q^\mu P_\mu$, $q_\nu = q^\mu g_{\mu\nu}$, etc. As P_ν, q_ν are independent and $q^2 \neq 0$, this gives

$$A + Cq^2 + D(q\cdot P) = 0,$$
$$B(q\cdot P) + Dq^2 = 0.$$

Solving for C and D and writing $A = W_1$, $B = W_2$, we have

$$D = -W_2 \frac{(q\cdot P)}{q^2}, \quad C = -\frac{W_1}{q^2} + \frac{W_2(q\cdot P)^2}{q^4}.$$

Hence

$$W_{\mu\nu} = W_1\left(g_{\mu\nu} - \frac{q_\mu q_\nu}{q^2}\right) + W_2\left(P_\mu - q_\mu \frac{q\cdot P}{q^2}\right)\cdot\left(P_\nu - q_\nu \frac{q\cdot P}{q^2}\right).$$

b. We are given

$$W_{\mu\nu} = \text{Tr}\left[(\not{P} - \not{q} + m)\gamma_\mu(\not{P} + m)\gamma_\nu\right]$$
$$= \text{Tr}\left[\not{P}\gamma_\mu\not{P}\gamma_\nu + \not{P}\gamma_\mu m\gamma_\nu - \not{q}\gamma_\mu\not{P}\gamma_\nu\right.$$
$$\left. - \not{q}\gamma_\mu m\gamma_\nu + m\gamma_\mu\not{P}\gamma_\nu + m\gamma_\mu m\gamma_\nu\right],$$

where $\not{P} = P_\alpha\gamma^\alpha$, $\not{q} = q_\alpha\gamma^\alpha$. The Dirac matrices satisfy the anticommutation relation

$$\{\gamma^\mu, \gamma^\nu\} \equiv \gamma^\mu\gamma^\nu + \gamma^\nu\gamma^\mu = 2g^{\mu\nu},$$

and so

$$\text{Tr}\left(\gamma^\mu\gamma^\nu\right) = 4g^{\mu\nu},$$
$$\text{Tr}\left(\gamma^{\mu_1}\gamma^{\mu_2}\cdots\gamma^{\mu_n}\right) = 0 \quad \text{for} \quad n=\text{odd},$$
$$\text{Tr}\left(\gamma^\mu\gamma^\nu\gamma^\lambda\gamma^\sigma\right) = 4\left(g^{\mu\nu}g^{\lambda\sigma} - g^{\mu\lambda}g^{\nu\sigma} + g^{\mu\sigma}g^{\nu\lambda}\right).$$

Hence in $W_{\mu\nu}$ the terms involving an odd number of γ vanish. Consider

$$\text{Tr}\left(\not{P}\gamma_\mu\not{P}\gamma_\nu\right) = \text{Tr}\left(P_\alpha\gamma^\alpha g_{\mu\lambda}\gamma^\lambda P_\beta\gamma^\beta g_{\nu\sigma}\gamma^\sigma\right)$$
$$= P_\alpha g_{\mu\lambda} P_\beta g_{\nu\sigma}\, \text{Tr}\left(\gamma^\alpha\gamma^\lambda\gamma^\beta\gamma^\sigma\right)$$
$$= 4P_\alpha g_{\mu\lambda} P_\beta g_{\nu\sigma}\left(g^{\alpha\lambda}g^{\beta\sigma} - g^{\alpha\beta}g^{\lambda\sigma} + g^{\alpha\sigma}g^{\lambda\beta}\right)$$
$$= 4\left(P_\alpha\delta^\alpha_\mu P_\beta\delta^\beta_\nu - P^\beta P_\beta\delta^\sigma_\nu g_{\nu\sigma} + P_\alpha g^\alpha_\nu P_\beta\delta^\beta_\mu\right)$$
$$= 4\left(P_\mu P_\nu - P^2 g_{\mu\nu} + P_\nu P_\mu\right),$$

$$\mathrm{Tr}\left(\gamma\!\!\!/\gamma_\mu P\!\!\!/\gamma_\nu\right)=4\left(q\!\!\!/_\mu P_\nu-\mathbf{q}\cdot\mathbf{P}g_{\mu\nu}+g_\nu P_\mu\right),$$

$$\mathrm{Tr}\left(m^2\gamma_\mu\gamma_\nu\right)=m^2 g_{\mu\alpha}g_{\nu\beta}\,\mathrm{Tr}\left(\gamma^\alpha\gamma^\beta\right)$$

$$=4m^2 g_{\mu\alpha}g_{\nu\beta}g^{\alpha\beta}=4m^2 g_{\mu\alpha}\delta^\alpha_\nu=4m^2 g_{\mu\nu}.$$

Then, as for an on-shell particle $P^2\approx m^2$, we have

$$W_{\mu\nu}=4\mathbf{q}\cdot\mathbf{P}g_{\mu\nu}+8P_\mu P_\nu-4\left(q_\mu P_\nu+q_\nu P_\mu\right).$$

A comparison with the given expression for $W\mu\nu$ we find

$$W_1=4\mathbf{q}\cdot\mathbf{P},\quad W_2=8.$$

Note that for an on-shell charge emitting a virtual photon, initially $P^2=m^2$, and finally $(\mathbf{P}-\mathbf{q})^2=P^2-2\mathbf{q}\cdot\mathbf{P}+q^2=m^2$, and so the two expressions for $W_{\mu\nu}$ are consistent.

8036

In order to account for the anomalous magnetic moments of particles, the Dirac equation given below can be used:

$$\left(i\nabla\!\!\!\!/-e A\!\!\!/+K\frac{e}{4m}\sigma_{\mu\nu}F^{\mu\nu}-m\right)\psi(x)=0.$$

Here e and m are the charge and mass of the particle, K is a dimensionless parameter, $A^\mu(x)$ is the four-dimensional potential and $F^{\mu\nu}$ is the electromagnetic field tensor, i.e.

$$F^{\mu\nu}=\frac{\partial A^\mu}{\partial x_\nu}-\frac{\partial A^\nu}{\partial x_\mu},\quad \sigma_{\mu\nu}=\frac{i}{2}\left[\gamma_\mu,\gamma_\nu\right],$$

where γ_μ is a Dirac matrix, $\gamma_0=\gamma^0=\beta,\gamma^i=-\gamma_i=\beta\alpha^i,i=1,2,3$.

a. It is well known that the above equation is covariant if $K=0$. We have

$$\psi'\left(x'\right)=S\psi(x),$$

where $x'^\mu=a^\mu_\nu x^\nu$ and $a^\mu_\nu\gamma^\nu=S^{-1}\gamma^\mu S$. Show that if $K\neq0$, the equation is still covariant.

b. Write the equation in the Hamiltonian form and show that the additional interaction does not destroy the Hermiticity of the original Hamiltonian.

(Buffalo)

Sol:

a. As

$$S^{-1}\gamma_\mu S = a_\mu^\nu \gamma_\nu$$

and a_μ^ν commutes with S and γ, we have

$$S^{-1}\gamma_\mu S a_\alpha^\mu = a_\mu^\nu a_\alpha^\mu \gamma_\nu = \delta_\alpha^\nu \gamma_\nu = \gamma_\alpha$$

Consider

$$\sigma'_{\alpha\beta} F'^{\alpha\beta}\psi'(x') = \frac{i}{2}\left(\gamma_\alpha\gamma_\beta - \gamma_\beta\gamma_\alpha\right) a_\mu^\alpha a_\nu^\beta F^{\mu\nu} S\psi'(x)$$

$$= \frac{i}{2}\left(SS^{-1}\gamma_\alpha SS^{-1}\gamma_\beta - SS^{-1}\gamma_\beta SS^{-1}\gamma_\alpha\right) a_\mu^\alpha a_\nu^\beta F^{\mu\nu} S\psi$$

$$= S\frac{i}{2}\left(S^{-1}\gamma_\alpha Sa_\mu^\alpha S^{-1}\gamma_\beta Sa_\nu^\beta - S^{-1}\gamma_\beta Sa_\nu^\beta S^{-1}\gamma_\alpha Sa_\mu^\alpha\right) F^{\mu\nu}\psi$$

$$= S\frac{i}{2}\left(\gamma_\mu\gamma_\nu - \gamma_\nu\gamma_\mu\right) F^{\mu\nu}\psi$$

$$= S\sigma_{\mu\nu} F^{\mu\nu}\psi$$

Hence

$$S^{-1}\sigma'_{\alpha\beta} F'^{\alpha\beta}\psi'(x') = \sigma_{\mu\nu} F^{\mu\nu}\psi(x).$$

Then as $\psi(x) = S^{-1}\psi'(x')$ and $\slashed{\nabla}$ and $\slashed{A}$ are invariant, under transformation the Dirac equation becomes

$$S^{-1}\left(i\slashed{\nabla} - e\slashed{A} + \frac{Ke}{4m}\sigma'_{\alpha\beta} F'^{\alpha\beta} + m\right)\psi'(x) = 0,$$

i.e. the equation is covariant.

b. As

$$\slashed{\nabla} = \gamma_\alpha\frac{\partial}{\partial x_\alpha} = \beta\frac{\partial}{\partial t} + \gamma\cdot\nabla,$$

$$\slashed{A} = \gamma_\alpha A^\alpha = \beta A^\circ + \gamma_j A^j = \beta A^\circ + \gamma_j g_{ji} A^i$$

$$= \beta A^\circ - \gamma\cdot\mathbf{A},$$

the Dirac equation can be written as

$$i\beta\frac{\partial}{\partial t}\psi = \left(-i\gamma\cdot\nabla + e\beta A^\circ - e\gamma\cdot\mathbf{A} - K\frac{e}{4m}\sigma_{\mu\nu} F^{\mu\nu} + m\right)\psi.$$

Note that we have used units such that $\hbar = c = 1$.

Mutiplying both sides by β from the left, as $\beta^2 = 1$, $\beta\gamma_i = \beta^2\alpha_i = \alpha_i$, we have the equation in the Hamiltonian form:

$$i\frac{\partial\psi}{\partial t} = H\psi$$

where

$$H = -i\alpha\cdot\nabla + eA^0 - e\alpha\cdot\mathbf{A} - K\frac{e}{4m}\beta\sigma_{\mu\nu}F^{\mu\nu} + m\beta,$$

with

$$\alpha = \begin{pmatrix} 0 & \sigma \\ \sigma & 0 \end{pmatrix}, \quad \beta = \begin{pmatrix} I & 0 \\ 0 & -I \end{pmatrix},$$

$\sigma_1, \sigma_2, \sigma_3$ being Pauli's matrices and I the unit matrix. By definition

$$\{\sigma_i, \sigma_j\} = \sigma_i\sigma_j + \sigma_j\sigma_i = 2I\delta_{ij}.$$

$$(i, j = 1, 2, 3)$$

It follows that

$$\{\alpha_i, \alpha_j\} = 2I\delta_{ij}, \{\beta, \alpha_i\} = 0$$

and so

$$\{\gamma_i, \gamma_j\} = 2g_{ij},$$
$$\{\beta, \gamma_i\} = 0.$$

Then

$$[\beta, \sigma_{ij}] = \left[\beta, \frac{i}{2}(\gamma_i\gamma_j - \gamma_j\gamma_i)\right]$$
$$= \frac{i}{2}(\beta\gamma_i\gamma_j - \beta\gamma_j\gamma_i - \gamma_i\gamma_j\beta + \gamma_j\gamma_i\beta) = 0$$

since $\gamma_i\gamma_j\beta = -\gamma_i\beta\gamma_j = \beta\gamma_i\gamma_j$, etc, and similarly,

$$\{\beta, \sigma_{0i}\} = \frac{i}{2}(\beta[\beta, \gamma_i] + [\beta, \gamma_i]\beta) = 0.$$

Since by definition σ_i and β are Hermitian, $\gamma_0 = \beta$ is Hermitian and $\gamma_i = \beta\alpha_i$ is anti-Hermitian. It follows that σ_{ij} is Hermitian and σ_{0i}, σ_{i0} are anti-Hermitian. Then

$$\sigma_{\mu\nu}^+\beta^+ = \sigma_{ij}^+\beta^+ + \sigma_{i0}^+\beta^+ + \sigma_{0i}^+\beta^+ = \sigma_{ij}\beta - \sigma_{i0}\beta - \sigma_{0i}\beta$$
$$= \beta\sigma_{ij} + \beta\sigma_{i0} + \beta\sigma_{0i} = \beta\sigma_{\mu\nu}$$

Hence the Hermitian conjugate of the additional interaction term is

$$\left(-K\frac{e}{4m}\beta\sigma_{\mu\nu}F^{\mu\nu}\right)^{+}=-K^{*}\frac{e^{*}}{4m^{*}}\sigma_{\mu\nu}^{+}\beta^{+}F^{\mu\nu^{*}}$$

$$=-K\frac{e}{4m}\beta\sigma_{\mu\nu}F^{\mu\nu}$$

noting that K,e,m and $F^{\mu\nu}$ are real numbers. Therefore the additional interaction is Hermitian, and it does not destroy the Hermiticity of the original Hamiltonian.

8037

Proton and neutron may be regarded as two "isospin" states of a single particle, the nucleon. Denote proton by $|+\rangle$ and neutron by $|-\rangle$ and define the following operators:

$$t_3|\pm\rangle=\pm\frac{1}{2}|\pm\rangle,\ t_{\pm}|\mp\rangle=|\pm\rangle,\ t_{\mp}|\mp\rangle=0.$$

The operators $t_1=\frac{1}{2}(t_++t_-),\ t_2=-\frac{i}{2}(t_+-t_-)$, and t_3 can be represented by one-half times the 2×2 Pauli matrices. Together they form a vector t in isospin space.

In a simple model the Hamiltonian for a system of N nucleons all in the same spatial state is the sum of three terms:

$$H=NE_0+C_1\sum_{i>j}t_i\cdot t_j+C_2Q^2,$$

where E_0,C_1 and C_2 are positive constants with $C_1>C_2$, $\mathbf{t}_j$ is the isospin of the j-th nucleon, and Q is the total electric charge in units of e. The sum is over all pairs of nucleons.

a. Show that $\sum_{i>j}\mathbf{t}_i\cdot\mathbf{t}_j=\frac{1}{2}\left[T(T+1)-\frac{3}{4}N\right]$, where T is the "total isospin" quantum number of the system.

In the rest of this problem it is essential to remember that neutrons and protons are spin-1/2 particles obeying Fermi statistics.

b. What are the energy eigenstates and eigenvalues of a 2-nucleon system? What is the total spin of each state?

c. What are the energy eigenstates and eigenvalues of a 4-nucleon system?

d. What are the energy eigenvalues of a 3-nucleon system?

(MIT)

Sol:

a. As $\mathbf{T}^2 = \left(\sum_i \mathbf{t}_i\right)^2$ has eigenvalue $T(T+1)$ and $\mathbf{t}_i^2$ has eigenvalue $\frac{1}{2}\left(\frac{1}{2}+1\right)$,

we have

$$\sum_{i>j} \mathbf{t}_i \cdot \mathbf{t}_j = \frac{1}{2}\left[\left(\sum_{i=1}^N \mathbf{t}_i\right)^2 - \sum_{i=1}^N \mathbf{t}_i^2\right]$$
$$= \frac{1}{2}\left[T(T+1) - \frac{3}{4}N\right]$$

b. A system of identical $-\frac{1}{2}$ particles must have an antisymmetric total wave function. Hence a system of two nucleons has the following possible structures:

Configuration	IsospinState	SpinState
(pp)	$\mid+\rangle\mid+\rangle$	$\frac{1}{\sqrt{2}}(\mid a\rangle\mid\beta\rangle - \mid\beta\rangle\mid a\rangle)$
(nn)	$\mid-\rangle\mid-\rangle$	$\frac{1}{\sqrt{2}}(\mid a\rangle\mid\beta\rangle - \mid\beta\rangle\mid a\rangle)$
(pn)	$\frac{1}{\sqrt{2}}(\mid+\rangle\mid-\rangle + \mid-\rangle\mid+\rangle)$	$\frac{1}{\sqrt{2}}(\mid a\rangle\mid\beta\rangle - \mid\beta\rangle\mid a\rangle),$
(pn)	$\frac{1}{\sqrt{2}}(\mid+\rangle\mid-\rangle - \mid-\rangle\mid+\rangle)$	$\frac{1}{\sqrt{2}}(\mid a\rangle\mid\beta\rangle + \mid\beta\rangle\mid a\rangle), \mid a\rangle\mid a\rangle, or \mid\beta\rangle\mid\beta\rangle$

The corresponding eigenvalues are as follows.

Configuration	T	S	Q	E
$(p\,p)$	1	0	2	$C_1 + 4C_2$
$(n\,n)$	1	0	0	C_1
$(p\,n)$	1	0	1	$C_1 + C_2$
$(p\,n)$	0	1	1	C_2

In the above, $\mid a\rangle, \mid\beta\rangle$ represent single-particle states with spin + and spin − respectively, and E is the energy above $\left(2E_0 - \frac{3}{4}C_1\right)$.

c. On account of Paul's principle, there can at most be 2 protons and 2 neutrons, each pair of opposite spins, in a given energy state. For the ordered combination $(pnpn)$ the spin states have four forms $(aa\beta\beta)$, $(\beta\beta aa)$, $(a\beta\beta a)$, $(\beta aa\beta)$. For other ordered combinations similar spin states apply. However, in this case the total wave function cannot be expressed as a simple product of the spin wave function and the isospin wave function. For the possible isospin values $T = 2, 1, 0$, the corresponding energy values are

$$E = 4E_0 + 4C_2 + \frac{3}{2}C_1, \qquad 4E_0 + 4C_2 - \frac{1}{2}C_1,$$

$$4E_0 + 4C_2 - \frac{3}{2}C_1.$$

But as the spatial wave functions of the four nucleons are the same and there are only two spin states for a nucleon, Pauli's principle requires the system's total isospin to be 0 and its energy state can only be the eigenstate

$$\psi(1,2,3,4) = \begin{pmatrix} |+\rangle_1 a_1 & |+\rangle_1 \beta_1 & |-\rangle_1 a_1 & |-\rangle_1 \beta_1 \\ |+\rangle_2 a_2 & |+\rangle_2 \beta_2 & |-\rangle_2 a_2 & |-\rangle_2 \beta_2 \\ |+\rangle_3 a_3 & |+\rangle_3 \beta_3 & |-\rangle_3 a_3 & |-\rangle_3 \beta_3 \\ |+\rangle_4 a_4 & |+\rangle_4 \beta_4 & |-\rangle_4 a_4 & |-\rangle_4 \beta_4 \end{pmatrix}.$$

d. The configurations for a three-nucleon system are (ppn) or (nnp), and the isospin can be $\frac{3}{2}$ or $\frac{1}{2}$

For (ppn):

$$E = 3E_0 + 4C_2 \pm \frac{3}{4}C_1,$$

for (nnp):

$$E = 3E_0 + C_2 \pm \frac{3}{4}C_1.$$

8038

A molecule in the form of an equilateral triangle can capture an extra electron. To a good approximation, this electron can go into one of three orthogonal states ψ_A, ψ_B, ψ_C localized near the corners of the triangle. To a better approximation, the energy eigenstates of the electron are linear combinations of ψ_A, ψ_B, ψ_C determined by an effective Hamiltonian which has equal expectation values for ψ_A, ψ_B, ψ_C and equal matrix elements V_0 between each pair of ψ_A, ψ_B, ψ_C.

a. What does the symmetry under a rotation through $2\pi/3$ imply about the coefficients of ψ_A, ψ_B, ψ_C in the eigenstates of the effective Hamiltonian? There is also symmetry under interchange of B and C ; what additional information does this give about the eigenvalues of the effective Hamiltonian?

b. At time $t=0$ an electron is captured into the state ψ_A. Find the probability that it is in ψ_A at time t.

(*MIT*)

Sol:

a. Under the rotation through $2\pi/3$, we have

$$R\psi_A = a\psi_B, R\psi_B = a\psi_C, R\psi_C = a\psi_A.$$

Then as

$$R^2\psi_A = aR\psi_B = a^2\psi_C,$$
$$R^3\psi_A = a^2R\psi_C = a^3\psi_A,$$

we have $a^3 = 1$ and hence

$$a = 1, e^{\frac{i2\pi}{3}} \quad \text{and} \quad e^{\frac{i4\pi}{3}}.$$

Suppose the eigenstate of the effective Hamiltonian is

$$\psi = a_1\psi_A + a_2\psi_B + a_3\psi_C.$$

Symmetry under the rotation through $\dfrac{2\pi}{3}$ means that $R\psi = \psi$, i.e.

$$a_1 a\psi_B + a_2 a\psi_C + a_3 a\psi_A = a_1\psi_A + a_2\psi_B + a_3\psi_C.$$

Then the orthogonality of ψ_A, ψ_B, ψ_C requires

$$a_1 a = a_2, \quad a_2 a = a_3, \quad a_3 a = a_1.$$

For $a = 1$ letting $a_1 = 1$, for $a = \exp\left(i\dfrac{2\pi}{3}\right)$ letting $a_1 = 1$, and for $a = \exp\left(i\dfrac{4\pi}{3}\right)$ letting $a_2 = 1$, we have the three combinations

$$\psi^{(1)} = \frac{1}{\sqrt{3}}\begin{pmatrix} 1 \\ 1 \\ 1 \end{pmatrix}, \quad \psi^{(2)} = \frac{1}{\sqrt{3}}\begin{pmatrix} 1 \\ e^{i2\pi/3} \\ e^{i4\pi/3} \end{pmatrix}, \quad \psi^{(3)} = \frac{1}{\sqrt{3}}\begin{pmatrix} e^{-i4\pi/3} \\ 1 \\ e^{i4\pi/3} \end{pmatrix}.$$

Let the equal expectation values of H for the eigenstates ψ_A, ψ_B, ψ_C be zero then the effective Hamiltonian can be represented by the matrix

$$H = \begin{pmatrix} 0 & V & V \\ V & 0 & V \\ V & V & 0 \end{pmatrix}.$$

As $H\psi^{(1)} = 2V\psi^{(1)}$, $H\psi^{(2)} = -V\psi^{(2)}$, $H\psi^{(3)} = -V\psi^{(3)}$, the energies corresponding to $\psi^{(1)}, \psi^{(2)}, \psi^{(3)}$ are $2V, -V, -V$ repectively.

There is symmetry in the interchange of B and C. Denote by P the operator for the interchange of the two atoms. As P does not commute with R, it cannot be represented by a diagonal matrix on the basis of $\psi^{(1)}, \psi^{(2)}, \psi^{(3)}$. However, $\psi^{(1)}$ is an eigenstate of P and $\psi^{(2)}, \psi^{(3)}$ are degenerate states, though not eigenstates, of P. So P imposes no condition on the eigenvalues of H.

b. At $t = 0$, we can expand the wave function ψ_A as

$$\psi_A(0) = \frac{1}{\sqrt{3}}[\psi^{(1)} + \psi^{(2)} + e^{-i2\pi/3}\psi^{(3)}].$$

At a later time t we have

$$\psi_A(t) = \frac{1}{\sqrt{3}}[e^{-i2Vt/\hbar}\psi^{(1)} + e^{+iVt/\hbar}\psi^{(2)} + e^{-i2\pi/3}e^{+iVt/\hbar}\psi^{(3)}].$$

Hence the probability of an electron, initially in state ψ_A, being in state ψ_A at time t is

$$|\langle \psi_A(t)|\psi_A(0)\rangle|^2 = \left|\frac{1}{3}\left(e^{i2Vt/\hbar} + e^{-iVt/\hbar} + e^{i2\pi/3}e^{-iVt/\hbar}e^{-i2\pi/3}\right)\right|^2$$

$$= \frac{1}{9}\left|e^{i3Vt/\hbar} + 2\right|^2 = \frac{1}{9}\left(5 + 4\cos\frac{3Vt}{\hbar}\right).$$

8039

The energy of a molecule is the sum of the kinetic energies of the electrons and of the nuclei and of the various Coulomb energies. Suppose that for a particular many-particle normalized wave function $\psi(x_1, \ldots, x_N)$, the expectation value of the kinetic energy is T and of the potential energy is $-U(U > 0)$.

a. Find a variational estimate of the ground state energy using a wave function $\lambda^{3N/2}\psi(\lambda x_1, \ldots, \lambda x_N)$ where λ is a parameter.

b. Suppose ψ is the true ground state wave function and that the true ground state energy $-B(B > 0)$. What are the true values of T and U?

(MIT)

Sol:

a. The mean kinetic energy T of the system is given by the sum of terms like

$$\frac{\dfrac{\hbar^2}{2m}\displaystyle\int \psi^*\left(x_1,\ldots x_N\right)\frac{\partial^2}{\partial x_i^2}\psi\left(x_1,\ldots,x_N\right)dx_1,\ldots,dx_N}{\displaystyle\int \psi^*\left(x_1,\ldots,x_N\right)\psi\left(x_1,\ldots,x_N\right)dx_1,\ldots,dx_N}.$$

When the trial wave function is used, the mean kinetic energy T' is given by the sum of terms like

$$\frac{\dfrac{\hbar^2}{2m}\lambda^{3N}\displaystyle\int \psi^*\left(\lambda x_1,\ldots,\lambda x_N\right)\frac{\partial^2}{\partial x_i^2}\psi\left(\lambda x_1,\ldots,\lambda x_N\right)dx_1,\ldots,dx_N}{\lambda^{3N}\displaystyle\int \psi^*\left(\lambda x_1,\ldots,\lambda x_N\right)\psi\left(\lambda x_1,\ldots,\lambda x_N\right)dx_1,\ldots,dx_N}.$$

$$=\frac{\dfrac{\hbar^2}{2m}\lambda^2\displaystyle\int \psi^*\left(\lambda x_1,\ldots \lambda x_N\right)\frac{\partial^2}{\partial\left(\lambda x_i\right)^2}\psi\left(\lambda x_1,\ldots,\lambda x_N\right)d\lambda x_1,\ldots,d\lambda x_N}{\displaystyle\int \psi^*\left(\lambda x_1,\ldots,\lambda x_N\right)\psi\left(\lambda x_1,\ldots,\lambda x_N\right)d\lambda x_1,\ldots,d\lambda x_N}.$$

$$=\frac{\dfrac{\hbar^2\lambda^2}{2m}\displaystyle\int \psi^*\left(y_1,\ldots y_N\right)\frac{\partial^2}{\partial y_i^2}\psi\left(y_1,\ldots,y_N\right)dy_1,\ldots,dy_N}{\displaystyle\int \psi^*\left(y_1,\ldots,y_N\right)\psi\left(y_1,\ldots,y_N\right)dy_1,\ldots,dy_N},$$

where $y_1=\lambda x_1$, etc. Hence $T'=\lambda^2 T$. Similarly, $-U$ is given by the sum of terms like

$$\frac{e_ie_j\displaystyle\int \psi^*\left(x_1,\ldots x_N\right)\frac{1}{\left|x_i-x_j\right|}\psi\left(x_1,\ldots,x_N\right)dx_1,\ldots,dx_N}{\displaystyle\int \psi^*\left(x_1,\ldots,x_N\right)\psi\left(x_1,\ldots,x_N\right)dx_1,\ldots,dx_N}$$

and so $-U'=-\lambda U$. Thus the mean value of the energy is

$$E(\lambda)=\lambda^2 T-\lambda U.$$

For the ground state, $\dfrac{dE(\lambda)}{d\lambda}=0$, giving $\lambda=\dfrac{U}{2T}$. Hence the variational estimate of the ground state energy is

$$E=-\frac{U^2}{4T}.$$

b. If ψ is the true ground state wave function, then $\lambda=1$. Hence

$$U=2T \quad\text{and}\quad E=T-U=-T.$$

As $E=-B$, we have

$$T=B,\,U=2B.$$

8040

In diatomic molecules, nuclear motions are generally much slower than are those of the electrons, leading to the use of adiabatic approximation. This effectively means that the electron wave functions at any time are obtained for the instantaneous positions of the protons. A highly idealized version of a "singly-ionized hydrogen molecule" is provided by the one-dimensional "electron" Hamiltonian

$$H = -\frac{\hbar^2}{2m}\frac{d^2}{dx^2} - g\delta(x - x_0) - g\delta(x + x_0),$$

where $\pm x_0$ are the proton coordinates.

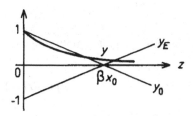

Fig. 8.10

a. What are the eigenfunctions and eigenvalues of all bound states for arbitrary x_0? You may specify eigenvalues in terms of a transcendental equation. Give the analytic results for the limiting cases $\frac{mgx_0}{\hbar^2} \gg 1$ and $\frac{mgx_0}{\hbar^2} \ll 1$.

b. Assume that the protons (mass $M \gg m$) move adiabatically and have a repulsive potential $V(2x_0) = g/200x_0$ acting between them. Calculate approximately the equilibrium separation of the protons.

c. Calculate approximately the frequency for harmonic vibrations of the protons about the equilibrium position. Is the adiabatic approximation justified?

(MIT)

Sol:

a. The Schrödinger equation can be written as

$$\frac{d^2\psi}{dx^2} + \beta\left[\delta(x - x_0) + \delta(x + x_0)\right]\psi = k^2\psi, \tag{1}$$

where $\beta = \dfrac{2mg}{\hbar^2}$, $k^2 = -\dfrac{2mE}{\hbar^2} = \dfrac{2m|E|}{\hbar^2}$ as E is negative for bound states.

For $x \neq \pm x_0$, the equation becomes $\dfrac{d^2\psi}{dx^2} = k^2\psi$. Furthermore as H is invariant under space inversion, its eigenstates are of two types, with odd and even parities as:

odd parity:

$$\psi(x) = \begin{cases} \sinh kx, & 0 \leq x \leq x_0, \\ ae^{-kx}, & x_0 \leq x, \end{cases}$$

even parity:

$$\psi(x) = \begin{cases} \cosh kx, & 0 \leq x \leq x_0, \\ be^{-kz}, & x_0 \leq x, \end{cases}$$

where a, b are constants. Integrating both sides of Eq. (1) from $x_0 - \varepsilon$ to $x_0 + \varepsilon$ and letting $\varepsilon \to 0$, we find

$$\psi'(x_0 + \varepsilon) - \psi'(x_0 - \varepsilon) + \beta\psi(x_0) = 0.$$

The continuity of ψ across x_0 requires

$$\psi(x_0 + \varepsilon) = \psi(x_0 - \varepsilon).$$

These two conditions give, for odd parity,

$$e^{-2kx_0} = 1 - \frac{2kx_0}{\beta x_0},$$

for even parity,

$$e^{-2kx_0} = -1 + \frac{2kx_0}{\beta x_0}.$$

As shown in Fig. 8.10, k and hence the eigenvalue E are given by the intercept of $y = e^{-z}$ with either

$$y_0 = 1 - \frac{z}{\beta x_0} \quad or \quad y_E = -1 + \frac{z}{\beta z_0},$$

where $z = 2kx_0$. When $\beta x_0 \ll 1$, as

$$\left|\frac{dy_0}{dz}\right| = \frac{1}{\beta x_0} \gg \left|\frac{dy}{dz}\right| \approx 1,$$

y and y_0 do not intercept and there is only solution for even parity. For this the interception occurs at small z given by

$$1 - z \approx -1 + \frac{z}{\beta x_0},$$

or $z \approx 2\beta x_0 (1-\beta x_0)$, i.e. $k \approx \beta(1-\beta x_0)$.

Hence

$$E \approx -\frac{\hbar^2 \beta^2}{2m}(1-\beta x_0)^2.$$

When $\beta x_0 \gg 1$, the interceptions occur near $z \approx \beta x_0$. Using this we have for the odd and even parities respectively

$$k \approx \frac{\beta}{2}\left(1 \mp e^{-\beta x_0}\right),$$

and hence

$$E \approx -\frac{\hbar^2 \beta^2}{8m}\left(1 \mp e^{-\beta x_0}\right)^2.$$

Note that for odd parity the energy

$$E \approx -\frac{\hbar^2 \beta^2}{8m}\left(1 - e^{-\beta x_0}\right)^2$$

decreases as x_0 increases, even before we consider the repulsive force between the protons. Thus the system is unstable and the state is not a bound state. Therefore, in both the limiting cases only the even parity solutions are valid.

b. The total energy of the system including the proton's is

$$\langle H \rangle = E_e + T_p + V_p,$$

where E_e is the electron energy obtained above for even parity, $T_p \approx 0$ in adiabatic approximation, and $V_p = \dfrac{g}{200 x_0}$.

The equilibrium separation $\bar{x}_0$ of the protons is given by

$$\frac{d}{dx_0}\langle H \rangle \bigg|_{\bar{x}_0} = 0,$$

which gives

$$100\left(\beta \bar{x}_0\right)^2 \left(1+e^{-\beta \bar{x}_0}\right) = e^{\beta \bar{x}_0}.$$

If $\beta x_0 \ll 1$, we have

$$\left(\beta \bar{x}_0\right)^2 \left(2-\beta \bar{x}_0\right) \approx \frac{1}{100},$$

or

$$\bar{x}_0 \approx \frac{1}{10\sqrt{2}\beta}.$$

If $\beta x_0 \gg 1$, we have

$$100\left(\beta \bar{x}_0\right)^2 \approx e^{\beta \bar{x}_0}.$$

Consider

$$h''\left(\bar{x}_0\right) \equiv \frac{d^2}{dx_0^2}\langle H \rangle \bigg|_{\bar{x}_0} = -\frac{g}{2}\beta^3\left(1+2e^{-\beta \bar{x}_0}\right)e^{-\beta \bar{x}_0} + \frac{g}{100\bar{x}_0^3}.$$

For $\beta x_0 \ll 1$, we have

$$h''\left(\bar{x}_0\right) \approx \frac{g}{100\bar{x}_0^3}\left[1-150\left(\beta \bar{x}_0\right)^3\right] \approx \frac{g}{100\bar{x}_0^3} > 0,$$

and the equilibrium is stable. For $\beta x_0 \gg 1$ we have

$$h''\left(\bar{x}_0\right) \approx -\frac{g}{200\bar{x}_0^3}\left(\beta \bar{x}_0 - 2\right) < 0,$$

and the equilibrium is unstable. Hence the equilibrium separation is

$$\bar{x}_0 \approx \frac{1}{10\sqrt{2}\beta}.$$

c.   Consider the case of stable equilibrium $\beta x_0 \ll 1$. The force constant is

$$K = h''\left(\bar{x}_0\right) \approx 20\sqrt{2}g\beta^3,$$

and so the vibrational frequency is

$$\omega = \sqrt{\frac{K}{m}} = \frac{4 \times 200^{1/4}mg^2}{\hbar^3}.$$

As the kinetic energy of protons is of the order

$$T_p = \frac{1}{2}K\bar{x}_0^2 \approx \frac{g\beta}{10\sqrt{2}}$$

while the electron has energy

$$\left|E_e\right| \approx \frac{\hbar^2\beta^2}{2m} = g\beta,$$

we have $T_p \ll \left|E_e\right|$ and the adiabatic approximation is valid, i.e. the protons may be taken to be stationary.

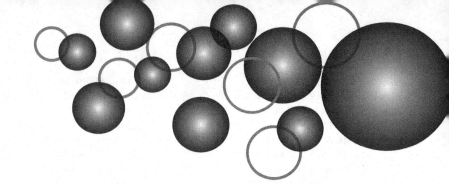

Index to Problems

690   *Index to Problems*

Printed in the United States
by Baker & Taylor Publisher Services